A Geology for Engineers

A Geology for Engineers

6th edition

F. G. H. Blyth Ph.D., D.I.C., F.G.S.

Formerly Reader in Engineering Geology,
Imperial College of Science and Technology, London

M. H. de Freitas B.Sc., F.G.S.

Lecturer in Engineering Geology,
Imperial College of Science and Technology, London.

Edward Arnold

© F. G. H. Blyth and M. H. de Freitas 1974

First published 1943
by Edward Arnold (Publishers) Limited
25 Hill Street, London W1X 8LL

Second edition 1945
Reprinted 1946 and 1949
Third edition 1952
Reprinted 1953, 1955, 1957, 1958 and 1959
Fourth edition 1960
Reprinted 1961, 1964 and 1966
Fifth edition 1967
Reprinted 1968, 1969 and 1971
Sixth edition 1974
Reprinted 1975
Reprinted with additional amendments 1976
Reprinted 1977
Reprinted 1979
Reprinted 1981
Reprinted 1982

Boards edition ISBN: 0 7131 2440 7
Paper edition   ISBN: 0 7131 2441 5

Photoset by The Universities Press, Belfast, Northern Ireland
Printed in Great Britain at The Pitman Press, Bath

Preface

For the Sixth Edition of *A Geology for Engineers* a complete revision has been made in order to present geology in a way that is appropriate to current engineering practice. The book is now divided into three sections: General Geology, Analytical Techniques, and Applications. The first section corresponds broadly with chapters 1 to 11 of earlier editions and provides a basic text in geology. Chapter 1 is new and outlines modern ideas of earth-structure and continental drift. Stratigraphy (Chapter 8) has been limited to a shortened geological history of the British Isles; while Structural Geology (Chapter 7) has been considerably extended to cover recent developments in this field. Attention has also been given to Mineralogy and Petrography (Chapters 3 to 6) as these subjects are considered an essential part of a geological education and necessary for an appreciation of the mechanical properties of rocks and soils.

The second section, Analytical Techniques, deals mainly with those geological investigations that are pertinent to engineering works. Site investigations, *in situ* testing and laboratory testing are considered, together with problems of sampling.

The third section, Applications, attempts to illustrate how the geology described in the first section can affect engineering schemes. The movement of water in the ground, the stability of slopes, and the excavation of materials are considered in successive chapters, as well as geological features which are relevant to dam and reservoir construction and to industrial and urban development of land. Numerous case histories are provided. There are also chapters on geological maps and materials for construction. Many references are given in each chapter to assist further reading. An Appendix sets out current sources of geological information.

New text-figures have been introduced throughout the book, bringing the total to over 200 illustrations, and in addition photographs for Plates have been drawn from many sources. In this connection the **Authors'** thanks for permission to reproduce photo-graphs are expressed to the following: Dr. G. P. L. Walker (Plate I); Dr. Brian Chadwick and the Director, Geological Survey of Green-land (Plate 3a); Dr. A. O. Fuller (Plates 4a and b); Institute of Geo-logical Sciences, London (Plates 6b, 8a and b); Messrs. Aerofilms Ltd. (Plates 7 and 10); Department of Geology, Exeter University (Plate 9a); Soil Mechanics Ltd. (Plates 11a and b); George Stow and Co. Ltd. (Plate 12a); Air Ministry, Crown Copyright (Plate 12b); Mr. G. Ellson (Plate 13); Sir Alexander Gibb and Partners (Plate 14); Foraky Ltd. (Plate 15a); Nobel's Explosive Co. Ltd. (Plate 15b); National Coal Board (Plates 16a and b).

The authors also thank Mary Pollard for her assistance with the preparation of this new edition.

It is hoped that the book will continue to be useful as a text for first degree Civil Engineering, Mining, and other students, and for references to professional papers on engineering geology topics and selected geological publications.

F. G. H. B.

1973   M. H. de F.

Note: In this book both British and S.I. units are used for lengths and weights throughout, except in quotations from Case Histories, where the original wording is retained.

Contents

(References are given at the end of each chapter)

General Geology

1 The Earth: Surface, Structure, and Age

Introduction The science of Geology is concerned with the earth and the rocks of which it is composed, the processes by which they were formed during geological time, and the modelling of the earth's surface in the past and at the present day. The earth is not a static body, but is constantly subject to changes both at its surface and at deeper levels. Rapid movements such as earthquakes frequently occur, as well as movements which take place slowly and are only susceptible to measurement over extended periods—for instance, the subsidence of areas such as the North Sea, or the geologically recent uplift in Scandinavia. Tides in the seas and oceans are generated by the gravitational pull of the moon, and coast-lines are being continually modified by the combined forces of waves and marine currents. Volcanic activity gives rise to eruptions of hot material from within the earth and builds new land forms—the island of Surtsey off Iceland in 1964–65 is an example—and disasters may be caused when inhabited areas are affected by the arrival of volcanic products at the surface (Plate 1).

During the very large span of geological time, processes such as those which operate at the present day have left their record in the rocks, sometimes clearly, sometimes partly obliterated by later events. The rocks therefore record events in the long history of the earth; and the remains of living organisms, animals or plants, when preserved as fossils in sediments, make their contribution to the geological record. In one sense geology is earth-history.

The term *rock* is used for the materials which form the thin outer shell, or *crust*, of the earth; it includes those which are relatively soft and easily deformed as well as those which are hard and rigid. They are accessible for observation at the surface and in mines and borings. Three broad rock groups are distinguished: *igneous rocks*, which have originated below the earth's surface and have solidified from a hot, molten condition (e.g. basalt, granite); *sedimentary rocks*, which are mainly formed from the breakdown products of older rocks, the fragments having been sorted by the action of water or wind and built up into deposits of sediment (e.g. sandstone, clay); some are chemically formed. Organic remains such as marine shells are found as *fossils* in many sediments. *Metamorphic rocks* are derived from either igneous or sedimentary rocks, but recrystallized from their original state by the action of heat and pressure (e.g. slate, schist, gneiss).

Rocks are made up of smaller crystalline units known as *minerals*; a rock can thus be defined as an assemblage of particular minerals. In later chapters the main rocks of the three groups are described (Petrography), and the main rock-forming minerals (Mineralogy). Geological processes such as those which give rise to topographical features at the earth's surface are discussed in Chapter 2, and folding and faulting in a later chapter (Structural Geology). The present chapter deals with the earth as a whole: its broad structure; its larger surface features—the oceans and continents; and its age and origin.

Dimensions and surface relief   The radius of the earth at the equator is 6378 km (3965 miles), and the polar radius is shorter by 22 km; thus the earth is not quite a perfect sphere. The surface area of the planet is 510×10^6 km^2 (197×10^6 square miles), and of this area 29 per cent is land. If however the shelf areas which surround the continuents are added, the total continental area is nearly 35 per cent. Surface relief is varied: mountains rise to several kilometres above sea level, with a maximum of 8.8 km (29 140 feet) at Everest; the ocean floors have a mean depth of about 3.7 km (12 000 feet). But the ocean floor descends to much greater depths in elongated areas or trenches, such as the Mariana Trench off New Guinea with a depth of 11.5 km (37 800 feet). These extremes of height and depth are however not extensive, and are small in comparison with the earth's radius. The average height of the land above sea level is barely one kilometre. The oceans, seas, lakes, and rivers are collectively referred to as the *hydrosphere;* the surrounding gaseous envelope is the *atmosphere.*

Features of the ocean floors   Until about 50 years ago the topography of the deep oceans was known only in broad outline, from soundings by lead-line. The advances in techniques which were brought about during the 1939–45 World War have made possible much more detailed surveys, particularly by depth-recording apparatus which draws a continuous profile of the ocean bottom. New methods of coring the sea floor at great depths have also yielded rock samples, so that the distribution and composition of the layers of sediment covering the sea floors is becoming known.

The topographical features of a continental margin are shown in Fig. 1.1. The *continental shelf* is a submerged continuation of the

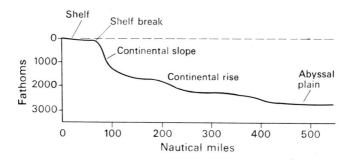

Fig. 1.1 Profile from continental margin to deep ocean floor. Based on data from the North Atlantic (after Heezen, Tharp and Ewing, 1959). Length of the profile is over 400 nautical miles.

land, with a gentle slope, 1 in 1000 or less, and is of varying width. It continues to an average depth of 183 m (100 fathoms), when there is a marked change in slope known as the *shelf break*, the gradient becoming 1 in 40 or more. The shelf break marks the beginning of the *continental slope*, which continues until the gradient begins to flatten out and merges into the *continental rise*, perhaps several hundred miles wide as in the North Atlantic, with a diminishing gradient. Eventually, at depths greater than about 5 km (2800 fathoms), the abyssal ocean floors are reached. The continental slopes at many points show erosional features known as *submarine canyons*. They are steep-sided gorge-like valleys incised in the sea floor (Fig. 1.2); some lie opposite the mouths of large rivers, as with the Hudson Canyon opposite Long Island. It is believed that many of the canyons have been excavated by turbidity currents, i.e. submarine currents carrying much suspended sediment and therefore denser than normal ocean water. In some instances they continue down to the continental rise.

Away from the continental margins, in the deepest parts of the oceans, the important *oceanic ridges* are located. These are submerged areas of high topography which have widths up to several hundred kilometres, with summits up to 4 km above the deep ocean floor. Their significance is discussed on p. 14.

Temperature gradient and density As is well known from the evidence of deep mining operations, temperature increases downwards from the surface at an average rate of 30°C per km. The rate is higher near an active volcanic centre. The pressure under which the rocks exist also increases with depth. Assuming that the temperature gradient persists at about the above rate, calculation shows that at a depth of 32 km the temperature would be such that most known rocks would begin to melt; but the high pressures prevailing at that

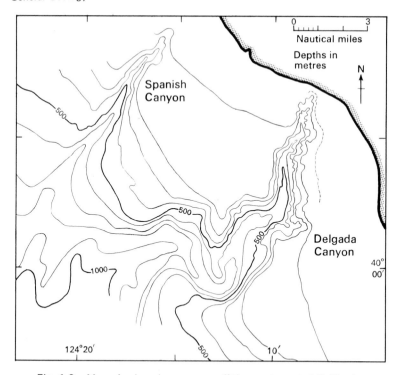

Fig. 1.2   Map of submarine canyons off the west coast of California.

depth result in the rock-material there remaining in a relatively solid condition.

The mean density of the earth, which is found from its estimated mass and volume, is 5.527 g per cm^3. This is greater, however, than the density of most rocks found at the surface, which rarely exceeds 3; sedimentary rocks for example average 2.3, and the abundant igneous rock, granite, about 2.7. In order to make up the mean density of 5.5 there must therefore be denser material at lower levels within the earth, i.e. below the crust. Our knowledge about the interior of the earth has come largely from the study of earthquakes, or seismology, which has shown that the earth has a *core* of heavy material with a density of about 8. Two metals, iron and nickel, have densities a little below and above 8 respectively, and the core is believed to be composed of iron or a mixture of iron and nickel. Surrounding the core is the region known as the *mantle* (Fig. 1.3) and overlying that the *crust*, which is itself composite. The mantle has a range of density intermediate between that of the core and the crust. In order to discuss further the seismic evidence for earth structure we now consider the subject of earthquakes.

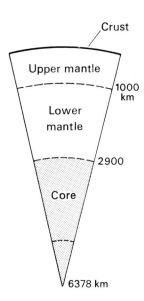

Fig. 1.3 Earth layers (after Bott, 1971).

Earthquakes The numerous shocks which continually take place are due to sharp movements along fractures (or faults, p. 228) which relieve stress in the crustal rocks. Stress accumulates locally from various causes until it exceeds the strength of the rocks, when slip along fractures occurs, followed sometimes by a smaller rebound. A small movement on a fault (perhaps a few cm or less) may produce a considerable shock because of the energy involved; and the fault 'grows' by successive movements. Earthquakes range from slight tremors which do little damage, to severe shocks which can initiate landslides, break and overthrow buildings, and sever supply mains and lines of transport. The worst effects are produced in soft ground where fissures may be opened and fault scarps formed.

Many earthquake centres are located along two belts of the earth's surface: one belt extends around the coastal regions of the Pacific, from the East Indies through the Philippines, Japan, the Aleutian Isles, and thence down the western coasts of North and South America; the other runs from Central Europe through the eastern Mediterranean to the Himalayas and the East Indies, where it joins the first belt (Fig. 1.4). Both these belts are in places parallel to the younger fold-mountain chains (p. 174), where much faulting is associated with the crumpled rocks; numerous volcanoes are situated along the earthquake belts. Many shocks continually occur also in zones of submarine fault activity such as the mid-Atlantic ridge (p. 15); and some in fault-zones on continents such as the Rift Valley system of

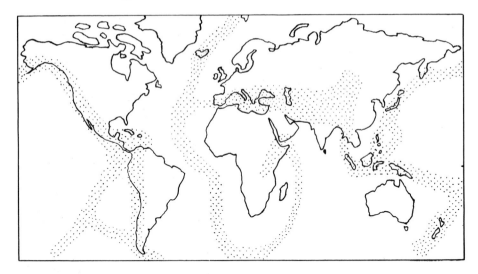

Fig. 1.4   Distribution of earthquakes; dotted areas are belts of active epicentres.

Africa. In Britain only small tremors and a few shocks of moderate intensity have been recorded; the latter include the Inverness earthquakes (1901, 1934), Nottingham (1957) and Dent, Westmorland (1970).

The intensity of an earthquake can be estimated from the effects felt by an observer or observed around him. Many observations when collected together are used to determine the centre of the disturbance. They are graded according to a *scale of intensity* such as the Mercalli Scale:

    I. Detected only by instruments.

    II. Felt by some persons at rest; suspended objects may swing.

    III. Felt noticeably indoors; vibration like the passing of a truck.

    IV. Felt indoors by many, outdoors by some; windows and doors rattle.

    V. Felt by nearly everyone; some windows broken, pendulum clocks stop.

    VI. Felt by all, many frightened; some heavy furniture moved, some fallen plaster; general damage slight.

    VII. Everyone runs outdoors; damage to poorly constructed buildings; weak chimneys fall.

    VIII. Much damage to buildings, except those specially designed (Note 1). Tall chimneys, columns fall; sand and mud flow from cracks in ground.

    IX. Damage considerable in substantial buildings; ground cracked, buried pipes broken.

    X. Disastrous; framed buildings destroyed, rails bent, small landslides.

    XI. Few structures left standing; wide fissures opened in ground, with slumps and landslides.

    XII. Damage total; ground warped, waves seen moving through ground, objects thrown upwards.

The intensity at points within the area affected can be marked on a map, and lines of equal intensity then drawn (*isoseismal lines*) to enclose all places where damage of a certain degree is done, giving an *isoseismal map*. For the Inverness earthquake of 1901 the innermost isoseismal line, within which buildings were damaged, was an ellipse 19 km long and 11 km wide, with its long axis at N 33°E. This direction is close to that of the Great Glen Fault, N 35°E, which passes near to Inverness (Fig. 8.7). The earthquake was caused by movement on this fault, which is part of an extensive fracture system that traverses Scotland and continues past Caithness and Orkney. Another, less severe, shock occurred in 1934. Frequent small movements also take place on the Highland Boundary Fault (Fig. 8.7).

A large displacement on the San Andreas Fault, California, in 1906 resulted in the disastrous earthquake that ruined San Francisco. The motion of the ground was mainly horizontal, one side of the fault moving relative to the other a distance of 4.6 m (15 feet) horizontally, parallel to the fault direction. Other shocks occurred in 1940 and subsequently, and surveys of the ground showed that the western side of the fault moves north relative to the eastern side (reference in Note 4). The NNW course of the San Andreas Fault system can be followed for over 1150 km (700 miles) and its trace at the surface is clearly seen from the air.

The amount of *energy* released in an earthquake is described by a number which is called the *magnitude* (*M*) (see Note 6).

Earthquake waves and seismograph records. Elastic waves are propagated in all directions from the centre of origin or *focus* of a shock. The point on the earth's surface immediately above the focus is called the *epicentre*. The elastic vibrations which travel out from the focus are of three kinds: (i) Compressional (or longitudinal) vibrations, denoted by the letter *P* (primary); these are the fastest and first to arrive at a recording station. (ii) Transverse or shear vibrations, called *S* (secondary), a little slower than the compressional waves. (iii) Surface or *L*-waves, transverse vibrations of long period which follow the periphery of the earth; they have a large amplitude and do the greatest damage (Note 2).

The vibrations can be detected and recorded by suitably placed *seismographs* (or *seismometers*). The principle on which these instruments work is that of a light beam, pivoted at one end to a frame and suspended, with a heavy weight near its free end. When the ground is shaken the frame of the instrument moves with the ground, but the lightly suspended weight remains relatively fixed in position owing to its inertia. Thus a motion is imparted to the beam, and its swing is recorded on a rotating drum by means of a ray of light reflected from a mirror attached to the end of the beam. Intervals of time are also marked on the record. Two seismographs are employed, to record the N–S and E–W components of the vibrations respectively, and a third to detect vertical movements. From the record or *seismogram* so obtained the times of arrival of vibrations are read off; and by assuming values for the velocities of transmission (see below) the

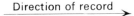

Direction of record

Fig. 1.5 Seismogram of distant earthquake (after Milne and Lee). *P* = longitudinal, *S* = transverse, *L* = surface waves.

distance of the epicentre from the recording station can be calculated. Explosions, also detected by seismographs, can be distinguished from earthquakes.

It is found that seismographs situated at distance up to 105° of arc from the epicentre record the onsets of *P*, *S*, and *L* waves (Fig. 1.5). But for greater distances than this the *P* and *S* waves are not recorded, until at stations 142° or more from the epicentre the *P* waves (only) are again received. They have, however, taken longer to arrive, and hence must have been slowed down over some part of their path through the earth. The region extending from 105° to 142° of arc away from an epicentre, in which no *P* or *S* waves are received, is called the 'shadow zone'. The interpretation of these facts was put forward by R. D. Oldham in 1906, in terms of an earth-core of different composition. Vibrations which penetrate to a greater depth than the 105° path (Fig. 1.6) enter the denser core and are slowed down by it. But since transverse vibrations are not received after 105°, the material of the core must have the properties of a fluid, which does not transmit shear vibrations. Later work has shown that the fluid core extends to within 2900 km of the earth's surface, i.e. its radius is rather more than half that of the earth. At the boundary of the core, which is determined from seismic data, there is a sharp discontinuity corresponding to the difference in physical properties of the core and the lower part of the mantle. The mantle transmits both *P* and *S* vibrations.

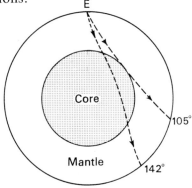

Fig. 1.6 Paths of earthquake waves through the earth (*E* = epicentre). Two paths only are shown, out of the many which radiate from the epicentre.

Records obtained relatively near an epicentre (within about 1000 km) yield information about the crust of the earth. It was noticed by the Serbian seismologist A. Mohorovičić, in 1909, that *two* sets of *P* and *S* waves were sometimes recorded, the two sets having slightly different travel times. He suggested that this indicated that one set of vibrations travelled by the direct path from the focus and the other set by a different route. In Fig. 1.7, P_g and S_g follow the direct route,

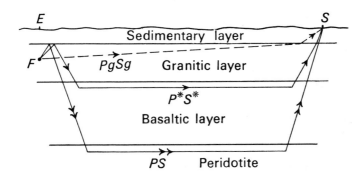

Fig. 1.7   Vibration paths in crustal layers. *E* = epicentre, *S* = seismograph. Focus (*F*) in granite layer. Each path represents a small fraction of the energy transmitted from the focus; refraction occurs at the interlayer boundaries.

while *P* and *S* are refracted at the boundary of a lower layer and travel there at a higher velocity. This boundary marks the base of the crust and is called the *Mohorovičić discontinuity* (colloquially 'the Moho'). Later a third set of vibrations was detected on some seismograms; they are called *P** and *S** and have velocities lying between those of the other two sets, suggesting that they follow a path in another kind of material in the crust (Fig. 1.7). The velocities for the three sets of waves, as determined by Jeffreys from European earthquake data, are as follows:

P_g	5.57 km s^{-1}	S_g	3.36 km s^{-1}
$P*$	6.65 km s^{-1}	$S*$	3.74 km s^{-1}
P	7.76 km s^{-1}	S	4.36 km s^{-1}

These values correspond to those derived from laboratory tests on the elasticity of the igneous rocks granite, basalt, and peridotite respectively. They are greater than velocities in average sedimentary rocks near the earth's surface (2 to 4 km s^{-1}).

Thus the fastest waves, *P* and *S*, travel for the greater part of their course in material approximating to peridotite of the upper part of the mantle, below the Mohorovičić discontinuity; their energy is dissipated by repeated refractions at the boundary. Above the M-discontinuity a *basaltic layer*, in which the *P** and *S** waves travel, forms the lower part of the crust (Fig. 1.7). *The granitic* layer, or upper part of the crust, transmits the P_g and S_g vibrations. The granitic layer is itself covered in places by a relatively thin veneer of sediments. The thickness of these crustal layers varies considerably from place to place; the average thickness of the whole crust in a continental area is about 35 km. Beneath a mountain mass, however, the crust is much thicker, as discussed below; outwards from a continental margin it becomes thinner, and beneath the oceans may have a thickness of only 5 km (Fig. 1.8).

Continental margin   Mountain range

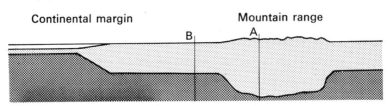

Fig. 1.8   Diagrammatic section through part of a continent. Sima shown by stipple, sial by light tint.

The picture we now have of earth structure is, therefore, that of a thin but composite outer shell or crust, overlying a solid mantle which extends to a depth of nearly half the earth's radius; and a heavy liquid core. Recent research suggests that within the core there is an inner solid core with a radius of about 1000 km.

Two conveniently short terms, sial and sima, can be used to denote the granitic layer (together with its discontinuous cover of sediments) and the basaltic layer respectively. *Sial* is coined from the initial letters of *si*lica and *al*umina, indicating the predominance of these two oxides in the composition of granitic rocks; and similarly *sima* from *si*lica and *ma*gnesia for the basaltic rocks. The sial has a density of about 2.7, and is crystalline and rigid; the sima, with a density of about 2.95, has lower rididity and yields slowly to long-continued stresses. The upper part of the mantle has a density of about 3.4, which increases with depth; recently, sub-divisions of the mantle have been proposed, and areas of modified mantle with a density of 3.15 may be present beneath some oceanic areas (Note 3).

Isostasy   The continental masses can be visualized as extensive blocks or 'rafts' of sial (i.e. of broadly granitic composition, although composite) supported by the underlying sima. The difference in density implies that the sial continuents are largely submerged in the heavier sima, rather like blocks of ice floating in water. Ideally, a state of balance—to which the term *isostasy* is given (from Greek, 'in equipoise')—tends to be maintained between large blocks of the sial crust and the sima. Thus the weight of a column of matter in a mountain region where the sial is thick, as at *A* in Fig. 1.8, equals that of Column *B*, where the sial is thinner and displaces less of the underlying sima. The columns are balanced at a depth (below sea level) where the weight of each is the same, and large topographical relief is said to be *compensated* by differences in rock density below it. The concept of isostatic balance has been tested by gravity surveys, which reveal excess or deficiency of density in the make-up of the crust for the area surveyed. When all the evidence is reviewed it appears probable that very large topographical features on the earth's surface are bounded by faults and supported by the upward pressure of the sima, i.e. they are isostatically compensated on a regional scale. The Alps, for example, are believed to be balanced in this way, their topographical mass above sea level being continued in a deep 'root' of granitic material (sial) below them. Local isostatic compensation for smaller masses is unlikely to be complete because their weight is partly supported by the strength of the surrounding crust, i.e. smaller mountains and valleys exist because of the crust's rigidity.

The transfer of load from one part of a continental area to another, by the processes of denudation and sedimentation, results in a local sinking of the area where sediment is deposited, accompanied by the slow outflow of sima from beneath it. Conversely there is a rise of a land surface undergoing denudation, as it is lightened, with an inflow of denser material below the area. Because of the difference in density, the amount of rise will not equal the depth of material denuded. If the densities of sial and sima are taken as 2.7 and 2.95 respectively, then the removal of 300 m of the former will be balanced by the inflow of about 276 m of the denser material, the final ground level when isostatic adjustment is complete being 24 m lower than before. It is thought that the height of the Himalayas, for example, has been maintained by this kind of mechanism during the erosion of their many deep gorges, involving the removal of great quantities of rock, much of which finds its way to the Ganges basin as sediment. Modern geophysical surveys have shown that continental margins are probably isostatically compensated at the present day.

Again, during the Glacial period, when thick ice sheets covered much of the lands of the Northern hemisphere (p. 65), the load of ice on an area resulted in the depression of the area. With the removal of the load as the ice melted, isostasy slowly restored the balance by re-elevating the area. In this way many raised beaches, such as those around the coasts of Scotland and Scandinavia, were elevated to their present positions after the melting of the ice (Fig. 8.16).

Ocean floors and continental drift   The idea of possible movement of the continents relative to one another, in the geological past, was first discussed at length by Alfred Wegener in 1912, and it soon became a matter of controversy. During the last twenty years, however, new evidence has come to light which gives strong support to the theory of continental drift. This evidence has arisen largely from the study of magnetism in rocks of the earth's crust, and from detailed surveys of the ocean floors, made possible by technical advances in depth sounding and the accurate fixing of positions at sea.

It had been noticed by Wegener and others that the shape of the coast lines of Africa and South America showed a marked resemblance on either side of the Atlantic. There are also many corresponding

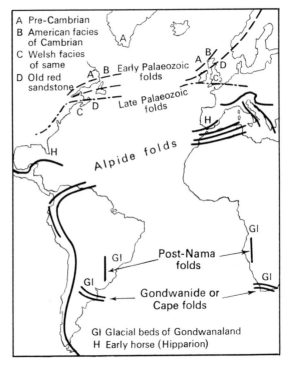

A Pre-Cambrian
B American facies
    of Cambrian
C Welsh facies
    of same
D Old red
    sandstone

B Early Palaeozoic
    folds

Late Palaeozoic
    folds

Alpide folds

Post-Nama
    folds

Gondwanide or
    Cape folds

GI Glacial beds of Gondwanaland
H Early horse (Hipparion)

Fig. 1.9   Geological resemblances across the Atlantic (after du Toit).

geological features between the two continents, such as belts of folded rocks in South Africa and North Africa, which run out to the coast, and have their counterparts in South America; and similarities in fossil faunas. These features (Fig. 1.9), which were set out in detail by A. L. du Toit (1937, see p. 20), would be simply explained if the two continents had originally been adjacent to one another. One example of the present-day distribution of certain fossil remains, the early horse (*Hipparion*), on either side of the Atlantic is shown in the figure. Within the last few years an accurate fit of South America against Africa along the continental shelves has been demonstrated (Fig. 1.10), and leaves little doubt that the two continents were once

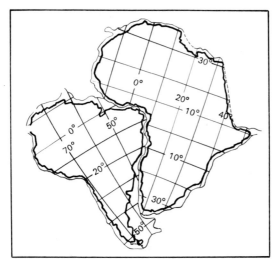

Fig. 1.10 Fit of Africa and South America at the 1000 fathom line (Bullard, 1964, *Q.J.G.S.* vol. 120; after Carey, 1958).

joined. A similar relationship can be found for the North Atlantic, where Canada and Greenland lay next to the Eurasian continent, with Greenland opposite to Scandinavia. When these positions are restored, mountain ranges such as the Appalachians in North America and the Caledonian folds of Scotland become continuous, and both have similar geological characters. Further, lands in the southern hemisphere, including South America, Africa, Australia, Antarctica and peninsular India, probably formed a very large continent in Carboniferous times, called *Gondwanaland* (Fig. 1.11), and have since moved apart to their present positions. It has been shown that

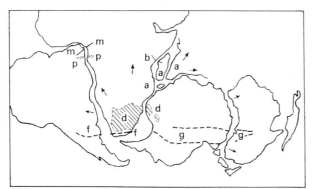

Fig. 1.11 Gondwanaland: re-construction of the fit of the southern continents (after Gilbert Smith and Hallam, 1970). Matchings include: *a*, Pre-Cambrian anorthosites; *b*, limit of Jurassic marine rocks; *d*, Mesozoic dolerites; *f*, fold-belt; *g*, geosynclinal (early Cambrian); *m*, mylonites; *p*, Pre-Cambrian geosyncline. Arrows show ice movement.

the Australian continent and Antarctica fit together with considerable accuracy with Tasmania lying near the Ross Sea area of the Antarctic, as in the figure. Certain geological features of the two continents then become aligned. The east side of peninsular India, with Ceylon, was probably joined to Antarctica, as is suggested when outlines of the land masses are taken at the 1000-fathom line. An extensive glaciation in Carboniferous times affected southern Africa, India, the southern margin of Australia, and parts of Brazil and Argentina, as evidenced by the glacial deposits found in all those areas. These and other geological facts would be readily explained if the glaciated lands were originally parts of one super-continent. When the latter broke up the separate parts began to move to their present-day positions; Africa and India moving northwards impinged on the southern margin of the Eurasian continent and ridged up the great fold-mountain systems of the Alps and Himalayas. It has been estimated that the rate of movement of the Indian block may have been as much as 20 cm per year.

The recognition of extensive fracture systems, with horizontal displacements of the order of hundreds of kilometres, has shown that very large fault movements form part of the architecture of the earth's crust. Such fractures include the San Andreas Fault near the west coast of California (p. 7), which is still active; the Great Glen Fault in Scotland, now known from recent geophysical surveys to extend north-north-east past Caithness; the Alpine Fault of New Zealand, and others. These are all transcurrent faults (p. 235) with large horizontal movements, as are probably those recently located in the deep ocean floor west of California (Note 4).

The modern study of the magnetism found in many rocks has yielded other independent evidence. Minerals which have magnetic properties are found in certain igneous rocks; when these small crystals were formed they acted as little magnets and became lined up in the earth's magnetic field at that time. Measurements show that the directions of magnetization in former geological periods vary considerably, after allowing for changes in the direction of the earth's field which take place periodically. An explanation was suggested in terms of relative movement between the continents. This study of *palaeomagnetism* has been important in stimulating the renewed investigation of continental drift.

Oceanic ridges. The submerged topography of the ridges is of great extent (p. 3). The best known example is the mid-Atlantic ridge, which was originally located from soundings along the course of trans-Atlantic telephone cables. It extends from Iceland southwards through the North Atlantic and, allowing for offsets due to east-west faults, continues into the South Atlantic (Fig. 1.13); thence it turns east and continues to the Indian Ocean. Other oceanic ridges lie below the East Pacific Ocean and between Australia and Antarctica. The mid-Atlantic Ridge, a profile of which is given in Fig. 1.12, rises nearly a mile and a half (2.3 km) above the ocean floor in its central belt, and to within 1200 fathoms (2.2 km) of the surface; it resembles a submerged mountain range. Along the central belt there is a deep narrow valley, the *median rift*, going down to 900 fathoms (1.64 km) or more below the heights on either side. Rock samples dredged up from the ridge are mainly of volcanic rocks such as basalt and related types, and appear to be material which has emerged from fissures, probably on the line of the median rift, and has spread

Flemish cap	Mid-Atlantic ridge	English Channel
Ocean basin	⌐Rift	Ocean basin

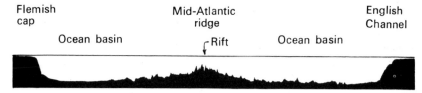

Fig. 1.12   Profile across the mid-Atlantic Ridge (after Heezen, 1959).

laterally over the ocean floor. The process is known as *ocean floor spreading*. Outflow of material from below the crust has moved in either direction away from the median area, and has resulted in the formation of new ocean floor in the region of the ridge.

The magnetization of the rocks of the ridge found from samples provides independent evidence for the lateral spreading. The originally molten material cooled and became magnetized in the direction of the earth's magnetic field. Surveys have revealed that on either side of the median rift the Atlantic Ridge is composed of alternate parallel strips of positively and negatively magnetized rocks. It is known that the magnetic field of the earth undergoes periodic reversals of polarity; the last reversal was about 700 000 years ago, and other reversals occurred at about $2\frac{1}{2}$ and $3\frac{1}{2}$ million years ago. The alternate strips of positive and negative magnetization would therefore be explained if they had been formed at successive times, and had moved in opposite directions by spreading from a central rift. The samples increase in age with distance from the median rift. Thus, rocks dredged up from 64 km to the west have an age of about 8 million years (found by potassium/argon dating, p. 18), whereas samples from near the rift itself are dated at only 13 000 years. On this interpretation, therefore, ocean floor spreading in the Atlantic has gone on for millions of years during the recent geological past, and it is believed to be the means by which the oceanic ridge has been formed. From figures such as those given above, a rate of spreading of between 1 cm and 3 cm per year can be calculated. Part of the adjustment to the spreading

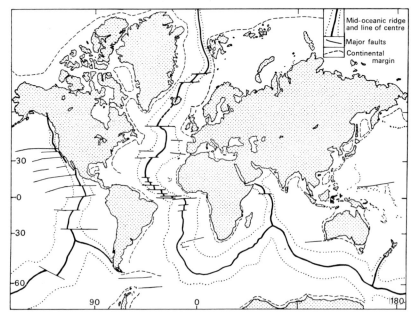

Fig. 1.13 Map of the oceanic ridges (after B. C. Heezen, 1963, in 'The Sea'). The heavy line shows the mean position of the centre of a ridge; thin lines show displacements by transcurrent faults.

is by horizontal slips along faults transverse to the oceanic ridges (Fig. 1.13); these faults are often marked by steep linear scarps on the ocean floor, which have been found by surveys.

Further, ocean floor spreading is thought to indicate the mechanism which has resulted in continental drift. The temperatures of rock of an oceanic ridge are found to be higher at the centre of the ridge than on either side of it. Material from the earth's mantle rises towards the surface below the hotter central part of a ridge; and it is believed that the cause of this upward flow is the existence of slow moving convection currents in the earth's mantle (Fig. 1.14). The currents rise towards the base of the crust and spread out horizontally, passing beneath the edges of the continents and descending again. The hotter rock in the rising current is less dense and possesses buoyancy, which is the driving force of the mechanism. As the current flows horizontally outwards beneath the oceanic ridge it exerts a drag on the overlying crust, and separates the crustal blocks on either side of the rising

Fig. 1.14 Diagram of a convection current (after Bullard, 1964, *Q.J.G.S.*, 120). Margins of continents are shown stippled.

current. Thus the oceanic ridges are zones that appear to mark the former junctions of adjacent continents which have drifted apart. Calculations show that for the mid-Atlantic Ridge the rate of separation is about 2 cm per year; and for the East Pacific Ridge the rate is twice as much. These rates may have varied in the past. The separation of the American continents from Eurasia and Africa probably began at the end of Jurassic times, about 200 million years ago (Notes 4 and 5).

Origin of earth and moon   As a member of the solar system the earth is one of the inner, smaller planets. Theories of its origin are broadly of two kinds, involving either the cooling and condensation of hot gaseous material or the formation of a cold body by accumulation of particles from a swarm in space. Of the former, the Tidal Disruption Theory proposes that the near approach of the primitive sun and another star, or a grazing contact between them, raised a tidal bulge on the sun which became detached as an elongated filament. This filament broke up into parts which in due course became the planets of the solar system, all moving in the same direction around the sun. Modifications of this theory have been introduced to meet mathematical objections concerning the speed of approach to the two stars and the angular momentum of the system.

The second kind of theory visualizes the growth of planetary bodies by the accumulation of solid fragments from swarms moving through space; clouds of silicate dust, for instance, are believed to exist in space. The earth, if formed in this way by accumulation, could then have acquired its layered structure during early geological time by gravitational settling of heavier elements towards its centre; meteorites which fall on the earth's surface may be fragments left over from the original swarm.

Much new information has come in, and is coming, from space satellites which have been launched since 1960, and the fresh data gives rise to new ideas about the earth's origin. A satisfactory theory which will take account of all the known facts lies in the future; the speed of progress of modern research may allow it to be formulated sooner than could have been contemplated a decade ago.

The moon may have originated from a cloud of matter, in the way indicated above for the earth, and at the same time; or, according to an earlier theory, it may have been spun off from the primitive earth when a tidal bulge was formed on the planet before its outer shell had become rigid (and when the earth had a higher speed of rotation). The separation would have occurred perhaps 3500 million years ago. The density of the moon is similar to that of the material forming the earth's crust. A third theory is that the moon was a small planet which was captured by the earth in orbit.

It has also been suggested that the earth, moon, and Mars (which has a similar density) once formed a single planet, and that both Mars and the moon were spun off from the earth when its rotation was faster than at the present day. Thus with several theories in the field, the way in which the moon was formed remains uncertain until further facts come to light. However, samples of rock brought back from one part of the moon's surface at the time of the Apollo spacecraft landings in 1969 have proved to be similar to certain terrestrial volcanic rocks, and their age is estimated to be between 3000 and 4000 million years. We now turn briefly to the question of the age age of the earth.

Age of the earth   Modern estimates of the age of rocks which form the earth's crust are based on determinations on radioactive minerals contained in the rocks. Before these *radiometric methods* were developed, early estimates of age had been made from data such as the amount of salt in the oceans and its rate of accumulation. These gave results which were too low because of inaccurate assumptions. The discovery of radioactivity came at the time when Pierre and Marie Curie, in 1898, first isolated compounds of radium. This element is found, together with uranium, in the mineral pitchblende, a blackish coloured pitch-like substance occurring in certain igneous rocks and veins. Uranium undergoes a transformation, with the emission of helium ions and other particles, into an isotope of lead, and radium is formed at one stage in the process. The rate at which this radioactive change takes place appears to be constant, no variation in it having been detected. Similarly the element thorium undergoes a transformation into another isotope of lead, at a constant rate. The rates being known, determinations of the amount of uranium and thorium in a mineral containing those elements, and the lead content of the mineral, give data for calculating the age of the mineral, i.e. the length of time that has elapsed during the formation of the lead. The helium content can also be used in a similar way, if all the helium generated is retained in the mineral.

Other radioactive changes which take place include potassium into argon, and rubidium into strontium, and methods based on these transformations are also used for estimating the age of minerals containing K and Rb. The potassium/argon method has become important because potassium is a constituent of many feldspars, a group of abundant rock-forming minerals.

The greatest age so far determined is about 3500 million years, for a mineral in a rock from the ancient group known as the Pre-Cambrian (p. 250). The age of formation of the earth's crust, of which the rock is a part, must be still greater, and the earth's age may therefore be of the order of 4000 to 4500 million years. The larger subdivisions of geological time, with radiometric ages in millions of years, are as follows:

Group	System
CAINOZOIC	{ Quaternary { Tertiary
	(70)
MESOZOIC	{ Cretaceous { Jurassic { Triassic
	(225)
NEWER PALAEOZOIC	{ Permian { Carboniferous { Devonian
	(400)
OLDER PALAEOZOIC	{ Silurian { Ordovician { Cambrian
	(570)
PRE-CAMBRIAN	(3500+)

The long periods of time implied by the above list (or 'geological column') may at first give the impression that geological processes go on at a very slow pace. But the time-scale is quite different when we consider individual items which come within the broad outline of events. An ice age may come and go in perhaps one or two million years; the erosion of coastal cliffs by the sea is sometimes rapid, producing marked changes in 100 years or even in a man's lifetime; and the sudden release of energy at the time of an earthquake can give rise to shocks lasting only a few seconds but causing great damage in the area affected. The earth is not a static body, but in a state of continual change.

NOTES AND REFERENCES

1. In engineering design of earthquake resisting structures, values for the acceleration of the ground are used, rather than a Scale of Intensity. And see Chapter 13, p. 379.
   References to the design of works, taking into account natural forces, are given by Hollins, E. P. 1971, in Bibliography of Earthquake Engineering, 3rd edition, *California Earthquake Engng. Res. Inst.*
2. Details of elastic vibrations can be found in a text-book such as Richter, C. F. 1958, Elementary Seismology. Freeman & Co., which also gives

isoseismal maps of the Californian earthquake of 1906, the Tokyo earthquake of 1923, and others; and in Milne, J. and Lee, A. W. 1950, Earthquakes and other earth-movements, 5th edition, Kegan Paul, London.

3. Discussed in The Earth's Mantle, 1967, Ed. T. F. Gaskell; in particular Chapter 13. Academic Press, London and New York.

4. A symposium on Recent Crustal Movements, edited by Collins, B. and Fraser, R. 1971, is given in Bulletin 9 of R. Soc. N.Z.; and includes papers on tectonic zones and earthquakes in New Zealand, Australia, and elsewhere; and a group of papers dealing with movements on the San Andreas Fault, California.

5. In the early 1960s, when the validity of continental drift was largely agreed, the idea was advanced that the earth's surface could be considered as a mosaic of about twelve large, rigid *plates*. The plates could move relative to one another and could also move over the material which underlay them. The study of this concept became known as Plate Tectonics. In oceanic areas the plates are covered by a thin crust essentially formed by the spreading of the sea floor at oceanic ridges, and are termed oceanic plate; in continental areas the plates are covered by thick crust (continental plate). Where two plates moved apart, new material emerged from below to build a ridge, as described on p. 15. Where two plates moved together, the leading edge of one was pushed down under the other, sinking into the earth's mantle at deep trenches in the ocean floor (such as the Aleutian, Japanese, and Mindanao trenches). Along lines of faulting (see Fig. 1.13) one plate could slide past another.

A recent account of these studies, with many references, is given by E. R. Oxburgh, 1974, in 'The Plain Man's guide to Plate Tectonics', *Proc. Geol. Assoc.,* **85,** p. 299.

6. A *scale of magnitudes* was devised by C. F. Richter (1952), based upon the maximum recorded amplitudes shown on seismograms made with a standard seismometer. It is a logarithmic scale of total energy released, ranging from magnitude 0 (8×10^9 ergs) to magnitude 9 (1.2×10^{25} ergs). The smallest felt shocks have magnitude $(M) = 2$ to $2\frac{1}{2}$; damaging shocks have $M = 5$ to 6; and any earthquake above $M = 7$ is a major disaster (e.g. Hawkes Bay, New Zealand, 1931; Niigata, Japan, 1964; Guatemala, 1976). Very great earthquakes have $M = 8$ or over, e.g. Chile, 1960; Alaska, 1964.

The Richter Scale of magnitudes and the Mercalli Scale of intensity (p. 6) are not strictly comparable; but $M = 5$ (damage to chimneys and plaster, etc.) corresponds roughly to grade VI on the Mercalli Scale.

GENERAL REFERENCES

BOTT, M. H. P. 1971. The Interior of the Earth. Edward Arnold, London.

DU TOIT, A. L. 1937. Our Wandering Continents. Oliver & Boyd, Edinburgh, an early statement of evidence then available for continental drift.

EIBY, G. A. 1967. Earthquakes (revised edition). Frederick Muller, London. A short, readable modern account.

Plate 1

Surtsey Volcano, Iceland, in eruption 1963. The dark cloud near the cone is a dense ash emission. (*Photograph by G. P. L. Walker*)

Plate 2

(a) Dune-bedding in Permian sandstone, Ballochmyle, Ayrshire.

(b) Horizontally bedded sandstones in sea-stack, Yescanby, Orkney.

2 Surface Processes

Introduction Land areas are continually being modified by processes which originate and act at the earth's surface; the general term for these processes is *denudation*, in which both weathering and erosion are included. Rocks exposed to the atmosphere undergo *weathering* from atmospheric agents. *Mechanical* weathering, or disintegration, is the breakdown of rocks into small particles by the action of agents such as rain, frost, and wind, all of which are helped by gravity; and in some climates changes of temperature produce flaking. *Chemical* weathering, or decomposition, is the breakdown of minerals into new compounds by the action of chemical agents, such as acids in the air and in rain and river water. By all these processes a covering layer of weathered rock is formed on a land surface. As parts of this cover are continually removed, fresh material comes under the influence of the weathering agents and the work of denudation continues; in some circumstances the weathered material may remain in position as a residual deposit.

As well as the atmospheric processes, agents of *erosion*—rivers, moving ice, water waves—contribute to the denudation of the land in their particular spheres of action; they also transport weathered and eroded material away from areas where it is derived, to form deposits of sediments elsewhere. According to the climatic conditions under which the denudation takes place a particular kind of activity may be enhanced: for example, the work of rain in tropical climates, frost and ice in glacial conditions, wind and insolation (changes of temperature) in semi-arid and desert climates. The following discussion is divided into sections which describe the main physical agents of weathering and erosion; but often two or more processes operate together and their interaction is then indicated.

Geological work of rain Except where bare rock is exposed, the surface on which rain falls consists of the soil which forms the upper part of the layer of weathered material. This top-soil ranges in depth from only a few inches to several feet, according to the type of rock from which it has been derived. It is in general a mixture of inorganic particles and vegetable humus, and has a high porosity, i.e. a high proportion of interstices in a given volume. The soil grades down into 'sub-soil', which is a mixture of soil and rock-fragments, with decreasing organic content, and then into weathered rock passing down into unweathered rock. In many places the soil (or A-horizon) and sub-soil (or B-horizon) have been derived *in situ* from the rock which underlies them (Note 1). A vertical column showing this sequence is called a soil-profile. The pore spaces in the soil are filled with air together with water from rainfall; some of the latter passes through the soil and percolates downwards into the rock below, where it is stored as *groundwater* in the zone of saturation, filling the interstices, joints, and other spaces in the rocks there. The infiltration of rain at the surface is discussed further in Chapter 12.

The *mechanical* action of rain, falling continually on such a surface, is to dislodge loose particles of soil and rock which are gradually washed down slopes to lower levels, a process known as *rain-wash*. It is the means by which rivers receive much of the sediment they carry in suspension (some is eroded from their banks), and the muddy appearance of a river after rainfall is evidence that the land surface is slowly but continually being lowered. The denuding effects of heavy showers and storms can be severe, especially in regions where a covering of vegetation is lacking. The rain may then be channelled into courses which collect into deeper channels or gullies, especially when heavy storms or cloudbursts occur. Cloud-bursts (a high rate of precipitation over a relatively small area) can produce deep gullies, entrenched in the surface, and cause local damage by the destruction of roads and live-stock and the undermining of farm buildings. The running water thus derived from rainfall then becomes an agent of erosion (p. 36). Gulleying is common in sub-tropical countries, but is occasionally found also in more temperate climates.

A cover of vegetation protects the ground from the immediate effects of the impact of rain, and the clearing of wooded areas has been followed in many places by considerable denudation of the bare ground surface. If this proceeds for some time, the entire covering of soil may be removed and the area rendered useless for cultivation. (Wind also plays a large part in soil erosion, as discussed on p. 31).

In areas of thick soil containing embedded boulders, *earth pillars* may be produced by rain denudation. These are more or less slender columns of soil each capped by a large stone or boulder which has preserved the material below it from being washed away. Groups of earth pillars are developed on valleys slopes, in earthy or clayey deposits with boulders, in the Tyrol and elsewhere. Occasional examples are found in Scotland; at Fochabers, pillars of soft conglomerate capped by boulders have been preserved. In soft, easily denuded formations of this kind the pillars stand for a time after the material between them has been removed by the action of rain. Deposits of unconsolidated volcanic ash have been carved into earth

Fig. 2.1   Earth-pillars at Urgüp, Turkey, eroded from a deposit of volcanic ash and capped by blocks of lava which were embedded in the ash.

pillars at Urgüp in Turkey (Anatolia) and form striking scenery, with volcanic 'bombs' embedded in the ash serving as cap-stones (Fig. 2.1).

Mud-flows are formed when rain, soaking into an upland area of deep soil over a long time, has built up a high water-content in the soil. The unstable, water-logged mass becomes mobile when sufficient pressure is developed to burst the lateral restraint at a weak point, and the mass then moves down slopes as a mud-flow. It may engulf or displace buildings and lines of transport before coming to rest at a lower level, and valleys have been temporarily blocked by such mud-flows, leading to local flooding.

Where *deep weathering* has taken place, by the action of ground-water which has soaked through the rocks, a thick cover of rotted material lies above an irregular surface which bounds the solid rock below. Chemical effects play an important part here. In areas of jointed igneous rocks such as granite, for example, the groundwater has become acid during its passage through the soil; then, penetrating along joints it has reduced the granite to a crumbly, rotted mass often many metres in depth. In parts of Australia, and elsewhere, depths of 60 to 90 m of weathered granite have been recorded. On Dartmoor, England, weathered granite up to 10 m deep has been encountered in excavations, especially in the proximity of faults, sometimes enclosing hard masses of the unweathered rock (corestones) (Note 2).

The formation of granite *tors* is related to the frequency of jointing in the granite, illustrated in Fig. 2.2; the tors are upstanding masses of more solid rock, preserved where the spacing of joints is wider, in contrast to the adjacent rock; the latter has been more easily and therefore more extensively denuded owing to the presence in it of closely spaced jointing, which has allowed more rapid weathering to go on.

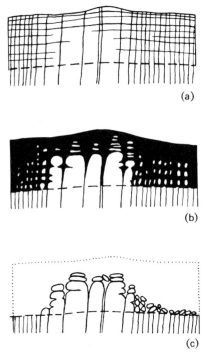

(a)

(b)

(c)

Fig. 2.2   Stages in the formation of granite tors (after Linton, 1955).

In sub-tropical climates, where ground covered with dense vegetation is subject to heavy rainfall—as in the monsoon areas of East Asia—very deep zones of weathered rock have been formed. Conditions of this kind were found in the Cameron Highlands, Malaya, where a hydro-electric scheme located in steep terrain was completed in 1963, and depths of weathered rock up to several hundred metres were encountered (Note 3).

Unloading.   One result of denudation is to reduce the load on an area as the removal of rock cover proceeds, leading to the relief of stress in the rock below. The unloading allows a small, mainly vertical, expansion which gives rise to the formation of sheets of rock by the opening up of joints parallel to the ground surface. This is frequently seen in igneous rocks such as granite (*cf.* Fig. 2.2), where sheet jointing is developed in the upper part of a granite mass; in valleys, for instance, the surfaces of parting lie parallel to the valley slopes and approach the horizontal at the bottom of a valley. The frequency of the sheet joints diminishes with depth, i.e. it is related to distance from the topographical surface. The production of smaller platy rock-fragments in a similar manner, sometimes seen in quarries, is known as spalling. The formation of other joints by the release of strain energy, in rocks which have been moved by some structural process from a deeper position to one where confining stress is lower, is discussed in a later chapter (p. 243).

The *chemical weathering* effects produced by rain are seen in its solvent action on some rocks, notably limestones; the process depends on the presence of feeble acids, derived from gases such as CO_2 and SO_2 which are present in the air in small quantities and which enter into solution in rain-water. An economic aspect of this chemical action is the decay of building stones, especially under the atmospheric conditions which prevail in cities, where a much higher content of impurities is present in the air than in rural districts.

While some other rocks are susceptible in a small degree to the solvent action of rain, limestones frequently show marked solution effects. The calcium carbonate of the limestone is dissolved by rain-water containing carbon dioxide, and is held in solution as calcium bicarbonate, thus:

$$CaCO_3 + H_2O + CO_2 \rightleftharpoons Ca(HCO_3)_2$$

The surface of a limestone area commonly shows solution hollows, depressions which may continue downwards as tapering or irregular channels. These may be filled with sediment such as sand or clay, derived from overlying deposits which have collapsed into the solution cavity after its formation had proceeded for some time. The upper surface of the Chalk shows many features of this kind, known as 'pipes'. In bare limestone districts, such as areas of the Carboniferous Limestone in the Pennines, the surface of the rock is channelled by runnels along which rain has trickled, and presents an irregular appearance. Vertical joints in the rocks are widened by solution as the rain passes down over their walls, and are then known as *grikes*. Solution may lead to the formation of *swallow holes*, rough shafts which communicate with solution passages at lower levels along which the water flows, especially where surface water is channeled to points at which major vertical joints intersect. Underground caverns are also formed by solution aided by the fall of loosened blocks of limestone from the roof of the cavern. Gaping Ghyll, on the slopes of Ingleborough in Yorkshire, is a swallow hole with a depth of 111 m (365 feet) leading to a long cavern system

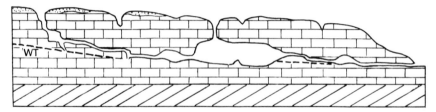

Fig. 2.3 Section through a limestone plateau to show solution features. *WT* = water table.

below, the water flowing thence into the surface drainage at Clapham (Fig. 2.3). Much larger systems of caves and solution channels are found, for example, in the French Pyrenees, the limestone plateau of Kentucky, and elsewhere. The Karst area of Istria and the Dalmatian coast of the Adriatic has given the name *karst topography* to land-forms which are characteristic of the chemical weathering of limestone; karst topography is also well seen in the Burren of County Clare, Ireland (Note 4).

The underground circulation of water helps to extend caverns and channels, particularly in hard limestone formations; streams which once flowed on the surface come to flow along bedding planes and joints below ground. In the Cheddar caves of the Mendips the former surface stream now runs 15 to 18 m below the floor of the caverns, which are dry. Cheddar Gorge, 128 m deep at one point, is thought to have been a large cave system formerly eroded, which has become exposed at the surface by the collapse of its roof. As water charged with calcium bicarbonate trickles over the walls and drips from the roofs of caves, part of it evaporates and calcium carbonate is slowly re-deposited as loss of carbon dioxide occurs (i.e. the equation given on p. 25 is reversible). In this way masses of *stalactite*, hanging from the roof or coating the walls of a cave, are formed, sometimes making slender columns where they have become united with *stalagmites* which have been slowly built up from the floor of the cave, on to which water has dripped over a long period of time. Sheet stalactite coats the walls of many caverns and may be coloured by traces of iron and lead compounds.

Weathering of rocks can also result, in some conditions, from pressures exerted by the crystallization of salts from solutions. This process operates particularly when the solutions occupy confined spaces, such as the pore spaces in rocks; the expansion which accompanies the growth of crystals then helps to disintegrate the fabric of the rock. The process can be prominent in desert areas (see p. 35).

The work of frost In cold climates repeated freezing breaks off flakes and angular fragments from exposed rock surfaces, a process referred to as the operation of the 'ice-wedge'. It is this which produces the serrated appearance of a high mountain sky-line, and aids in the formation of screes on mountain slopes. Water enters rocks along pores, cracks, and fissures; the ice formed on freezing occupies nearly 10 per cent greater volume, and exerts a pressure of about 13.8×10^6 N/m^2 (2000 psi) if the freezing occurs in a confined space. It is therefore like a miniature blasting action and brings about the disintegration of the rock. The loosened fragments fall and accumulate as heaps of *scree* or *talus* at lower levels, and this material may later be consolidated into a deposit known as *breccia*. By the removal of the fragments the surface of the rock is left open to further frost action, and so the process goes on. Some well-known screes include those in the English Lake District along the eastern side of Wastwater, where the mountain slopes fall steeply to the water's edge. Joints and cleavage planes in rocks assist the action of frost and to some extent control the shape of the fragments produced.

Where the ground is flatter, frost-shattered rock may lie at the surface, forming an uneven blanket of angular stones. These *block-fields*, as they are called, move only slowly if at all, as individual fragments are lifted by the growth of ice crystals beneath them and lowered when thawing occurs. Such deposits can occasionally be observed in upland areas in Great Britain.

Experimental work on rock samples has shown that frost-shattering depends more on the number of freeze-and-thaw cycles to which they are subjected than on the intensity of the freezing. The presence of planes of weakness in a rock increases the rate of shattering, and rocks such as shale, built of very small mineral constituents, disintegrate to a great extent; a coarse igneous rock such as granite is more resistant. It was found that very little material of smaller size than 0.6 mm (the upper limit of the silt grade, p. 177) is produced by the frost action (Note 5).

Permafrost. The term 'permafrost' is used to denote perennially frozen ground, which has remained below 0°C for many years and in most cases for thousands of years (Note 6). Conditions of this kind at the present day are found within the Arctic Circle, but extend well south of it in continental areas, in some instances as far south as 60°N, as in North America. It is estimated that one-fifth of the earth's land surface is underlain by permafrost. The impervious permafrost layer is formed in the ground a little below the surface, and may range in thickness from less than 1 m to hundreds of metres; thicknesses of over 400 m have been reported from boreholes in Alaska and Spitzbergen. Permafrost requires a cold climate with a mean annual air temperature of -1°C (30°F) or less for its formation and maintenance. The frozen ground conditions are primarily due to the loss of heat from the surface exceeding the amount received from solar radiation. In the Arctic the period of summer warming is short, and for many months of each year heat is dissipated from the ground into the atmosphere.

The upper limit of continuous permafrost in the soil is generally situated between about 10 and 60 cm from the surface, but well-drained soils may thaw annually to a depth of one or two metres during the milder season. This shallow zone which freezes in winter and thaws in summer is known as the *active layer*. The thawing of near-surface soils means the release of large volumes of water from within their mass, and this water cannot drain away through the still frozen ground below. It accumulates in the soil, which then flows readily down slopes, even those as low as 3 degrees. The amount of material involved in the movements is large and flow is comparatively rapid. A landscape under such conditions is reduced to long smooth slopes and gently rounded forms.

The frozen ground may contain much ice, especially where the soil texture is of fine grain and poorly drained; the ice occurs in thin films, grains, veinlets, vertical wedge-shaped masses, layers, and irregular masses of all sizes. In coarse-grained sediments the ice binds the grains together. Some kinds of ground contain little or no ice (dry permafrost). Unfrozen zones within perennially frozen ground may occur near the surface, and may mark permeable water-bearing layers or indicate past climatic fluctuations. Permafrost has been responsible for the preservation of many fossils, including complete mammals such as the mammoths unearthed at Siberian localities.

Near the margins of permafrost areas, where nearly flat slopes are covered by mud and rock fragments, the slow movement of this surface material over the frozen ground beneath it takes place by repeated freezing and thawing. The process is known as *solifluction* (= soil flow).

Frost-heaving occurs when the freezing of the soil results in the formation of layers of segregated ice at shallow depths. Each lens of ice is separated from the next by a layer of soil, whose water content freezes solid. The ice lenses vary in thickness from a few millimetres to about 30 mm. The total heave of the surface is approximately equal to the aggregate thickness of all the ice layers.

Many engineering problems arise from this cause in areas of permafrost in connection with bridge foundations and rail and road construction and maintenance, as in Alaska (Note 7). The disposal of sewage from large camps is difficult and contamination of surface water supplies may easily occur. Normal percolation of water downwards from the surface is prevented by the impervious frozen layer, and a concentration of organic acids and mineral salts in solution is built up in the shallow depth of soil above it. Removal of swamps, which was carried out on stretches of Alaskan highways, resulted in the formation of a deeper active zone, because the swamp layer of matted decayed vegetation when *in situ* formed a good insulation against frost.

The frost-heaving of piles used to support structures is promoted by forces originating in freezing of the active layer and is a common problem in the Arctic. Structures built on the piles are heaved up with them; and in fine-grained sediments, such as silts, the upward-acting forces increase as the frozen layer becomes thicker by seasonal penetration into the underlying silt.

The construction of buildings which are heated has the effect of inducing a rise of temperature in the ground beneath the structure, which may then undergo differential settlement. In order to avoid this, a structure can be built a little above ground level with large air spaces beneath it. Cold air in winter then circulates under the building and counteracts the heat from it (Note 8).

Geological work of wind   Wind is one of the two natural agents which transport rock material against gravity; its denuding action is seen most prominantly in regions which have a desert climate. Blowing over weathered surfaces it removes small loose particles of dry and decayed rock, both in the deserts and in more temperate regions. In desert areas sand particles are carried along near the ground in vast quantities, giving rise to repeated dust-storms. The grains may be driven against rock surfaces standing above the level of the desert, and then act as an abrading sand-blast. Rocks are smoothed by the wind-blown sand, sharp corners are rounded, and softer bands of rock become more deeply etched than the harder layers. These features are typical of surfaces which have been subject to wind erosion over a long period of time (Note 9). Upstanding masses of rock on the floor of a desert are undercut by eddying sand near the ground, so that they become 'mushroom' shaped, with the base narrower than the top (Fig. 2.4). The blown sand accumulates in suitable situations to form sand dunes.

Evidence of wind action in earlier geological times is sometimes seen in areas which are no longer arid; at Mountsorrel near Leicester, where granite is covered by Triassic red marl, the surface of the granite when excavated was found to be smoothed and to show unmistakable signs of wind erosion, pointing to a former desert climate in England. In making this deduction we are interpreting past events in terms of processes seen at work today; on this principle (Hutton's 'uniformit-arianism', 1795) events recorded in the rocks can be understood by reference to the present-day activity of geological agents in different places and climates.

Wind-blown (eolian or aeolian) grains become worn down to

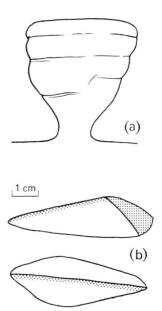

Fig. 2.4 (*a*) Mushroom rock (height about 8 to 10 m); (*b*) dreikanters.

well-rounded, nearly spherical forms with frosted surfaces, and from their shape are called 'millet seed' grains. This rounding is more perfect than for water-worn sand, and the grains are also well graded, i.e. of nearly uniform size, since wind of a given velocity cannot move particles larger than a certain diameter (or weight). Particles of quartz 1 mm diameter need a wind velocity of 8 metres per second to move them; a wind of 4 metres per second would lift grains up to about 0.35 mm diameter, (see Fig. 10.2, p. 303). Most wind-blown grains of the deserts are between 0.15 and 0.3 mm in size. Wind also exerts a winnowing action: the finer particles of dust are separated from the sand and carried over larger distances, to be deposited far away from their source (see *loess*, below).

The sand is swept along until it accumulates against obstacles to form dunes. Pebbles and boulders lying on the hard floor of a desert (and too large to be lifted by the wind) become smoothed and faceted by exposure to sand-storms, and frequently present three-cornered shapes whence they are known as *dreikanters* (Fig. 2.4*b*). The surface of a desert may be worn down locally until a level is reached where groundwater is present in the rocks. Many hollows in which *oases* are situated have been eroded down to the level of the water-table. The wet sand is the lower limit of wind action, as in the deep depressions of the Egyptian desert (p. 34). Lines of communication may be affected by wind-blown sand in arid countries. To avoid the accumulation of sand alongside railway embankments, as in the Sudan, culverts can be constructed to allow for easy passage of the wind and its load of sediment.

Perhaps the most serious damage caused by the action of wind is the removal of vast quantities of dry soil in many regions of the world. The 'Dust Bowl' areas of Kansas and Nebraska provide an example of widespread *soil erosion* in a district of low rainfall. A century ago, this region of the Great Plains, between the Missouri River and the Rocky Mountains, was a short grass country supporting large herds of buffalo. Settlers used the country as a cattle belt and later (during the 1914–18 war) ploughed up large areas for wheat, obtaining big crops from the fertile soil. Over-cultivation followed, trees were cut down and swamps drained, and the soil became dry and exhausted in less than twenty years. A series of droughts followed, and wind began to scatter the loose dusty soil, reducing large areas to desert. The district became a wind-eroded waste, bare of vegetation and useless for cultivation until sufficient time had elapsed for new soil to be formed. Measures to promote the formation of new soil were put in hand and have slowly taken effect, and the area is again yielding crops. Other instances of soil erosion, for example in Kenya and Nigeria, have arisen largely through the overstocking of sparse grasslands with cattle and sheep. Depletion of the scanty cover of grass soon follows and erosion of the soil begins. On a minor scale, soil erosion or 'blowing' has affected light soils in East Anglia from time to time.

The *deposits* formed by the action of wind include (i) coastal sand-hills in temperate regions; (ii) desert dunes, such as *barchans* (crescentic dunes), and *seifs* (ridge dunes), and the larger sand accumulations of desert areas (e.g. the Sand Sea of the Egyptian Desert); and (iii) *loess* deposits.

(i) Mounds of blown sand, or dunes, are piled up by the prevailing wind in some coastal regions of more humid climate as well as in arid countries. Thus the western coasts of Europe are liable to dune formation because of the prevailing south-west winds, for example on the west coast of France south of the Gironde estuary. Accumulations of blown sand occur on the East Anglian coast and in other parts of Britain. The dunes are not stationary but move continually and may overwhelm land areas; stages in the advance of a sand-dune are illustrated in Fig. 2.5. Sand is blown from the back of the dune, over the crest, and dropped on the leeward slope at its natural angle of repose, 30° to 35° according to the size of grains. The dune thus migrates in the direction of the prevailing wind until arrested by some obstruction.

Evidence for the movement of coastal dunes is sometimes provided by old records and maps. At Eccles on the coast of Norfolk, the

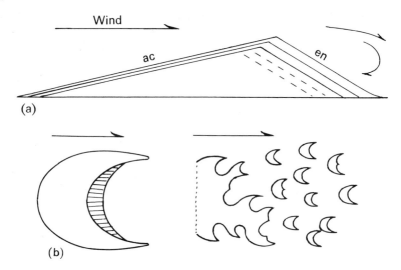

Fig. 2.5 (*a*) Structure of a sand dune. *ac* = accretion layer, *en* = encroachment layer; broken lines indicate successive positions of sand-slopes. (*b*) Barchan dune, slip face shaded; group of barchans (right) advancing in direction of wind.

church was almost completely buried by sand in 1839; twenty-three years later the tower was uncovered as the sand-hills moved inland, and later its foundation was seen on the shore after further advance of the dunes. The rate of movement of the sand here can be estimated at about one metre per year; it varies in different localities, and at times may be rapid enough to warrent steps being taken to arrest the process if valuable land areas are threatened. The stabilization of coastal dunes is frequently effected by the planting of marram grass (*Ammophila arenaria*), or rice grass, whose long roots bind the surface layers of sand and so prevent its removal by wind. Afforestation is a larger scale method of dealing with the same problem as, for example, the planting of pines at Culbin on the Moray Firth. In the Culbin area great storms at the end of the 17th century resulted in a large tract of farmland and houses being completely buried by advancing sand. The area remained a waste until the present century, when reclamation was begun in 1928; protective mats were used to stabilize the ground surface prior to the planting of grasses and then trees. The area has now been largely reclaimed and is again under cultivation. On a smaller scale, rubbish dumped on an area of blown sand at Wallasey, Cheshire, resulted in the stabilization of the ground in a short time and provided land for a recreation ground.

(ii) *Barchan dunes* in desert regions have crescentic shapes (Fig. 2.5) and are found in areas where the wind blows from a nearly constant direction, with a limited supply of sand. They are mounds of sand up to about 30 m in height and several times as wide, with the horns of the crescent pointing forward in the direction of the wind. Groups or colonies of barchans move slowly forward across the rocky desert floor where the wind and supply of sand are constant. They are found in desert areas of California, in the Libyan desert, and elsewhere.

Most dunes have a structure composed of inclined layers of sand (Fig. 2.5), known as *dune-bedding* (Plate 2a). Sand blown onto the windward slopes is compacted more than on the leeward side, giving accretion layers (*ac* in the figure). Accumulation is at a maximum rate on the summit of the dune; when the upper part of the leeward slope becomes too steep and unstable, the sand there shears along a *slip-plane* (or slip surface) at an angle of about 30 degrees. More sand from the windward builds up, until another slip-plane is formed just ahead of the first, and the process continues. Each inclined layer of sand between two slip-planes is called an encroachment layer (*en* in the figure), and the dune moves forward by the addition of successive layers. Dune-bedding is seen in some sandstones formed in past geological ages, as in the Nubian Sandstone of North Africa; and in the Permian sandstones of Ayrshire, Scotland (Plate 2a). The eolian grains in these rocks are rounded and frosted. Model dunes made in a wind tunnel have been studied in the laboratory, and the different packings and porosities of the sand on the windward and leeward slopes measured (Note 10).

The larger ridge dunes, or *seifs*, have the form of long ridges of coarser sand several hundred metres high, which are surmounted by crests of finer sand at intervals. Their formation has been discussed by R. A. Bagnold (Note 11). The ridges extend in straight lines for great distances, sometimes 160 km (100 miles) or more, as in the region south-east of the great Qattara Depression on the northern margin of the Sahara, west of Cairo (Fig. 2.6). The predominant winds blow from the north-north-west and the flat floor of the desert is seen between the ridges. The Qattara, one of a group of deep depressions excavated by wind action, extends in places to -128 m (-420 feet) below sea level, the local level of the water-table, and contains salt marshes at that level. Part of the material removed from the depressions has contributed to the ridge dunes and other large sand accumulations to the south. It has been estimated that the Sahara desert is advancing southward at the rate of a kilometre per year.

(iii) The wind-blown deposit known as *loess*, a fine-grained calcareous clay or loam, extends from Central Europe through Russia into Asia and covers large areas in China, where it reaches its greatest thickness. It consists of the finer particles which are blown out from the deserts to distant areas, forming a porous deposit traversed by networks of narrow tubes which once enclosed the roots of grasses; these during their growth bound the particles of dust and

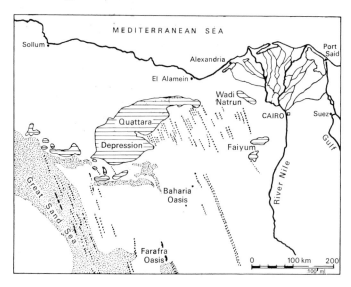

Fig. 2.6 Map of part of the Egyptian desert, showing sand accumulations (dots), depressions, and lines of dunes (after Holmes).

silt in their grip. Loess is resistant to weathering and supports steep slopes, and where dissected by the action of streams, stands as low vertical cliffs. The deserts of North Africa have contributed much of this material to Europe and the steppes of Russia. Generally of buff colour, loess is darkened by admixture with vegetable matter and in this condition forms the 'black earth' of the Russian steppes. An essentially similar deposit developed in the semi-arid regions of the western U.S.A. and in the pampas country of South America is called *adobe*.

Insolation   When a rock surface is exposed to a considerable daily range of temperature, as in arid and semi-arid regions, the expansion which occurs during the day and the contraction at night, constantly repeated, weaken the texture of the rock. The outer heated layers tend to pull away from the cooler rock underneath and flakes and slabs split off, a process known as *exfoliation*. This kind of weathering is prominent in climates where high day and low night temperatures are prevalent; it may also be observed to a lesser degree in more temperate lands.

In areas where an arid cycle of erosion has reduced an old landscape to a nearly level surface, residual peaks of hard rock are left upstanding above the general level and are termed *inselbergs* (or 'island mounts'). Resistant rocks such as granite or gneiss generally compose the larger inselbergs; they have undergone long periods of exfoliation, during which successive shells of heated rock have fallen away.

Spectacular examples occur in parts of Africa, including Mozambique and South-West Africa (Plate 4a), and others in Northern Australia. The level ground from which they stand out is probably the result of denudation in alternate dry wind erosion and wet periods.

A large range of temperature occurs daily in deserts, commonly 30°C and sometimes as high as 50°C; the daily temperature range for rock *surfaces* is often higher than for air. Strain is set up in a rock by the unequal expansion and contraction of its different mineral constituents and its texture is thereby loosened. Laboratory experiments have shown that this effect is enhanced if the rock surface is subjected to frequent wetting and drying. A more homogeneous rock, made up largely of one kind of mineral, would not be affected so much as a rock containing several kinds of minerals having different rates of expansion (e.g. the quartz and feldspar in a granite).

Under natural conditions, insolation of rock-faces results in the opening of many small cracks (some of hair-like fineness) into which water enters, and so both the decomposition of the rock and its disintegration by the help of frost are promoted. The crystallization of salts from solution in confined spaces, such as cracks and interstices (pore spaces) has been studied. Rocks having large pores (called macropores) are more liable to be affected by such crystallization than rocks having mainly very small pores (micropores). The process may be pronounced in deserts and in coastal areas. The thermal expansion of certain salts in confined spaces as they are heated is also a contributory factor (Note 12).

Weathering by organic agents   Effects which are small in themselves, but noticeable in the aggregate, are due to plants and animals (*biotic weathering*). Plants retain moisture, and any rock surface on which they grow is kept damp, thus promoting the solvent action of the water. The chemical decay of rock is also promoted by the formation of vegetable humus, the organic products of the decay of plants, which is aided by the action of bacteria and fungi; organic acids are thereby added to percolating rain-water and increase its solvent power. Some bacteria are active in reducing conditions and contribute to the making of sulphides. Others can convert nitrogen to NH_4 compounds and affect the pH of soils.

The mechanical break-up of rocks is brought about by the roots of plants which penetrate into cracks and wedge apart the walls of the crack. Earth-worms and other burrowing animals bring to the surface large quantities of finely divided soil; in their casts, worms are estimated to turn over 2.5 kg per m^2 (10 tons per acre) per year in the upper soil, and this repeated mixing exposes fresh material to air and water near the surface. The general result of all these processes is to assist in the production of the weathered mantle of soil (p. 21), which is then open to the action of transporting agents such as rain-wash, running water, and wind.

The work of rivers   The work of erosion performed by rivers results in the widening and deepening of their valleys during their course of development. River erosion is greatly enhanced in times of flood. Rivers are also agents of transport; they carry much material in suspension and re-deposit some part of it further downstream, the rest being transported to the sea. Some matter goes into solution in the river water and ultimately helps to increase the salinity of the oceans. The energy which is imparted to sediment moved by a stream, the finer particles in suspension and the coarser (including boulders) rolled along the bed during floods, performs work by abrading the channel of the river.

Valleys.   A drainage system is initiated when, for example, a new land surface is formed by uplift of the sea floor. Streams begin to flow over it and excavate their valleys, their courses mainly directed by the general slope but also controlled by any irregularities which it may possess. In many instances present day valleys have been cut by the streams which occupy them except in so far as they have been modified by the action of ice or other agents. In course of time valleys become deepened and broadened, and the rivers are joined by tributaries. Stages of *youth, maturity*, and *old age* may be distinguished in the history of a river, and topographical forms characteristic of these stages can be recognized in modern landscapes. Thus there is the steep-sided valley of the youthful stream; the broader valley and more deeply dissected landscape of the mature river system; and the flat, meandering course of a river in old age.

Youthful rivers cut gorges in hard rocks and V-shaped valleys in softer rocks (profile 1, Fig. 2.7*a*), and are characteristic of many mountainous regions. Debris, loosened by frost and rain, falls from the valley sides, to be carried away by the stream and assist in its work of abrasion. Rain-wash and soil-creep (p. 51) also contribute material from the slopes; this is especially the case in the more mature stages of valley development, when a mantle of soil has been formed on the land surface. Gradually as the headwaters of a river cut back into the land, its valley is deepened and widened into a broader V, and small scree slopes form at the bottom of the valley sides (the flatter inner slopes of profile 2). The deepening and widening, if unin-terrupted, continue as shown by profiles 3 and 4 in the figure, until the stage of maturity is reached, when the maximum topographical relief (i.e. the difference in height between valley floor and adjacent ridge tops) has been produced. Later, as denudation reduces the height of the ridges and the river enters its old age, the valley comes to have a wide, flat floor over which the river follows a winding course;

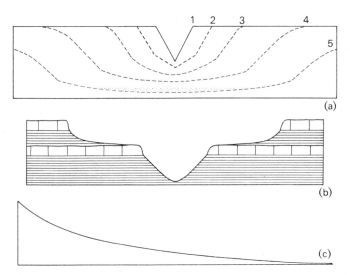

Fig. 2.7 Valley profiles. (*a*) Successive valley profiles during widening. (*b*) Valley eroded in hard and soft beds. (*c*) Profile along the course of a mature (graded) river. The horizontal line is base-level.

the upper slopes are convex (profile 5) and depth of relief is less. Such a sequence would be followed by a river in a temperate climate, if uninterrupted. Under other conditions of valley development, as in a drier climate, the slopes stand at a steeper angle because less affected by atmospheric weathering, and the ridges between valleys are then less rounded. Flat-topped hills may be formed in horizontal strata where a layer of hard rock such as sandstone, or a dolerite sill, forms the top of the hill; the term *mesa* (= table) is used for this topographical form.

The shape of a valley slope also depends on the nature of the rocks which have been eroded; Fig. 2.7*a* illustrates a valley excavated in rock of uniform character. When alternate hard and soft layers are present, as in Fig. 2.7*b*, erosion of the softer rock is more rapid than of the harder, and terraced slopes are developed. If the rock layers or strata are inclined and the river flows parallel to their upturned edges, instead of across them, the land surface will be carved into long hollows or vales along the softer beds, separated by ridges of the harder rock which form escarpments (Fig. 2.8; and see Fig. 7.2).

Grading and rejuvenation. The profile taken *along* the course of a river also changes during the river's evolution. For a young stream, actively eroding, this profile is an irregular curve which is steeper where the river crosses more resistant rocks, perhaps forming rapids or waterfalls, and flatter where it flows over more easily eroded rocks. If left undisturbed by movements such as uplift, or other factors causing change, the river continues to reduce irregularities of gradient

NW   SE
Wenlock Edge

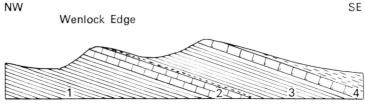

Fig. 2.8   Section through Silurian strata, Shropshire, to illustrate escarpments formed by the Wenlock Limestone (*2*) and Aymestry Limestone (*4*). (*1*, Wenlock Shales. *3*, Lower Ludlow Mudstones).

and to smooth out its bed, until in maturity it is said to be *graded* or at grade. Its longitudinal profile is then independent of the kind of rock over which it passes, and tends towards a smooth curve (Fig. 2.7*c*) of the type expressed by the equation:

$$y = a - k \cdot \log (p - x),$$

where *y* is the height of a point above datum, *x* its distance from the river's mouth, *p* the total length of the river, and *a* and *k* are constants. It was shown by O. T. Jones that for the River Towy, in Wales, a curve fitting part of the longitudinal profile of the river was given by this formula. For the upper course of the short River Mole, England, the constants found were *a* = 241.5 and *k* = 65, when *y* is measured in feet and *x* and *p* in miles (Note 13). Similar studies have been made for large rivers, including the Colorado in Arizona.

The *base-level* of a river is the level of the sea or lake into which it discharges, for clearly it cannot cut down below this. For a tributary, base-level is the level of the main stream at its point of entry, and as this changes in the course of time the tributary is continually adjusting its grade to a new level. The cutting power of a river which has reached maturity or old age may, however, be revived by uplift or tilting of the land, by recession of the coast-line due to marine erosion (p. 57), or from other causes; as a result the stream is given a new fall to the sea. It is said to be *rejuvenated*, and begins to cut back again and lower its bed by the newly acquired energy. Owing to such interruptions in their cycle of activity not all streams become completely graded.

An example of rejuvenation due to a local uplift is seen in the valley of the River Greta near Ingleton, Yorkshire. In its upper course this stream has a wide, mature valley; but about 2 km above Ingleton the river plunges downwards in a series of waterfalls, cutting into hard rocks with youthful energy in striking contrast to the placid flow of middle age which is evident only a short distance upstream. Below Ingleton the river again resumes a meandering course. The rejuvenation of a river leaves its mark on the longitudinal profile, where

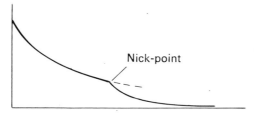

Fig. 2.9   Nick-point (see text).

the break in slope at the junction of the old course with the new (deeper) cut is called the *nick-point* (Fig. 2.9).

As a river grows older it remains vigorous only in its upper reaches, where the flow is swiftest and the gradient steepest (Fig. 2.7c). In its lower course the speed and carrying power are reduced, and the river begins to meander from side to side of its valley. It becomes subject to seasonal floods and under these conditions much sediment is deposited (p. 43). We first consider some characteristics of youthful rivers.

Waterfalls and gorges.   Waterfalls and rapids are formed where a stream in a youthful stage flows over rocks of differing hardness. A hard layer or band is worn away less rapidly than softer rock, with the result that the river's gradient is locally steepened where the outcrop of the harder rock is crossed. If the river crosses a layer of more resistant rock, less resistant material below it is undercut by the eddying of the water and an overhanging ledge is formed over which the stream falls (Fig. 2.10a). The weight of the overhang in time becomes greater than can be supported by the strength of the rock, the ledge breaks away, and the fall gradually recedes upstream. The Whin Sill, in the north of England, is a hard igneous rock (dolerite) forming a nearly horizontal sheet which gives rise to numerous waterfalls in this way; two examples are High Force in Teesdale and High Cup Nick in Westmorland. Hollows known as *pot-holes* may be worn in the rock of a river-bed by the motion of pebbles which are swirled round by eddies, especially near a waterfall. Such a water-worn rock surface is easily recognized, and if observed near but above an existing stream, it marks a former course at a higher level.

At Niagara on the Canada–U.S.A. border, the waters of Lake Erie flowing to Lake Ontario cross a hard limestone formation, the Niagara Limestone, which lies above softer shales, and the river here makes a drop of 55 m. The limestone forms a broad ledge which is undercut by the river, and the falls retreat upstream as the ledge collapses from time to time. Below the falls the river flows through a gorge 11 km long. Formerly the falls receded at a rate of nearly 1.2 m per year; taking this as an average, the time taken to erode the gorge

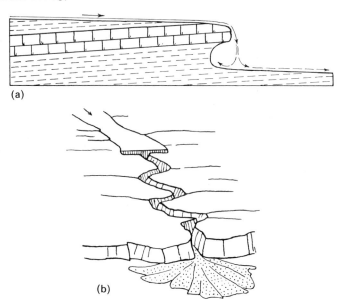

(a)

(b)

Fig. 2.10 (*a*) Waterfall formed at a hard layer, below which softer beds are eroded. (*b*) Sketch of gorge eroded in jointed lavas below the Victoria Falls on the Zambesi River (after H. Cloos).

would be about 10 000 years—i.e. it has been formed since the end of the Pleistocene glaciation (p. 65). The rate of recession of the Niagara falls is now reduced to about a quarter of its former amount, since hydro-electric power generation has greatly lessened the flow of water over them.

The 110 m high Victoria Falls on the Zambesi River, Rhodesia, are situated at a point where the river leaves the gently undulating surface of a lava plateau and plunges into a deep gorge from a ledge 1.58 km (nearly a mile) wide. The gorge has been cut in basalt lavas, eroded along lines of weakness formed by intersecting vertical joints and fractures, which cause it to have a zig-zag course (Fig. 2.10*b*). At the exit from the vertical-sided gorge 96 km long, the debris carried by the river is dumped as a wide delta.

Gorges in general are cut by young streams which erode rapidly downwards; the weathering of the sides of the gorge takes place only slowly during its formation if the rock is hard, or if the area is one of low rainfall. In some instances slow uplift of the area has helped to produce a deep gorge by maintaining the fall of the stream during downcutting. Where prominent joints or other lines of weakness in the rocks are present they help to control the shape of the eroded channel. If the rock is sufficiently strong it will stand with steep or vertical walls; in less strong rock the gorge will become widened out after a time, particularly if erosion is aided by solution. In the Carboniferous Limestone of the Pennines and other areas, solution effects have played a part in shaping a gorge, as at Dovedale in Derbyshire.

The Grand Canyon of the Colorado, a great erosional feature 480 km (300 miles) long and with a maximum depth of 1830 m, has been formed in relatively soft rocks by a young river flowing from the Rockies on entering an arid region. The river was rejuvenated by a geologically recent uplift, rapid downward erosion being assisted by a rising land surface. The upper part of the Canyon has become widened out, but denudation of its walls by atmospheric agents has proceeded slowly on account of the arid climatic conditions and low rainfall prevailing, so that nearly vertical cliffs with scree slopes form the upper walls. At the bottom of the Canyon, a steep-sided inner gorge some 305 m deep has been cut in harder, older crystalline rocks (Pre-Cambrian).

River capture. A river which is cutting back vigorously in an easily eroded formation and extending its catchment, may approach the course of a neighbouring stream and meet and divert the headwaters of the latter into its own channel. This process is known as *river capture;* the stream which has lost its headwaters is said to be beheaded, and dwindles in size. It is then too small for the valley in which it flows and is called a 'misfit'. The River Blackwater near Farnham, Surrey, a tributary of the Thames, is an example. The course of the now short stream once extended much farther to the south-west, whence it transported distinctive gravels containing chert, to deposit them north of the gap in the Chalk ridge at Farnham (Fig. 8.10). The source rocks of the gravels prove the former extent of the river. An eastward flowing stream, the Wey, cutting back rapidly in soft strata to the south-east of Farnham, tapped the upper reaches of the Blackwater and diverted their flow into its own channel. The point of the capture is marked by a right-angle bend which the river makes just south of Farnham (Note 14).

In north-east England three rivers, the Nidd, Ure, and Swale in Yorkshire, flowing east off the Pennines to the North Sea, were captured by the River Ouse as it flowed south to the Humber along a belt of soft Triassic rocks. Rapid erosion under these conditions effected successive captures as the Ouse extended its headwaters to the north. And in the Tyne drainage system the present short rivers Blyth and Wansbeck, which flow to the east coast, had longer courses of which the upper reaches were captured by the North Tyne river, working back northwards in soft sediments of Carboniferous age.

Meanders. When a river has cut down nearly to base-level it flows more slowly with a reduced gradient and begins to swing from side to side of its valley. The energy imparted to the load of sediment which it carries is expended in the widening of the valley by lateral erosion, and the course of the river develops big loops called *meanders*. Stages in the development of a meander are shown in Fig. 2.11*a*. The length of a loop when fully formed is about sixteen times the width of the stream. On the concave side of a curve the bank is undercut and eroded, while detritus is deposited on the convex side, with the formation of banks of sand and gravel (Fig. 2.11*b*). The flow of the river at a bend moves from one side of its channel to the other, as indicated by the arrows in the figure; the water near the stream bed flows obliquely under that nearer the surface. The scouring effect of the bed current often results in a deepening of the channel on the concave side of the bend.

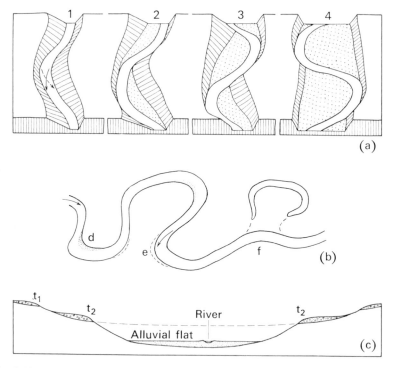

Fig. 2.11   Meanders and river terraces. (*a*) Stages in the widening of a valley floor and development of meanders (after Cotton). (*b*) Fully developed meanders, showing deposition (*d*) and erosion (*e*) at bends. (*c*) Section across a valley to show alluvial flat and older river terrace deposits at $t_1 t_2$.

As the meanders progress, a *flood-plain* or *alluvial flat* is eventually developed (stage 4, Fig. 2.11*a*), and the river flows through its own deposits in the widened valley. When the meanders reach their full size they tend to shift bodily downstream, cutting into the sand and gravel of the flood-plain and rebuilding it behind them as they move forward. The river thus works in time over a large area of land, re-sorting its earlier deposits. A comparison of old and new maps of the same river will often show this shifting and change in shape of the meanders.

If a river in the meandering stage breaks through the neck of land at the end of one of the loops, as at f in Fig. 2.11b, and thus shortens its course, it will leave a crescent-shaped lake called an 'ox-bow' or 'mortlake' in the abandoned meander. Examples may be seen from a topographic map of the River Wye, e.g. near Ross, Herefordshire, where contours show the form of the old channel. The Wye is also remarkable in possessing a meandering course although the river flows in a gorge-like valley below Kerne Bridge. This phenomenon, known as *incised meanders*, is due to the rejuvenation of the river after it had reached a meandering stage in an earlier drainage system. Fresh energy imparted to the stream enabled it to cut downwards once more and so carve out the gorge along the great loops of its old channel.

Other rivers which have developed meanders are the Forth, Thames, and Seine, and large parts of their lower courses have now been trained between walls and embankments, thus preventing further changes in position. The Mississippi is discussed in Chapter 16 in connection with flood control.

The chief factors which control the formation of meanders are a low gradient and a moderate load of sediment. It has been demonstrated by means of models that if the gradient is increased beyond a limiting value, or if the load of sediment becomes too great, no meanders are developed.

River deposits   The general term given to deposits laid down by rivers is *alluvium*, though this is often restricted to the finer material such as silt and mud, as distinct from gravels and larger fragments (Note 15). The transporting power of a stream increases at the rate of the fifth or sixth power of its velocity. Thus, if the normal velocity is trebled, as may be the case after heavy rains, the carrying capacity is increased several hundred times. The result is seen in boulder-strewn torrent tracks in hill country. Large boulders which would not be moved under normal conditions of flow, are shifted with an intermittent motion by the stream in spate, and become partly rounded by the buffeting they receive. Large quantities of smaller boulders, gravel, and sand are also transported at such times, the coarser particles being rolled along on the stream bed and the finer carried in suspension. The principles underlying the transport of small particles by a current of water are utilized in the process of elutriation for separating particles of different sizes in a sample sediment.

Transported sediment is dropped by a stream whenever its velocity is checked. A river emerging from a mountain valley on to flatter ground, such as the edge of a plain, builds up a heap of detritus known as an *alluvial cone* where the change of gradient occurs. Since this pile of sediment affects the flow, the river changes its course and so the cone spreads outwards. Deposition at places along a river's course where the flow is locally slackened, as at bends, has been noted above in connection with a meandering stream. The material deposited is mainly sand and fine gravel, partly rounded during transport, the finer mud particles being carried on down to the sea. Owing to the cushioning effect of the water, particles are only partly rounded by impact; they are less well-rounded than wind-blown sands.

In the lower course of a mature river the finer alluvium is spread out to form an *alluvial flat* (Fig. 2.11c); this is subject to periodic flooding, and a fresh layer of alluvium is deposited at each flood. The coarser particles are dropped nearest the stream, and gradually build up a bank or *levee* on each side of it, which is only overtopped by the water in time of flood. In some cases the formation of an alluvial flat may involve the burial of the lower slopes of the valley sides, and the alluvium then abuts directly on to the convex upper slopes (*cf.* Fig. 2.7a). Such alluvial deposits may have only a small thickness, but may go up to 10 m or more; they are very porous, and in excavations the zone of saturation may lie at small depths below the surface of the alluvium. Running sands may occur under these conditions.

The rapid variation in the nature of alluvial deposits is well illustrated by exposures in dock excavations in East London and at other places; layers of mud, silt, and sometimes peat alternate and thin out in irregular fashion, and may fill hollows (wash-outs) which have been scoured out by floods in earlier deposits.

If a river, after having formed a flood plain of alluvium, is rejuvenated and cuts down its channel to lower levels, remnants of the earlier deposits may be left on the old valley slopes as *terraces* at different levels (Fig. 2.11c). Thus, in the valley of the River Thames three such terraces are found (see Fig. 8.15, p. 268), known as the Boyn Hill, Taplow, and Flood Plain terraces respectively, the first two being named after localities near Maidenhead. Valuable deposits of gravel and sand which occur in these terraces are excavated on a large scale for supplies of aggregate and sand for concrete and mortar, as for example in Middlesex, around Ashford and Staines.

Alluvial mud may be used as one of the raw materials in the manufacture of cement, as on the River Medway and at Lewes in Sussex, where in both cases river mud is mixed with Chalk in the required proportions (see Chapter 17).

Drainage of pore-contained water from river-deposited sands and gravels may result when water is taken from wells sunk in the alluvium, and when proper replenishment by rainfall is prevented, as in a built-up area. This removal of the interstitial water causes a slight compaction of the deposit, leading to slow subsidence. In the London area, precise measurements of levels made over a period of 68 years (to 1932) revealed a fairly general subsidence, the maximum amount being nearly 7 inches (178 mm) at the Bank of England. Traverses across the area showed that the subsidence begins where the London Clay gives place at the surface to the gravels and sands which overlie it, and is greatest over the buried channels of old streams such as the Fleet and Walbrook, which were tributaries of the Thames at the time the deposits were laid down (Note 25). The Bank is built over the old deposits of the Walbrook. The compaction, although small in itself, could cause unequal settlement of large buildings erected on the alluvium and resulting strain on the structures; this happened in the case of St. Paul's Cathedral, where extensive remedial measures were called for.

Current bedding.   Shallow-water deposits (p. 64) and river deposits frequently show a structure known as current-bedding or cross-bedding, which is contemporaneous with their deposition (Plate 2b). Successive heaps of sand or gravel deposited by the sudden checking of a current lie one above another, as shown in Fig. 2.12a. A change

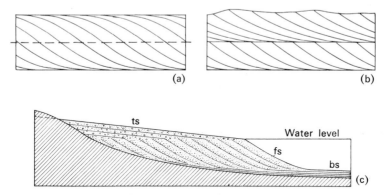

Fig. 2.12   Current-bedding and delta deposits. (*a*) Current-bedded deposits, the upper part of which is later washed away down to the broken line. (*b*) Further material laid down on the truncated edges of the lower deposits. (*c*) Section through a small lake delta; *ts* = topset beds; *fs* = foreset beds; *bs* = bottomset beds.

in the direction or intensity of the current may wash away the upper part of these layers, and more sediment may be laid down upon their truncated edges (Fig. 2.12b). Coarse and fine layers may be intercalated. When such current-bedding is preserved through a sequence of shallow-water deposits, after the sediment has been consolidated the structure is seen in the resulting sandstone. The top or bottom of a current-bedded series of sediments is always indicated by the fact that the tops of the layers are truncated, while the bottoms are asymptotic to the surface on which they were deposited. Current-bedding by wind is also found in eolian sands and is called *dune-bedding* (Plate 2a).

Deltas.   A river entering a body of still water, such as a lake or the sea, drops much of its load of sediment as its velocity is reduced, and in many cases forms a delta which is gradually built forward into the water. An example of a lake delta is the area of sedimentation at the head of Derwentwater, in the Lake District; and in many Scottish lochs, lateral streams entering build up flat cones of debris, which in time become grassed over and stand as small promontories. At a coast, where a tidal estuary is present no delta is formed but sand- and mud-flats are laid down where the river's flow is checked by the tidal water; a fresh-water channel, which may branch into a number of subsidiary channels, is scoured out through the deposits by the discharge of the river at low tide. Not only the coarser sandy sediment in suspension but also much of the finer muddy material settles near a shore-line, though some is carried farther out by weak currents. Settlement of the very fine particles of mud is promoted by flocculation (the aggregation of fine particles into clusters or flocks).

The building of a *delta* proceeds as a fan of sediment is deposited at a river's mouth; the stream then divides and flows through the water on either side of the obstacle which it has made. Further deposition of sediment takes place along these distributaries, and after further barriers have been formed the streams branch again. In this way, by repeated bifurcation and sedimentation, the deltaic deposits come to cover a large area which may have a roughly triangular shape (like the Greek letter Δ).

A section through the deposits of a typical small delta is given in Fig. 2.12c; as each flood brings down its load of sediment the coarser material is dropped on the seaward side of the growing pile and comes to rest at its angle of repose in water, building up the sloping *foreset* beds. Observations show that their slope varies from about 12° to 32°, larger particles standing at the higher and smaller at lower angles. Ahead of the foreset beds and continuous with them, the finer material is deposited as the *bottomset* beds. As the delta is built forward, foreset deposits come to rest on earlier bottomset deposits, as shown in the figure. The upper surface of the delta is composed of gently sloping *topset* beds of coarse material, which are a continuation of the alluvial plain of the river and cover successive foreset deposits.

The sediments of an ancient delta can be recognized in the sandstones of the Millstone Grit formation of Yorkshire, which were laid down by a large river discharging in a southerly direction (p. 256).

A *large delta* such as that of the Nile, Niger, Ganges, or Indus, all of which are over 160 km across and have a large under-water extent beyond the present coast-lines, is a very thick pile of sediment in which land-derived materials may be present as well as shallow-water deposits. The Nile delta-front extends northwards under the sea for some 100 km, and its slope seaward increases from about 1 in 1000 near the shore to 1 in 40 at its outer edge, this steeper slope representing the foreset beds. The deposits consist of Nile silts and clays overlying sands with layers of gravel. Borings have indicated thicknesses of up to 700 m for these deposits, which are all geologically recent.

The low-lying Indo-Gangetic plains which separate the Himalayas to the north from peninsular India on the south are also an area where hundreds of metres of alluvial deposits, brought down by rivers from the mountains, have accumulated. Deep borings in this alluvium, the floor of which has not been reached, have penetrated many successive beds of sand and clay with intercalated layers of peat and kankar, and the deposits are of similar type throughout (Note 16). This, and the fact that a pile of sediment of this order of thickness constitutes a considerable load on the crust below it, can be explained by assuming that the area of sedimentation was sinking as the deposits accumulated, i.e. under isostatic control, as referred to on p. 11. Alternatively, the area could be sinking for other, tectonic, reasons and thus act as a basin in which shallow-water sediments would accumulate, sedimentation keeping place with the sinking. (Tectonic structures are discussed in Chapter 7).

The Mississippi delta, which has an area of about 30 000 km², is another deltaic area which is slowly extending seaward, but differs in some respects from those discussed above. It possesses an unusual feature, a 'bird's foot' projection (called the Balize sub-delta) which has been formed in recent times. River water with much mud and silt in suspension flows out by long channels along whose sides deposition of the coarser particles takes place, the finer material building up swamps between the channels.

River discharge problems in the Fens of East Anglia Where rivers traverse low-lying areas before reaching the sea, problems of flooding frequently arise, sometimes on a large scale. We consider here the geological conditions which have given rise to difficulties of this kind in East Anglia; engineering works in connection with river training and flood control are further discussed in Chapter 15.

The flood problems of the River Great Ouse where it flows through the Cambridgeshire and Norfolk fenlands before discharging into the tidal waters of the Wash, had become serious by about 1940. The fens are a flat expanse of silt and clay, with beds of peat at certain levels. The area was once a bay or estuary, of which the Wash is now a remnant, formed by the submergence of a broad valley which was hollowed out in soft Jurassic clays. This bay became silted up, mainly by material brought in by the sea, and layers of peat grew at times, especially near the landward margin. The Fen deposits are sandy silts (called *warp*) and smooth clays (the *buttery clay*); they have accumulated as sea level has risen slowly, with several oscillations in level, since the end of the glaciation of the Pleistocene period (p. 269). Slight emergence of the area resulted in an extension of the growth of peat seawards; submergence caused the silts and clays to be deposited *over* the peat as the sea re-advanced. The deposits reach a thickness of about 18 m and rest on boulder clay. Much of the low-lying fenland is now protected from the sea by artificial embankments.

The fens around Ely are a sunken area, and water draining off them is pumped up into a system of artificial channels (or drains) maintained at a level high enough to give gravity flow to the sea. Shrinkage of the fen deposits through drainage and cultivation has resulted in a sinking of the land surface. It is on record that a post driven through 22 feet of peat into underlying clay at Denton Fen in 1848, with its top then level with the ground, had become exposed over its upper 11 feet by 1932; the depth of peat had thus been halved in the interval. This represents an average rate of lowering of the surface of 40 mm per year. River embankments built on this surface had, of course, sunk with it and had to be constantly raised, so that in places they were over twice their original height; the necessary width for further raising of the banks was not everywhere available and at many points the limit had been reached. The more recent rate of lowering of the surface has been estimated at about 20 mm per year.

The water from the upper Ouse catchment above Earith flows north through two artificial channels known as the Old Bedford River and the New Bedford or Hundred Foot River, down to Denver Sluice, thus short-circuiting the course of the river past Ely (see Fig. 2.13). Below Denver the river is tidal. The above two

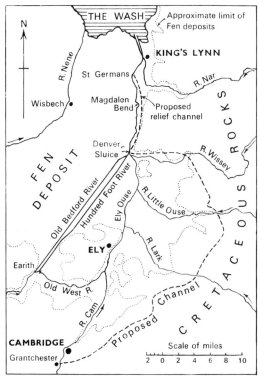

Fig. 2.13   Map of Great Ouse drainage area.

channels have high banks on their east and west sides; the area between them, some 5000 acres (2023 hectares), is known as the Washlands and can store flood water when the sluice gates at its northern end are closed by the tide. The drainage from the Fens east of the Hundred Foot River, which is pumped into the high level system already mentioned, can also only be discharged at low tide, and in times of high flood dangerously high water levels occurred upstream of Denver.

After serious floods in 1937 and 1940, several schemes were put forward for remedying the situation, particularly by reducing the silting in the tidal part of the channel below Denver. The proposal adopted by the Great Ouse Catchment Board involved the cutting of a new relief channel 11 miles (18 km) long between Denver sluice and a point downstream near Kings Lynn, to the east of the river. This channel would be used only in times of flood and would discharge when the tide was low, intermittent storage of water thus being provided for; it was completed in October 1959. In another part of the proposal, tidal water was to be impounded by a barrage constructed in the loop of the river known as Magdalen Bend, and used for scouring out the lower channel of the river into the Wash (Fig. 2.13), thus removing silt deposited by incoming tides and preventing a reduction of the capacity of the natural channel. The effect of such a barrage was tested by means of a model and the scour was found to extend to the end of the existing training walls at the mouth of the river (Note 17).

With the Fenland surface continuing to sink, a further relief channel has been constructed (the broken line east of Cambridge in the figure). This cut-off channel passes round the margin of the Fens, and takes water from the rivers Cam, Lark, Little Ouse, and Wissey northwards to Denver sluice. It is located on Cretaceous rocks and is thus independent of the Fen deposits and the compaction of peats in them. With its construction, the threat of serious flooding in the Fens has receded.

Landslides Movements of rock masses known as landslides or landslips (the two terms can be taken as synonymous) arise from the action of gravity on unstable material. They are important factors in denudation, and when they occur at a coast-line help the sea in its attack on the cliffs. Landslides may be divided into two broad groups:

(i) *slides*, in which a surface of sliding is present, separating the moving mass from the stable ground; and

(ii) *flows*, where movement takes place by continuous deformation, without a surface of sliding. The motion in flows is generally less rapid than for slides and may be very slow indeed; however, their acceleration on failure can be considerable (Note 18).

(i) Included in the first group is the commonly occurring type of slide where movement takes place on surfaces such as inclined bedding planes (the parallel surfaces which bound the beds or layers of sedimentary rocks). Beds of rock (or strata) which lie on one another may, in a line of cliff, have a slope down towards the sea; they are then said to *dip* seaward (Fig. 2.14*a*), and are prone to slipping especially

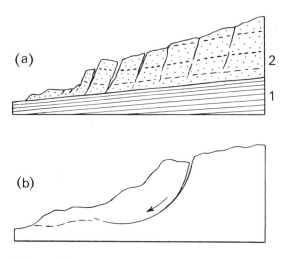

Fig. 2.14   Landslides. (*a*) Section showing a lubricated clay (*1*) on which slipping of overlying beds (*2*) has taken place. (*b*) Rotational slip in clay.

when a bed through which water can percolate rests on clay, as shown in the figure. The water is hindered from draining, and is stored in the rock; it exerts a hydrostatic pressure which reduces the strength of the rock, and masses break away from the cliff and move down the inclined clay surface under the action of gravity. This is seen, for example, at Ventnor and other localities in the Isle of Wight, where many falls of cliff due to the Chalk slipping over the Gault (clay) have taken place. At Rousdon, East Devon, the Lias clay has led to similar landslide conditions; movements here during the past 100 years, along several kilometres of coast, have given rise to chasms where large slipped masses (Chalk and Greensand) have moved seaward from the stable ground behind them, leaving inland cliffs. Ground which has moved in this way has a characteristic tumbled appearance.

Rotational slips (Fig. 2.14*b*) are of common occurrence in clays and unconsolidated materials; they develop in natural banks or cliffs and in railway and road cuttings where excavation has created conditions leading to slipping, in particular an increased water content of the clay. They are discussed in Chapter 13 (Plate 13).

Rock-falls and *rock-slides* occur when masses of unstable rock break off along joints and bedding planes and move either vertically or on inclined surfaces. In the great rock-slide at Turtle Mountain,

Alberta, in 1903 millions of cubic yards of rock which formed a projecting shoulder of the mountain broke off on inclined joints which traversed the mass, and slid rapidly with great force down the side of the mountain (see Chapter 13, p. 392).

(ii) Flows, in the second group, move by continuous deformation of the mass of sediment. They include mud-flows (p. 23), and the slumps in clay cuttings which arise where the clay has acquired greater plasticity from an increased water content. This is promoted by rain entering the surface of the clay through cracks formed after a period of drying. Bulging of the sides of clay cuttings is sometimes due to this cause, and may be associated with slipping; such movements add to the difficulties of maintenance of the cuttings (see also p. 395).

Under the same general heading come the very slow movements of surface layers on slopes, such as *soil creep* (Plate 3b) and *rock creep* on hillsides, which continue over a long period and may ultimately displace trees, fences, and lines of communication. Although these effects are not called landslides they arise from like causes and are included in the general category of *mass-movements*. The cumulative effect of creep in bringing soil and rock fragments within the sphere of action of transporting agents such as rivers is an important contributory factor in the processes of denudation.

The work of the sea   The waves which break on a shore erode the land by the force of their impact and, especially in storms, by the impact of the debris they carry forward. The debris itself comprises rock fragments derived either from local cliffs or brought to an area from adjoining beaches. The fragments become rounded and reduced by the battering they receive from the waves, and make up the familiar pebbly and sandy beach deposits.

The continental shelf around the margins of a large land mass has been referred to in Chapter 1. A distinction must be drawn between relatively shallow seas, such as the North Sea or the Baltic, which lie on the margins of the continents and are called *epicontinental* seas, and the deep *oceans* such as the Atlantic and Pacific.

The terms used by different writers in describing the parts of a region where land and sea meet vary; those employed here may be defined as follows. The margin of the land is called the *coast-line*, and the land-zone adjacent to it is the *coast*, which is frequently bounded by a line of cliffs. The *shore* is the zone extending from the base of the cliffs down to low-water mark (Fig. 2.15); it may be sub-divided into the *fore-shore*, which is that part lying between ordinary high and low tide marks, and the *back-shore* or area between high tide level and the foot of the cliffs. When no back-shore is present, high tide then extends up to the cliff base.

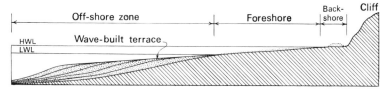

Fig. 2.15   Nomenclature of shore zones.

The shore is a wave-cut platform (p. 56) and on it lie the *beach* deposits, which are moved by waves and covered and uncovered daily by the tides. In the *off-shore zone* beyond low tide level, thicker deposits of land-derived sediment form a terrace whose seaward slope continues that of the wave-cut platform (Fig. 2.15). The deposits which form this terrace grade from coarser to finer material with progress outwards from the shore.

Tides and currents. The periodic rise and fall of the sea, or *tide*, is due essentially to the pull exerted by the sun and moon on the water; the bulge of water thus produced moves round the earth as the latter rotates. The highest ordinary tides are called Spring tides; these occur at the times of full and new moon, when the sun and moon and earth are in line. The smallest or Neap tides occur at the moon's first and third quarters, when its pull is at right angles to that of the sun. At the equinoxes exceptionally high (equinoctial) tides are generated; when these are accompanied by heavy seas in severe storms the coarser deposits of the back-shore, which are normally above the level of spring tides, may be piled up to form a ridge or storm-beach.

With the rise of the tide the water in shallow seas and narrow channels is heaped up so that a *tidal current* is generated; in the English Channel, for instance, the tidal current flows eastwards at about 3 km per hour as the tide rises, and westwards on a falling tide. The tidal current in the Mersey estuary reaches 12 km per hour. In estuaries which shallow and narrow rapidly, the incoming water advances with a steep wave at its front called a *bore*, as in the case of the Severn below Gloucester. In estuaries which are artificially deepened the bore disappears. Tidal currents are in many cases strong enough to move the less coarse sediment in the shore zone, but more rarely to move shingle.

Waves. Wave motion is produced when a water surface is swept by wind. It is an oscillatory motion, and in deep water each particle near the surface moves in a vertical circular orbit, as may be observed by watching the movement of a floating object. Particles below the surface move in phase with the surface particles but in smaller orbits. Waves within an area where they are being driven by wind are known as *forced waves;* when they move out into water where they are unaccompanied by wind they are called *free waves* or *swell.*

The height of waves from trough to crest normally varies up to about 9 m, but is much more than this in heavy seas. The wave-length (distance from crest to crest) of forced waves is from 60 to 180 m in the open ocean, but the wave-length of a swell may be greater. Wave motion diminishes with depth from the surface, and at a depth equal to the wave-length the movement ceases altogether. At greater depths, even the finest mud on a sea floor is not stirred by water movement. Hence the lower limit of wave action varies, and may be as much as 180 m, a depth corresponding to the edge of the continental shelf.

As a wave runs into shallow water (i.e. water less deep than a wave-length), friction retards the movement of water in contact with the sea bottom, while the top of the wave runs on. The wave-front thus increases in steepness and ultimately the crest falls over in front of the wave, and the wave is said to break. This forward movement of the water ('wave of translation') constitutes the means by which a wave becomes an agent of erosion. The breaking of the wave is followed by the *backwash*, which combs down the components of the beach, especially the finer material, towards deeper water where they are temporarily deposited.

With an onshore wind blowing at right angles to the shore-line, water is heaped up against a coast; this is compensated by a return current away from the land, called the *undertow*, which may be concentrated into narrow channels. The undertow can transport finer sediment out to sea, and in storms a strong undertow can remove large quantities of beach.

Waves meeting a steep coast, such as a line of cliffs fronting relatively deep water, are partly reflected and an up-and-down movement is imparted to the water at the cliff face. On the other hand, where waves run in a long distance in very gradually shallowing water, much of their energy is absorbed by friction, and they become reduced in size (and hence in erosive power) before reaching the shore. A flat, sandy fore-shore is therefore to a large extent its own protection against erosion. When the slope of the fore-shore is steeper, low frequency waves may have a constructive effect, resulting in aggradation of the beach deposits.

The pressures exerted by waves were measured by Stevenson, who found that they varied from about 600 up to 2000 pounds per square foot in winter, while in one gale as much as 6000 pounds per square foot was recorded (Note 19). Experiments made with models by R. A. Bagnold and C. M. White showed that waves breaking against a wall produced shock pressures ranging from 1440 lb/sq ft ($68\,950\ \text{N/m}^2$) to more than three times that amount. The effectiveness of storm waves in breaking up massive stone and concrete structures has been demonstrated many times around the shores of Britain, and it should be noted that storm waves do far more damage in a short time than normal seas acting over longer intervals. Experiments on the action of waves in moving and modifying the form of beach deposits have been made with the aid of models, in the U.S.A., Great Britain, and other countries (Note 20).

Littoral drift. Waves which approach a coast obliquely carry beach material forward up the shore as they advance and break, but the backwash which flows down the beach as each wave recedes drags back the pebbles and sand along a path nearly at right angles to the shore-line, thus: An alongshore movement is therefore imparted to the fragments by the incoming waves, and the cumulative effect is known as *littoral drift* or 'long-shore' drifting. It may be supplemented to some extent by coastal currents, but these can carry only the finer sediment, and it is the action of the waves which is the chief cause of the drifting. Transport of coarse material alongshore by wave action in one direction and of fine material by currents in the opposite direction has been observed.

On the east coast of England the prevalent drift of shingle is southward, and along the south coast it is eastward; but there are local exceptions to this general statement, as on the Norfolk coast west of Sheringham where the drift is westward towards the Wash. Since a good shingle beach affords the land a considerable measure of protection from the waves, it is generally an advantage to preserve beach deposits and to encourage their accumulation by suitably placed groynes, a matter which is discussed later. On the other hand, interference with the normal process of littoral drift by engineering works may lead to undesired results. For example, the construction of harbour works at Lowestoft arrested the southward travel of beach material there, and much shingle was piled up north of the harbour. Serious erosion resulted at Pakefield, a mile to the south, where the cliffs have been cut back rapidly. The shore lost its protective beach when the normal drift was interrupted, and became open to the attack of the sea.

Spits and bars. The growth of these accumulations of shingle and sand is brought about by longshore drift in situations where there is a change in direction of the coast-line. A *spit* is a ridge of sand and shingle formed by longshore drift which extends out into open water from a bend in the coast, as at the mouth of a river or bay or from the leeward side of a headland. If the latter, erosion of the headland may contribute rock material to the building of the spit. On the east coast of England Spurn Head, Blakeney Point, Orfordness, and Great Yarmouth, are spits built by the southward movements of beach material; and the Hurst Castle and Mudeford spits on the south coast derive their material from the west. The curved ends of some spits are developed by the sweeping round of the dominant waves, as at Hurst Castle.

Orfordness, in Suffolk, deflects the mouth of the River Alde; this river once entered the sea at Aldeburgh but now turns southwards there, just before reaching the sea, and flows parallel to the coast for some ten miles (16 km) separated from the sea by the shingle barrier. It is estimated that the spit grew $5\frac{1}{2}$ miles (9 km) in length, from near Orford to its termination, during 700 years from the founding of Orford Castle in the 12th century; more recently it has advanced further but with periods of recession.

When a spit grows across a bay it then forms a *bay-bar*, as at Looe Pool on the south coast of Cornwall. The Dovey estuary in Wales, also, is shut off from the sea by a shingle bar except for a small channel. (For a small river much of its flow can percolate through a shingle barrier, which is permeable.) The formation of a bar results in the shortening of the coast-line and protects the bay or estuary from wave attack. The bay then becomes silted up with river-trans-ported sediment, which forms mud-flats lying between high and low water levels. The mud and silt trapped behind a bay-bar in this way may be converted gradually into dry land by the growth of *salt-marsh* vegetation, as at Scolt Head Island on the Norfolk coast, or Gibraltar Point (Nature Reserve) on the south Lincolnshire coast. In the upward growth of a salt-marsh, the rate of accumulation of sediment has been found by spreading a layer of coloured material on the surface and measuring the thickness of silt deposited above it. At Scolt Head Island this varied from about $\frac{1}{3}$ to 1 cm per year (Note 21).

The long shingle barrier known as the Chesil Bank, Dorset, extends for about 22.5 km from near Abbotsbury to the Isle of Portland (Fig. 2.16) and encloses a lagoon behind it (the Fleet). There is a

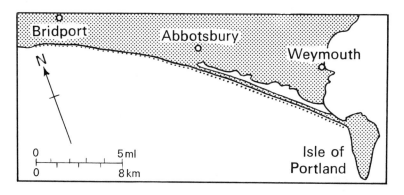

Fig. 2.16   Chesil Bank, Dorset, linking the Isle of Portland to the mainland.

gradation in the size of pebbles which form the Bank, from small at the north-western end to very large at the south-eastern. It is essentially a storm beach, to which longshore drift from the west has contributed some material at the western end; but at the other end there is much locally derived material from the Portland Limestone, and the storm beach there reaches a height of 13 m. Its length is about at right angles to the mean direction of the dominant waves which reach that part of the Dorset coast.

Besides the forms described above, other prominences which are in effect compound spits may develop on a coast and are known as *cuspate forelands*. They are gradually built out to sea, as at Dungeness in Sussex (Fig. 8.10). This large shingle structure now extends some 16 km beyond the old coastline and its outer part consists of storm shingle ridges which face the dominant waves. Storm waves run in mainly from two directions, south and east, the nearness of the French coast preventing large waves from arriving head-on from the south-east. Seas approaching the coast from the two main directions have built up successive storm beaches, now seen as parallel shingle ridges, in the course of centuries; a similar explanation would account for other forelands of this type, which are situated where there is a change of exposure of the coast-line. Sea level off Sussex in Roman times was probably nearly 2 m lower than at present (cf. p. 61), and the Romney Marshes which lie behind Dungeness were probably drained then.

Types of shore-line.   Many shore-lines are the result of *submergence*, i.e. a sinking of the land relative to the sea. When this happens, water-level rests against the hill and valley slopes of the former land surface; the hills stand out as headlands and cliffs are then formed around them by erosion, as described below. A shore-line may also result from the *emergence* of a land area from beneath the sea, in which case the coast is straight and featureless, bordering an area of newly exposed sea-floor. Other types of shore-line are formed, for example, by the growth of coral reefs or by volcanic activity.

Coastal erosion   Where a cliff marks the coast-line, a *wave-cut platform* is usually formed at its base. The elevation of the platform approximates to low-water level, and it slopes gently seaward. In many instances the rocks of the fore-shore may be seen exposed on this platform, in others they are covered by beach deposits. At the base of the cliffs the sea exerts a 'sawing' action, cutting a horizontal notch into and gradually undermining the cliff face. While this goes on, atmospheric denudation is wearing away the upper part of the cliff, tending all the time to reduce its steepness, and debris which falls to the foot of the cliff is broken up by the waves and removed into the off-shore zone, where it is spread as a bank or terrace of sediment (Fig. 2.15). Thus the cliff recedes as it is undercut by waves and at the same time worn down by rain and frost; the action of the sea is gradually reduced as the wave-cut platform is widened. These combined processes, if left to run their course, would in the end stabilize the position of the coast and reduce the land to a nearly level surface—results rarely achieved because of interruptions which occur in the cycle of erosion. As cliffs are cut back, streams which drain to the coast may be rejuvenated by the steepening of their lower courses; if the cutting back takes place faster than the streams can achieve a new grade, waterfalls may be formed where the truncated valleys meet the cliff.

In hard rocks a cliff may stand with a vertical face; if the rocks are jointed, marine erosion uses the joints and other planes of weakness to remove loosened joint-blocks; in this way caves and tunnels are carved out. The process is greatly assisted by the repeated compression of air in the joints by incoming waves, since the pressure acts at the back and sides of joint-blocks, which become loosened. When the roof of a cave collapses, a long narrow inlet or *geo* is formed. A cave sometimes communicates with the surface by a vertical shaft or *blow-hole* on rocky coasts, as in Cornwall. Erosion controlled by a joint-system (Fig. 2.17), may in time leave isolated pillars of rock, or

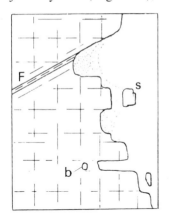 

Fig. 2.17   (*Left*) Joint-control of erosion, with formation of stacks; joint system shown by broken lines (plan). (*Right*) Arch in coastal cliff; when crown of arch falls a stack is formed. (*b*) = blow-hole, (*s*) = stack, (*F*) = fault, controlling local direction of cliff.

sea-stacks, sometimes of fanciful shape; stacks are formed in relatively soft rocks such as the Chalk of the Dorset coast and the Isle of Wight, and notably in harder rocks, as around the northern coasts of Scotland and the Orkney and Shetland Islands (Plate 2b).

Where belts of harder and softer rock alternate along a length of coast, the former stand out as headlands and the latter are hollowed out into intervening bays in the early stages of shore-line development. The Welsh coast near St. Davids well illustrates the way in which steeply dipping bands of resistant rock, running at an angle to the coast, reinforce the softer strata and project as headlands.

Adjacent hard and soft beds may run parallel to the coast, as at Lulworth Cove, Dorset, where the breaching of a hard outer layer (Portland Beds) on the seaward side has allowed the sea access to softer rocks (Purbeck and Wealden Beds), and these have been hollowed out into the nearly circular cove behind the barrier of harder strata.

Hard beds lying horizontally over softer strata give rise to over-hanging ledges, produced by unequal rates of erosion in the two kinds of rock.

In a mature stage of erosion the headlands themselves are reduced, an effect promoted by *wave-deflection*. Waves, as they approach the shore, run into shallow water off the headlands before they reach water of similar depth in the bays. Each wave-front thereby becomes deflected, the line of crests being bent so that it is concave towards a headland and convex towards a bay. This results in a concentration of wave attack on the headlands from both sides, while the bays are to a considerable extent protected. Erosion of the headlands thus goes on vigorously and at the same time there is comparatively smooth water at the heads of bays, where beaches are accumulated. In course of time the headlands are worn down and the bays filled in, and the coast-line tends to become more nearly a smooth line.

Where softer rocks such as sands and clays form low cliffs they offer less resistance to erosion and are usually worn back more rapidly. The Norfolk cliffs between Sheringham and Happisburgh, for example, are built of glacial clays, sands, and gravels, arranged in irregular layers which may be bent and sometimes contorted. Where clay layers slope seawards, rain soaking through overlying gravels and sands saturates the clays, which then locally become surfaces of sliding (*cf*. Fig. 2.14). Frequent slips have been caused in this way along the coast; the cliffs become more stable when erosion has cut them back to a position in which the folded layers of sediment have a landward dip, i.e. slope away from the sea. Wind erosion of dry sandy beds also produces small falls of cliff locally.

Rate of erosion.   The rate of coastal erosion may vary considerably from place to place, according to the characters of the rocks and the structures they contain, and the presence or absence of protective beach. In England erosion is greater, in general, on the east and south coasts than on the west. Exact figures relating to the progress of past erosion are not easy to obtain and usually more measurements are needed. An estimate of the change that has occurred for a particular locality can, however, sometimes be made by comparing old and new editions of the Ordnance Survey maps of the district. The Sussex coast, for example, east of Beachy Head, is estimated to have receded by 4 m per year over a period of 110 years, a high rate of recession. On the Holderness coast of Yorkshire, losses by erosion of the relatively soft boulder clay cliffs have probably amounted to some 2.7 m in a year, about 1 acre of land (4046 m²) having been lost annually; a strip of coast between three and four kilometres wide has been lost here since Roman times (Note 22).

The stretch of coast on the east side of Selsey Bill, Sussex, has probably suffered more erosion during the past 50 years than any area in Britain. The mean wave approach here is from the south-south-west, oblique to the shore, giving rise to a north-easterly littoral drift (Fig. 2.18). Old sea defences were smashed in successive storms, and

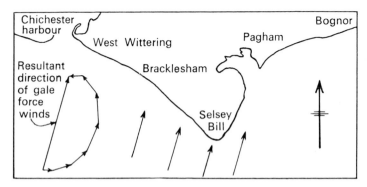

Fig. 2.18   Map of Selsey Bill, Sussex, with mean direction of wave approach (after Duvivier).

houses and roads at Selsey were being lost to the sea. Erosion along the east side of the promontory had reached a rate of 7.6 m (25 feet) per year; the lifeboat station, which was built next to the cliff in 1923, was by 1950 some 230 m (750 feet) from the shore. (It was rebuilt in 1960.) Erosion along the west side of the Bill was slower because of better protection by groynes there. With the passing of the Coast Protection Act in 1950, which eased financial difficulties, new works including articulated groynes (see p. 63) were constructed to secure Selsey Bill against further erosion (Note 23).

Tidal surges in the North Sea   On the night of January 31, 1953, a severe storm swept south past Scotland, where high northerly winds did much damage to forests, and into the North Sea; it caused extensive flooding and damage at many places on the east coast of England, from the Humber to Kent. It also affected the coasts of Holland, with disastrous results. The spring tides expected at the time were not exceptional, but superimposed on them was a tidal surge, which travelled from north to south and produced high-water levels up to about 8 feet above those predicted (2 feet at Aberdeen, 7 feet at the mouth of the Tees, and over 8 feet at Southend). A surge is defined, simply, as 'a water movement which is quickly generated and whose effects are soon over'; in this instance it was a rapid rise of sea level which moved down the North Sea with the incoming tide, rather like the bore of a river.

The high-water levels, combined with a northerly gale, resulted in the overtopping of many lines of coastal protection, the sea also breaking through dunes, as between Mablethorpe and Sutton in Lincolnshire, and inundating low-lying areas behind them. Other wider belts of dunes, e.g. on the north coast of Norfolk, were eroded but some withstood the storm. At Horsey, on the east coast of Norfolk, the new wall built in 1938 helped greatly to maintain that stretch of coast. But at places where the beach was steeper, more severe erosion occurred.

At Lowestoft large areas of the town were flooded by water which came in both behind and over the sea-walls (p. 62), which themselves were unbroken. At Aldeburgh the river Alde kept to its course, but the outer shingle bank at the bend of the river just south of the town, referred to on p. 55, was flattened and much shingle was displaced landwards by the heavy seas. On the Essex coast near Clacton, and at Canvey Island, extensive flooding of areas lying below high-water mark was combined with serious damage to property and loss of life. The railway system of north Kent was seriously disrupted by flooding. The main sea-walls along the Thames estuary held, but water flooded over the lower walls at the Benfleet and other creeks, and once the walls were overtopped they were eroded from inside (Note 24). Much new construction of defence works has been put in hand since the events described, and also measures to restore to productivity the agricultural land which was flooded. A few years earlier, in January 1949, a smaller tidal surge affected the east coasts and had many points in common with that of 1953; but water levels in 1949 were somewhat lower.

Submergence in south-east England   There is evidence that a slow downward movement of the land relative to the sea has been going on in south-east England for some centuries. Recent changes in level were found by the Ordnance Survey when the second geodetic levelling in Great Britain was completed in 1921, the first levelling having been made over 60 years earlier. When the two sets of precise levels were compared considerable differences between them were found, the greatest being a fall of 2 feet (61 cm) at Harwich (Note 25). The changes were not distributed haphazardly, but the country as a whole showed a fall in level in the east and south, and a somewhat smaller rise in the west and north. It was considered unlikely that these differences could be explained in terms of instrumental or other errors, and they therefore appeared to show that during the period of 60 years there had been a small tilt of the land to the south-east. The validity of this result was questioned on the ground that the older levelling did not possess an accuracy of the same order as that of the second levelling. But more recently, measurements of sea levels at tidal gauges have indicated that, taken over a period of years, the average height of high tides in southern England has gradually increased, thus supporting the suggestion that a real lowering of the land has gone on. At Newlyn, Cornwall, the rate of the lowering is estimated at about 9 inches per century (or 2.3 mm per year)—a figure which agrees with the average rate at which the land surface in the Thames Valley has subsided since Roman times.

Valentin has published a map of the British Isles on which isobars in mm per year are drawn for the movement from the known data. The − 2 mm isobar runs from London westwards to Avonmouth, the − 1 isobar from Pembroke north-eastwards to the Humber, and the zero line in Anglesey curves northwards to near Dunbar on the Scottish coast. West of the zero isobar a slow rise appears to be taking place with a possible maximum, according to Valentin, of + 4 mm per year in the Scottish Highlands (Note 26). Changes of this kind are not confined to the British Isles; a longer discussion, with references, is given by Jelgersma and others (Note 27).

The slow submergence in the south-east of England appears to be related to the accumulation of a large thickness of sediment in the North Sea Basin, into which many present day rivers discharge silt and mud. An estimate by Collette in 1967 put the thickness of sediment at 6000 m (nearly 20 000 feet). He suggests that this has accumulated from Jurassic times onwards, at intervals, and is mainly shallow-water sediment. The crust locally has subsided as the load has increased, and there are no indications that the subsidence has come to an end (Note 28).

Much low-lying land along a North Sea coast or estuary at the present day is subject to the risk of flooding should another tidal surge comparable to that of 1953 (p. 59) coincide with high tide levels. Great damage was done by flooding in Hamburg in 1962. Along the River Thames, flood protection for London has involved the raising of the river walls to withstand a flood height of 5.3 m ($17\frac{1}{2}$ feet), corresponding to the level reached in 1953, and it is proposed to raise the flood defence level a further 1.8 m (6 feet). For greater protection in the future, a tide-control barrier across the Thames near or below Woolwich is to be built.

Coast protection works   Marine erosion may be arrested locally by the building of defences such as sea-walls, or by the construction of groynes which serve to accumulate some of the sand and shingle drifting alongshore and so build up a protecting beach. On sandy coasts, the erosion of dunes may be offset by the planting of grasses and shrubs to form a stable surface layer which in time becomes grassed over.

Sea-walls are used to protect cliffs from wave-action, the shock of which is taken by the wall instead of by the cliff foot. One form of construction may be a low bank of earth, sand, or other material; on account of its relatively low cost a bank of this kind is often used for protecting agricultural land. It usually has a gently sloping seaward face covered with a pitching of stone blocks or other resistant material. At Winchelsea, Sussex, a shingle bank which had been breached by the sea was successfully repaired with a filling of chalk, which was tipped and consolidated in successive layers and built up into a bank closing the gap. Timber stakes driven into the beach in front of a bank have been used to break the force of the waves, as at Walcheren in Holland. At Rye Bay, west of Dungeness, a double line of piles with a filling of shingle between them, forming a continuous barrier, was constructed about 1936 along a length of fore-shore to protect the existing shingle bank; in front of this barrier was placed a single line of closely spaced timber piles to act as a further wave-screen (Note 29).

Walls of concrete or stone are more durable but more costly; they are commonly employed in residential areas, generally in conjunction with a system of groynes. In the past massive concrete sea-walls were built with a vertical or nearly vertical face; the wave-impact and undertow, however, sometimes induced scour at the toe of the wall, leading to instability of the structure when deep scour occurred. A different form of wall which has been employed more recently in many situations is of reinforced concrete with a stepped or gently sloping seaward face. The sloping face forms a protective apron whose function is to absorb much of the force of the waves as they move up it; sheet piles driven into the beach at the toe of the wall can be used to prevent scour there. At Lowestoft, after the breaking of earlier vertical walls by heavy seas, a concrete wall with a 15 m (50 ft) wide nearly horizontal apron and steel piling at the toe was constructed about 1936 (Note 30). This new wall was broken in severe storms early in 1946; the position is an exposed one on the east coast, subject to north-easterly gales, which change the configuration of protective sandbanks off-shore. Subsequently another defence wall was built behind the line of the previous structure and withstood the tidal surge of 1953.

Groynes. A good beach is itself a protection against the erosion of the land, as it absorbs the impact of the waves which break on it. But a beach may be considerably modified by storms and sometimes changed rapidly; instances are on record of the lowering of beaches by 4.27 m or more in a single storm. Groynes can be used to retain or accumulate beach deposits along a shore. In general they are placed normal to a coast-line or to the prevailing longshore current, and at a distance apart approximately equal to their length; there is much difference of opinion as to the dimensions and spacing of groynes, and conditions vary at each locality. Groynes were commonly made of heavy timbers or of stonework; but concrete structures are now used in many places, and need to be founded sufficiently deep to guard against scour at low tides. Drifting beach deposits accumulate on the weather side of each of a series of groynes, which can be made high enough to produce a uniform accumulation of beach, but not so high as to give rise to a large drop in level on their leeward sides. As the profile of a beach changes it may be necessary to adjust the height of the groynes, for example by adding longitudinal planks between the posts of timber structures; adjustable screw-piles which can be raised or lowered by turning, to suit the prevailing beach level, have also been used instead of fixed posts.

Efforts to stabilize the seriously eroded headland at Selsey Bill (Fig. 2.18) began in 1950 with the construction of an *articulated groyne*. This consisted of massive concrete blocks weighing $15\frac{1}{2}$ tons (15.75 metric tonnes) each, linked transversely in pairs by hinged steel bars, and pinned to the underlying gravel by vertical steel rails. Successive pairs of blocks formed the first groyne, 190 feet long and 16 feet wide (58 m and 4.9 m), which was built on the west side of the debris at Selsey village. This groyne functioned well for several years, taking the impact of heavy seas from the south-south-west; a second articulated groyne was then built to the east of it, and the original groyne converted into a rigid break-water in 1960. These structures, together with a new sea-wall and system of groynes along the east side of the promontory, succeeded in controlling the erosion.

By analogy with natural conditions, from which it is seen that erosion proceeds most rapidly at points where rocks of one kind adjoin others of different hardness (and therefore of different resistance), it is evident that defence works such as a length of sea-wall along a coast may be of local value, but end-effects have to be considered. The effect of a wall is like that of a cliff of hard rock; where it ends erosion goes on rapidly, possibly cutting back behind the end of the wall, so that one part of a coast is protected at the expense of adjacent areas. A comprehensive plan dealing with a whole coast-line is therefore needed. In default of this ideal, provision can sometimes be made for suitable terminations to coastal works. At the leeward end of a series of groynes serious erosion may develop to the lee of the last groyne; where the new works do not join on to existing works it may be possible to carry on the groyning along the shore to a region where there is greater stability of beach deposits, or to deal with the matter in some other way. The problem of erosion is aggravated when shingle is removed in large quantities from beaches, e.g. for local construction, and in places it has been necessary to regulate or prohibit this.

Marine deposits   Much material derived by denudation from land areas is deposited near a coast, as discussed in Chapter 5. After sorting by waves and currents, the sediments are built up into new deposits mainly in shallow water. The fragments are distributed according to size, ranging from the coarse shingle of the beach, through sands and muds, to the finest particles which are carried by feeble currents out to sea and slowly accumulated. When compacted and hardened the deposits become sedimentary rocks such as sandstones and shales. These *shallow-water deposits* are laid down on the continental shelf (p. 2) below low-tide level, to depths of about 100 fathoms (183 m), the lower limit of wave action. They include the sands, silts, and muds, and are deposited over wider areas than the coarser material of the beaches. The latter comprise boulders, pebbles, and coarse sands. Marine shells and other organic remains which ultimately are preserved as fossils may be embedded in the muds and sands. In some environments much calcareous organic matter such as shell sand may be present, or fine calcareous muds laid down, eventually to become beds of limestone (Note 33).

There is a gradation in shallow-water sediments from coarse to fine with increasing distance from a coast (Fig. 2.19). They form lens-shaped layers, thicker at the coarser end and thinning out to seaward.

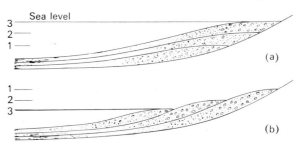

Fig. 2.19   Deposition of sediments near a shore-line, (*a*) area submerging, (*b*) area rising relative to sea-level. *1, 2, 3* denote successive water-levels. Vertical scale is exaggerated.

If the sediments are deposited on a sinking sea-floor, as must often have been the case in past geological times, the lenses of material overlap successively; as in the figure (*a*), and finer grades come to be deposited above coarser. Thus sands may grade into silts and muds in a vertical as well as a horizontal direction. Under these conditions a large thickness of deposit can accumulate, the sedimentation keeping pace with the sinking area and the depth of water remaining much the same. When deposits are laid down on a slowly *rising* sea-floor, successive lenses of sediment are related as shown in Fig. 2.19(*b*), and coarser grades may then come to lie on the finer parts of older layers. If the area of sedimentation is intermittently raised and lowered, giving varying depths of water at different times, then the deposits oscillate from one kind to another, e.g. coarser layers may be intercalated between finer, and vice versa.

It is evident that at any given time different kinds of deposit may be forming in different environments. The term *facies* is used to express this conception, and is applied to the sediments or to the rocks resulting from their consolidation. There is, for instance, a transition from a sandy facies near a shore to a contemporaneous muddy facies in deeper water. Not only the nature of the sediments, but also other features such as colour and the different kinds of fossils preserved in them, are understood in the term facies, which includes all the characters which go to distinguish a deposit. Sediments formed on land, such as wind-blown sands, constitute a continental facies as distinct from a marine facies. Structures such as bedding planes, which divide a thickness of similar material into beds or *strata*, and generally represent small interruptions in the course of sedimentation, are discussed in Chapter 5.

The work of ice   A land surface whose topographical features have been fashioned by the action of rivers and atmospheric agents is considerably modified when it becomes covered by an ice-sheet or by glaciers. Valleys are deepened and straightened, and rock surfaces smoothed by erosion, and when the ice melts away it leaves behind a variety of deposits which mark its former extent. The main features of these processes are discussed here; glacial deposits formed in Britain during the Pleistocene glaciation, when large areas of the British Isles, north-west Europe, North America, and other northern lands were under a load of ice, are summarized on pp. 266–7.

Ice is formed by the compaction of snow in cold regions and at high altitudes, where the supply of snow exceeds the wastage by melting. In an intermediate stage between snow and ice the partly compacted granular mass is called *névé*. Ice of sufficient thickness will begin to move down a slope, and such a moving mass is called a *glacier*. It may occupy a valley, as a *valley glacier*, of which many examples are found in the Alps, the Rockies, and other mountain regions; they are the relics of larger ice-caps. Where several valley glaciers meet on low ground in front of a mountain range, a stagnant accumulation of ice, or *piedmont glacier*, is formed, e.g. the Malaspina Glacier of Alaska. The much larger accumulations of thick ice constitute the *ice-sheets*, and cover great areas. The Greenland ice-sheet extends over about 1.3×10^6 km^2 at the present day, and the Antarctica ice-sheet is six times greater. 'Islands' of rock standing up through an ice-sheet are called *nunataks*. When a glacier meets the sea it begins to float and break up into ice-bergs; any land-derived debris in the ice is carried by the bergs, and dropped as they become reduced by melting.

Valley glaciers. Extensive studies have been made in Switzerland and in other countries where valley glaciers are accessible. The ice breaks away from the parent snow-field by a big crevasse known as the *bergschrund*, and in moving over the ground it behaves rather like a very viscous body, flowing over the irregularities of its course. It has a granular texture similar to the crystalline structure of metals. Movement occurs by melting and re-freezing at the surfaces of crystals, along gliding-planes within the crystals, and also along shear planes in the mass, the ice breaking as it rises over an obstruction, on its way down to lower levels; tension crevasses are formed especially near the margins of the ice (Plate 3a). While the upper part of the mass undergoes movement on inclined shear fractures, the lower part of the ice moves more by flowage under pressure.

An average rate of movement of an Alpine glacier is 0.6 m per day, but this figure varies considerably. It depends on the steepness of the slope over which the ice is moving, on the thickness of ice, and on the air temperature. Rates up to 18 m per day have been measured in Greenland. At lower levels, where melting balances the supply of ice descending a valley, the size of a glacier is gradually reduced until it ends in a 'snout,' from which issues a stream fed by the melting ice. Such glacial streams may supply reservoirs which store water for hydro-electric power generation; the glacier is in effect a frozen reservoir.

Transport by ice. A glacier carries along boulders and stones of all sizes which fall on to its surface from the valley walls on either side, and superficial debris of this kind is called *moraine*. It frequently lies in two marginal bands, or *lateral moraines*, parallel to the sides of the glacier; in some places the ice is completely covered by stones and dirt. The confluence of two glaciers, as when a tributary enters the main valley, results in the formation of a *medial moraine* from the two laterals which become adjacent where the ice-streams meet. The united glaciers below the point of junction thus possess one more line of moraine than the number of tributaries.

Rock debris which falls into crevasses is carried forward within the ice, and blocks under these conditions may penetrate by their own weight to the sole of the glacier. Other fragments are 'plucked' from rocks over which the ice moves, and are held in the lower part of the mass, which becomes packed with dirt and fragments. This assorted

Plate 3

(a) Glaciers in south-west Renland, Scoresby Sound, East Greenland. The ice-dammed lake is 1.5 km wide and lies 820 m above sea level; the cliffs on the left rise a further 550 m above the lake level. (*Courtesy of the Director, Geological Survey of Greenland, Photograph by Dr. Brian Chadwick*)

(b) Steeply dipping sandstones with soil and rock-creep.

Plate 4

(a) Large inselberg at Spitzkop, South West Africa. (*Courtesy of Dr. A. O. Fuller*)

(b) Basic dyke cutting folded Damara rocks, Brandberg West area, South West Africa. (*Courtesy of Dr. A. O. Fuller*)

material, much of which is protected from wear as it is surrounded by ice during transport, is known as *englacial material;* fragments may become partly rounded at a subsequent stage by the action of glacial streams. Later, when the ice has melted, large transported blocks are left stranded wherever they happen to be. They are called indicator boulders, or *erratics,* since they are generally different from the local rock on which they come to rest. A study of the distribution of erratics has made it possible to draw conclusions as to the extent and direction of travel of the ice which, during the Pleistocene glaciation of the British Isles, moved out from the higher mountain regions in the north. Blocks of the distinctive microgranite from Ailsa Craig in the Firth of Clyde, for example, are found as far south as North Wales; and boulders of laurvikite (*q.v.*), a Norwegian rock, are found along the Yorkshire coast, indicating ice-transport across what is now the North Sea. (Fig. 8.14).

Glacial erosion. Ice moving over a land surface removes soil and loose material, exposing the bed-rock below, and acts as a soft abrasive. The surface debris, together with material which has worked its way down through the ice, is held in the lower part of the glacier as explained above. It is rubbed over the rock floor, which becomes smoothed in the process and lubricated by the melting of the ice under pressure; thus a *glaciated surface* is formed. Blocks held in the sole of the glacier may cut and scratch the surfaces over which they are carried, producing grooves or *striae* which point in the direction of motion; the blocks themselves may become striated. Hollows ground out by the ice form *rock basins* and are often occupied by lakes or tarns at the present day.

At the head of a valley glacier, the 'plucking' action of the ice at the bergschrund forms an armchair-shaped hollow called a *corrie* or *cirque* (the Welsh *cwm*), which is sometimes occupied by a lake as at Glaslyn, Snowdon. Rock obstructions in the path of a glacier are smoothed on their iceward slopes and 'plucked' on the leeward side; when exposed to view after the removal of the ice, they show the rounded forms known as *roches moutonnées* (Fig. 2.20*a*), which are typical of glaciated mountain regions. The profile of a valley down which a glacier has flowed is changed from a **V**- to a **U**-shaped form. Such a valley presents a rougher, rocky aspect when viewed upstream than when looking down the valley, since in the latter case it is the ice-smoothed slopes that are seen. The valley is deepened by the passage of the ice, and spurs which formerly projected into it are truncated so that sharp bends have been removed. Tributary valleys which were once graded to the main valley are left cut off at some height above the new floor; they are known as *hanging valleys,* and the streams from them form waterfalls as they drop down to the new

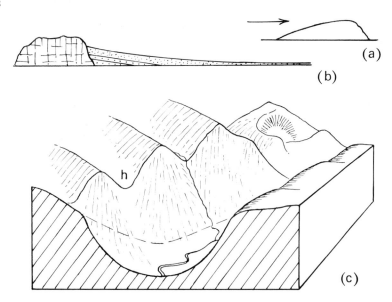

Fig. 2.20 Ice erosion forms. (*a*) Roche moutonnée. (*b*) Crag and tail (to smaller scale), ice movement from left to right. (*c*) **U**-shaped profile of glaciated valley, with hanging valleys (*h*) and truncated spurs between tributaries to main valley. Corrie in distance. Original valley profile shown by broken line.

level (Fig. 2.20*c*). Alluvium subsequently deposited on the floor of the main valley may level up the bottom of the **U**-shaped profile.

A large rock mass in the path of advancing ice, in low-lying country, may preserve a 'tail' of drift on its lee side, the ice dividing and passing on either side of the 'crag'; a well-known example of *crag and tail* is provided by Castle Rock, Edinburgh, from which High Street and Canongate running eastwards to Holyrood follow the gentle slope of the tail. And in Co. Down, Northern Ireland, a fine 'tail' extends south-south-east from Slieve Gullion (the 'crag') for several kilometres.

Glacial deposits When a glacier or ice-sheet has retreated, it leaves behind it characteristic deposits of sediment to which the term *glacial drift* is given. This is irregularly distributed, generally thickest in valleys, and shows a lack of sorting and arrangement which distinguishes it from water-borne sediment. Debris dropped at the end of a glacier as the ice melts forms a hummocky ridge called a *terminal moraine;* in the British Isles these end-moraines are found, for example, in north Yorkshire, South Wales, and southern Ireland.

Streams of melt-water flow in tunnels in the base of the ice and move much material, which becomes partly rounded in the process. When these streams emerge along the ice-front their speed is checked and the transported debris is spread out on land as an *outwash fan* of gravel and sand. If the debris is discharged into still water it builds up a delta. Ridges and mounds of gravel deposited by such sub-glacial streams, with their length roughly at right angles to the ice-front, are termed *eskers;* one broad belt of eskers extends east and west across the Central Plain of Ireland, between Galway and Co. Dublin (Note 31). Other ridges or sheets of sand and gravel which were derived from the ice along its margins during pauses in its retreat, and lie parallel to the ice margin, are called *kames.* Good examples of these glacial gravels can be seen near Carstairs in Lanarkshire, and in the Clyde valley to the south, and at many places in the north of England.

Fine laminated clays known as *varves* may also be deposited in still water impounded in front of a glacier, or in glacial lakes (p. 70); they show alternate coarser (silty) and finer (muddy) layers. The mud particles, which settle slowly in water, lie on the more quickly sedimented silt, and each pair of layers (coarse and fine) represents a season's melting of the ice. Varved clays thus provide a time scale, in which a pair of layers counts as one year. In an engineering excavation at Lake Ragunda, Sweden, a thickness of 24 metres of undisturbed varves lay above a deposit of moraine. From the number of laminae in this deposit it was estimated that 7000 years had elapsed since the ice margin retreated past the region; other counts near by, when added in, showed that in all 9000 years had passed since Stockholm became free from ice. This method of dating has also been applied extensively to varves in Canada, Scotland, and elsewhere.

The deposit known as *boulder clay* is formed underneath an ice-sheet; it is a more or less stiff clay composed of the ground-up debris (fine rock-flour) contributed by the melting out of englacial material and by the abrasion of the rocks over which the ice has passed, and embedded in this clay are boulders and stony fragments of all sizes. It is deposited as an irregular layer over the surface of the ground, and is unstratified except where modified by water. The colour of the clay depends on that of the rocks from which the ice obtained the raw materials during its passage. Thus a red boulder clay is formed when the ice has passed over red rocks, as with the Irish Sea ice which crossed the red Triassic rocks of Cheshire on its way into the Midlands, carrying also far-travelled boulders from the Lake District and southern Scotland (see Fig. 8.14). An old name for hardened boulder clay is *till.*

A *drumlin* is a smooth oval-shaped mound, commonly 30 to 45 metres in height, which is built of boulder clay or englacial debris, sometimes moulded over a roche moutonnée. The long axis of the oval is parallel to the direction taken by the ice. Drumlins are well developed in the country bordering the Lake District, as in the Vale of Eden, and in parts of Galloway. Clew Bay, in N.W. Ireland, is a remarkable drumlin archipelago formed by partly submerged drumlin country (Note 32).

Boulder clays may give rise to difficulties in engineering construction because of their extreme variability. They may consist of a high proportion of stones and boulders, or pockets of sand and gravel in a clay matrix, or may be nearly entirely of clay. When very large boulders are present excavation is impeded; the large boulders hinder the use of excavating machinery and may have to be broken by blasting and removed piecemeal. Boulder clay also masks the true nature of the underlying rock surface, concealing hollows and sometimes deep valleys; buried channels have been proved by borings through drift in the valleys of the rivers Clyde, Forth, and Tyne, and in many other valleys. The old course of the River Mersey, now a buried drift-filled channel, was encountered during the construction of the first Mersey road tunnel (1936) and explored by inclined drill-holes from the workings. The presence of such buried valleys can often be detected by a geophysical survey (p. 280). An early instance of this was the gravity survey used to find the extent of a buried channel at Drumry, Glasgow: the method was employed there because of the difference in density between the drift (1.72) and the bed-rock (2.38).

Trial borings through glacial drift to find the depth to the rock floor beneath it provide more precise data on which design can be based, so long as the number of boreholes used is adequate. It is false economy to have too few boreholes. A large boulder partly penetrated by a borehole has sometimes, in the past, been mistaken for the solid rock below the drift. Boulder clay which may form a useful water-tight layer at a reservoir site, or other locality, needs to be sampled and tested to ensure that it is of suitably low permeability (see Chapter 9).

Glacial lakes   When glaciers descending from higher ground encroach upon an area, streams normally flowing through the valleys may be dammed up by the ice, and a lake thus formed (Plate 3a). In Glen Roy, north-east of Ben Nevis, three shore-terraces—the Parallel Roads—run at successive levels along the slopes of the glen; they mark the beaches of a former glacial lake. The waters from the glen drain into Glen Spean and, when the latter was occupied by ice, were ponded up and found outlet by a col at the head of Glen Roy. This level was maintained long enough for a well-marked beach to be formed; then as the ice in the main valley diminished, a lower outlet for the glacial lake was established and the second terrace was formed, the third being made later at a still lower level.

Overflow channels from old glacial lakes have sometimes been cut deeply if they acted as a drainage outlet for a long time, and are then prominent topographical features at the present day. The Ironbridge Gorge, Shropshire (p. 268), is one example. The Grand Coulee, a canyon over 64 km long and 3 to 5 km wide, in the State of Washington, U.S.A., provides an instance of an old glacial channel being utilized for engineering works of considerable size. The canyon was cut by the overflow waters of a glacial lake which was impounded by ice blocking the gorge of the Columbia River. The waters overflowed past the margin of the ice at a high level, for a long enough time to cut a large natural spillway (the Grand Coulee). This spillway was eroded in basaltic lavas, and is in places over 183 m deep, with a large waterfall (now dry) at one point. After the recession of the ice the river recovered its old channel. The Grand Coulee has now been made the site of a reservoir, water for irrigation purposes being impounded between two low dams built across it. The reservoir was filled in 1952, water being pumped up to it from the main dam on the Columbia River, adjacent to the Coulee.

NOTES AND REFERENCES

1. The term *soil* is used in another connection in engineering geology, as in soil mechanics (p. 177).
2. Examples of this are seen in exposures made about 1965 for road-widening south-east of Moretonhampstead. The local term 'growan' is used for the weathered granite.
3. See Dickinson, J. C. and Gerrard, R. T. November 1963. The Cameron Highlands H.E. Scheme, in *Proc. Inst. C.E.* **26**, 387.
4. For an illustrated account of the weathering of limestones see the essay by Sweeting, Marjorie M. 1966, in *Essays in Geomorphology*, Ed. G. H. Dury. Heinemann, London.
5. Potts, A. S. 1970. *Trans. Inst. Brit. Geographers*, **49**, 109.
6. The term *permafrost* is due to Muller, S. W. 1945. *U.S.G.S. Spec. Rep.*
7. Described by Ferrians, O. J., Kachadoorian, R., and Greene, G. W. 1969. Permafrost and related engineering problems in Alaska, *U.S.G.S. Prof. Paper 678*, Washington.
8. See Brown, R. J. 1970. Permafrost in Canada: its influence on northern development. University of Toronto Press and Oxford University Press, London.
9. Abrasion by a sand-blast is used in commercial processes for frosting glass and for roughening a metal finish.
10. Unpublished thesis by Morton, J. A. 1964. Birmingham University.
11. A full account is given in The Physics of Blown Sand and Desert Dunes by Bagnold, R. A. Methuen, London, 1941.
12. Cooke, R. U. and Smalley, I. J. Dec. 21, 1968. Salt weathering in deserts, *Nature*, **220**, 1226. See also Skinner, B. J. 1966. Handbook of Physical Constants, Mem. 97, *Geol. Soc. Amer.*
13. The Upper Towy drainage system, 1924, *Q.J.G. Soc.*, lxxx, p. 568. The River Mole is discussed by Green, J. F. N. *et al.*, 1934, in *Proc. Geol. Assoc.*, **45**, 35–69.
    For a fuller discussion of the development of drainage systems and

river patterns see Small, R. J. 1970. Study of Landforms. Cambridge University Press.

14. An illustrated account of this river capture is given in *British Regional Geology:* The Wealden District, H.M.S.O., London, 1954.

15. By contrast with alluvial deposits, the term *colluvium* is used for rock-debris transported by gravity. Colluvial deposits include screes, avalanches, landslip deposits, and similar materials, and are usually of limited extent as they are situated near the cliffs or slopes from which they are derived.

16. The thickness of the deposits has been estimated by Glennie, E. A. as 6500 feet (1.98 km). A depth of 1306 feet was bored through them at Calcutta in 1938; an earlier borehole at Ambala had reached 1612 feet (0.5 km).

17. A general description of the proposed works was given by Doran, W. E. Apl. 1941, in *Geogr. Journ.* xcvii, 217. The hydraulic model referred to on p. 49 was constructed for the Great Ouse Catchment Board.

18. The classification by Sharpe, C. F. S. in Landslides and Related Phenomena (New York, 1938) includes three groups: (i) Flows, (ii) Slides, and (iii) Subsidences (in which the moving mass is entirely surrounded by unmoved material).

19. T. Stevenson, 1849 in *Trans. Roy. Soc. Edinburgh*, xvi, 25. White, C. M. June 1939, in *Jour. Inst. Civil Engrs.*, **12**, 202.

20. R. A. Bagnold, Nov. 1940, *Jour. Inst. Civil Engrs.*, **15**, 27. See also publications of U.S. Beach Erosion Board; and U.S. Army Board Report on Sand Movement and Beach Erosion, 1937 (reviewed 1937, in *The Engineer*, **163**, 719).

21. J. A. Steers, May 1939. *Geogr. Journ.*, 399; and 1948, *Geol. Mag.* **85**, 163. See also Beaches and Coasts, by King, C. A. M. 1959, Edward Arnold, London; and Evans, G. 1965. Intertidal flat sediments in the Wash, *Q. J. Geol. Soc.* **121**, 209–246, who records accretion of 1.2 to 1.5 cm in a year.

22. Extensive data relating to the changes in the coast-line of Gt. Britain were published in the Report of the Royal Commission on Coast Erosion (3 vols.), 1909–11.

23. J. Duvivier, Dec. 1961. The Selsey Coast Protection Scheme, *Proc. I.C.E.* **20**, 481–506.

24. Many other details are given in the account by Steers, J. A. 1953. *Geogr. Journ.*, **119**, 280.

25. The Subsidence of London by Capt. T. E. Longfield, 1932. *Ordnance Survey* Prof. Paper no. 14 (N.S.).

26. Valentin, H. Sept. 1953. Present vertical movements of the British Isles, *Geogr. Journ.*, **119**, 299–305.

27. Jelgersma, S. 1966. Sea level changes during the last 10 000 years. *Proc. Int. Symp. on World Climate from 8000 to 0 B.C. Roy. Meteorological Soc.*, London. Also, The Subsidence of South-eastern England. *Phil. Trans. Roy. Soc. London.* 1972. **A272**, 79–274.

28. Collette, B. J. 1968. The Subsidence of the North Sea Area: a summary,

Canad. Journ. of Earth Sciences, **5,** 1123. (Paper presented at *Int. Symp. on Continental Margins,* Zurich, 1967).
29. Described in *Engineering,* Jan. 10 1936, p. 27 and p. 81.
30. Deane, H. J. and Latham, E. Sept. 25, 1936. *Engineering,* p. 329 and p. 377.
31. The term 'esker' is taken from the Irish *eiscir.* A map of glacial deposits in Ireland, including eskers and drumlins, is given by Nevill, W. E. in Geology and Ireland, Figgis, Dublin, 1963.
32. A map of Clew Bay is given in The Geology of Ireland by Charlesworth, J. K. 1953. Oliver & Boyd, Edinburgh.
33. For a helpful review of the engineering problems in recent marine sediments, see Offshore Soil Mechanics, 1976, 468pp., Lloyd's Register of Shipping, London.

GENERAL REFERENCES

CHARLESWORTH, J. K. 1958. The Quaternary Era (Vol. 1). Edward Arnold, London
COTTON, C. A. 1958. Geomorphology. 7th edition. Whitcombe & Tombs, New Zealand.
EMBLETON, C. and KING, C. 1976. Glacial Geomorphology. Edward Arnold, London.
FLINT, R. F. 1957. Glacial and Pleistocene Geology. J. Wiley & Sons, New York.
GREGORY, K. and WALLING, D. 1976. Drainage Basin Form and Process. Edward Arnold, London.
HOLMES, A. 1965. Principles of Physical Geology. 2nd edition. Nelson, London.
SPARKS, B. W. 1972. Geomorphology. 2nd edition. Longmans Green, London.
WASHBURN, A. L. 1973. Periglacial processes and environments. Edward Arnold, London.

3 Minerals

The three broad groups of rocks—igneous, sedimentary, and metamorphic—together with their mineral constituents, were referred to on p. 1. A *mineral* can be defined as a natural inorganic substance having a particular chemical composition or range of composition, and a regular atomic structure to which its crystalline form is related. Before beginning the study of rocks it is necessary to know something of the chief rock-forming minerals. This chapter discusses the identification of minerals from their *physical characters*, including their *crystalline form*, and from their *optical characters*. First we notice some general facts about the composition of the earth's crust.

The average composition of crustal rocks has been calculated from many chemical analyses to be as follows:

	%		%	
SiO_2	59.26	Na_2O	3.81	
Al_2O_3	15.35	K_2O	3.12	
Fe_2O_3	3.14	H_2O	1.26	
FeO	3.74	P_2O_5	0.28	
MgO	3.46	TiO_2	0.73	
CaO	5.08	rest	0.77	Total: 100.0

The last item includes the oxides of manganese, barium, zirconium, lithium, strontium, copper, zinc, nickel, and other metals; and also gases such as carbon dioxide, sulphur dioxide, chlorine, fluorine, and others; and trace elements, which occur in very small quantities. In terms of *elements* the order of abundance is: oxygen (46.60%), silicon (27.72%), aluminium (8.13%), iron (5.00%), calcium (3.63%), sodium (2.83%), potassium (2.59%), and magnesium (2.09%). Thus, silicon and oxygen together make up nearly 75 per cent of crustal rocks, and the above eight elements over 98 per cent. They are combined in various ways as metallic silicates and oxides, to give the bulk of the rock-forming minerals; other minerals include carbonates, phosphates, and sulphates. Some rare metals, for example gold and silver, found in small amounts, sometimes occur uncombined or 'native'.

Since silicon and oxygen preponderate in the rocks, the chief rock-forming minerals are silicates. Although over three thousand different minerals are known to the mineralogist, the commonly occurring silicates are relatively few. They are described below, together with the main non-silicate minerals. For a fuller classification the student is referred to textbooks on Mineralogy (Note 1).

PHYSICAL CHARACTERS

Included under this head are properties such as *colour, lustre, form, hardness, cleavage, fracture,* and *specific gravity.* Not all of these properties would necessarily be needed to identify any one mineral; two or three of them taken together may be sufficient, apart from optical properties (p. 83). Other characters such as *fusibility, magnetism,* and *electrical conductivity* are also useful in some cases as means of identification, and will be referred to as they arise in the descriptions of mineral species. In a few instances *taste* (e.g. rock-salt) and *touch* (e.g. talc, feels soapy) are useful indicators.

Colour   Some minerals have a distinctive colour, for example the green colour of chlorite. On the other hand, most naturally occurring minerals contain traces of substances which modify their colour. Thus quartz, which is colourless when pure, may be white, grey, pink or yellow, when certain chemical impurities or included particles are present.

Much more constant is the colour of a mineral in the powdered condition, known as the *streak.* This may be produced by rubbing the mineral on a piece of unglazed porcelain, called a streak-plate, or other rough surface. Streak is useful, for example, in distinguishing the various oxides of iron; hematite (Fe_2O_3) gives a red streak, limonite (hydrated Fe_2O_3) a brown, and magnetite (Fe_3O_4) a grey streak.

Lustre   Lustre is the appearance of a mineral surface in reflected light; it depends upon the amount of reflection that occurs at the surface. It may be described as *metallic,* as in pyrite or galena; glassy or *vitreous,* as in quartz; *resinous* or greasy, as in opal; *pearly,* as in talc; or *silky,* as in fibrous minerals such as asbestos and satin-spar (fibrous gypsum). Minerals with no lustre are described as *dull.*

Form   Under this heading come a number of terms which are commonly used to describe various shapes assumed by minerals in groups or clusters; the crystalline form of minerals is discussed on page 78.

Acicular—in fine needle-like crystals, e.g. schorl, natrolite.

Botryoidal—consisting of spheroidal aggregations, somewhat resembling a bunch of grapes; e.g. chalcedony. The curved surfaces are boundaries of the ends of many crystal fibres arranged in radiating clusters.

Concretionary or nodular—terms applied to minerals found in detached masses of spherical, ellipsoidal, or irregular shape; e.g. the flint nodules of the Chalk.

Dendritic—moss-like or tree-like forms, generally produced by the deposition of a mineral in thin veneers on joint planes or in crevices; e.g. dendritic deposits of manganese oxide.

Fibrous—consisting of fine thread-like strands; e.g. asbestos and satin-spar.

Granular—in grains, either coarse or fine; the rock marble is an even, granular aggregate of calcite crystals.

Reniform—kidney-shaped, the rounded surfaces of the mineral resembling those of kidneys; e.g. kidney iron-ore, a variety of haematite.

Scaly—in small plates; e.g. tridymite.

Tabular—showing broad flat surfaces; e.g. the 6-sided crystals of mica.

Note that the above terms do not apply to rocks.

Hardness, or resistance to abrasion, is measured relative to a standard scale of ten minerals, known as Mohs' Scale of Hardness:

1. Talc.	6. Orthoclase Feldspar.
2. Gypsum.	7. Quartz.
3. Calcite.	8. Topaz.
4. Fluorspar.	9. Corundum.
5. Apatite.	10. Diamond.

These minerals are chosen so that their hardness increases in the order 1 to 10. Hardness is tested by attempting to scratch the minerals of the scale with the specimen under examination. A mineral which scratches calcite, for example, but not fluorspar, is said to have a hardness between 3 and 4, or H = 3–4. Talc and gypsum can be scratched with a finger-nail, and a steel knife will cut apatite (5) and perhaps feldspar (6), but not quartz (7). Soft glass can be scratched by quartz. The hardness test, in various forms, is simple and easily made and extremely useful; it is a ready means for distinguishing, for example, between quartz and calcite.

Cleavage   Many minerals possess a tendency to split easily in certain regular directions, and yield smooth plane surfaces called *cleavage planes* when thus broken. These directions depend on the arrangement of the atoms in a mineral (p. 92), and are parallel to definite crystal faces. *Perfect, good, distinct,* and *imperfect* are terms used to describe the quality of mineral cleavage. Mica, for example, has a perfect cleavage by means of which it can be split into very thin flakes; feldspars have two sets of good cleavage planes. Calcite has three directions of cleavage.

Fracture   The nature of a broken surface of a mineral is known as *fracture*, the break being irregular and independent of cleavage. It is sometimes characteristic of a mineral and, also, a fresh fracture shows the true colour of a mineral. Fracture is described as *conchoidal*, when the mineral breaks with a curved surface, e.g. in quartz and flint; as *even*, when it is nearly flat; as *uneven*, when it is rough; and as *hackly* when the surface carries small sharp irregularities. Most minerals show uneven fracture.

Specific gravlty   Minerals range from 1 to over 20 in specific gravity (e.g. native platinum, 21.46), but most lie between 2 and 7. For determining this property a steelyard apparatus such as the Walker Balance can be used, for crystals or fragments which are not too small. The mineral (or rock) is weighed in air and in water, and the specific gravity, sp. gr., is calculated from the formula: sp. gr. $= w_1/(w_1 - w_2)$, where w_1 is the weight in air and w_2 the weight in water. Other apparatus is described in appropriate textbooks (Note 1).

The specific gravity of small mineral *grains* is estimated by the use of heavy liquids, of which the chief are:

bromoform ($CHBr_3$)   sp. gr. $= 2.80$
methylene iodide (CH_3I_2) sp. gr. $= 3.33$ }(dilute with benzene);

and Clerici's solution (4.25), a mixture of thallium salts (dilute with water).

The separation of light and heavy grains in sands can be carried out by the use of heavy liquids. A sand which consists mainly of quartz grains generally contains a small proportion of heavy grains, the identification of which may yield information concerning the derivation of the sand, and also serves for comparison between samples. Minerals commonly found in sediments, with their specific gravities, include:

glauconite	2.3	tourmaline	3.0–3.2	rutile	4.2
feldspar	2.56–2.7	sphene	3.5	zircon	4.7
quartz	2.65	topaz	3.6	ilmenite	4.8
muscovite	2.8–3.0	kyanite	3.6	monazite	5.2
apatite	3.2	staurolite	3.7	magnetite	5.2
hornblende	3.2 (av.)	garnet	3.7–4.3	cassiterite	6.9

Specific gravity is a property used in Mineral Technology for separating ore from gangue (p. 166) and for concentrating residues.

Heavy mineral separation.   (1) Using heavy liquids. A small sample of the sediment is placed in a funnel fitted with a short length of rubber tubing and with a clip to close the tubing, and is stirred in the liquid to exclude bubbles of air. Grains which are denser than the fluid sink and can be withdrawn by opening the clip for a moment, the lighter grains still floating; they are then washed, dried, and mounted (if required) for microscope examination. A temporary mount on a glass slip may be made with cedar wood oil (refractive index, 1.52). The heavy liquid is recovered for further use.

To determine the specific gravity of small grains (of one kind), a liquid in which they float can be slowly diluted until the grains just begin to sink, and its density then determined.

(2) *Panning*. This method is useful for testing a deposit in the field, as in prospecting. A sample (e.g. of a stream sand) is placed in the pan, a shallow metal dish with preferably not too smooth a surface, which is agitated under water. The heavier grains gradually work their way to the bottom of the sample, and the lighter material (quartz, feldspar, etc.) can be allowed to escape over the edge of the pan, the grains possessing considerable buoyancy in water. When the operation is carried out in a stream the running water is made to assist the panning. A rough separation of the heavy fraction in the sample is obtained in this way.

(3) Further separation of heavy grains into magnetic and non-magnetic varieties is sometimes useful and is carried out by means of an electromagnet. In the above list of minerals, magnetite is strongly attracted, ilmenite and iron-rich garnets are moderately magnetic, tourmaline and monazite weakly magnetic, while cassiterite (tin-stone), topaz, sphene, rutile, and zircon are non-magnetic.

CRYSTALLINE FORM

Minerals occur as *crystals*, i.e. bodies of geometric shape which are bounded by *faces* arranged in a regular manner and related to the internal atomic structure (p. 92). When a mineral substance grows freely from a fused, liquid state (or out of solution, or by sublimation), it tends to assume its own characteristic crystal shape; *the angles between adjacent crystal faces are always constant for similar crystals of any particular mineral*. Faces are conveniently defined by reference to *crystallographic axes*, three or four in number, which intersect in a common origin within the crystal and form, as it were, a scaffolding on which the crystal faces are erected. The arrangements of faces in crystals possess varying degrees of symmetry; according to their type of symmetry, crystals can be arranged in seven Systems of crystalliza-tion, which are summarized below and illustrated in Fig. 3.1. A *plane of symmetry* divides a crystal into exactly similar halves, each of which is the mirror image of the other; it contains one or more of the crystallographic axes. The number of planes of symmetry stated in the tabulation is that for the highest class of symmetry in each of the Systems (Note 2).

System	Axes	Planes of Symmetry (Max.)	Mineral Examples
Cubic (or Isometric)	3 equal axes at right angles to one another	9	Garnet, leucite, fluorite, rocksalt, zinc-blende, pyrite
Hexagonal and Trigonal	4 axes: three equal and horizontal, and spaced at equal intervals; one vertical axis	7	Beryl, nepheline, apatite Tourmaline, calcite, quartz
Tetragonal	3 axes at right angles: two equal and horizontal, one vertical axis longer or shorter than the others	5	Zircon, cassiterite (tin-stone), idocrase
Orthorhombic	3 axes at right angles, all unequal	3	Olivine, enstatite, topaz, barytes
Monoclinic	3 unequal axes: the vertical axis (c) and one horizontal axis (b) at right angles, the third axis (a) inclined in the plane normal to b	1	Orthoclase feldspar, hornblende, augite, biotite, gypsum
Triclinic	3 unequal axes, no two at right angles	none	Plagioclase feldspars, axinite

Crystal faces   In the Cubic System many crystals are bounded by faces which are all similar; such a shape is called a *form*. Two forms are illustrated in Fig. 3.1(*a*), the cube (six faces) and the octahedron (eight faces); other forms are the dodecahedron (twelve diamond-shaped faces, Fig. 3.22*b*), the trapezohedron (24 faces, Fig. 3.22*c*), and with lower symmetry the pyritohedron (12 pentagonal faces, Fig. 3.25*c*); among others the tetrahedron has four triangular faces, the smallest number possible for a regular solid. Crystals may grow as one form only, or as a combination of two or more forms; for example, Garnet (Fig. 3.22*b*, *c*) occurs as the dodecahedron, as the trapezohedron, or as a combination of the two.

In the Orthorhombic, Monoclinic, and Triclinic systems, where the axes are all unequal, faces are named as follows: A face which (when produced) cuts all three axes is called a *pyramid;* there are eight such faces in a complete form, one in each octant formed by the axes. Faces which cut two lateral axes and are parallel to the vertical axis are known as *prisms,* and make groups of four, symmetrically placed

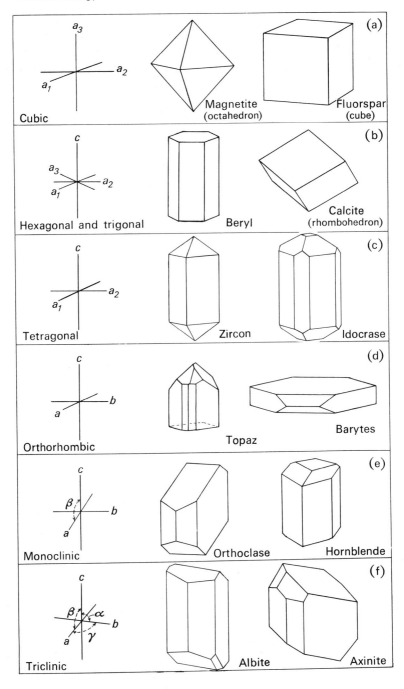

Fig. 3.1   Crystal systems. Crystal are drawn with the c-axis vertical (the reading position).

about the axes. A *pinacoid* is a face which cuts any one axis and is parallel to the other two. A *dome* cuts one lateral and the vertical axis, and is parallel to the other lateral axis. These faces are illustrated in Fig. 3.2. In the Tetragonal system the terms pyramid and prism are

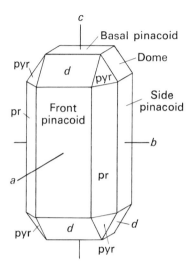

Fig. 3.2   Olivine crystal. To show crystal axes and faces in the orthorhombic system. *pr* = prism. *d* = dome. *pyr* = pyramid.

used as above; but as the two horizontal axes are equal, the term *2nd order prism* is used instead of pinacoid, and *2nd order pyramid* instead of dome.

Lastly, in the Hexagonal and Trigonal systems the names pyramid and prism are used as before for faces which cut more than one lateral axis; but since there are now three lateral axes instead of two, six prism faces (parallel to the *c*-axis) are found in the Hexagonal system (Fig. 3.1, beryl), and 12 pyramid faces. In Trigonal crystals, some faces are arranged in groups of three, equally spaced around the *c*-axis. This is seen, for example, in a rhombohedron of calcite (Fig. 3.1*b*), a form having six equal diamond-shaped faces. Calcite also occurs as 'nail-head' crystals, bounded by the faces of the rhombohedron combined with the hexagonal prism (Fig. 3.3*a*); and in 'dog-tooth' crystals (Fig. 3.3*b*) where the pointed terminations are bounded by the six faces of the scalenohedron, each face of which is a scalene triangle. In quartz crystals the six-sided terminations, which sometimes appear symmetrical, are a combination of two rhombohedra whose faces alternate, as shown by shading in Fig. 3.3*c*. This fact is shown by etching a quartz crystal with hydrofluoric acid, when two different sets of etch-marks appear on the alternate triangular faces. The

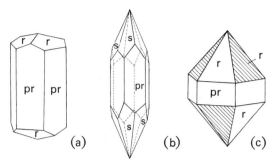

Fig. 3.3 Calcite and quartz. (*a*) Nail-head spar. (*b*) Dog-tooth spar. (*c*) Quartz, with negative rhombohedron shaded. *r* = faces of rhombohedron. *pr* = prism. *s* = scalenohedron.

trigonal (or three-fold) symmetry of quartz is also shown by extra faces which are sometimes present (Note 3).

Symbols are given to crystal faces and are based on the lengths of the intercepts which the faces make on the crystallographic axes, or the reciprocals of the intercepts (Miller symbols). The Miller notation is set out fully in textbooks of Mineralogy.

Twin crystals When two closely adjacent crystals have grown together with a crystallographic plane or direction common to both, but one reversed relative to the other, a *twin crystal* results. In many instances the twin crystal appears as if a single crystal had been divided on a plane, and one half of the crystal rotated relative to the other half on this plane. If the rotation is 180°, points at opposite ends of a crystal are thus brought to the same end as a result of the twinning and re-entrant angles between crystal faces are then frequently produced and are characteristic of many twins. Examples are shown in Fig. 3.4.

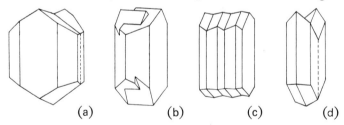

Fig. 3.4 Twin crystals. (*a*) Augite, showing re-entrant angle; twin plane parallel to front pinacoid. (*b*) Carlsbad twin of Orthoclase. (*c*) Multiple twin of plagioclase (Albite twinning). (*d*) Arrow-head twin of Gypsum.

The feldspar twins are important; in orthoclase the plane on which the two halves of the twin are united may be parallel to the side pinacoid (Carlsbad twin, Fig. 3.4*b*), or to a diagonal plane (Baveno twin), or to the basal pinacoid (Manebach twin). These are all *simple twins*. The plagioclase feldspars develop repeated or *multiple twinning* on planes parallel to the side pinacoid (Fig. 3.4*c*); the twin lamellae are generally narrow and are seen as regular parallel stripes on certain faces of a plagioclase crystal such as the basal plane (see also p. 108).

Twin crystals are a special case of *crystal aggregates;* the latter range from irregular groups of crystals to perfectly parallel growths. A crystal aggregate is made up of two or more individuals, and may contain crystals of different kinds or all of one kind.

The foregoing short account of crystals is necessarily incomplete, but the features which have been indicated enable the shapes of many crystals to be understood on inspection.

OPTICAL PROPERTIES OF MINERALS

Rock slices   The use of the petrological microscope in the study of minerals was made possible by a device due by H. C. Sorby of Sheffield, in 1850. Sorby, applying the method used earlier by W. Nicol for sectioning fossil wood, prepared slices of minerals and rocks so thin that light could be transmitted through them and their contents studied under the microscope.

In making a rock slice, a chip of rock (or slice cut by a rotating steel disc armed with diamond dust) is smoothed on one side and mounted on a strip of glass 75 × 25 mm (or for mineral grains 48 × 25 mm).

Fig. 3.5   Thin section of rock prepared for the microscope.

The specimen is cemented to the glass strip by means of Canada balsam, a gum which sets hard after being heated. The mounted chip of rock is then ground down with carborundum and emery abrasives to the required thinness, generally 30 micrometres (1 micrometre = $\frac{1}{1000}$ millimetre). It is now a transparent slice, and is completed by being covered with a thin glass strip fixed with balsam. Surplus balsam is washed off with methylated spirit. The surfaces of the specimen have been smoothed in making the slice, and they are free from all but very small irregularities. The effects observable when light is transmitted through such a slice of crystalline material are described in the following pages.

Refraction and refractive index   A ray of light travelling through one medium is bent or *refracted* when it enters another medium of different density. Figure 3.6 shows the path of such a ray (RR), which makes angles i and r with the normal (NN) to the surface separating the two media. The angle between the ray and the normal to the surface is smaller in the optically denser medium, i.e. a ray is bent towards the normal on entering a denser medium, and conversely.

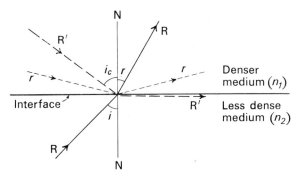

Fig. 3.6 Refraction of rays of light at the interface between two media.

If the angle of incidence is measured for air, as is usual, then the ratio sin i/sin r is called the *refractive index*, n, for the other medium, and is constant whatever the angle of incidence. It can be shown that sin i and sin r are proportional to the velocities of light (v_1, v_2) in the two media, i.e. $n = \sin i/\sin r = v_1/v_2$. The refractive index of a substance is therefore inversely proportional to the velocity of light through the substance.

When a ray (R′R′, Fig. 3.6) passes from one medium (n_1) into another of lower refractive index (n_2), at an angle of incidence known as the *critical angle* (i_c), it is refracted along the interface between the two media. If the angle of incidence exceeds the critical angle, the ray will not cross the interface but will be totally reflected from it, as shown for the ray rr in the figure (Note 4).

For Canada balsam, the gum in which rock slices are mounted, $n = 1.54$. Minerals with a much higher or lower refractive index than this appear in stronger outline, in a thin section under the microscope, than those which are nearer in value to 1.54 (this is illustrated by some of the minerals shown in Fig. 3.18); garnet ($n = 1.83$) is an example of a mineral which shows a very strong outline. Minerals whose refractive index is near to 1.54 appear in weak relief, as in the case of quartz ($n = 1.553$ to 1.544). Some minerals have an index less than 1.54, e.g. fluorspar ($n = 1.43$).

Becke's test The refractive index of a crystal can be compared with that of an adjacent crystal or of the mounting medium by means of this test. For a mineral which is in contact with balsam at the edge of a slice, a bright line of light appears along the margin of the mineral. This is best seen with a high-power objective, and is due to the concentration of light rays refracted at the boundaries of the mineral. On racking *up* the focussing screw of the microscope a short distance, the bright Becke line moves into the substance with the *higher* refractive index. Conversely, on racking *down*, the bright line moves into the medium of *lower* refractive index. It is thus possible to determine whether the refractive index of a mineral is above or below 1.54, or, in the case of two adjacent minerals, which of them has the higher value. Best results are obtained from crystal margins which are approximately vertical in the slice, i.e. not overlapping obliquely.

Polarized light   According to the Wave Theory of Light, a ray is represented as a wave motion, propagated by vibrations in directions at right angles to the path of the ray. In ordinary light these vibrations take place in all planes containing the direction of propagation; in plane polarized light the vibrations are confined to one plane (Fig. 3.7*a*). Light which passes through a crystal is, in general, polarized.

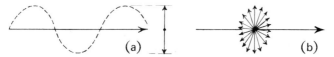

Fig. 3.7   (*a*) Polarized light, consisting of vibrations in one plane. (*b*) Ordinary light (diagrammatic), composed of many vibrations in planes containing the direction of the ray. Double-headed arrows represent double amplitude of vibrations.

While much can be learned from a microscopic examination of minerals in ordinary transmitted light, polarized light enables minerals to be identified with certainty, reveals twin crystals, and shows the nature of mineral alterations.

Double refraction   Crystals other than those in the Cubic system have the property of splitting a ray of light which enters them into two rays, one of which is refracted more than the other. A cleavage rhomb of clear calcite does this effectively, as is shown if it is placed over a dot on a piece of paper; two dots are then seen on looking down through the crystal, i.e. two images are produced (1 and 2 in Fig. 3.8).

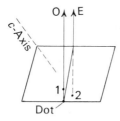

Fig. 3.8   Double refraction of calcite. O = ordinary ray. E = extraordinary ray. 1, ordinary image. 2, extraordinary image.

On turning the crystal, one dot appears to move round the other. The light passing through the calcite is split into two rays, called the ordinary ray (O) and the extraordinary ray (E). The two rays travel at different velocities in the crystal, the E-ray being the faster, and both rays are plane polarized.

A mineral which has this property of dividing a ray of light into two is said to be *doubly refractive* or *birefringent*. Since the two rays within the mineral travel at different velocities, there are two values of refractive index, one for each ray. The difference between these two values is known as the *birefringence* of the mineral; it is spoken of as 'strong' or 'weak' birefringence according as its amount is great or small. Calcite is an example of a strongly birefringent mineral; its maximum and minimum indexes of refraction are 1.658 and 1.486, and its birefringence is therefore 0.172. Quartz is an example of a mineral with a weak birefringence, 0.009.

Minerals which have the same refractive index for light which enters in any direction are called *isotropic;* they do not divide a ray entering them and are therefore *singly refracting.* All Cubic crystals are isotropic, and also all basal sections of Hexagonal and Tetragonal crystals (Note 5).

The nicol prism   It is possible to isolate one of the two rays passing through a doubly refracting calcite crystal by means of an invention due to W. Nicol (1828). A long rhomb of clear calcite, the variety known as Iceland Spar, of rhombic cross-section, is cut diagonally along its length and the two halves cemented together with balsam. The ends of the crystal are ground down to make an angle of 68 degrees with the long edge of the prism. A ray of light entering the prism parallel to its length is divided into ordinary and extraordinary rays as described above. But the film of balsam is at such an angle to the path of the O-ray that the latter is totally reflected from it, and absorbed in the black mount of the prism. Only the E-ray passes through the diagonal film and emerges from the opposite face of the calcite prism; the plane in which the vibrations of the E-ray take place is called the vibration plane of the Nicol prism.

Petrological microscope   Two polaroid discs (or two Nicol prisms in older instruments) are mounted in the microscope (Fig. 3.9), one (the *polarizer*) below the stage, the other (the *analyser*) above the stage. The polarizer can be rotated against a spring-loaded catch which normally holds it in one position. The analyser moves in a slot and can be slid into and out of the tube of the microscope. When the polarizer is held by its catch, the two polaroids (or nicols) are so set that the vibration directions for their E-rays are at right angles. This setting is known as 'crossed polars' (or 'crossed nicols'). The stage of the microscope can be rotated and is graduated in degrees.

Passage of light through the microscope   Light is reflected from the mirror (Fig. 3.9) up through the polarizer, where it is plane polarized. If no mineral or rock section is placed on the stage, the light from the polarizer passes up through the objective and so enters the analyser, vibrating as it left the polarizer. Since the analyser only transmits vibrations at right angles to those from the polarizer, the polars being 'crossed,' no light emerges from the eyepiece of the

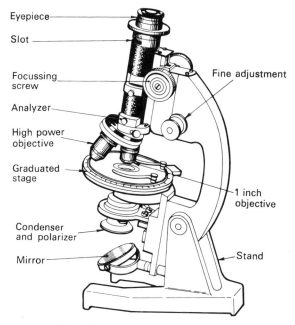

Eyepiece

Slot

Focussing screw

Analyzer

High power objective

Graduated stage

Condenser and polarizer

Mirror

Fine adjustment

1 inch objective

Stand

Fig. 3.9 A petrological microscope.

microscope. This fact should be tested as follows: With polarizer in, analyser out, look through the eye-piece and adjust the mirror so that the light is reflected up the tube. Slide in the analyser; a completely dark field of view should result if the polaroids are in adjustment and of good quality.

But when a slice of a doubly refracting mineral is placed on the stage of the microscope, and polaroids crossed, a coloured image of the mineral is usually seen. A thicker or thinner slice of the same mineral produces a different colour. These effects are explained as follows:

The ray of light which leaves the polarizer vibrating in one plane enters the thin section of the mineral on the stage of the microscope, and in general becomes resolved into two rays (since the mineral is birefringent), an O-ray and an E-ray. One of these travels faster than the other through the mineral slice; they can therefore be called the 'fast' ray and the 'slow' ray. They emerge from the slice vibrating in two planes at right angles (Fig. 3.10), and there is a phase difference between them, since they have travelled at different speeds and by different paths through the mineral slice.

The two rays enter the analyser, where each is again resolved into two. Of these *four* rays now travelling through the analyser, the two O-rays are eliminated, and only the two E-rays emerge from it *vibrating in the same plane* but out of phase with one another. This

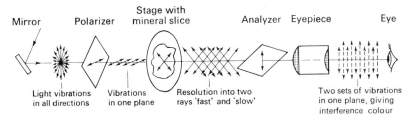

Mirror   Polarizer   Stage with
mineral slice   Analyzer Eyepiece   Eye

Light vibrations   Vibrations   Resolution into two   Two sets of vibrations
in all directions   in one plane   rays 'fast' and 'slow'   in one plane, giving
interference colour

Fig. 3.10   Passage of light through microscope (diagrammatic).

results in interference between the emergent vibrations, and the eye sees an interference colour, also called the *polarization colour*, which is of diagnostic value.

Polarization colours   White light is made up of waves of coloured light, from red at one end of the spectrum to violet at the other. Each colour has a different wave-length; that of red light, for example, is 0.00076 mm, and of violet light 0.00040 mm. These lengths are also expressed as 760 and 400 nanometres respectively. (1 millimetre = 1000 micrometres = 10^6 nanometres, or nm).

The interference of the two rays leaving the analyser of the microscope results in the suppression of some colours and the enhancement of others. A colour is intensified when its two waves are additive; another is eliminated when the two waves for that colour cancel one another. Whether the two waves for any particular colour are additive or the reverse depend on the amount of the phase difference (acquired in the mineral slice) in relation to the wave-length for the colour in question. The complete or partial suppression of some components of the white light leaves a balance which is the interference or polarization colour.

The polarization tints shown by minerals between crossed nicols form the series known as Newton's Scale of Interference Colours; a familiar example of Newton's Scale is provided by the colours seen in a thin film of oil resting on water. They can be demonstrated by means of a tapering slice of a single mineral; a *quartz wedge* is such a slice, cut from a quartz crystal parallel to its *c*-axis and ground so that it increases in thickness from zero at one end to about 0.25 mm at the other. It is mounted in Canada balsam between glass strips, and can be inserted in a slot in the tube of the microscope in a direction at 45° to the plane of the nicols. The wedge when viewed between crossed nicols shows a series of colour bands across its length (Fig. 3.11). At the thin end come greys, which pass into paler grey, then white, yellow, orange, and red as the thickness increases. This red marks the end of the 1*st Order* of colours. Next come violet, blue, green, yellow, orange, and a second and brighter red which marks the end of the 2*nd Order*. The colours of the spectrum are repeated in the 3rd and 4th Orders, but

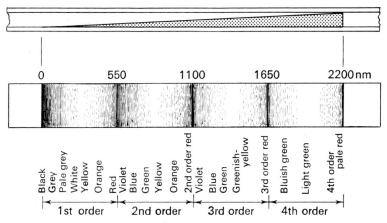

0 550 1100 1650 2200 nm

Black
Grey
Pale grey
White
Yellow
Orange
Red
Violet
Blue
Green
Yellow
Orange
2nd order red
Violet
Blue
Green
Greenish-yellow
3rd order red
Bluish green
Light green
4th order pale red

|←——— 1st order ———→|←——— 2nd order ———→|←——— 3rd order ———→|←——— 4th order ———→|

Fig. 3.11 Quartz wedge and (below) Newton's Scale of colours.

these contain progressively more delicate tints, and in the 5th and higher orders (at the thick end of the wedge) the colours become so pale that they ultimately merge into white. The phase difference between two waves of white light which gives a rise to a 1st Order red is 550 nm, that for the 2nd Order red 1100 nm, and so on, a red colour appearing at intervals of 550 nm.

The polarization colour obtained with a particular mineral slice depends on (1) the birefringence of the slice, which in turn depends on the refractive indexes of the mineral and the direction in which it has been cut; and (2) the thickness of the slice. These facts are illustrated by the quartz wedge and are discussed further in the next paragraph.

Polarization of quartz A crystal of quartz has a maximum refractive index of 1.553, for light vibrating parallel to the c-axis of the crystal, and a minimum value of 1.544 for light vibrating in a direction perpendicular to the c-axis (Fig. 3.12). The difference in these two values is 0.009, which is the maximum birefringence for the mineral. This birefringence holds only for a longitudinal slice of a quartz crystal; the polarization colour of such a slice is a pale yellow (if the slice is of standard thickness). As explained above, the polarization

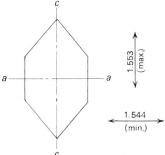

Fig. 3.12 Vibration directions and refractive indices in a vertical section of quartz. (cc = direction of c-axis, aa = direction of a lateral axis.)

colour is produced by the interference of two rays which have acquired a phase difference in traversing the mineral; the birefringence of 0.009 represents a phase difference of 0.009 micrometres, or 9 nanometres, per micrometre thickness of slice. Since the usual thickness is 30 micrometres, the full phase difference between the two rays after passing through the slice of quartz is 30 × 9 = 270 nanometres, which corresponds to a pale yellow in the 1st Order of Newton's Scale (Fig. 3.11).

Consider now a basal section of quartz, i.e. one cut perpendicular to the c-axis (the direction aa in Fig. 3.12). Such a slice has only one value of refractive index for light vibrations traversing it. It has therefore no birefringence, i.e. it is isotropic and appears completely black between crossed nicols.

Between the above extremes, the pale yellow colour (maximum) and black (or nil), a slice of quartz cut obliquely to the c-axis will show a white or grey polarization colour, the tint passing from dark grey to pale grey and white as the orientation of the slice approaches parallelism with c-axis.

Thus, in general, a crystal slice gives a characteristic colour between crossed nicols according to the direction in which it has been cut from the mineral. The maximum birefringence and polarization colour of some common minerals, for a 30-micrometre thickness of slice, are given below:

Mineral	Max. Biref.	nm	Colour
Muscovite	0.043	1290	= 3rd order delicate green
Olivine	0.035	1050	= bright 2nd order red
Augite	0.025	750	= 2nd order blue
Quartz	0.009	270	= 1st order pale yellow
Orthoclase	0.007	210	= 1st order grey

Extinction   When a birefringent mineral slice is rotated on the microscope stage between crossed polars, one or other vibration direction in the mineral can be brought parallel to the vibration plane of the polarizer. This occurs four times in each complete rotation. In such positions the light vibrations from the polarizer pass directly through the mineral slice to the analyser, where they are cut out (because it is set at right angles to the polarizer), so that no light emerges; the mineral thus appears completely dark at intervals of 90° during rotation. This effect is known as *extinction*. Half-way between successive extinction positions the mineral appears brightest.

The polars (or nicols) of the microscope are so set that their vibration planes are parallel to the cross-wires of the diaphragm, one 'east-west,' the other 'north-south.' If now a mineral is found to be in extinction when some crystallographic direction such as its length or a prominent cleavage is brought parallel to a cross-wire, the mineral is said to have *straight extinction* with regard to that length or cleavage. If extinction occurs when the length of the mineral makes an angle with the cross-wire, it is said to have *oblique extinction;* the extinction angle can be measured by means of the graduations around the edge of the stage.

Pleochroism   This is the name given to the change of colour seen in some minerals when only the polarizer is used and the mineral is rotated above it on the stage of the microscope. Pleochroism is due to the fact that the mineral absorbs the components of white light differently in different directions, as is well seen in the mica biotite. When this mineral is oriented with its cleavage direction parallel to the vibration plane of the polarizer it appears a much darker brown than when at 90° to that position. Hornblende and tourmaline are two other strongly pleochroic minerals.

Twinkling   Certain minerals have one value of refractive index much higher than, and the other nearer to, that of Canada balsam (1.54). The values for calcite, for example, are 1.66 and 1.49. When a slice containing crystals of calcite is viewed through the microscope and the polarizer rotated below it (the analyser not being used), the crystals show a change in relief and strength of outline as the vibrations from the polarizer coincide alternately with the directions of maximum and minimum refractive index in the mineral. The result is a characteristic *twinkling* effect. It is also seen in the mineral dolomite.

Summary of observations with a thin slice of mineral or rock.
1. Ordinary light. Look for form, colour, inclusions, and alteration products. Estimate refractive index relative to the mounting medium (balsam) using Becke's Test.
2. Polarized light. (*a*) Polarizer only: test for pleochroism, twinkling. (*b*) Polarizer and analyser: Cubic minerals are isotropic. In other (anisotropic) minerals observe the polarization colour, extinction, twinning, and alteration (if any).

   Examine opaque minerals by reflected light (or 'top light'), i.e. light reflected from the surface of the mineral.
3. The magnification of the microscope, for the combination of lenses used, may be estimated by observing a scale with small graduations; the apparent diameter of the microscope field divided by the intercept seen on the scale gives the approximate magnification.

ATOMIC STRUCTURES

The atomic structure of crystals can be investigated by methods of X-ray analysis (Note 6). The arrangement and spacing of the atoms of which a crystal is composed control its regular form and properties. For example, the atoms of sodium and chlorine in a crystal of common salt (NaCl) are arranged alternately at the corners of a cubic pattern (Fig. 3.13), which is repeated indefinitely in all directions. Salt crystals

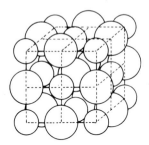

Fig. 3.13   Arrangement of atoms in unit cell of NaCl (small circles, Na; large circles, Cl).

grown from solution are cubes, a shape which echoes the internal structure.

When the silicate minerals were investigated it was found that they could be placed in a very few groups, according to the arrangement of the silicon and oxygen atoms. These groups corresponded broadly to the existing mineral families which had long before been worked out from a study of their shape and symmetry. A silicon atom is tetravalent, and is always surrounded by four oxygen atoms which are spaced at the corners of a regular tetrahedron (Fig. 3.14a). This SiO_4-tetrahedron is the unit of silicate structure, and is built into the different structures as follows:

1. *Separate* SiO_4-*groups* are found in some minerals, e.g. olivine, garnet. In olivine the tetrahedra are closely packed in regularly spaced rows and columns throughout the crystal structure, and are linked together by metal atoms (Mg, Fe, Ca, etc.) situated between the tetrahedra. Since each oxygen has two negative valencies and silicon four positive valencies, the SiO_4-group has an excess of four negative valencies; these are balanced when it is linked to metal atoms contributing four positive valencies, as in olivine, Mg_2SiO_4. Some Mg atoms may be replaced by Fe. (*Note:* In this account the word atom is used throughout instead of ion, which is more strictly correct.)

2. *Single Chain Structures* (Si_2O_6) are formed by SiO_4-groups linked together in linear chains, each group sharing two oxygens with its neighbours (Fig. 3.14e). This structure is characteristic of the Pyroxenes, e.g. diopside, $CaMg(Si_2O_6)$. Similar considerations to those mentioned above as regards valencies hold good here. The chains lie parallel to the c-axis of the mineral, and are bonded together by Mg, Fe, Ca, or other atoms, which lie between them. An end view

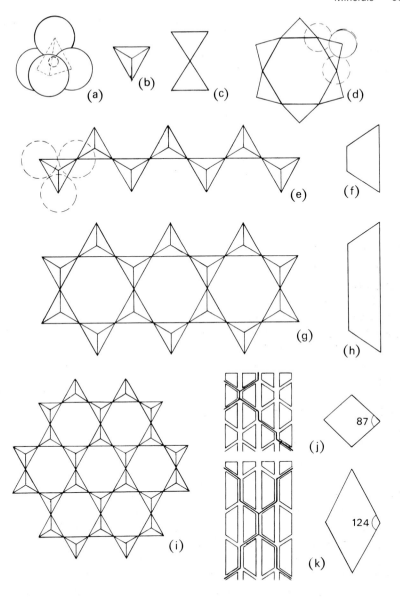

Fig. 3.14 Silicate structures. (a) SiO_4-tetrahedron: large circles represent oxygen ions, small circle silicon. (b) The tetrahedron in plan, apex upwards (circles omitted). (c) Two tetrahedra sharing one oxygen, Si_2O_7. (d) Ring of six tetrahedra (the centres of the two outer oxygens lie one above another), Si_6O_{18}. (e) Single chain, Si_2O_6; tetrahedra share two oxygens each; (f) conventional end-view of single chain. (g) Double chain, Si_4O_{11}, and (h) end view of double chain. (i) Sheet of tetrahedra, Si_4O_{10}; each shares three oxygens with adjoining tetrahedra. (j, k) Stacking of single and double chains respectively (viewed along c-axis); resulting cleavage directions shown by heavy lines, cleavage pattern on right.

of a chain is given at (f) in the figure. The vertical cleavages in the mineral run between the chains, as shown at (j), and intersect at an angle of 87°. Aluminium atoms, since they have nearly the same size as silicon, may replace silicon in the structure to a limited extent, and may also occur among the atoms which lie between the chains; in this way aluminous pyroxenes such as augite are formed.

Ring structures are built up by groups of three, four, or six tetrahedra, each of which shares two oxygens with its neighbours. Figure 3.14d shows a ring of six linked tetrahedra, which is found in the mineral beryl, $Be_3Al_2Si_6O_{18}$.

3. *Double Chain Structures* (Si_4O_{11}), in which two single chains are joined together side by side (Fig. 3.14g), are found in the Amphiboles, e.g. tremolite, $Ca_2Mg_5(Si_4O_{11})_2(OH)_2$. The double chains run parallel to the c-axis of the minerals, and are linked laterally by metal atoms lying between the chains. The cleavage directions are as shown at (k), and intersect at an angle of 124°. In hornblende, aluminium replaces part of the silicon (cf. augite above). Hydroxyl groups, (OH), are always present in the Amphiboles to the extent of about one to every eleven oxygens.

4. *Sheet Structures* (Si_4O_{10}) are formed when the SiO_4-tetrahedra are linked by three oxygens each, and lie with their bases in a common plane (Fig. 3.14i). Sheets of this kind are found in the Micas and other flaky minerals (e.g. chlorite, talc, the clay minerals), whose perfect cleavage is parallel to the silicon-oxygen sheets. In the mica muscovite, $KAl_2(AlSi_3)O_{10}(OH)_2$, aluminium replaces about one quarter of the silicon, and hydroxyl is always present. The Si_4O_{10}-sheets are arranged in pairs, with aluminium atoms between them, and each pair is separated from the next pair by a layer of potassium atoms. Other structures involving silicon-oxygen layers are discussed under Clay Minerals, p. 119.

5. *Three-dimensional Frameworks* (SiO_2) are formed when each tetrahedron is linked by all four corners, each group sharing its four oxygens with adjacent groups. The mineral quartz, SiO_2, has a framework in which the SiO_4-groups form a series of linked spirals (Fig. 3.21). In the Feldspars another type of framework is found, and Al replaces part of the Si. Thus in orthoclase feldspar, one silicon in four is replaced by Al; the substitution of trivalent aluminium for a tetravalent silicon releases one negative (oxygen) valency, which is satisfied by the attachment of a univalent sodium or potassium atom, thus:

Orthoclase $= KAlSi_3O_8$; Albite $= NaAlSi_3O_8$; and Anorthite $= CaAl_2Si_2O_8$. The K, Na, or Ca atoms are accommodated in spaces within the frameworks.

The above results are summarized in the following table:

Type of Structure	Repeated Pattern	Mineral Group
Separate SiO_4-groups	SiO_4	Olivine
Single Chain	Si_2O_6	Pyroxenes
Double Chain	Si_4O_{11}	Amphiboles
Sheet	Si_4O_{10}	Micas
Framework	$\begin{cases} (Al, Si)_n O_{2n} \\ SiO_2 \end{cases}$	Feldspars Quartz

Progress from the simple structure of separate tetrahedra to the three-dimensional framework involves increasing complexity of atomic structure; this is related in a general way to physical properties such as density and refractive index, which decrease as the atomic structures become more complex. Also, those minerals of simple structure (e.g. olivine) crystallize out early from magma, at a high temperature, whereas those of more complex structure tend to form at successively lower temperatures. The descriptions which follow are given in the order of the groups above.

THE ROCK-FORMING MINERALS

It is convenient to distinguish between minerals which are *essential* constituents of the rocks in which they occur, their presence being implied by the rock name, and others which are *accessory*. The latter are commonly found in small amount in a rock but their presence or absence does not affect the naming of it. *Secondary* minerals are those which result from the decomposition of earlier minerals, often promoted by the action of water in some form, with the addition or subtraction of other material, and with the formation of by-products.

Identification of minerals in hand specimen.   Many minerals can be identified in a rock specimen with some degree of certainty, but it is not always possible to do this, for two reasons: (i) the mineral may be too small to be identifiable easily; (ii) only one or two characteristic features may be seen, and these together may be insufficient to determine the mineral. In such cases, low magnification (six or eight diameters) with a pocket lens may be enough, but failing that recourse must be had to the microscope for identification.

It should be possible to identify the common rock-forming minerals in the hand specimen with a pocket lens where one dimension of the mineral grain is not less than about 1 mm. With practice much smaller grains can be determined. The most useful characteristics for this purpose are:

1. General shape of grains, depending on the crystallization of the mineral; the faces of well-formed crystals can often be observed, but where grains have been modified (e.g. by rounding) other properties must be used.
2. Colour and transparency.
3. Presence or absence of cleavage.
4. Presence or absence of twinning, and type of twinning.
5. Hardness.

In the following descriptions of minerals, notes are included to aid identification in the hand specimen on the above lines. They are followed by notes on the simpler optical properties of the mineral; the abbreviations *R.I.* (refractive index) and *Biref.* (birefringence) are used throughout, and in stating the polarization colours of minerals it is assumed that sections are of normal thickness (30 micrometres).

SILICATE MINERALS

THE OLIVINE GROUP

Olivine

Common olivine has the composition $(MgFe)_2SiO_4$, in which Fe replaces part of the Mg. The pure magnesium silicate is called *forsterite* (Mg_2SiO_4), a mineral found in some metamorphosed limestones (p. 205); the corresponding iron silicate *fayalite*, Fe_2SiO_4, is rare in nature.

Crystals: Orthorhombic (Fig. 3.2); pale olive-green or yellow; vitreous lustre; conchoidal fracture. H = $6\frac{1}{2}$. Sp. gr. = 3.2 to 3.6.

Olivine occurs chiefly in basic and ultrabasic rocks. Since it crystallizes at a high temperature, over 1000°C, it is one of the first minerals to form from many basic magmas. (*Magma* is the molten rock-material from which igneous rocks have solidified.) The crystals thus grow in a largely fluid medium and develop their own characteristic shape. Though small they may sometimes be seen with a hand lens in rocks such as olivine-basalt, because of their colour and glassy appearance. Alteration to green serpentine is common (p. 116).

Properties in thin section

Porphyritic crystals[1] commonly show 6- or 8-sided sections, the outline generally somewhat rounded. Cleavage rarely seen; irregular cracks common (Fig. 3.18*a*).

Colour: None when fresh. Alteration to greenish serpentine is very

---

[1] Porphyritic crystals are large compared with the grain size of the matrix in which they are set; they are generally well-formed crystals.

characteristic, this mineral being often developed along cracks and around the margins of olivine crystals. Some olivines have been entirely converted to serpentine; or relics of original olivine may be preserved as isolated colourless areas in the serpentine. Magnetite (Fe_3O_4) may be formed, during the alteration, from iron in the original olivine, and appears as small black specks in the serpentine. *Mean R.I.* = 1.66 to 1.68, giving a bold outline in Canada Balsam. *Biref:* Strong (max. = 0.04), giving bright 2nd and 3rd order polarization colours. Twinning is usually absent.

THE PYROXENE GROUP

The pyroxenes belong to two systems of crystallization:

1. Orthorhombic, e.g. *enstatite, hypersthene.*
2. Monoclinic, e.g. *diopside, augite, aegirite.*

They possess two good cleavage directions parallel to the prism faces of the crystals (*prismatic cleavage*); the intersection angle of the cleavages is nearly 90°, a characteristic feature of the group (see Fig. 3.14*j*). They form 8-sided crystals, and being silicates of Fe and Mg they are dark in colour (except diopside, CaMg). In the crystal structures, single chain units (Si_2O_6) lie parallel to the *c*-axis (Fig. 3.14), with metal ions between the chains in regular positions.

Orthorhombic pyroxenes

Enstatite, $MgSiO_3$. **Hypersthene,** $(MgFe)SiO_3$

Enstatite generally contains a small amount of Fe which replaces part of the Mg; when the proportion of $FeSiO_3$ exceeds 15 per cent of the whole, certain optical properties are changed and the mineral is called hypersthene.[2]

Crystals: Usually dark brown or green (hypersthene nearly black), 8-sided and prismatic. In addition to the prismatic cleavages mentioned above there are poorer partings parallel to the front and side pinacoids; lustre, vitreous to metallic. H = 5 to 6. sp. gr. = 3.2 (enstatite), increasing with the iron content to 3.5 (hypersthene).

The minerals occur in some ultrabasic rocks and in basic rocks such as norite (*q.v.*), as black lustrous grains interlocked with the other constituent minerals; also in some andesites, usually as small black porphyritic crystals.

Properties in thin section

8-sided cross-sections are shown when crystals are idiomorphic (i.e. having its own form, well-developed), with two good cleavages

---

[2] The names 'enstatite' and 'hypersthene' have Greek derivations which refer to the colour change in pleochroism: *enstates*, weak; *sthene*, strong.

nearly at right angles; traces of a third cleavage are sometimes seen, parallel to the side pinacoid.

Colour: Enstatite is nearly colourless and not pleochroic; hypersthene shows pleochroism from pale green to pink.

Mean R.I. = 1.65 (enstatite) to 1.73 (hypersthene); cf. augite.

Biref: Weak; max. = 0.008 (enstatite) rising to 0.016 (hypersthene), giving grey, white, or yellow 1st order polarization colours.

Extinction: Straight with reference to the cleavage in longitudinal sections parallel to the *c*-axis. Twinning rare.

Enstatite and hypersthene are distinguished from augite by their weaker birefringence and straight extinction, and relative scarcity of twinning.

Monoclinic pyroxenes

Augite, $(CaMgFeAl)_2(SiAl)_2O_6$

A complex aluminous silicate whose formula can be written as above in conformity with the Si_2O_6 pattern of the atomic chain structure. The relative proportions of the metal atoms (Ca, Mg, Fe, Al) are variable within limits, giving a range of composition and therefore different varieties of the mineral. Some Al is substituted for Si in augite.

Crystals: Commonly 8-sided and prismatic, terminated by two pyramid faces at each end; brown to black in colour, vitreous to resinous lustre. Twin crystals (Fig. 3.4*a*) show a re-entrant angle. H = 5 to 6. sp. gr. = 3.3 to 3.5. The two vertical cleavages may be observed on the end faces of suitable crystals (Fig. 3.15).

Augite occurs chiefly in basic and ultrabasic rocks; e.g. in gabbro, where it appears as dark areas intermingled with the paler feldspar. In fine-grained basic rocks it is not distinguishable in the hand specimen unless it is porphyritic. Augite is also a constituent of some andesites and diorites, and occasionally of granites.

Properties in thin section

Idiomorphic crystals show characteristic 8-sided transverse sections, bounded by prism and pinacoid faces, with the two prismatic cleavages intersecting at nearly 90°. Longitudinal sections show only one cleavage direction (Fig. 3.18*b*).

Colour: Pale brown to nil; except a purplish variety containing titanium (titaniferous augite). Zoning (see p. 109) and 'hour-glass structure' are shown by colour variations. Pleochroism generally absent.

Mean R.I. = about 1.70, giving strong relief in balsam.

Biref: Strong (max. = 0.025). Polarization in bright 1st and 2nd order colours, in sections of normal thickness. Weak biref. shown by some sections.

Extinction: Oblique, up to 45° in longitudinal sections, except those parallel to the front pinacoid, which show straight extinction. In transverse sections extinction is symmetrical with the intersecting cleavages. Simple and 'strip' twins are frequent.

Alteration: Augite may change to chlorite (p. 116) by hydration, with the loss of some constituents; epidote and calcite may form as by-products.

Diallage A variety of augite possessing an extra parting parallel to the front pinacoid, which gives the mineral a laminated structure. Colour, green or greenish-brown; lustre, metallic. In thin section the prominent parting appears as closely spaced parallel lines, and the mineral is easily recognized by this feature. Found in some gabbros and peridotites.

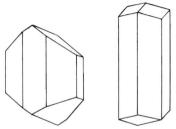

Fig. 3.15 Crystals of augite (left) and common hornblende (right).

Diopside, $CaMg(Si_2O_6)$ A non-aluminous pyroxene forming 8-sided crystals, usually pale green in colour. It occurs in some basic rocks and in metamorphic rocks such as contact-altered impure limestones. In thin section it is colourless or very pale green, but similar in polarization colour to augite. It differs from augite in having a smaller extinction angle (38°) and somewhat lower R.I.

Aegirite, $NaFe^{3+}(Si_2O_6)$, usually with some Ca, Mg, and Al in small amounts. (The name *acmite* is used for the pure silicate.) Crystals are usually green or brown in colour, 6-sided and prismatic; vitreous lustre. H = 6. sp. gr. = 3.5. Occurs chiefly in Na-rich rocks, such as nepheline-syenite and phonolite, where (like augite) its dark colour is in contrast with the other constituents. It is not always identifiable without an examination in thin section. The name *aegirite-augite* is used for minerals intermediate between augite and aegirite (i.e. with some content of $NaFe^{3+}$).

Properties in thin section

Cross-sections often 6-sided, longitudinal sections elongated; cleavages as in augite.

Colour: Green; pleochroism marked, in shades of green and greenish-brown.

Mean R.I. = about 1.77.

Biref: Strong (max. = 0.04), giving 3rd order polarization colours which are largely masked by the green body-colour of the mineral.

Extinction: Oblique in most longitudinal sections, at a small angle (about 6° for acmite): this and the colour and pleochroism are means of distinction from augite. Aegirite-augite has a larger extinction angle, from 15 to 30° in side pinacoid sections.

THE AMPHIBOLE GROUP

This large family of minerals contains mainly monoclinic members. The crystals are elongated in the c-direction and usually bounded by six vertical faces, of which the prism faces intersect at an angle of 124° (Fig. 3.15). This is also the angle between the two cleavage directions, parallel to the prisms. It is determined by the internal atomic arrangement (Fig. 3.14g, h, k) in which double chain units (Si_4O_{11}) lie parallel to the c-axis and are linked by Ca, Mg, Fe, Na, and other ions between them. Only the most common monoclinic amphibole, *hornblende*, is given here in detail, together with *tremolite* and *actinolite*. Rarer varieties include *glaucophane* and *anthophyllite* (orthorhombic), which are found in certain metamorphic rocks.

Hornblende, $(CaMgFeNaAl)_{3-4}(AlSi)_4O_{11}(OH)$

A complex aluminous silicate whose formula can be written as above. The relative proportions of the metal atoms vary within the limits shown, giving a range of composition; the (OH)-radicle is found in all amphiboles. Al is substituted for Si in about one in four positions.

Crystals: Monoclinic, dark brown or greenish black; usually 6-sided, of longer habit than augite, with three dome faces at each end (Fig. 3.15). Twinning on the front pinacoid results in a crystal of similar appearance, but having four faces at one end and two at the other. Vitreous lustre. H = 5 to 6. sp. gr. = 3 to 3.4.

Common hornblende is found in diorites and some andesites as the dark constituent; in andesite it is porphyritic and may be recognized by its elongated shape, the length of the crystals being often several times their breadth. It is also found in some syenites and granodiorites, and in metamorphic rocks such as hornblende-schist.

Properties in thin section

6-sided transverse sections, bounded by four prism and two pinacoid faces, are very characteristic (Fig. 3.18c), and show the prismatic cleavages intersecting at 124°. Longitudinal sections are elongated and show one direction of cleavage parallel with the crystal's length.

Colour: Green to brown; pleochroism strong in shades of green, yellow, and brown. A brown variety, often with a dark border, is called 'basaltic hornblende' and contains ferric iron.

Mean R.I. varies from 1.63 to 1.72 in different hornblendes.

Biref: Strong (max. $= 0.024$), giving 2nd order polarization colours which are somewhat masked by the body-colour of the mineral.

Extinction: Oblique in most longitudinal sections, at angles up to $25°$ with the cleavage; sections parallel to the front pinacoid show straight extinction. Twinning is common.

Alteration to chlorite, with the formation of by-products, is often seen.

Hornblende is to be distinguished from augite by its colour and lower maximum angle of extinction; and from biotite, which it may resemble in sections showing one cleavage direction, by the fact that biotite always gives straight extinction.

Tremolite, $Ca_2Mg_5(Si_4O_{11})_2(OH)_2$ }
Actinolite, $Ca_2(MgFe)_5(Si_4O_{11})_2(OH)_2$ } Non-aluminous amphiboles

Crystals: Elongated, often in bladed aggregates; colour white (tremolite) or green (actinolite), sometimes translucent; vitreous lustre. $H = 5$ to 6. sp. gr. $=$ about 3.

These minerals occur mainly in metamorphic rocks, e.g. tremolite in impure contact-altered limestones, actinolite in metamorphosed basic rocks such as actinolite-schist.

Properties in thin section

Elongated crystals, sometimes in radiating clusters; diamond-shaped cross-sections with two good cleavages intersecting at $124°$.

Colour: Tremolite is colourless; actinolite, pale green and weakly pleochroic. *Mean R.I.* $= 1.62$ to 1.64.

Biref: $= 0.027$ (max.), giving bright 2nd order polarization colours.

Extinction: Oblique in longitudinal sections, up to $18°$.

Asbestos   The fibrous form of tremolite, in which crystals grow very long and are flexible. In commerce the term 'asbestos' also includes other fibrous minerals such as *chrysotile* (a form of serpentine, *q.v.*) and *crocidolite* (a soda-amphibole). These minerals are useful because of their resistance to heat and because of their fibrous nature, which enables them to be woven into fireproof fabrics, cord, and brake-linings, and made into boards, tiles, and felt.

THE MICAS

The micas are a group of monoclinic minerals whose property of splitting into very thin flakes is characteristic and easily recognized. It is due to the perfect cleavage parallel to the basal plane in mica crystals (Fig. 3.16), which itself results from the layered atomic structure of the minerals. Each layer of Si_4O_{10} composition is strongly bonded, by Al ions, to the similar layer of a pair; and successive pairs are separated by layers of potassium ions, as shown in Fig. 3.17. It is along the K-layers that the cleavage directions lie.

The common micas include *muscovite* (colourless or slightly tinted), *paragonite* (Na-mica), *lepidolite* (pale pink lithium mica), *phlogopite* (light brown Mg-mica), and *biotite* (dark brown to nearly black). Two varieties, muscovite and biotite, are described below. Muscovite, phlogopite, and lepidolite have important economic uses as insulators in electrical apparatus on account of their dielectric properties. Mica crystals are six-sided, with pseudo-hexagonal symmetry. Their cleavage flakes are flexible, elastic, and transparent. A six-rayed percussion figure is produced when a flake is struck with a pointed instrument, and one of the cracks thus formed is parallel to the plane of symmetry of the monoclinic mineral.

Muscovite, $KAl_2(Si_3Al)O_{10}(OH)_2$

Form and cleavage as stated above. White in colour, unless impurities are present to tint the mineral; pearly lustre. H $= 2$ to $2\frac{1}{2}$ (easily cut with a knife). sp. gr. $=$ about 2.9 (variable).

Muscovite occurs in granites and other acid rocks as silvery crystals, from which flakes can be readily detached by the point of a penknife; also in some gneisses and mica-schists. It is a very stable mineral, and persists as minute flakes in sedimentary rocks such as micaceous sandstones. The name *sericite* is given to secondary muscovite, which may be produced by the alteration of orthoclase (see p. 106). The mica of commerce comes from large crystals found in pegmatite viens (p. 160).

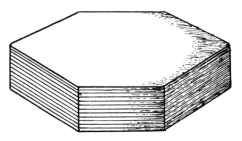

Fig. 3.16   Mica crystal.

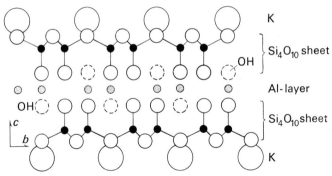

Fig. 3.17 The structure of muscovite (diagrammatic). The silicon-oxygen tetrahedra are linked to form 'sheets', two of which are shown, with aluminium ions lying between them. Layers of potassium ions are situated between successive pairs of the Si_4O_{10} sheets.

Properties in thin section

Longitudinal sections (i.e. across the cleavage) are often parallel-sided and show the perfect cleavage (Fig. 3.18d); basal sections appear as 6-sided or irregular colourless plates. Alteration uncommon.

Mean R.I. = 1.59.

Biref: Strong (max. = 0.04), giving bright 3rd order pinks and greens in longitudinal sections. Basal sections have a weak double refraction.

Extinction: Straight, with reference to the cleavage.

Biotite, $K(MgFe)_3(Si_3Al)O_{10}(OH)_2$

Crystals are brown to nearly black in hand specimen; single flakes are pale brown and have a sub-metallic or pearly lustre. Form and cleavage as stated above. H = $2\frac{1}{2}$ to 3. sp. gr. = 2.8 to 3.1.

Biotite occurs in many igneous rocks, e.g. granites, syenites, diorites, and their lavas and dyke rocks, as dark lustrous crystals, distinguished from muscovite by their colour. Also a common constituent of certain gneisses and schists.

Properties in thin section

Sections showing the cleavage often have two parallel sides and ragged ends (Fig. 3.18d). In some biotites, small crystals of zircon enclosed in the mica have developed spheres of alteration around themselves by radioactivity. These spheres in section appear as small dark areas or 'haloes' around the zircon and are pleochroic.

Colour: Shades of brown and yellow in sections across the cleavage, which are strongly pleochroic; the mineral is darkest (i.e. light absorption is a maximum) when the cleavage is parallel to the vibration direction of the polarizer. Basal sections have a deeper tint and are only feebly pleochroic.

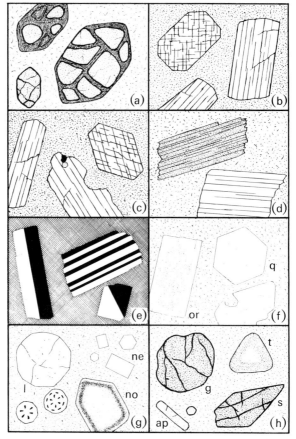

Fig. 3.18   Minerals in thin section. (*a*) Olivine, with serpentine-filled cracks. (*b*) Augite. (*c*) Hornblende. (*d*) Mica (biotite, dark; muscovite, light). (*e*) Feldspars (crossed nicols): simple twins of orthoclase, multiple twin of plagioclase. (*f*) Quartz (*q*) and kaolinized orthoclase (*or*). (*g*) Feldspathoids: leucite (*l*); nepheline (*ne*), nosean (*no*). (*h*) Garnet (*g*), sphene (*s*), apatite (*ap*), and tourmaline (*t*).

Mean R.I. = about 1.64.

Biref: Strong, about 0.05 (max.) in sections perpendicular to the cleavage, but the 3rd order polarization colours are obscured by the body-colour of the mineral. Basal sections are almost isotropic.

Extinction: Parallel to the cleavage. Alteration to green chlorite is common, when the mineral loses its strong birefringence and polarizes in 1st order greys (see under Chlorite).

Phlogopite, $KMg_3(Si_3Al)O_{10}(OH)_2$ Less deeply coloured than biotite, sometimes a coppery red; often shows a 6-rayed star pattern (asterism) when a cleavage flake is held up to the light. Found in certain igneous rocks and in metamorphosed impure limestones. *R.I.* and birefringence are similar to those of biotite.

FELDSPARS

The feldspars form a large and abundant group of monoclinic and triclinic minerals, and are essential constituents of most igneous rocks. They have atomic *framework* structures (p. 94) in which the silicon-oxygen tetrahedra are linked by sharing all their oxygens, in a 3-dimensional arrangement. Some Al is substituted for Si in the frameworks, and potassium, sodium, and calcium ions lie in the spaces between the linked tetrahedra. The chief minerals of the feldspar family are: *orthoclase*, $KAlSi_3O_8$, *albite*, $NaAlSi_3O_8$, and *anorthite*, $CaAl_2Si_2O_8$; they are shown in Fig. 3.19. Members containing

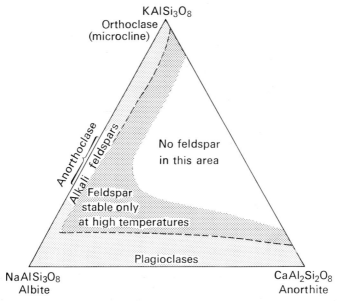

Fig. 3.19   Feldspar composition diagram (after Chudoba). For a feldspar represented by a point within the triangle, the proportions of the three quantities Orthoclase, Albite, Anorthite are given by the lengths of the perpendiculars from the point on to the sides of the triangle. Each corner represents 100 per cent of the component named there.

essentially K and Na, in varying proportions, are called the *alkali feldspars*. They include orthoclase, soda-orthoclase (containing some Na), and anorthoclase (in which the proportion of Na is greater than K). Feldspars having mainly Na and Ca are the *plagioclases*, which range from albite to anorthite in a series of solid solutions (p. 107) (Note 7).

Orthoclase (Fig. 3.1*e*) is the common potassium feldspar; a high temperature form, *sanidine*, is found in lavas which have been rapidly chilled; and a low temperature form, *microcline*, which has cooled slowly and developed twinning, also has the same composition as orthoclase. In this account, only orthoclase, microcline, and the plagioclases are described; other members of the group are discussed in text-books of Mineralogy (Note 1). The low temperature feldspar *adularia*, of restricted occurrence in hydrothermal veins, has a distinctive habit (Fig. 3.20) and a composition near orthoclase (approximately $Or_{90}Ab_9An_1$, the suffixes indicating percentages). The rare barium feldspar, *celsian*, has the composition $BaAl_2Si_2O_8$.

Orthoclase, $KAlSi_3O_8$. Potassium feldspar

Crystals: Monoclinic; white or pink in colour, vitreous lustre; bounded by prism faces, basal and side pinacoids, and domes (Fig. 3.20). Simple twins are frequent: Carlsbad twins unite on the side

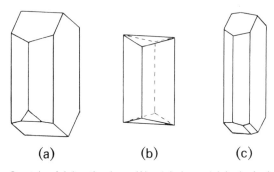

(a)   (b)   (c)

Fig. 3.20   Crystals of (*a*) orthoclase, (*b*) adularia, and (*c*) plagioclase (albite).

pinacoid (Fig. 3.4*b*), Manebach twins on the basal plane, and Baveno twins on a diagonal plane (dome) parallel to the *a*-axis (Fig. 3.18*e*). Two good cleavages, parallel to the side and basal pinacoids, intersect at 90°; the name of the mineral is due to this property (from the Greek *orthos*, straight or rectangular, and *klasis*, breaking). H = 6. sp. gr. = 2.56. The colourless glassy variety, *sanidine*, is a high temperature form found in quickly-cooled lavas.

Orthoclase occurs in granites and syenites as hard, cleaved white or pink crystals, generally constituting the greater part of the rock. In the dyke rocks related to granites and syenites, orthoclase may occur as porphyritic crystals (see p. 142). Found also in some gneisses and felspathic sandstones.

Properties in thin section

Shape often nearly rectangular if crystals are idiomorphic, irregular if interlocked with other minerals. Cleavage not always seen.

Colour: None when fresh, but shows frequent alteration to kaolin (p. 117), when the mineral appears 'cloudy' and looks white by top light (reflected from the surface of the slice).

Mean R.I. = 1.52; this is below that of Balsam, and may provide a useful means of confirming identification, using Becke's test.

Biref: Weak (max. = 0.008), giving 1st order grey and white polarization colours, in sections of normal thickness.

Extinction: Usually oblique, up to 21° on the side pinacoid; sections perpendicular to the side pinacoid show straight extinction. Simple twins common; distinguished from plagioclase by the absence of multiple twinning. *Alteration* to kaolin is common (p. 117); sometimes alters to an aggregate of very small flakes of *sericite* (secondary white mica), revealed by their bright polarization colours.

Microcline, $KAlSi_3O_8$. Triclinic potash-feldspar

Crystal form is similar to that of orthoclase, but the two cleavages (parallel to the basal and side pinacoids) meet at $89\frac{1}{2}°$ instead of exactly 90°.

Colour: White, pink, or green.

R.I. and *Biref:* Similar to orthoclase. Multiple twinning, with two intersecting sets of gently tapering twin-lamellae; in thin sections this is seen as a characteristic 'cross-hatching' effect between crossed polars, and distinguishes microcline from other feldspars. Found, for example, in granites which have been subjected to stress.

Anorthoclase Triclinic alkali-feldspar (Fig. 3.19), with more Na than K. One specimen analysed had the composition: albite 65 per cent, orthoclase 35 per cent, with mean *R.I.* = 1.525, and sp. gr. = 2.58. Properties in thin section similar to those of microcline. Anorthoclase is the beautiful grey schillerized feldspar which occurs in *laurvikite* (a Norwegian syenite, p. 159); it is also found as rhomb-shaped crystals in *rhomb-porphyry* in the Oslo district.

Schiller is the play of colours seen by reflected light in some minerals, in which minute platy inclusions are arranged in parallel planes; the schiller effect is produced by the interference of light from these included plates.

The Plagioclases

Feldspars of this series are formed of mixtures (solid solutions) of albite, $NaAlSi_3O_8$, and anorthite, $CaAl_2Si_2O_8$, in all proportions (Note 7). The range of composition is divided into six parts, which are named as follows:

From 100 to 90%	Albite, and 0 to 10%	Anorthite .	. *Albite*
From 90 to 70	Albite, and 10 to 30	Anorthite .	. *Oligoclase*
From 70 to 50	Albite, and 30 to 50	Anorthite .	. *Andesine*
From 50 to 30	Albite, and 50 to 70	Anorthite .	. *Labradorite*
From 30 to 10	Albite, and 70 to 90	Anorthite .	. *Bytownite*
From 10 to 0	Albite, and 90 to 100	Anorthite .	. *Anorthite*

A plagioclase, for example, containing 40 per cent albite and 60 per cent anorthite would be called labradorite (written $Ab_{40}An_{60}$).

Crystals: Triclinic; white or colourless (albite) to grey (labradorite), bounded by prisms, basal and side pinacoids, and domes (Fig. 3.20c). Vitreous lustre. Cleavages are parallel to the basal and side pinacoids, and meet at an angle of about 86° (hence the name, from the Greek *plagios*, oblique; *klasis*, breaking). Multiple twins parallel to the side pinacoid (Albite twinning, Fig. 3.4c) are characteristic of the plagioclases; the closely spaced twin-lamellae can often be seen with a lens as stripes on the basal cleavage and other surfaces. Another set of twins is sometimes developed on planes, parallel to the b-axis, which usually make a small angle with the basal pinacoid (Pericline twinning. H = 6 to $6\frac{1}{2}$. sp. gr. = 2.60 (albite), rising to 2.76 (anorthite).

Plagioclase feldspars occur in most igneous rocks, and in some sedimentary and metamorphic rocks. In the coarser grained igneous rocks the plagioclase appears as white or grey cleaved crystals, and is often distinguishable by its multiple twinning.

Properties in thin section

Idiomorphic crystals (e.g. in lavas) commonly show rectangular sections; parallel-sided 'laths,' with their length several times as great as their breadth, are seen when the crystals sectioned are tabular (i.e. flat and thin) parallel to the side pinacoid (see Fig. 4.5d). Cleavage not often visible. The minerals are normally colourless but may be clouded with alteration products.

The characteristic multiple twinning appears as light and dark grey parallel stripes between crossed polars (Fig. 3.18e), and sets of alternate stripes extinguish obliquely in different positions.

Mean R.I. and other optical properties vary according to the composition of the minerals:

	Mean R.I.	Maximum Biref.	Max. Symmetrical Extinction Angles for twin-lamellae, in sections perpendicular to side pinacoid (see below)
Albite	1.53	0.011	from 18° to 12°
Oligoclase	1.54	0.008	from 12 to 0, 0 to 13
Andesine	1.55	0.007	from 13 to 27
Labradorite	1.56	0.009	from 27 to 39
Bytownite	1.57	0.010 ⎫	over 39°
Anorthite	1.58	0.012 ⎭	

Albite and andesine, which may have the same angle of extinction, have values of *R.I.* which are respectively below and above 1.54 (the *R.I.* of Balsam), and Becke's test may be used to decide a particular case.

The values of extinction angle given above apply only to those sections in which the twin planes are perpendicular to the plane of the slice. This condition is fulfilled, (1) if a crystal shows even illumination when the twin-lamellae are parallel to one of the cross-wires, and (2) if the sets of alternate twin-lamellae extinguish symmetrically on either side of a cross-wire. To test this, select a likely crystal and note the angles of extinction of alternate lamellae by turning the stage of the microscope to the left and to the right, starting each time from the mid-position where the twin-lamellae are parallel to a cross-wire. The two angles thus obtained should be equal, or very nearly so. Measure a number of suitable crystals in the slice and take the *maximum* value as the criterion. It is important to follow this rule closely.

As a first approximation, it should be noted that a symmetrical extinction angle of a few degrees (or straight extinction in a microlith) indicates an oligoclase, and a wide angle one of the basic plagioclases such as labradorite.

A description of other methods of distinguishing the different plagioclases in thin section is beyond the scope of this book.

Zoning: (1) A plagioclase crystal may contain numerous inclusions which became locked up in it at some period during its growth; these inclusions are often arranged in zones parallel to the crystal faces. The core of a crystal, for instance, may contain many inclusions while the marginal zone is clear. This is known as zoning by inclusions.

(2) Zoning also results from differences in the composition of successive layers of material acquired during a crystal's growth. For example, a plagioclase may have begun to grow as anorthite; but because of changes in the relative concentrations of constituents in the melt which was breeding the crystal, further growth may have made use of material containing progressively less anorthite and more albite. A slice of such a crystal seen between crossed polars does not show sharp extinction in one position, but the zones of different composition extinguish successively at slightly different angles as the slice is rotated. Augite, tourmaline, may also show this zoning.

Alteration products in plagioclase are kaolin in Na-rich, and epidote in Ca-rich varieties.

FELDSPATHOIDS

Minerals of this group resemble the feldspars chemically, and have 3-dimensional framework structures; they differ from the feldspars in their lower content of silicon. Stated in another way, the Al:Si ratio is higher in the feldspathoids than in the corresponding feldspars. The two chief minerals of the group, which are only briefly discussed here because their occurrence is somewhat limited in nature, are:

Leucite, $K(AlSi_2)O_6$ (cf. Orthoclase)
Nepheline, $Na(AlSi)O_4$ (cf. Albite)

Nepheline generally contains a small amount of potassium. Derivatives from nepheline by the addition of a sulphate, chloride, or carbonate, are:

Sodalite, $3(NaAlSiO_4).NaCl$
Nosean, $6(NaAlSiO_4).Na_2SO_4$
Hauyne, $3(NaAlSiO_4).CaSO_4$
Cancrinite, $4(NaAlSiO_4).CaCO_3.H_2O$ (approx.)

Leucite, $K(AlSi_2)O_6$

Crystals: Cubic, commonly in the form of the trapezohedron (Fig. 3.22*c*), with an imperfect cleavage. Colour, dull white (hence the name, from the Greek *leucos*, white). H = 5 to 6. sp. gr. = 2.5.

Leucite occurs in certain undersaturated lavas which have a low silica- and high alkali-content, such as the leucite-basalts from Vesuvius, and leucitophyre.

In thin section, 8-sided or rounded colourless sections, often much cracked, are common. The small 'sago grain' variety found in some lavas generally has symmetrically arranged inclusions (Fig. 3.18*g*).

R.I. = 1.51 (i.e. below balsam). Crystals may be isotropic, or may show anomalous double refraction in patchy 1st order greys, due to lamellar twin structures which develop as the crystals cool.

Nepheline, $Na(AlSi)O_4$

Crystals: Hexagonal; short 6-sided prisms when idiomorphic (as in lavas). Colourless or white. The variety *elaeolite*, which occurs in plutonic rocks, has no regular form, and is grey with greasy lustre. The latter, together with its lower hardness, distinguish it from quartz. H = $5\frac{1}{2}$ to 6 (can just be cut with knife). sp. gr. = 2.6.

Nepheline occurs in undersaturated rocks such as phonolite (*q.v.*) and nepheline-syenite; weathered surfaces of the latter may be pitted where the nepheline crystals have been worn down below the level of the more resistant feldspars.

In thin section, hexagonal (basal) and rectangular (longitudinal) sections are shown in lavas (Fig. 3.18*g*). Colourless when fresh, the mineral is frequently altered and then appears clouded (hence the name, from the Greek *nephele*, a cloud). Yellow cancrinite is also an alteration product.

R.I. = 1.54 (mean), slightly higher than for orthoclase.

Biref: Weak (0.004), giving low grey polarization tints. Basal sections are isotropic.

Extinction: Straight in longitudinal sections. Nepheline does not twin; this and its usual lack of cleavage, help to distinguish it from orthoclase.

FORMS OF SILICA

Silica is found uncombined with other elements in several crystalline forms of which *quartz*, one of the most common minerals in nature, is of special importance. Other forms of silica include the high temperature *tridymite* (see below); *chalcedony*, aggregates of quartz fibres (p. 118); and the cryptocrystalline forms *flint* and *opal*, which are described under Secondary Minerals.

Quartz, SiO_2

In the structure of quartz the silicon-oxygen tetrahedra build up a three-dimensional framework in which each oxygen is shared between two silicons. The tetrahedra form spiral structures (Fig. 3.21) which

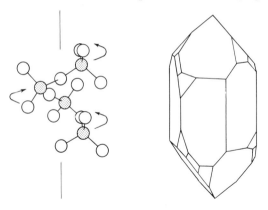

Fig. 3.21   *Left,* spiral structure in atomic framework of quartz; *right,* quartz crystal with extra (trigonal) faces.

extend in the direction of the *c*-axis and are linked laterally by sharing oxygens; they twist either clockwise or anti-clockwise, giving right-handed or left-handed crystals. There are no substitutions of other ions in the silicon positions.

Crystals: Trigonal (p. 82); 6-sided prisms and rhombohedral terminations (Fig. 3.21); faces sometimes unequally developed; occasionally other faces belonging to trigonal forms are present. Vitreous lustre; conchoidal fracture. Colourless when pure (e.g. *'rock crystal'*), but many coloured varieties occur, the colour being due to traces of impurities, e.g. *rose quartz* (pink), *smoky quartz* (grey), *milky quartz* (white), *amethyst* (violet). Some quartz contains minute inclusions or liquid-filled cavities, which may be arranged in regular directions in a crystal. No cleavage; twins rare. H = 7, cannot be scratched with a knife. sp. gr. = 2.66.

Quartz is an essential constituent of granites, and can be recognized in the rock as hard, glassy grains of irregular shape and without cleavage. It occurs in small amount in granodiorite and quartz-diorite, and is present as well-shaped porphyritic crystals in acid dyke-rocks and lavas. *Vein quartz* is an aggregate of interlocking crystals of glassy or milky appearance, filling fractures in a rock; the boundaries of the crystals may be coated with brown iron oxide. Well-shaped quartz crystals are found in cavities (*druses*) in both veins and granitic rocks. Most sands and sandstones have quartz as their main constituent; the grains have a high resistance to abrasion (p. 177) and thus persist over long periods during erosion and transport. The mineral is also found abundantly in gneisses, quartzites, and in some schists and other metamorphic rocks.

Properties in thin section

Basal sections are regular hexagons (Fig. 3.18) when the crystals are well-formed; see Fig. 3.12 for longitudinal section. When the mineral has crystallized among others, as in granite, its shape is irregular.

Colourless. Never shows alteration, but crystals in lavas sometimes have corroded and embayed margins.

R.I. = 1.553 (max.), 1.544 (min.); weak outline in Canada balsam.

Biref: = 0.009 (max.) in sections parallel to the *c*-axis, giving pale yellow polarization colour (see p. 89); basal sections are isotropic. Oblique sections give 1st order greys or whites. *Extinction:* Straight in longitudinal sections.

Twinning occurs, but is rarely seen in thin sections. Quartz is distinguished from orthoclase by the absence of twinning and by its entire lack of alteration; it always appears fresh, although inclusions may be present.

Tridymite   This form of silica crystallizes at temperatures above 850°C, as small hexagonal plates, and appears in thin sections as a series of minute scales. It is found in some acid lavas, and also in artificial furnace linings such as silica bricks, where its presence is an index to the temperature reached during the burning of the brick.

ACCESSORY MINERALS

Silicates included under this head are tourmaline, garnet, sphene, zircon, and, for convenience, the metamorphic minerals andalusite and cordierite. Other accessory minerals, such as apatite and magnetite, are described under Non-silicate Minerals (p. 122).

Tourmaline   Complex boro-silicate of Na, Mg, Fe, and Al, with a structure having Si_6O_{18} rings (Fig. 3.14d). There are many varieties, according to the relative proportions of the metal atoms; common tourmaline is black; boron makes up nearly 2 per cent of the whole, and is an essential part of the mineral.

Crystals: Trigonal; commonly long prismatic forms (Fig. 3.22a),

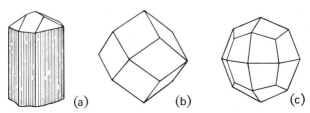

Fig. 3.22   (a) Crystal of tourmaline. (b) Garnet, rhomb-dodecahedron (12 faces). (c) Garnet, trapezohedron (24 faces).

which show striations parallel to their length. Vitreous lustre. Red, green, and blue transparent varieties, sometimes zoned, are cut for gems.

Schorl is an iron-rich tourmaline which commonly grows in radiating clusters of needle-like crystals, e.g. in pneumatolysed granites (p. 208). H = 7. sp. gr. = 3 to 3.2.

Tourmaline occurs as an accessory in some granites; also in vein rocks such as pegmatites, which fill fissures where steam and vapours containing boron and fluorine have been active. It is frequently idiomorphic and black in colour, and is then easily recognized.

In thin section, colour is yellow to greenish-grey (Fig. 3.18h); frequently shows colour zoning; strongly pleochroic, maximum absorption occurring when the length of a longitudinal section is at right angles to the vibration direction of the polarizer, and the mineral then appears darkest.

R.I. = 1.65 (max.), 1.62 (min.).

Biref: Moderately strong, but polarization colours are often masked by the body-colour. Basal sections are isotropic. Extinction parallel to length.

Garnet   The garnets form a group having the general composition $R_2^{2+}R_3^{3+}(SiO_4)_3$, where $R^{2+} = Ca$, Mg, Fe, or Mn, and $R^{3+} = Al$, Fe^{3+}, or Cr. Common garnet is the claret-red *almandine*, $Fe_3Al_2(SiO_4)_3$, or the dark brown *andradite*, $Ca_3Fe_2(SiO_4)_3$. Among the other species may be mentioned *pyrope*, $Mg_2Al_2(SiO_4)_3$, precious garnet; and *grossular*, $Ca_3Al_2(SiO_4)_3$ in some metamorphosed limestones.

Crystals: Cubic, in the form of the trapezohedron (Fig. 3.22c), the rhombdodecahedron (Fig. 3.22b) or combinations of the two. H = $6\frac{1}{2}$ to $7\frac{1}{2}$. sp. gr. = 3.5 to 4.2, for the different species.

Garnet commonly occurs in metamorphic rocks such as mica-schist, as well-formed crystals (porphyroblasts, p. 212); also as an accessory mineral in granites and other igneous rocks.

In thin section, garnet stands out in bold relief by reason of its high R.I. (1.7 or more), and often presents a somewhat rounded outline (Fig. 3.18h); no cleavage; generally broken by irregular cracks. Colourless or pale pink. Garnet is isotropic (i.e. transmits no light between crossed polars).

Sphene (Titanite), $CaTiSiO_5$

Crystals: Monoclinic; flat and wedge-shaped (named from the Greek *sphene*, a wedge). Colour: Generally yellow, brown, or grey. Prismatic cleavage. Frequently twinned. H = 5. sp. gr. = 3.5. Sphene occurs as small well-formed crystals in many granites, diorites, and syenites.

In thin section, sphene stands out in strong relief on account of its high *R.I.* (1.9 to 2.0), and surface irregularities are emphasized. Pleochroism is often marked. *Biref:* Strong, giving delicate polarization tints which are usually masked by the body colour. The wedge shape is characteristic (Fig. 3.18h).

Zircon, $ZrSiO_4$

The tetragonal crystals, usually very small, are bounded by prism and pyramid faces (Fig. 3.1c). Some zircons are colourless, other varieties yellow or reddish-brown. H = 7.5. sp. gr. = 4.7.

Zircon contains traces of radio-active elements which give rise to pleochroic haloes, as in mica enclosing zircon (see p. 103). It occurs in granites and syenites as an original constituent (of early crystallization and therefore often enclosed in other minerals); it is very resistant to abrasion and hence frequently found as water-worn grains in the heavy residues of sands derived from the weathering of granite areas (p. 179).

In thin section, zircon appears as small colourless grains. *R.I.* = 1.93, giving a strong outline in Balsam. *Biref:* Very strong; polarization is in high order colours. Recognized chiefly by its shape and occurrence.

Andalusite, Al_2SiO_5, or $Al_2O_3SiO_2$

Crystals: Orthorhombic, bounded by four long prism faces and two basal pinacoids; a transverse section is nearly square. Colour: Pink or grey. The variety *chiastolite* contains inclusions of carbon which are clustered mainly along the edges where two prism faces meet and down the centre of the crystal (see cross-section, Fig. 3.23). H = 7.5. sp. gr. = 3.2.

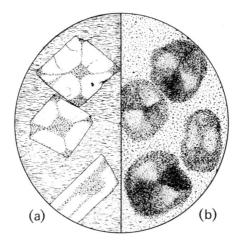

Fig. 3.23   (*a*) Chiastolite in chiastolite-slate. (*b*) Cordierite in spotted rock (crossed polars, to show sector twinning).

In thin section, andalusite is distinguished by its patchy pink pleochroism, the coloured areas not extending to the boundary of the mineral.
Mean R.I. = 1.64. *Biref:* Weak (0.009 max.). Chiastolite is distinguished by its carbon inclusions and shape of cross-sections.
Andalusite is found in contact-metamorphosed shales and slates (e.g. andalusite-hornfels, see p. 204), and sometimes as an accessory in granite.

Cordierite, $Mg_2Al_3(AlSi_5)O_{18}$

Crystals: Orthorhombic; nearly hexagonal shape; groups of three crystals often grow together as sector twins (Fig. 3.23*b*). Occurs in metamorphic rocks such as spotted-rock and hornfels (p. 204) and occasionally in granites near a contact zone.
In thin section: Colourless, resembles quartz in *R.I.* and birefringence. May be distinguished from quartz by the sector twinning, by the presence of alteration products, and by its manner of growth (p. 204), which results in imperfectly formed crystals containing many inclusions; yellow pleochroic haloes are sometimes present.

SECONDARY MINERALS

Described under this head are the minerals *chlorite, serpentine, talc, kaolin, epidote, zeolite,* and *chalcedony,* all of which result from the alteration of pre-existent minerals.

Chlorite, $(MgFe)_5Al(Si_3Al)O_{10}(OH)_8$, variable

The chlorites (named from the Greek *chloros*, green) form a family of green flaky minerals which are hydrous silicates of magnesium and aluminium. Some Fe replaces Mg and gives colour to the chlorite. Like the micas, they have a perfect cleavage, due to the atomic sheet structure (p. 94); they differ from the micas in containing no alkalies, and in other properties noted below. Different kinds of chlorite are given distinctive names (e.g. penninite, clinochlore); these are not distinguished in the following general description.

Crystals: Monoclinic, frequently 6-sided in shape, with a perfect cleavage parallel to the basal plane; the mineral splits into hexagonal flakes which are flexible but not elastic (cf. mica). $H = 2$ to $2\frac{1}{2}$ (often soft enough to be scratched by the finger-nail). sp. gr. $= 2.65$ to 3.0.

Chlorite is found in igneous rocks, as described below, and in metamorphic rocks such as chlorite-schist.

Properties in thin section

Chlorite occurs as an alteration product of biotite, augite, or hornblende; it may replace these minerals completely, forming a *pseudomorph* ($=$ 'false form') in which the aggregate of chlorite flakes and fibres retains the shape of the original mineral. Together with other minerals such as calcite, chlorite also forms an infilling to cavities in basalts (*q.v.*); the variety called *vermicular chlorite* appears as small rounded and 'worm-like' areas.

Colour: Shades of bluish-green and yellowish-green, sometimes very pale; noticeably pleochroic; cleavage often seen.

Mean R.I. = about 1.58.

Biref: Weak; grey polarization tints, but abnormal interference colours known as ultra-blue and ultra-brown are shown by some chlorites (penninite).

Serpentine, $Mg_6Si_4O_{10}(OH)_8$. Some Fe replaces Mg, in part

Serpentine is an alteration product of olivine, of orthorhombic pyroxene, or of hornblende. One possible equation of the change from olivine to serpentine may be written thus:

$$4(Mg_2SiO_4) + 4(H.OH) + 2CO_2 \rightarrow Mg_6Si_4O_{10}(OH)_8 + 2MgCO_3.$$
$\quad$ Olivine $\qquad\quad$ Water $\qquad\qquad\qquad\qquad$ Serpentine

This reaction takes place in an igneous rock while it is still moderately hot (*hydrothermal action*), the source of the hot water being magmatic; it is thought that the change from olivine to serpentine may also be brought about by the action of water and silica.

Serpentine grows as a mass of green fibres or plates, which replace the original mineral as a pseudomorph. The fibrous variety is called *chrysotile*, and is worked in veins for commercial asbestos (see p. 101); the name *bastite* is given to fibrous pseudomorphs after orthorhombic pyroxene, and a platy variety of serpentine is called *antigorite*. In the mass, serpentine is rather soapy to the touch, and may be coloured red if iron oxide is present. H = 3 to 4. sp. gr. = 2.6. Serpentine is found in basic and ultrabasic rocks (p. 150), and in serpentine-marble.

Properties in thin section

As a pseudomorph after olivine, serpentine appears as a matte of pale green fibres, weakly birefringent, and having a low R.I. (1.57). Specks of black magnetite, the oxidized by-product from iron in the original olivine, are often present. The change to serpentine involves an increase in volume, and this expansion may fracture the surrounding minerals in the rock, fine threads of serpentine being developed in the cracks so formed.

Talc, $Mg_3Si_4O_{10}(OH)_2$ A soft, flaky mineral, white or greenish in colour, which occurs as a secondary product in basic and multrabasic rocks, and in talc-schist (p. 213). It is often associated with serpentine. Flakes are flexible but not elastic, and are easily scratched by the finger-nail. H = 1. Talc is used as a filler for paints and rubber, as an absorbent, toilet powder, etc. *Steatite* (*soap-stone*) is a massive form. In thin section, talc resembles muscovite in its polarization, but the rocks in which it occurs are not those which would contain muscovite, a feature which enables the distinction to be made.

Kaolin (china clay)

This substance is largely made up of the mineral *kaolinite*, $Al_4Si_4O_{10}(OH)_8$, one of the group of Clay Minerals (p. 119) which, like the micas, are built up of silicon–oxygen sheets.

Kaolin is derived from the breakdown of feldspar by the action of water and carbon dioxide; the chemical equation for the change is given and the kaolinization of granite masses is described on p. 208. It is white or grey, soft, and floury to the touch, with a clayey smell when damp. sp. gr. = 2.6. In thin section it is seen as a decomposition product of feldspar, which when altered appears clouded and looks white by top light (i.e. by light reflected from the surface of the slice and not transmitted through it).

Epidote, $Ca_2(AlFe)_3(SiO_4)_3(OH)$

The monoclinic crystals of this mineral are typically of a yellowish-green colour, and are elongated parallel to the *b*-axis. Often in radiating clusters; vitreous lustre. H = 6 to 7. sp. gr. = 3.4.

Epidote occurs as an alteration product of calcic plagioclases or of augite; also as infillings to vesicles in basalts, and as pale green veins traversing igneous and metamorphic rocks.

In thin section, grains of epidote are pale yellow and pleochroic; the basal cleavage is sometimes seen. *Mean R.I.* = about 1.75, giving a strong outline in Balsam; polarization is in bright 2nd and 3rd order colours, a distinctive feature of the mineral.

Zeolites

These form a group of hydrous aluminous silicates of calcium, sodium, or potassium; they contain molecular water which is readily driven off on heating, a property to which the name refers (Greek *zein*, to boil). They occur as white or glassy crystal clusters, filling or lining the cavities left by escaping gases (amygdales, p. 143) in basic lavas, or filling open joints, and are derived from feldspars or feldspathoids by hydration. Artificial zeolites are manufactured and used in water-softening processes, which depend on the *base-exchange* properties of the minerals: a sodium zeolite takes up calcium ions from the calcium bicarbonate in the water being treated, in exchange for its sodium ions, which go into solution as sodium bicarbonate. The mineral is restored or 'regenerated' by treating with brine, from which it again acquires sodium ions in exchange for calcium ions, and is then ready for further use.

Two commonly occurring natural zeolites are:

Analcite, $NaAlSi_2O_6H_2O$

Cubic; crystallized as trapezohedra, white in colour. Sp. gr. = 2.25. Occurs as a primary mineral in undersaturated rocks such as analcite-dolerite (p. 148), as well as in the amygdales of basalts. In thin section, analcite appears colourless and isotropic, with *R.I.* = 1.487, i.e. well below that of Balsam.

Natrolite, $Na_2Al_2Si_3O_{10}2H_2O$

Forms white, acicular orthorhombic crystals, generally in radiating clusters. sp. gr. = 2.2. In thin section, crystals show straight extinction and weak birefringence. *R.I.* = 1.48.

Chalcedony, SiO_2

Radiating aggregates of quartz fibres, their ends often forming a curved surface; white or brownish colour and of waxy appearance in the mass. Chiefly found in layers lining the vesicles of igneous rocks. In thin section, such layers show a radiating structure, of which the crystal fibres have straight extinction and give an extinction 'brush' which remains in position as the stage is rotated. *R.I.* = 1.54; polarization colours are 1st order greys.

Flint Cryptocrystalline[1] silica, possibly with an admixture of opal, representing a dried-up gel; occurs in nodules in the Chalk (see p. 263). Often black in colour on a freshly broken surface, with conchoidal fracture. Split flints were much used in the past as a decorative facing to buildings, but the flint industry which once flourished in parts of England is now nearly extinct.

Opal Hydrated silica, SiO_2nH_2O; amorphous. White, grey, or yellow in colour, with a pearly appearance (opalescence), and often displaying coloured internal reflections. Conchoidal fracture. H = about 6. sp. gr. = 2.2. Occurs as a filling to cracks and cavities in igneous rocks. When it replaces woody tissues it preserves the original textures and is known as *wood opal*. Opal is an undesirable constituent in rocks used for concrete aggregates, owing to the possibility of reaction occurring between it and alkalis in the cement (p. 512).

In thin section, colourless and isotropic, with low *R.I.* (1.44).

CLAY MINERALS

In common with other flaky minerals such as the micas, chlorites, and talc, clay minerals are built up of two-dimensional layers or sheets (p. 93) which are stacked one upon another in the *c*-direction. The layers are of two kinds:

(i) a silicon–oxygen sheet, formed by the linking together of tetrahedral SiO_4-groups as described on p. 94, and generally referred to as a *tetrahedral layer*. The composition of this layer is a multiple of Si_2O_5, or with attached hydrogen, $Si_2O_3(OH)_2$.

(ii) an *octahedral layer*, in which a metal ion (Al or Mg) lies within a group of six hydroxyls which are arranged at the corners of an octahedron (Fig. 3.24b). Adjoining octahedra are linked by sharing hydroxyls (Fig. 3.24c). Such an octahedral layer has the composition $Al_2(OH)_6$ or $Mg_3(OH)_6$ (the minerals *gibbsite* and *brucite* respectively); in addition, substitution may take place for the metal ion, for example Fe^{3+} or Mg^{2+} for Al^{3+}.

Different arrangements of the above layers build up the units of which the clay minerals are composed, the flat surfaces of the minute crystals being parallel to the layers. Some clay minerals have two-layer units, as in Fig. 3.24 (kaolinite); others (e.g. montmorillonite) have three-layer units in which an octahedral layer lies between two tetrahedral layers, one of which is inverted relative to the other so that

---

[1] *Cryptocrystalline:* made of a large number of minute crystals which are too small to be distinguished separately except under very high magnification.

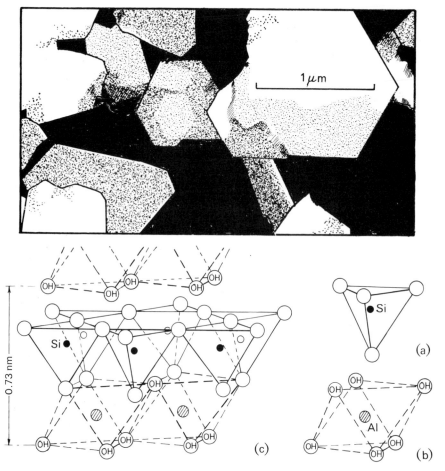

Fig. 3.24 Kaolinite. *Upper.* Electron-microscope photograph (× 35 000). *Lower.* Atomic structure of kaolinite (open circles represent oxygen atoms). (*a*) tetrahedral group; (*b*) octahedral group; (*c*) unit of structure, consisting of a silicon-oxygen layer of linked tetrahedral groups, combined with an aluminium-hydroxyl layer. The *c*-axis of mineral is vertical.

apexes of the tetrahedra point inwards in the unit (*cf.* Fig. 3.17). Layers of molecular water may lie between these units, as in montmorillonite.

The commoner clay minerals include:

(*a*) *Kaolinite*, $Al_4Si_4O_{10}(OH)_8$, made up of alternate silicon- and aluminium-layers; each pair, $Si_2O_3(OH)_2 + Al_2(OH)_6$, with loss of water becomes $Al_2Si_2O_5(OH)_4$. Kaolinite occurs in hexagonal flakes of minute size, and forms the greater part of kaolin (china clay) deposits; it is also found in soils and sedimentary clays, of which it forms a variable and often small proportion. It is the main constituent of fire-clays (*q.v.*) (Note 8).

Dickite has the same composition as kaolinite, but the layers in the structure have a different relative arrangement to one another. *Halloysite*, $Al_2Si_2O_5(OH)_4 \cdot 2H_2O$, may be included in a group with dickite and kaolinite; it occurs as minute tubes, the rolled-up 'sheets' of silicon-oxygen and Al-hydroxyl composition. Certain clays having a high content of halloysite possess special properties, which are discussed on p. 525.

(*b*) *Montmorillonite*, which has important base-exchange proper-ties, is built up of 3-layer units comprising two silicon-layers separated by an aluminium-layer, and has the ideal formula $Al_4Si_8O_{20}(OH)_4$. Some aluminium is usually replaced by magnesium or iron, and small amounts of sodium or calcium are then attached, as ions lying between the 3-layer units or around the edges of the minute crystals. These alkali ions are exchangeable, giving rise to the high base-exchange capacity of the mineral. In addition, layers of molecular water may occur between the 3-layer units. A typical Ca-mont-morillonite would be represented by the formula: $Ca_{0.5}(MgAl_3)Si_8$-$O_{20}(OH)_4 \cdot xH_2O$; the calcium is replaced by sodium in Na-montmorillonite. The proportion of water is variable, and water absorption between the 3-layer units gives rise to the considerable swelling properties possessed by clays containing much montmor-illonite.

The mineral occurs sparsely in soils together with kaolinite, but is the chief component of clays such as fuller's earth and bentonite, which are described below, with a note on their uses. The related minerals *saponite* and *nontronite* are formed when the aluminium in montmorillonite is all replaced by magnesium or ferric iron respec-tively.

(*c*) The mineral *illite* (named after Illinois by R. E. Grim, 1937) has come to be recognized as a distinct species. It is similar in many respects to white mica, but has less potassium and more water in its composition, and gives a distinct X-ray pattern. It has a much lower base-exchange capacity than montmorillonite. Illite is built up of units comprising two tetrahedral layers separated by an octahedral layer, and forms minute flaky crystals in a similar way to montmor-illonite. Some of the silicon is replaced by aluminium, and atoms of potassium are attached, giving a general formula of the type: $K_xAl_4(Si_{8-x}Al_x)O_{20}(OH)_4$, the value of x varying between 1.0 and 1.5. (By comparison with the white mica muscovite, illite thus contains about half as much K.) The OH-content may exceed 4, out of total $(O + OH) = 24$.

According to Grim, sedimentary clays are mostly mixtures of illite and kaolinite, with some montmorillonite, and shales have illite as the dominant clay mineral. Illite is probably the most widely dis-tributed clay mineral in marine argillaceous sediments.

The property of base-exchange (see above and p. 108) was used e.g. to render impervious a leaky clay lining to the artificial freshwater lake constructed at Treasure Island, San Francisco. The material used for the lining was a sandy clay, having a small content of calcium which was probably attached as ions to aggregates of colloidal particles, by virtue of which the clay was 'crumby' and to some extent permeable. By filling the lake with salt water the clay was enabled to take up sodium in exchange for the calcium; this resulted in a considerable decrease in its permeability and a 90 per cent reduction in the seepage losses. The colloidal aggregates were dispersed by the exchange of bases, thus changing the physical properties of the clay and filling the voids with a sticky gel which rendered it largely impervious to water. This treatment is the reverse of the common agricultural process of adding calcium (in the form of lime) to a heavy, sticky soil in order to improve its working qualities.

NON-SILICATE MINERALS

Only the more commonly occurring non-silicate minerals have been selected for description, fuller treatment being beyond the scope of this book; they are here listed in the order in which they are described:

Oxides: Magnetite, Hematite, Limonite, Ilmenite, Cassiterite.
Carbonates: Calcite, Dolomite, Siderite.
Phosphate: Apatite.
Sulphates: Gypsum, Barytes.
Sulphides: Pyrite, Marcasite, Pyrrhotite, Galena, Zinc-blende.
Fluoride: Fluorspar.
Chloride: Rock-salt.

Magnetite, Fe_3O_4. Magnetic oxide of iron.

Crystallizes in black octahedra (Fig. 3.1a, Cubic system); also occurs massive. Metallic lustre; opaque. Black streak. H $= 5\frac{1}{2}$ to $6\frac{1}{2}$. sp. gr. $= 5.18$. Occurs in small amount in many igneous rocks, not usually visible in hand specimen; when segregated into large masses it forms a very valuable ore of iron, as in North Sweden and the Urals. Grains of magnetite are commonly found in the heavy residues obtained from sands.

In thin section, black and opaque when viewed by transmitted light; by reflected light, steely grey.

Haematite, Fe_2O_3

Crystals are rhombohedral (Trigonal), black and opaque, with steely metallic lustre; this variety of hematite is called *specular iron ore*. *Kidney ore* is a massive, reniform variety with a radiating structure. Earthy haematite is called *reddle*. Red streak. H = $5\frac{1}{2}$ to $6\frac{1}{2}$. sp. gr. = about 5.2. When it occurs in large deposits, as in Cumberland, haematite is an important ore of iron; the great Lake Superior deposits contain 50–60 per cent of iron. At Hamersley, Western Australia, banded haematite and chert form an important economic deposit. Haematite is a cementing material in many sandstones, is soluble in dilute acids, and gives the red staining seen in many rocks (p. 182). In thin section it appears red by 'top light' or, if thin enough, by transmitted light.

Limonite. $2Fe_2O_33H_2O$. Hydrated iron oxide.

Amorphous; occurs in masses having a radiating fibrous structure (similar to that of hematite), in concretions, and in an earthy state. Colour, brown to yellow; brown streak. H = 5. sp. gr. = 3.8. Limonite is a very common colouring medium in rocks, and results from the hydration of other iron oxides and from the alteration of other minerals containing iron (e.g. pyrite). The mineral *goethite*, $FeO(OH)$, is often associated with limonite; it occurs as yellow-brown orthorhombic crystals, often tabular, and is pleochroic.

Ilmenite. $FeOTiO_2$

Forms platy Trigonal crystals which are black and opaque. Moderately magnetic; black streak. H = 5 to 6. sp. gr. = 4.5. Found in basic igneous rocks in small grains; also occurs massive, and may form large segregations, as in Norway and Canada. Ilmenite is the chief ore of titanium, which is used in the manufacture of white paint.

In thin section ilmenite is black and opaque when unaltered; but it changes by hydration to a whitish substance called *leucoxene*, which can be detected by its white appearance when viewed by reflected light. The presence of the alteration product therefore helps to distinguish ilmenite from magnetite.

Cassiterite (tin-stone), SnO_2

Crystals are tetragonal, dark brown to black in colour, opaque and heavy. Knee-shaped twins often seen. H = 6 to 7. sp. gr. = 6.8. Found in veins of quartz, often associated with tourmaline and fluorspar, in granite areas (e.g. Cornwall). Water-worn grains of cassiterite can be recovered from many stream sands ('placer' deposits) in granite areas; deposits of this type were worked extensively for supplies of tin in Malaya, but are now virtually exhausted.

Calcite, $CaCO_3$

Crystals: Trigonal (p. 79); a cleavage rhombohedron is shown in Fig. 3.1*b*. 'Dog-tooth spar' and 'nail-head spar' (Fig. 3.3) are varieties formerly named from their resemblance to those objects. Generally colourless or white, but may have various tints. Cleavage is perfect, parallel to the rhombohedral faces, and twinning is common on rhombohedral planes. H = 3; the ease with which the mineral can be scratched with a knife affords a useful index to identification. sp. gr. = 2.71. Calcite dissolves in dilute acids, with effervescence.

Limestones are essentially composed of calcium carbonate, of which the crystalline form, calcite, may constitute a large part. Calcite also occurs as a secondary mineral in igneous rocks, e.g. in the amygdales of basalts. Veins of calcite often fill fractures in many rocks. Calcite is commonly associated with sulphide ores such as blende and galena in mineral veins. Vein calcite may be recognized by its white or pink colour, and is distinguished from quartz by its lower hardness.

Properties in thin section

Crystals generally irregular in shape, colourless, the cleavage apparent in some sections.

R.I. = 1.658 (max.), 1.486 (min.); these values are respectively above and below the *R.I.* of Balsam. In consequence, a *twinkling* effect is seen when the polarizer is rotated below the slice (the analyser not being used): the mineral shows changes in relief and strength of outline as the plane of the polar coincides alternately with each of the two vibration directions of the calcite, with their different refractive indexes. Sections transverse to the *c*-axis do not show twinkling.

Biref: = 0.172 (max.); polarization colours are whites of the higher orders, or delicate pinks and greens for sections whose birefringence is less than the maximum. Lamellar twinning appears as bands of colour along the diagonals of rhomb-shaped sections.

Aragonite, orthorhombic, is another crystalline form of calcium carbonate; it is less common, and less stable, than calcite, and is often associated with gypsum.

Dolomite, $CaCO_3MgCO_3$

Crystals: Trigonal; crystals occur as rhombohedra, sometimes with curved faces. Cleavage perfect as in calcite. Colour: White, yellow or brown. Vitreous to pearly lustre. H = $3\frac{1}{2}$ to 4 (cf. calcite). sp. gr. = 2.85. Dolomite is not readily dissolved by *cold* dilute HCl; this provides a useful test for the mineral, which may otherwise be difficult to distinguish from calcite in a rock specimen. Occurs chiefly in dolomitic limestones (see p. 191).

Properties in thin section

Similar to those of calcite, but dolomite develops more prominent rhomb-shaped crystals and is thereby recognizable (Fig. 5.3c). Crystals are sometimes zoned by inclusions near their borders.

Siderite (chalybite), $FeCO_3$

Found as brown rhombohedral crystals, and also massive. White streak, pearly to vitreous lustre. H = $3\frac{1}{2}$ to $4\frac{1}{2}$. sp. gr. = 3.8. Siderite is important in ironstone deposits as a source of iron, e.g. the Cleveland iron ore, Yorks. It may replace calcium carbonate in a limestone and may also be precipitated direct from solution to form a sedimentary deposit (p. 192).

Apatite, $Ca_5F(PO_4)_3$ fluor-apatite; in chlor-apatite, Cl is present instead of F.

Crystals: Hexagonal, pyramids and prisms terminated by basal plane (Fig. 3.25*a*). Colour: Usually a brown or greenish tint, but

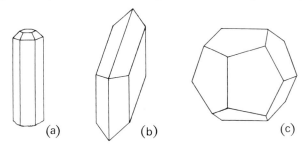

(a)   (b)   (c)

Fig. 3.25   (*a*) Apatite. (*b*) Gypsum. (*c*) Pyrite (pyritohedron).

varies greatly. H = 5. sp. gr. = 3.2. Apatite occurs as a common accessory mineral in many igneous rocks, the crystals being usually visible only with the microscope; large crystals occur in coarse-grained veins (pegmatites), which yield commercial supplies, from which phosphate is obtained for use as fertilizer.

Properties in thin section

Small, colourless elongated prisms and hexagonal cross-sections are typical (Fig. 3.18*h*); sometimes in long, needle-like crystals.

R.I. = 1.63; the mineral shows moderately strong relief in Balsam.
Biref: Very weak, giving low grey polarization colours. Straight extinction.

Gypsum, $CaSO_4 2H_2O$

Forms colourless or white monoclinic crystals, diamond-shaped and flat parallel to the side pinacoid (Fig. 3.25*b*); also 'arrow-head' twins (Fig. 3.4*d*). Cleavage is perfect parallel to the side pinacoid. Pearly lustre. H = $1\frac{1}{2}$ to 2 (easily scratched by the finger-nail). sp. gr. = 2.3. *Selenite* is the crystalline variety. *Alabaster* is a white or pink massive variety, and the form known as *satin-spar* is made of silky fibres occurring in veins.

Gypsum is formed chiefly by the evaporation of salt water in shallow inland seas, the calcium sulphate in solution being precipitated, as at the southern end of the Dead Sea; extensive deposits of Permian age, hundreds of feet thick, are worked at Stassfurt, Germany (p. 194). Gypsum is also formed by the decomposition of pyrite (FeS_2) in the presence of calcium carbonate, e.g. crystals of selenite found in the London Clay are due to this reaction. Gypsum is much used in the building industry in the manufacture of plasters and plaster board, and as a retarder of cement. *Anhydrite*, $CaSO_4$, is frequently associated with gypsum in evaporites (p. 194), and also occurs in the cap-rock of salt domes (p. 227). It forms white or greyish tabular orthorhombic crystals, H $= 3-3\frac{1}{2}$, sp. gr. $= 2.93$. It expands slightly on hydration (see note 12 of Chapter 2).

Barytes, $BaSO_4$

Forms colourless, white, or brown orthorhombic crystals which are elongated parallel to the *b*-axis (Fig. 3.1*d*). Perfect cleavages are developed parallel to the prism faces and to the basal plane, so that when broken the mineral yields flat diamond-shaped cleavage fragments. Also occurs massive and in other forms. Vitreous lustre. H $= 3$ to $3\frac{1}{2}$. sp. gr. $= 4.5$ (hence known as 'heavy spar'). Barytes, generally associated with calcite (and sometimes with the barium carbonate, *witherite*), occurs chiefly in ore-veins carrying zinc-blende and galena, e.g. in the north of England. It is used in the manufacture of paint and paper.

Pyrite, FeS_2

Occurs as Cubic crystals, commonly in the form of the cube or the pyritohedron (a form with twelve pentagonal faces, Fig. 3.25*c*), pale brassy-yellow in colour, with metallic lustre. The faces of the cube are often striated. Pyrite strikes fire with steel (a point of distinction from marcasite), and was formerly used instead of flint in fire-arms; it is not readily soluble in hydrochloric acid. H $= 6$ to $6\frac{1}{2}$. sp. gr. $=$ about 4.8.

Pyrite occurs in ore-veins, in shales and slaty rocks, and as an accessory mineral in some igneous rocks. It sometimes contains small quantities of copper and gold; large deposits of pyrite are worked for sulphur, as at Rio Tinto, Spain, and copper is also obtained from them. By decomposition and oxidation, pyrite gives rise to sulphuric acid (H_2SO_4), which is recovered in some smelting processes. The iron released during natural decomposition helps to form limonitic coatings on rock surfaces (p. 123).

Marcasite, FeS_2

The orthorhombic iron sulphide, paler and slightly softer than pyrite, and decomposed readily by acids. H $= 6$. sp. gr. $= 4.9$.

Found in concretions in sedimentary rocks, such as the Chalk; nodules of pyrite with a radial crystalline structure are also found in the Chalk.

Pyrrhotite (magnetic pyrite), Fe_nS_{n+1} (where n = 6 to 11).

Forms tabular hexagonal crystals with a brownish or coppery colour. Magnetic (a distinction from pyrite); soluble in HCl (pyrite is not). H = 4. sp. gr. = 4.5. Pyrrhotite often contains a small amount of nickel (up to 5 per cent), and is then worked as a source of that metal. The extensive nickel deposits at Sudbury, Canada, which yield the greater part of the world's nickel supply, contain a large proportion of pyrrhotite.

Galena.   Pbs; generally with some content of silver sulphide ('argentiferous galena').

Forms Cubic crystals, with faces of the cube and octahedra, lead-grey in colour; perfect cleavage parallel to cube faces. Also found massive. Metallic lustre, even fracture. H = $2\frac{1}{2}$. sp. gr. = 7.5. Galena occurs in lodes associated with blende, calcite, quartz, and other minerals, often filling fracture zones, as at Broken Hill, New South Wales. It is also deposited in joints and fissures in limestones, as in the Derbyshire occurrences. It is the chief ore of lead and an important source of silver.

Blende (zinc-blende; sphalerite), ZnS.

Forms Cubic crystals of tetrahedral type, usually brown or black in colour, with perfect cleavage parallel to crystal faces. Some varieties are transparent. Resinous lustre; brittle, conchoidal fracture. H = $3\frac{1}{2}$ to 4. sp. gr. = about 4. Blende is the principal ore of zinc, and occurs often with galena, as at the big Sullivan mine, British Columbia.

Fluorspar (fluorite), CaF_2

Cubic crystals are commonly cubes, occasionally octahedra, and may be colourless, white, green, yellow, or purple. Perfect cleavage parallel to octahedron faces; vitreous lustre; often transparent. H = 4. sp. gr. = 3.2. Sometimes zoned, from green (at centre) through white to purple (outermost zone), the latter being probably formed at a lower temperature than the white and green varieties. Fluorspar occurs in veins and is frequently associated with blende and galena, or with tinstone (see p. 123). It is used in enamels, as a flux in steel-making, and in the manufacture of certain kinds of glass. A massive purple or blue variety from Derbyshire is called 'blue john' and is used for ornaments.

Rock Salt (halite), NaCl.

Forms Cubic crystals which may be colourless, white, or yellow cubes, soluble in water and having a characteristic taste. Perfect cubic cleavage. H = 2. sp. gr. = 2.2. Rock salt occurs, with gypsum and other salts, as a deposit from the evaporation of enclosed bodies of salt water (see p. 193). Deposits are worked in the Triassic beds of Cheshire, at Stassfurt, Germany, in Ontario and Michigan, and in other countries. The salt is extracted either as brine pumped up from the salt-beds and then evaporated, or by mining by means of shafts and galleries. It is used in many chemical manufacturing processes, and for preserving and domestic purposes.

NOTES AND REFERENCES

1. Books include 'Mineralogy', by Berry, L. G. and Mason, B. 1959, (new edition 1970), Freeman & Co. and Rutley's Mineralogy, by Read, H. H. 26th Edition, 1971. Murby & Co., London.
    A good summary of rock-forming minerals, illustrated, is given in Chapter 3 of Introduction to Geology (Vol. 1), by Read, H. H. and Watson, J. 1967. Macmillan, London.
2. There are 32 Classes of symmetry, which are subdivisions of the seven Systems of crystallization.
    A more fundamental classification of crystals is based on the shapes of *unit cell*, determined by X-ray analysis of the three-dimensional patterns of atoms in crystals. The unit cell is the smallest complete unit of pattern which, when repeated in three dimensions, would build up a crystal.
3. There are high-temperature and low-temperature forms of quartz, distinguished as β-quartz and α-quartz respectively. Low-temperature quartz has Trigonal symmetry and is the form commonly found in many rocks. At 573°C it changes into β-quartz, with Hexagonal symmetry; in rapidly chilled rocks the β form may be preserved as crystals at normal temperatures.
4. The property of total reflection is used in the Nicol prism for obtaining polarized light, as described in earlier editions of this book. The Nicol prism has now been largely superseded by 'polaroid', a synthetic substance consisting of a plastic film in which are embedded minute needle-shaped crystals which are oriented parallel to one another. This material polarizes the light passing through it.
5. There is one direction in, for example, a calcite crystal along which light entering it is not split into two rays, but passes through the crystal undivided and unpolarized. This direction is known as the *optic axis* of the crystal, and in calcite it coincides with the crystallographic *c*-axis. Such a mineral is called *uniaxial:* all Hexagonal and Tetragonal minerals are uniaxial. On the other hand Orthorhombic, Monoclinic, and Triclinic crystals all have *two* optic axes, i.e. two directions along

which light can pass without being doubly refracted. They are therefore called *biaxial*.

6. See for example Crystal Structures of Minerals by Bragg, W. L. and Claringbull, G. F. 1965. Bell, London.

   An earlier work by Bragg, W. L. Atomic Structure of Minerals, published in 1937 and reprinted in 1957 by Cornell University Press, New York, gives the structures of a large number of minerals including the main silicate minerals.

7. The complete range of mixtures in the plagioclases is possible because the Na and Ca ions have nearly the same size, and either can enter the same atomic framework without distorting it. But the K-ion is larger, and cannot be built so readily into one crystal structure with the smaller Na and Ca ions.

8. Kaolinite of very fine particle size (less than 1 μm), with a disordered lattice structure, is formed under low temperature conditions of kaolinization as in zones of weathering of suitable sediments. This type of kaolinite occurs, for example, in ball clays (p. 188), where it may form over 50 per cent of the deposit.

   See Keller, W. D. 1964. Processes of origin and alteration of clay minerals, in *Soil Clay Mineralogy* (Ed. Rich and Kunze).

   The process is discussed in connection with ball clays in North Devon by Bristow, C. M. 1968. The derivation of the Tertiary sediments in the Petrocstow Basin, North Devon, in *Proc. Ussher Soc.*, **2**, 29–35 (Redruth, Cornwall).

Additional Reference: Grim, R. E. 1962. Applied Clay Mineralogy. McGraw-Hill, New York.

4 Igneous Rocks

Geological processes due to the natural agents which operate at the earth's surface have been discussed in Chapter 2. Other processes, however, originate below the surface and these include the action of volcanoes, or volcanicity. Molten rock material which is generated within or below the earth's crust reaches the surface from time to time, and flows out from volcanic orifices as *lava*. Similar material may, on the other hand, be injected into the rocks of the crust, giving rise to a variety of igneous *intrusions* which cool slowly and solidify; many which were formed during past geological ages are now exposed to view after the removal of their covering rocks by denudation. The solidified lavas and intrusions constitute the *igneous rocks*.

The molten material from which igneous rocks have solidified is called *magma*. Natural magmas are hot, viscous siliceous melts with a gas content; the chief elements present are silicon and oxygen, and the metals potassium, sodium, calcium, magnesium, aluminium, and iron (in the order of their chemical activity). Together with these main constituents are small amounts of many other elements, and gases such as CO_2, SO_2, and H_2O. Magmas are thus complex bodies and the rocks derived from them have a wide variety of composition. Cooled quickly, a magma solidifies as a rock-glass, without crystals; cooled slowly, rock-forming minerals crystallize out from it. Studies made on artificial melts have yielded much information on the course of crystallization in natural magmas.

The content of silica (SiO_2) in igneous rocks varies from over 80 per cent to about 40 per cent. Magmas and rocks containing much silica (regarded as an acid-forming oxide) were originally called *acid*, and those with less silica and correspondingly more of the basic oxides were called *basic*. This broad distinction is a useful one. Basic magmas are less viscous than acid magmas; the temperatures at which they exist in the crust are incompletely known, but measurements at volcanoes indicate values in the neighbourhood of 1000°C. for basic lavas, a figure which may be considerably lowered if fluxes are present. (A flux lowers the melting point of substances with which it is mixed; the gases in magma, for example, act as fluxes).

Volcanoes are essentially conduits between the earth's surface and bodies of magma within the crust; we now consider some of the phenomena which they exhibit.

VOLCANOES

Different styles of volcanic action may be distinguished, as follows:
(i) *Fissure eruptions*, where lava issues quietly from lines of fracture
at the surface of the earth, with little gas emission;
(ii) *Shield volcanoes*, with large flat lava cones (Hawaiian Type);
(iii) *Central Type volcanoes*, which build cones around a central
orifice with the emission of much gas, and are sometimes violently
explosive. In the waning stages of this type of vulcanicity *fumaroles*
(gas vents), *geysers*, and *hot springs* may be formed.

(1) **Fissure eruptions** These represent the simplest form of
extrusion, in which lavas issue quietly from linear cracks in the ground.
The lavas are basaltic and mobile; they have a low viscosity and
spread rapidly over large areas. In past geological times vast floods of
basalt have been poured out in different regions, and are attributed to
eruptions from fissures. These *plateau-basalts* at the present day
include the Deccan Traps (*trap* is an old field term for a fine-grained
igneous rock), which cover 10^6 km^2 in peninsular India; built up of
flow upon flow of lava they reach a thickness in places exceeding
1800 m.

The plateau-basalts of the Snake River area in North America
cover 500,000 km^2; and the basalts of Antrim and the Western Isles
of Scotland, including the hexagonally-jointed rocks of the Giant's
Causeway and of Staffa, are the remnants of a much larger lava-field
which included Iceland and Greenland. The basalt flows of the world
are estimated to cover in all 2.5 million km^2. The source of these
widespread lava flows is the basaltic layer situated beneath the earth's
outer crust (p. 9).

Fissure eruptions have occurred in historic times in Iceland, some
of the fissures reaching lengths of the order of 15 km or more. The
Laki fissure, nearly 32 km long, was active in 1783, and extruded a
very mobile basalt which covered an area of 670 km^2. Often small
craters due to subordinate explosive activity are built at intervals
along the course of a fissure. The Laki fissure carries a line of such
cones and was opened twice. Recently active fissures in which the
rock was still hot have been observed in eastern Iceland, an island
which lies across the mid-Atlantic rift and is built of volcanic rocks.
It has been suggested that Iceland has grown as the walls of the rift
have moved apart, with the emergence of successive lava flows (Note
1). The building of the new island of Surtsey around a volcanic vent
off the south coast of Iceland, in 1964, was observed and photographed
from the air in all its stages (Plate 1).

(2) **Shield volcanoes** Shield volcanoes are characterized by
large, flat cones of lava with gentle slopes; they are built up of many

lava flows, mainly basaltic in nature, which are emitted from a caldera
or from fissures on the slopes of the cone. It has been suggested that
they represent volcanic activity localized at points along original fissures. The lavas are emitted quietly with little explosive activity, and in
great volume.

Mauna Loa is a shield volcano situated in Hawaii in the South
Pacific, and is the largest known. The island of Hawaii is composed
almost entirely of lava and rises over 9000 m from the ocean floor,
with a summit 4440 m above sea level (Fig. 4.1a). The caldera at the

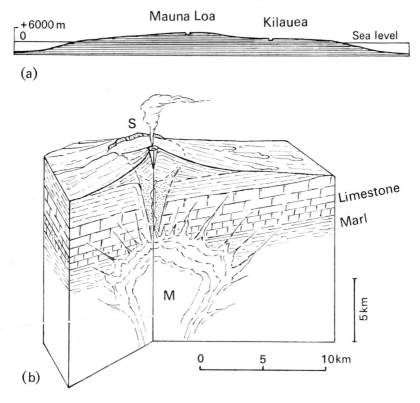

Fig. 4.1   (a) Flat lava-cone of Hawaii; (b) block diagram of Vesuvius; S = Somma
(old crater), M = magma below volcanic pipe (after Umbgrove).

summit of Mauna Loa is a great elliptical pit 5.6 km long and 2.8 km
across, and was probably formed by subsidence. Fountains of lava
play within it, the jets of molten rock sometimes reaching a height of
hundreds of metres. Spray from these fountains is blown by the wind
to form glassy threads ('Pele's hair') which fall on the surface. Eruptions from fissures on the slopes of the volcano occur at intervals of a
few years, with flows of ropy lava. Sometimes when the surface of a

flow has solidified the mobile lava below the crust of the flow drains away, leaving an empty tunnel with roof and walls of lava.

A more accessible crater is that of Kilauea, situated on the eastern slope of the main mountain of Hawaii, about 1.2 km above sea-level. Within it is a lava lake, the top of the volcanic conduit, a vertically sided pit in which the level of the molten rock at times rises and falls with great rapidity. At intervals the lake becomes crusted over, and at other times lava fountains are ejected, due to the discharge of gases. The temperature of the lava at the surface has been measured and found to be 1050°C or more; but 100° less at 6 m below it, since burning gases keep the basic lava hotter at the surface.

Shield volcanoes are found on other islands of the Sandwich group, in Java, and in Iceland, where they were active in geologically recent (Pleistocene) times.

(3) **Central type volcanoes** With their stratified cones and central funnel-shaped orifices, these approach more closely to the popular conception of a volcano. Vesuvius, Etna, and Stromboli in the Mediterranean region, Popocatepetl in the Andes, and many others active at the present day belong to this type. Where the dissected cones of volcanoes which were active in past geological ages are now exposed to view by the processes of denudation, their structure can be studied in detail; Arthur's Seat, Edinburgh, and Largo Law in Fife, are good examples of such 'fossil volcanoes.'

In the upper part of the magma chamber beneath a volcano gases accumulate and exert pressure, which forces magma up the volcanic *pipe* or conduit. The chief gas concerned in this activity is steam, though others are present, such as CO_2, N_2, H_2S, and SO_2. In some instances magma is quietly extruded from a vent as lava; in others, where the accumulated gas pressure is sufficiently high, it may be discharged upwards at intervals with explosive violence. The expanding gases burst the lava into countless small fragments which ultimately fall around the vent, or are blown to a distance by the wind. Thus deposits of fragments called *volcanic dust* or *ash* are formed; larger fragments, termed *lapilli*, and still larger lumps of ejected magma, known as *bombs*, may also be ejected. With these magmatic products may be mingled fragments and blocks of rock torn by the force of the eruption from the walls of the volcanic vent, or from the rocks underlying the volcano.

An eruption is frequently spectacular and often disastrous to life and property, so that it is an event that attracts much attention; descriptions of past volcanic outbursts date back to Pliny's account of the violent eruption of Vesuvius which overwhelmed Pompeii and Herculaneum in A.D. 79. Modern studies have been made from observatories situated on volcanoes, as at Vesuvius or Kilauea; more recently aerial reconnaissance using photographic, radar, and geophysical equipment has made possible more comprehensive studies.

A volcanic cone such as that of Vesuvius is built up mainly of layers of ejected material—ash, lapilli, bombs, and blocks; the layers dip radially outwards on the outer slopes of the cone, and inwards near the central pipe or neck (Fig. 4.1*b*). The latter forms the main conduit for the emission of gases and lava, but during an eruption the sides of the cone may be breached by fissures connected with the volcanic pipe, and lava streams flow down the slopes of the volcano from these orifices. As the eruption progresses, the gases which have accumulated under pressure in the magma chamber beneath the volcano are released, and magma is blown out in fragments by their sudden expansion. These ejected fragments fall on and around the volcano as a rain of dust and ash; if they become mixed with rainfall they form a fluid mud which flows down the slopes as a mud-stream, and may overwhelm buildings and crops, as at Herculaneum (A.D. 79). The volcanic pipe itself is cleared by the tremendous uprush of gas, but after the eruption has subsided the vent ultimately becomes choked again with debris or filled with a solid plug of lava.

The debris in and around the vent contains the largest ejected masses of lava (bombs), which are embedded in dust and ash; a deposit of this kind is known as *agglomerate*. The layers of ash and dust which are formed for some distance around the volcano, and which build its cone, become hardened into rocks which are called *tuffs*. (Note 2). When eruptions take place on the sea floor, as may happen if a submarine vent is opened, the ejected dust, ash, and lapilli may form deposits interbedded with normal aqueous sediments. This is the case with many of the ancient rocks of North Wales, Shropshire, and the Scottish Lowlands, where considerable thicknesses of bedded tuffs are found, pointing to extensive volcanic activity in the past. All these deposits comprise the *pyroclastic rocks* (literally, 'fire broken'), a name which refers to their mode of origin.

Poorly consolidated tuffs from the Naples area were used by the Romans for making 'hydraulic' cement, and were called *pozzolana;* mixed with lime they harden under water. Similar material from the Eifel volcanic district of Germany, has also been used: tuffs known as *trass*, when mixed with an equal amount of limestone, form a cement. Artificial pozzolans, as they are now called, are used at the present day (p. 527).

The famous diamond mines of Kimberley are located in old volcanic necks, which are filled with the diamond-bearing breccia known as 'blue ground.' The breccia has the composition of serpentine (*q.v*); it is easily weathered, and is crushed for the extraction of the diamonds which are sparsely distributed through it.

Paroxysmal eruptions. Some volcanoes, notably a group in the West Indies, are characterized by sudden and rapid activity which consists mainly in the discharge of large quantities of deadly gases, with very little outflow of lava. One such volcano, Mont Pelée on the island of Martinique, has a disastrous history. In May and December 1902 two violent eruptions occurred, each time with the emission of a hot cloud (*nuée ardente*) of heavy gases, loaded with incandescent dust particles. In the May eruption, the hot cloud rushed as an avalanche down the mountainside, completely destroying the town of St. Pierre at its foot. Buildings were obliterated and the 30 000 inhabitants perished, except for one survivor. A similar eruption later in the same year, with violent gas discharge through the side of the cone at a weak point, was observed by an expedition afloat. A plug of almost solid lava—the 'spine' of Mont Pelée—was pushed up through the volcanic pipe by gas pressure within, to a height of 213 m above the crater. It crumbled and almost disappeared by weathering after a few months. The tuffs formed on land from a *nuée ardente* typically have their component fragments welded together by heat at the time of their deposition, and are called *ignimbrites* (from Latin *ignis*, fire, and *nimbus*, a cloud).

Krakatoa, an island of the East Indies, was the scene of four tremendous explosions of the paroxysmal type in August 1883. The island formed part of the rim of a former large crater, much of which had become submerged. After two centuries of quiescence, the eruption began with the emission of steam, culminating with the great explosions which blew away most of the island, estimated at 18 km^3 of rock. Shocks were felt virtually the world over, and tidal waves generated near the volcanic centre did great damage to low coasts. The dust from Krakatoa floated round the earth in the upper atmosphere and caused remarkable sunsets for a year after.

Other volcanoes belonging to the same class are Papandayang in Java, and Bandai-San in Japan, both of which have in modern times lost a large part of their bulk by great explosions.

Fumaroles, geysers, and hot springs. In areas of dying volcanic activity, the emission of steam and gases (HCl, CO_2, H_2S, and HF) at high temperature from *fumaroles* or gas-vents may continue for a long time. Sulphur is deposited around some of the lower temperature gas-vents, called *solfataras*, as in the Vesuvian area, and commercial supplies of sulphur were obtained from these deposits.

Geysers (Icelandic, = roarer) are eruptive springs of boiling water and steam; jets of water are blown into the air at intervals by steam pressure generated in fissures in hot rocks below the surface. The Yellowstone Park region in Wyoming, U.S.A., is famous for its geysers and hot springs. The geyser called Old Faithful, which erupts regularly at intervals of about an hour and shoots a column of water to a height of 46 m, has recently been investigated and it was found that the geyser tube is much deeper than was formerly thought. Heat from surrounding rocks raises the temperature of the water, which was measured at different levels. It was shown that an eruption begins when the water becomes superheated by several degrees above boiling point. The upper layers of water begin to boil, and the boiling spreads downwards in the tube until enough steam has accumulated to eject the water in the upper part of the tube. The time interval between successive eruptions at Old Faithful decreased from an average of 64 minutes to about 60 minutes just before the earthquake at Hogben Lake in August 1959. After the earthquake the time between eruptions quickly increased to 67 minutes; it fell again before the onset of the more distant Alaska earthquake of March 1964. The possibility of a link between such an earthquake and the behaviour of the geyser has still to be evaluated (Note 3).

Geysers and hot springs are common in Iceland, where the hot water is utilized for domestic heating and cooking, and for laundry supplies. The temperature of hot springs is generally lower than that of geysers, and the former flow constantly, whereas geysers are intermittent. Around the orifice of a spring or geyser there is often deposited a cone or mound of *sinter*, a siliceous substance which is either a sublimate (Note 4) or is thrown out of solution from the hot spring. Sinter is usually white but may be a variety of colours; it is developed on a large scale at Taupo, New Zealand, and pink sinter deposited by hot springs formed the well-known terraces of Rotomahana, which were largely obliterated by a later eruption.

INTRUSIVE ROCK-FORMS

A body of magma which is under pressure in the earth's crust may be forced to higher levels, penetrating the upper rocks of the crust; it is then said to be *intrusive*. It may, during the process of intrusion, incorporate within itself some of the rocks with which it comes into contact, a process known as *assimilation* (see p. 164). In some cases it may also give off mobile fluids which penetrate and change the rocks in its immediate neighbourhood. If the intrusive magma cools under cover at some depth below the surface, the rocks which result are called *plutonic rocks* and are coarsely crystalline; a large mass of this kind constitutes *a major intrusion*. When magma rises and fills fractures or other lines of weakness in the crust, it forms *minor intrusions*, i.e. smaller igneous bodies. These include *dykes*, which are wall-like masses, steep or vertical, with more or less parallel sides; and *sills*, which are sheets of igneous rock whose extent is more or less horizontal, and which lie parallel to the bedding planes of sedimentary rocks into which they are intrusive (Fig. 4.2). Dyke and sill-rocks commonly have a fine-grained texture. *Veins* are smaller and irregular bodies of igneous material, filling cracks which may run in any direction.

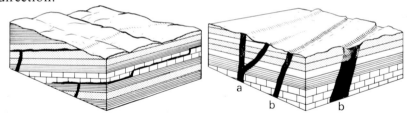

Fig. 4.2   Sills (left) and dykes (right). (*a*) Dyke more resistant to weathering, (*b*) less resistant, than the country-rock.

Magma which rises to the earth's surface and flows out as a lava, as described earlier in this chapter, is called *extrusive*, and under these conditions it loses most of its gas content. The rocks formed when it solidifies are termed *volcanic rocks*, and since they have cooled comparatively quickly in the atmosphere they are frequently glassy (i.e. non-crystalline), or very fine-grained.

Dykes and sills   Of the minor intrusions, *dykes* vary in size from a few centimetres to many metres in width. Most dykes, however, are less than 3 m across. They tend to outcrop in nearly straight lines, and may extend for a short distance or may run for many miles across country (see Fig. 11.7). The Armathwaite-Cleveland dyke of dolerite in the north of England, for instance, can be traced for 209 km. The dyke from Lochgoilhead to near Perth, one of many east-west dykes in Scotland, has a length of 113 km. If the dyke-rock is harder than the country-rock into which it was intruded, after weathering and denudation it will stand up above the level of the surrounding rocks as a wall-like mass (Fig. 4.2).

A group of many parallel or radial dykes is called a *dyke swarm*. A good example of the former is found in Mull, where 375 basaltic dykes running in a general N.W.–S.E. direction form a parallel swarm and cut the south coast of the island within a distance of 20 km. Some of the dykes extend across to the Scottish mainland, where they are traceable for a considerable distance farther. The stretching of the crust which was necessary to open up the dyke fissures points to the operation of tensile stresses across the area, at about the time that the Tertiary basalts of Antrim were being extruded. Another swarm of parallel dykes, which trend east and west, crosses the Midland Valley of Scotland. These dykes are of quartz-dolerite, and are closely related to the Whin Sill and other intrusions of the north of England (p. 138). A radial swarm occurs in the island of Rum, off the west coast of Scotland, where 700 basic dykes have been recorded and are grouped about a centre in the south of the island.

Basaltic dyke swarms of Tertiary age in eastern Iceland, which are associated with volcanic centres comparable with that of Mull, have been described (Note 5). They are believed to be the feeders of basalt lava flows.

In the Karroo Region of South Africa nearly horizontal sandstones and shales, which cover a vast area, are invaded by many dolerite sills and dykes (Plate 4b). Narrow dykes, from 2 to 9 m in width and nearly vertical, run for great distances in almost straight lines. Some cut the dolerites of the sills, indicating a slightly later age of intrusion, but others merge with the sills (Note 6).

The great dyke of Rhodesia is a composite intrusion several kilometres wide and 500 km (312 miles) long; situated west of Salisbury, it trends nearly NNE towards the Zambesi valley.

Sills, as distinct from dykes, have been intruded under a flat cover or 'roof', against a vertical pressure due to the weight of the cover. A columnar structure is frequently imparted to the igneous sheet by the presence of sets of joints lying at right angles to its roof and floor. This structure is well seen, for example, in the dolerite sill which forms Salisbury Craigs, Edinburgh. The sediments above and below a sill are baked by the heat of the intrusion and the columnar jointing develops during the cooling and contraction of the sill-rock. A sill may sometimes be stepped up from one level to another, the two flat parallel sections being connected by a short length of dyke (Fig. 4.2). If the rocks into which the sill is injected are later tilted or folded, the igneous mass will also partake of those movements.

Sills, and also dykes, are sometimes *composite*, i.e. having contrasted margins and centres due to successive injections of different material; successive injections of similar material produce a *multiple* sill or dyke. Fine-grained *chilled margins* are formed by rapid cooling where an intrusion has come into contact with the colder 'country-rock' into which it was injected.

As with dykes, sills vary greatly in thickness and extent. The Whin Sill of the North of England is a sheet of basic rock (quartz-dolerite) which has an extent of more than 3890 km². Its average thickness is about 30 m, and it is exposed along the sides of many valleys and escarpments, where denudation has cut down through it, and in coastal cliffs. It yields a high quality road-stone, locally called 'whinstone' (see p. 520).

The large Palisades Sill of the Hudson, New Jersey, has a thickness of 303 m and is mainly an olivine-bearing dolerite. Its mineral composition varies from top to bottom as a result of the settling of early-formed crystals (olivine and pyroxene), and fine-grained chilled margins of basalt are present at the upper and lower contacts.

The thick basic sill of the Shiant Isles, 32 km north of the Isle of

Skye, and many others, similarly show a layering due to the settling of heavier crystals such as olivine during the cooling of the igneous body.

Ring structures *Ring dykes* are intrusive masses filling curved fractures, sometimes appearing as a complete circle or loop in plan. Other fractures which are shaped like the surface of a cone with its apex downwards, and are filled with igneous material, are called *cone-sheets* (Fig. 4.3). These two types of intrusion are well seen in the

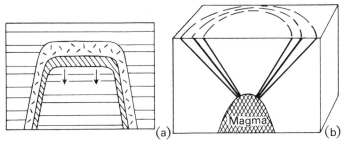

Fig. 4.3 Ring structures. (*a*) Cauldron subsidence: magma rises to fill annular fracture as central plug moves downwards. (See also Glen Coe, p. 270.) (*b*) *Cone-sheets:* conical fractures produced by vertical pressure over a small area, and filled by rising magma.

Ardnamurchan peninsula of the west of Scotland, and a stress theory to account for the origin of the two sets of nearly circular fractures from vertical pressures has been worked out (Note 7).

Cone-sheets are frequently of basic rock, i.e. dolerite or basalt. Acid cone-sheets, however, and an acid ring-dyke have been described from south-eastern Rhodesia, in the Nuanetsi area, where they cut a large sill-like intrusion of basic rock (gabbro) which is 3 km thick and over 6 km in diameter. This igneous centre and others of the surrounding region are believed to have been located at the intersections of major fractures, which cut the Karroo basalts in which the intrusions are emplaced. (Note 8).

Granitic ring-complexes are found in Northern Nigeria, in the Jos Plateau and adjoining areas. In these complexes the extrusion of rhyolitic lavas was followed by the emplacement of granites, which was controlled by ring-fracturing and the subsidence of the blocks within the ring fractures. Most of the granites are steeply dipping ring-dykes, stocks, and plutons, and in many of the intrusions there is evidence of stoping during the rise of the granite magma (Note 9).

Laccoliths and phacoliths A *laccolith* is a 'mushroom-shaped' body (Fig. 4.4) of relatively small size, up to several kilometres in diameter, having a flat floor and up-arched roof. The roof-rocks were lifted or arched by the pressure of the incoming magma, and their form is therefore due to the process of intrusion. Laccoliths were first described by G. K. Gilbert from the Henry Mountains, Utah, where

Fig. 4.4 Cross-sections of an ideal laccolith (left) and a batholith (right). Igneous rock stippled. *s* = stock, *x* = detached masses of country-rock enclosed within the igneous body during intrusion. Scales of the two diagrams are not the same.

they are intruded into mainly horizontal strata and are now exposed in all stages of erosion. They are found in many regions, including Iceland and Skye (gabbro laccoliths); a laccolith of porphyrite occurs in the Tinto Hills, Lanark, from which the red rock is quarried for road surfacing material.

The *phacolith* is a somewhat similar intrusive form, but with both floor and roof curved, a shape due to the magma having been intruded into rocks which were already folded. The dolerite phacolith of Corndon, Shropshire, occupies the crestal region of a dome, over 2 km in length, in folded mudstones. Both laccoliths and phacoliths are classed here as minor intrusions (p. 142).

Batholiths, stocks, and sheets Of these major intrusions the name *batholith* (literally 'depth-rock') is given to very large igneous masses, bounded by steep walls of sedimentary or other rocks across which the igneous body cuts without having any apparent floor (Fig. 4.4). The largest known igneous masses of the world are among the batholiths and, where exposed at the surface, may cover thousands of square kilometres. Projections from the upper parts of such masses are called *stocks* and *bosses:* stocks have cross-sectional areas up to 100 km^2 (40 square miles as defined by R. A. Daly), while bosses are smaller and roughly circular in cross-section. Both have steep walls and rise from the upper part of a batholith, cutting across the structures in the country-rocks (*i.e.* the rocks penetrated by the igneous body). Batholiths are found in regions of folded rocks, especially in the cores of mountain fold-belts and are dominantly acid (granitic) in composition. The Cordillera of North America, for example, contain very large batholiths, elongated parallel to the length of the folds. Others, such as the Coastal Batholith of Peru, are emplaced in the Andean folds of South America (reference in Note 10).

One important process by which magma rises in the crust, to become emplaced as an igneous body, is known as *stoping*, or magmatic stoping. As 'fingers' of magma penetrate outwards into cracks and joints in the surrounding country-rocks, blocks of rock are wedged off and if the density of the blocks or fragments is greater than that of the magma they will sink in the viscous fluid. Shales and many metamorphic rocks, for example, are heavier than granitic magma and would sink in this way. As the stoped rocks remain for some time in the hot environment they become softened and may be streaked out by viscous flow; some may be partly or wholly assimilated or dispersed in the magma, thereby locally changing its composition, while others remain as solid enclosures or *xenoliths*, often with their constituents recrystallized. Such enclosures are frequently seen in the marginal parts of granite bodies. Many examples of stoping are on record, and bosses and stocks several hundred feet in height have been observed in granite areas, whose presence can be largely attributed to the process (Note 10).

The wedging apart of sedimentary layers by magma penetrating along bedding-planes and other surfaces, in the roof-rocks of an intrusion, is an allied process which gives rise to thin sheet-like intercalations of the igneous rock in the sediments, eventually seen after consolidation and subsequent denudation.

The Leinster granite, which lies to the south-west of Dublin and occupies 1000 km² of surface, is the largest granite mass in the British Isles and may be batholithic. It is elongated in a north-north-east direction (Fig. 8.7), not quite parallel to the folds which affect the slatey rocks of the area. Recent studies have shown that its western contact is steep, and in places faulted, while the eastern contact dips more gently under the country-rocks. (Note 11). Large bodies of granite of similar age, the forms of which are not completely known, are found in Galloway, and others in the Scottish Highlands (see Chapter 8). Among the granites of Donegal several different styles of intrusion are represented (Note 12).

The term *pluton* is used to denote any large igneous mass (other than a batholith), irrespective of its size and shape. The granites of Devon and Cornwall are a group of plutons of related composition, and are probably connected underground at a lower level. The granites are now exposed where their sedimentary roof-rocks have been removed by denudation. Other buried plutonic masses probably exist, whose cover has not yet been denuded sufficiently to reveal the igneous rock beneath. One such mass underlies the northern part of the Pennines between Alston and Stanhope (Co. Durham), where mineral veins are found injected into the sedimentary rocks of the district; geophysical surveys suggested the presence of a buried granite, which was reached by a boring at Rookhope, in 1961, at a depth of 389 m (Note 13). Certain ores which are commonly associated with the roof- and wall-rocks of granite masses are discussed on page 164.

Large *sheets* of igneous rock, which are much thicker in proportion to their extent than sills, and often basic in composition, may occupy many square kilometres of ground although not everywhere exposed at the surface. Some of the gabbro masses of Aberdeenshire are probably irregular sheets of this kind, e.g. the Huntly intrusion. The Duluth gabbro of Minnesota, which is a sheet of basic rock with an extent of about 6000 km² and a thickness estimated at 6 km, and the great norite sheet of the Bushveld Complex in the Transvaal, are examples on a larger scale.

We can now construct the following grouping for igneous rocks, based on their mode of occurrence as described above, and leading to the classification on page 145.

EXTRUSIVE   (Volcanic) . . **Lavas** . . . . Glassy or very finely
                                                crystalline.

INTRUSIVE {
  Minor . . . **Dykes and sills,**   Mainly fine-grained
                 laccoliths . . . rocks.

  Major . . . **Batholiths, stocks,**   Coarsely crystalline
  (Plutonic)   **and sheets** . . rocks.

THE IGNEOUS ROCKS

Textures   The texture, or relative size and arrangement of the component minerals, of an igneous rock corresponds broadly to the rock's mode of occurrence. Plutonic rocks, which have cooled slowly under a cover perhaps several miles thick, are entirely crystalline or *holocrystalline:* their component crystals are large (2 to 5 mm or more) and can easily be distinguished with the naked eye. Rocks of fine grain generally have crystals less than 1 mm across and of medium grain between about 1 and 2 mm. When the texture is so fine that individual crystals are indistinguishable without the aid of a microscope it is called *microcrystalline,* and when even the microscope fails to resolve the rock into its component minerals, yet its crystalline nature is apparent between crossed polars, it is said to be *cryptocrystalline.* These textures are all even-grained, i.e. composed mainly of crystals of much the same size; but, in contrast to this, some rocks show the *porphyritic* texture, in which a number of larger crystals are set in a uniformly finer base or ground-mass. The large, conspicuous crystals are called porphyritic crystals (or megacrysts); porphyritic feldspars in some granites may be 5 to 10 cm long.

Extrusive rocks, which have cooled more rapidly at the earth's surface, are often entirely glassy or *vitreous* (without crystals), or partly crystalline and partly glassy. Within a single lava flow the outer layers may be glassy, because rapidly chilled, and the inner part crystalline. Expanding gases in magma during its extrusion as lava give rise to cavities or *vesicles*, resulting in a vesicular texture; the vesicles are frequently elongated and somewhat almond-shaped, and may subsequently become filled with secondary minerals, when they are called *amygdales* (amygdaloidal texture). Basalts are frequently amygdaloidal. Banding or *flow-structure* is produced by differential movement between layers in viscous lava during the process of flow, as in rhyolite. Another kind of banding is formed by crystal-settling (p. 149).

Composition of igneous rocks   The mineral composition and colour of rocks are related to their chemical composition. When the chemical analyses of an acid rock like granite and of a basic rock (e.g. basalt) are compared, important differences are seen, such as the greater proportion of silica and alkalies (Na_2O and K_2O) in the acid rock, and the higher content of lime, magnesia, and iron oxide in the basic. This is shown by the following figures, which are averages of a large number of analyses:

	Average Granite %	Average Basalt %	
SiO_2	70.2	49.1	
Al_2O_3	14.4	15.7	
Fe_2O_3	1.6	5.4	
FeO	1.8	6.4	
MgO	0.9	6.2	
CaO	2.0	9.0	
Na_2O	3.5	3.1	
K_2O	4.1	1.5	
H_2O	0.8	1.6	
rest	0.7	2.0	Total: 100.

The higher alkali content in granite corresponds to a greater proportion of feldspar in the rock; conversely, basalt has more dark minerals containing Fe, Mg, and Ca.

During the cooling of a magma, the different constituents unite to form crystals of silicate (and other) minerals. In a basic magma, for example, minerals like olivine and magnetite may be the first to crystallize, and they take up some of the silica, magnesia, and iron oxide; the remainder of the magnesia and iron (with some alumina and other oxides) is used up later in augite, hornblende, and perhaps dark mica. On account of their composition, such minerals are called *ferromagnesian* or *mafic* (coined from *ma* for magnesium and *fe* for iron). In contrast to these dark and relatively heavy mafic minerals, the alkalies and lime, together with alumina and silica, form light-coloured or *felsic* minerals, which include the feldspars, feldspathoids, and quartz. Most of the calcium in a basic magma would enter a feldspar such as labradorite, a little going into augite. In acid rocks felsic minerals predominate and give the rocks a paler colour, in contrast to the darker basic rocks. Between *acid* and *basic* types there are rocks of *intermediate* composition.

When a magma contains enough silica to combine fully with all the metallic bases and still leave some over, it is said to be *oversaturated;* the excess silica crystallizes as quartz, and an igneous rock which contains quartz is called an oversaturated rock, e.g. granite. Minerals which can exist in a rock in the presence of free silica are said to be *saturated*. Certain minerals with a low silica content, in particular olivine and the feldspathoids, are not normally found in association with quartz and are termed *undersaturated;* thus olivine (Mg_2SiO_4) with the addition of silica would become enstatite ($Mg_2Si_2O_6$), a saturated mineral. It is convenient, therefore, to speak of igneous rocks as oversaturated (i.e. containing quartz or its noncrystalline equivalent), saturated (containing only saturated minerals), and undersaturated (containing olivine or feldspathoids or both).

Classification   A simple scheme of classification can be constructed here for the more common varieties of igneous rocks. The scheme does not include all igneous types; some of the less common ones are mentioned in the descriptions which follow; but more than this is not within the scope of the book.

In the table below, the rocks are arranged in three columns headed Acid, Intermediate, and Basic, forming the Granite, Diorite, and Gabbro groups respectively. A transitional type, Granodiorite, between Granite and Diorite, is also indicated. Rocks of the Syenite family fall outside this grouping and are treated separately on p. 158, for the reasons stated there. The range from left to right in the table corresponds to a decreasing silica content (over-saturated to saturated); an extension to the right of the table would include the Ultrabasic rocks, most of which contain olivine and are undersaturated. In each column there are three divisions, the lowest containing the name of the coarse-grained plutonic member of the group, the middle division the dyke or sill equivalent of the plutonic type, and the top division the extrusive or volcanic rocks. The main minerals which make up the plutonic rocks are shown in a separate diagram below the table. The columns therefore give a grouping based on mineral composition (which in turn expresses the chemical composition), and the horizontal divisions are based on mode of occurrence, which largely governs the rock texture.

Silica percentage decreases from Granite to Gabbro. In the lower diagram, intercepts on any vertical line give the approximate mineral composition for the corresponding plutonic rock in the table.

Table of Igneous Rocks
(excluding alkaline rocks)

	Acid	Intermediate	Basic	Ultra-basic
Extrusive (Volcanic)	Rhyolite Dacite Andesite		Basalt	
Minor intrusive (Dykes and Sills)	Felsite Quartz-porphyry	Porphyrite	Dolerite	
Major intrusive (Plutonic)	Microgranite Granite Granodiorite Diorite		Gabbro	Picrite Peridotite

Oversaturated. . . . .Saturated. . . . .Undersaturated

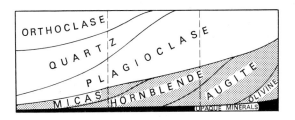

Generalized mineral composition

DESCRIPTIONS OF IGNEOUS ROCKS

GABBRO, DOLERITE AND BASALT

These basic rocks form a large group and are of considerable economic importance on account of the value of certain types as road metal. They have a large content of ferromagnesian minerals, which give the rocks a dark appearance and a range of specific gravity from about 2.9 to 3.2. On account of the relative fluidity of basic lavas, basalt is frequently completely (though finely) crystalline, and locally grades uniformly into dolerite; similarly dolerite may grade into gabbro as the texture grows coarser. Undersaturated basic types include olivine-gabbro, olivine-dolerite, olivine-basalt, and feldspathoidal varieties.

Gabbro

Minerals

Essential minerals are plagioclase (labradorite to anorthite) and a monoclinic pyroxene, e.g. augite or diallage. The lime-rich nature of the plagioclase is related to the high CaO and low Na_2O content in a normal gabbro (see analysis, p. 161). Other minerals which may be present (not all in one rock) include enstatite or hypersthene, olivine, hornblende, biotite, and feldspathoids. Ilmenite and apatite are common accessories. There is no quartz or orthoclase in a normal gabbro, but a very small amount of interstitial quartz may be present in some varieties. Mafic minerals form about 50 per cent of the rock.

Texture

Coarsely crystalline, rarely porphyritic, sometimes with finer modifications. The hand specimen appears dark in colour, grey to black or greenish black, owing to the high proportion of mafic minerals and the grey tint of the plagioclase. Under the microscope the rock shows interlocking crystal plates (Fig. 4.5). Olivine if present often shows well-formed crystals, because of its early separation from the magma. Grains of iron oxide may be enclosed in olivine and other minerals. Frequently decomposition products such as serpentine after olivine or chlorite after pyroxene are present.

Varieties

Some are named after the chief mafic mineral present other than augite, thus: enstatite- or hypersthene-gabbro; olivine-gabbro; hornblende-gabbro. *Norite* is a gabbro containing enstatite or hypersthene instead of augite. *Troctolite* is an olivine-gabbro without augite. *Quartz-gabbro* (e.g. at Carrock Fell, Cumberland) has a little interstitial quartz, which represents the last siliceous liquor to crystallize from a slightly oversaturated gabbro magma. Feldspathoidal varieties include nepheline-gabbro (*essexite*).

Gabbros in Britain are found in Skye and Ardnamurchan (Plate 5, *upper*), in northern Ireland, the Lake District, and at the Lizard, Cornwall (Note 14). The large basic sheets of Aberdeenshire (p. 141), including the Insch, Haddo, and Huntly masses, contain much norite and hypersthene-gabbro.

The important nickel-bearing intrusion of Sudbury, Ontario, is largely composed of norite, in which the nickel-sulphide ores are concentrated near the base of the sheet.

The basic igneous complex of Skaergaard, east Greenland, consists of a layered series of gabbros, with olivine-bearing gabbro below and

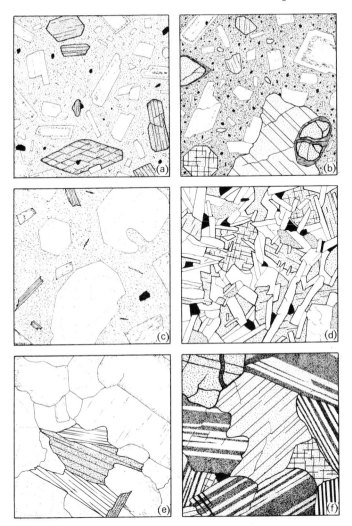

Fig. 4.5 Igneous rocks in thin section. (*a*) Hornblende-andesite. (*b*) Olivine-basalt. (*c*) Quartz-porphyry. (*d*) Ophitic dolerite. (*e*) Muscovite-biotite-granite. (*f*) Gabbro (crossed nicols). (Magnification about × 12.)

iron-rich gabbro above a middle gabbro without olivine. Olivine-gabbro probably represents the original magma before differentiation. The complex has an area of 50 km² and extends downwards as a funnel-shaped mass (Note 15).

Dolerite

Dolerite occurs in dykes and sills, and is dark grey, sometimes nearly black in colour, except when the proportion of feldspar is high;

the texture is usually of medium to fine grain. Dolerite is included in the trade name 'basalt' (as are also andesite, basalt, and other rocks). It is important as a road metal because of its toughness, and its capacity for holding a coating of bitumen and giving a good 'bind,' (See Chapter 17).

Minerals

As for gabbro.

Texture

When the lath-shaped plagioclase crystals are partly or completely enclosed in augite the texture is called *ophitc* (Fig. 4.5). This interlocking of the chief mineral components gives a very strong, tough rock. The augite may, however, occur as granules between the plagioclase laths, when the texture is described as *intergranular*. The nearer dolerite approaches gabbro, as the texture increases in coarseness, the less emphatic will be the lath-like form of the feldspars. On the other hand, dolerite merges by degrees into basalt as the texture becomes finer.

Varieties

Normal dolerite = labradorite + augite + iron oxides; olivine-dolerite; hypersthene-dolerite; quartz-dolerite, with accessory quartz (cf. quartz-gabbro), is a common dyke-rock in the Midland Valley of Scotland and elsewhere, and forms the Whin Sill; analcite-dolerite (= *teschenite*) is an undersaturated type. Much-altered dolerites, in which both the feldspars and the mafic minerals are decomposed, are sometimes called *diabase*. Many American authors use the term synonymously with the British usage of dolerite.

Working quarries in dolerite are numerous; English localities include Rowley Regis, near Birmingham; the Clee Hills, Shropshire (olivine-dolerite); many workings in the Whin Sill and dykes in northern England; and in Scotland the quartz-dolerites and Tertiary dykes referred to on p. 266.

Basalt

Basalt is a dense-looking, black rock, often weathering to a brown colour, and is the commonest of all lavas. It is estimated that the basalt flows of the world have five times the volume of all other extrusive rocks together.

Minerals

Essential minerals are plagioclase and augite. The normal feldspar of basalts is labradorite, but andesine, oligoclase, or albite may occur in different varieties. Magnetite and ilmenite are common accessories; olivine occurs in many basalts and commonly shows alteration to serpentine; calcite, chlorite, zeolites, chalcedony, and other secondary minerals may fill vesicles. Nepheline, leucite, and analcite are found in undersaturated types.

Texture

None seen in the hand specimen, unless the rock is porphyritic or vesicular. Seen under the microscope (Fig. 4.5) the texture is microcrystalline to cryptocrystalline or glassy, often with porphyritic crystals of olivine or augite which are too small to be visible without magnification. Basalt glass is called *tachylite* and is found as a chilled base to flows of basalt lava, or as the chilled margins of dykes. Vesicular and amygdaloidal textures are common.

Varieties

Basalt and olivine-basalt are the commonest varieties; others include quartz-basalt (cf. quartz-dolerite), and feldspathoidal types like nepheline-basalt and leucite-basalt (e.g. the lavas from Vesuvius). Soda-rich basalts in which the plagioclase is mainly albite are called *spilites;* these rocks often show 'pillow-structure.' All varieties may be amygdaloidal; weathered types are sometimes called *melaphyre.*

Some of the great flows of basalt in different parts of the world are referred to on p. 131, and the list given there could be extended. Olivine-basalt lavas are extruded from volcanoes like Kilauea, Hawaii. Plateau-basalts, so called from the topographical forms which they build, are prominent in the west of Scotland and Antrim, in the volcanic region of Auvergne, Central France, and in the Deccan, India, to quote only a few.

PICRITE AND PERIDOTITE

These rocks consist essentially of mafic minerals and contain little or no feldspar. They are coarse-grained, holocrystalline, mostly dark in colour, and have a high specific gravity (3.0 to 3.3). As their silica content is in many cases about 40 per cent or less they are known as *ultrabasic* rocks, and in the classification of p. 145 they form an undersaturated group next to the gabbros. Ultrabasic rocks have relatively small outcrops at the earth's surface. They sometimes form the lower parts of basic intrusions, the heavy crystals of which they are composed having sunk through the body of magma before it consolidated, resulting in a *layering* of the mass (by composition, p. 138). In other situations, ultrabasic rocks such as bodies of serpentinite, derived from a deep-seated source, have been upthrust towards the surface in fault-zones and are now seen in fault-bounded outcrops. (Note 16). Occurrences of this kind are found at the great Alpine Fault of New Zealand (Fig. 1.13), and other rifts.

Picrite

Picrite contains little feldspar (not more than about 10 or 12 per cent), the bulk of the rock being made of olivine and augite or hornblende, generally with some ilmenite. The olivine crystals are sometimes enclosed in the augite or hornblende. By increase of the feldspar content and corresponding decrease in the other constituents, picrite grades into olivine-gabbro and gabbro. The Lugar sill, Ayrshire, contains a thickness of some 7.6 m of picrite, which merges downwards into peridotite and upwards into a feldspathoidal gabbro.

Peridotite

Olivine is the chief constituent of this rock, which is named from the French *peridot*, olivine. Other minerals may include enstatite, augite, hornblende, biotite, and iron oxides. Felsic minerals are virtually absent. A variety composed almost entirely of olivine is called *dunite*, from the Dun Mountains, New Zealand; it is used on a small scale as a decorative stone.

Serpentine-rock or *serpentinite* may result from the alteration of peridotite by the action of steam and other magmatic fluids, while the rock is still hot (p. 116). Large masses of red and green serpentinite occur in the Lizard district, Cornwall (see Note 14). The fibrous variety of serpentine, *chrysotile*, furnishes one source of commercial asbestos when it occurs in veins of suitable size.

Other bodies of ultrabasic rock consist almost entirely of one kind of mafic mineral, such as *pyroxenite* (all pyroxene), and *hornblende rock* (all hornblende). They are usually associated with basic rocks like gabbro, and are of small volume.

DIORITE, PORPHYRITE AND ANDESITE

The intermediate rocks forming the diorite group are saturated with regard to silica, and typically contain little quartz. But by increase of the silica content and the incoming of orthoclase they grade into the acid rocks, thus: *diorite—quartz-diorite—granodiorite—granite*. They are quarried for roadstone and for kerbs and setts, and used as stone for rough walling. The average specific gravity of diorite is 2.87.

Diorite

Minerals

Essential minerals are plagioclase (normally andesine) and hornblende. Accessories include iron oxides, apatite, and sphene. Biotite and quartz and a little orthoclase are frequently present. The mafic minerals may form from 15 to 40 per cent of the rock. The higher soda and lower lime content as compared with gabbro (p. 161) are reflected by the change in the nature of the plagioclase, from labradorite to andesine; some of the lime, together with magnesia and iron, goes to form hornblende.

Texture

Holocrystalline, of coarse to medium grain, rarely porphyritic; diorites on the whole are rather less coarse than granites, but in the hand specimen the different minerals can generally be distinguished with the aid of a pocket lens. The rock is normally less dark in colour than gabbro, on account of the smaller proportion of mafic minerals and the paler tint of the plagioclase. Under the microscope the minerals show interlocking outlines, with a tendency for the mafic mineral to be idiomorphic.

Varieties

Diorite (= andesine feldspar + hornblende); augite-diorite, forming a link with gabbro; biotite-diorite; quartz-diorite, perhaps a more common type than normal diorite, and grading into grano-diorite as explained above. Fine-grained varieties are called *micro-diorite*.

Diorites have a somewhat restricted distribution, and frequently form local modifications to granodiorite and granite intrusions due to assimilation (see p. 164), as in the case of many of the 'newer granites' of Scotland (p. 252). Small masses of diorite occur at Comrie and Garabal Hill, Perthshire.

The microdiorite of Penmaenmawr, North Wales, is an intrusive mass which is exposed in large quarries along the coast between sea-level and 457 m above sea level, and becomes increasingly basic from the top downwards. Quartz and orthoclase are present in the upper part of the intrusion; they decrease in amount at lower levels, and are practically absent in the lowest (visible) part of the mass, while at the same time the plagioclase varies from oligoclase to labradorite and the proportion of mafic minerals increases. Parts of the intrusion would more properly be called porphyrite.

Porphyrite (diorite-porphyry).

Minerals

Similar to diorite.

Texture

Porphyritic crystals of plagioclase and hornblende in a micro-crystalline groundmass consisting mainly of feldspar, with some hornblende or biotite. In some varieties the groundmass contains patches of feldspar and quartz in micrographic intergrowth (p. 154).

Varieties

When mafic minerals other than hornblende occur as porphyritic crystals their name is prefixed to the rock name, giving varieties thus: augite-porphyrite, mica-porphyrite. Minor intrusions of these rocks are found in the Charnwood Forest area, Leicestershire; sills of mica-porphyrite occur on Canisp, Sutherland, and many porphyrite dykes in the Southern Uplands.

Andesite

The andesites form a large family of rocks occurring mainly as lava flows and occasionally as small intrusives. They are compact, sometimes vesicular, and often brown in colour, and in extent are second only to the basalts. It is estimated that basalts and andesites together have fifty times the volume of all other extrusive rocks combined. The name is taken from the Andes of South America, where many volcanoes have emitted lavas and ash of andesitic composition. Many andesites are useful as road-metal when quarried.

Minerals and texture

The essential constituents are plagioclase (generally andesine) and a mafic mineral (hornblende, augite, enstatite, or biotite), which occur as porphyritic crystals in a base which may be glassy, crypto-crystalline, or microlithic (Fig. 4.5). (Microliths are very small, elongated crystals, generally of feldspar; a texture in which large numbers of them are present is said to be microlithic). Such varieties frequently show flow-structure, in which the microliths have a roughly parallel arrangement (cf. the trachytic texture, p. 160). Hornblende may show resorption borders. Grains of iron oxide are nearly always present as accessories, and quartz may be formed in oversaturated types.

Varieties

Hornblende-andesite; augite-andesite; enstatite-andesite; biotite-andesite; quartz-andesite (= dacite). The term pyroxene-andesite is used if both orthorhombic and monoclinic pyroxenes are present; these pyroxene-bearing varieties grade into basalts, and are very abundant.

Andesites which have been altered by hot mineralizing waters of volcanic origin, with the production of secondary minerals, are called propylites.

In Britain, andesite lavas are found in many areas of volcanic rocks, such as the Pentland Hills, Edinburgh; the Glencoe and Ben Nevis districts, and the Lorne volcanic plateau, Argyll; the Cheviot Hills; and south Shropshire. The Borrowdale Volcanic Group of the Lake District (p. 211) is largely composed of andesitic lavas and tuffs, some of which have been subjected to stresses which induced a slaty cleavage and form the green Cumberland slates.

GRANITE AND GRANODIORITE

These two oversaturated plutonic types form most of the very large acid bodies, the batholiths, which occur in the cores of folded mountain ranges; they also form many plutons in the upper levels of the earth's crust (p. 141), and are the most abundant of all the plutonic rocks. Granite is the main structural stone from among the igneous rocks, because of its good appearance, its hardness and resistance to weathering, and its strength in compression; the crushing strength of sound granite ranges from about 135×10^6 to 240×10^6 N m^{-2} (20 000 to 35 000 lbs per sq. inch). Its strength and rough fracture are also valuable properties when it is used as concrete aggregate. The average specific gravity of granite is 2.67, and of granodiorite, 2.72. The trade name 'granite' is used for many rocks which are not granite in the geological sense, e.g. diorite or gneiss (p. 507).

Granite

Minerals

Quartz and feldspar are the essential minerals. The latter includes both orthoclase and plagioclase (albite or oligoclase), and in some rocks microcline is present, as in certain Scottish granites. Quartz may form 25 to 40 per cent of the rock, and feldspar up to 50 per cent or more. Mica of some kind is commonly present and may be a dark variety like biotite, or the light mica muscovite, or both. Other minerals which may be present in different granites (not all in any one rock) are hornblende, augite, and tourmaline; soda-rich minerals such as reibeckite and aegirite appear in alkaline types of granite. Accessory minerals include apatite, zircon, sphene, garnet, and magnetite.

The average composition of granite calculated from many chemical analyses is given on p. 161. The large feldspar content of the rock is reflected in the high percentages for soda and potash, though some of the latter helps to form mica; the low ferromagnesian content is to be noted. The high value for silica results in the formation of free quartz after the metal bases have been fully combined with silica, and the rock is therefore oversaturated. Small constituents such as titanium (for sphere) and phosphorus (for apatite) are included in the last item of the analysis.

Texture

Hand specimens of the rock are, on the whole, light in colour, with a white or pink tint according to the colour of the feldspar, and are coarsely crystalline. Individual mineral grains can be distinguished by eye, flaky micas contrasting with cleaved feldspar, and both with the glassy quartz crystals. The texture may be porphyritic (e.g. the Shap granite, with large pink twinned orthoclase crystals). Sometimes small or large pieces of country-rock have been caught up by the granite during its intrusion and recrystallized by the heat, to appear as dark inclusions or xenoliths.

Under the microscope the component crystals are seen to interlock at their contacts with one another, producing the granitic texture (Fig. 4.5); this texture is not confined to rocks of granitic composition.

Granites, though generally coarse-textured, include less coarse varieties; the fine-grained type called *microgranite* is frequently found as a chilled margin to a larger mass, or as a vein rock. The reibeckite-bearing microgranite from Ailsa Craig is an example; it is the Scottish rock from which curling stones are made.

The *graphic* texture is due to an intergrowth of quartz and feldspar, in which oriented angular pockets of quartz have crystallized within the feldspar (orthoclase or microcline) in parallel positions. The resulting appearance may have a resemblance to Hebrew writing, and for this reason the texture was called 'graphic' (Fig. 4.6); it is also developed on a fine-grained scale (micrographic) in rocks such as *granophyre* (= graphic microgranite).

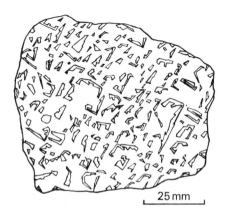

25 mm

Fig. 4.6   Graphic granite, showing angular pockets of quartz in orthoclase.

The proportions of the two minerals in such rocks are nearly constant, about 70 per cent feldspar and 30 per cent quartz; the quartz intergrowths run in regular directions through the feldspar crystals. In a thin section, groups of adjacent quartz areas show simultaneous extinction between crossed nicols, indicating that each group forms part of a single crystal structure which extends through the feldspar. According to one view, this texture is explained as due to simultaneous crystallization of the two minerals, which form a mixture in the eutectic proportions. The *eutectic* mixture of two solids (which do not form solid solutions) is the one which has the lowest freezing point; this principle has considerable applications in connection with metallic alloys. The proportions for an orthoclase-quartz eutectic are, orthoclase 72.5 and quartz 27.5 per cent—nearly the same values as are commonly found in graphic granite.

An alternative view is that there has been a replacement of parts of the feldspar by quartz; at a late stage in the consolidation of the rock, silica-bearing fluids penetrated the feldspar, and potassium and aluminium were exchanged locally for silica. Such replacement could give rise to the 'graphic' appearance. Other quartz replacements are also known.

Varieties

Varieties of granite are named according to the chief mineral present, other than quartz and feldspar, and include: muscovite-granite; biotite-granite (Plate 5, *lower*); muscovite-biotite-granite (or 'two-mica granite'); hornblende-granite; tourmaline-granite (see also p. 208).

The wide extent of granitic rocks has already been indicated. The main granites of the British Isles are shown in Fig. 8.7. In Scotland, as well as in the south, the rock has been quarried extensively for structural stone; in the neighbourhood of Aberdeen a group of quarries yield grey granites, one being the muscovite-biotite-granite (with microcline) from Rubislaw, much used for city buildings and bridges. The pale grey muscovite-granite from Kemnay was employed for the piers of the Forth Bridge. Red biotite-granite from the Peterhead quarries, north of Aberdeen, is a structural stone which has been widely exported.

The porphyritic granite from Shap, Westmorland, has two varieties which differ in the pink or white colour of the groundmass feldspar. The Merrivale quarries near Princetown, Dartmoor, yield a grey slightly porphyritic biotite-granite (Plate 5) which is a good structural stone, also employed as dressed or polished facing slabs (p. 518).

The Cornish granites from the Penryn, Carnsew, and other quarries are mainly muscovite-biotite-granites, grey in colour and sometimes porphyritic. They have provided dimension stone for many dock and harbour works (e.g. Swansea, Belfast, Colombo, Ceylon) and for bridges, buildings, and river walls.

Grey biotite-granites in Ireland are worked at Newry and Castle-wellan, Co. Down, and other localities. The Leinster granite (p. 141) is quarried at Ballyknockan and Ballyedmonduff, Co. Dublin.

Some of the very numerous granite occurrences have been mentioned on p. 140, including the great batholiths of the North American Cordillera. Among other large masses in the U.S.A. the Vermilion batholith (biotite-granite) of Minnesota covers an area of some 4000 km² and has a belt of foliation along its northern margin. And at the northern end of the Appalachian fold-belt, on the eastern seaboard the granites of Maine have been extensively quarried and have many associated pegmatites and aplites.

The conception of a Granite Series is discussed in Note 19 (p. 169).

Granodiorite

In rocks of this type the proportions of orthoclase and plagioclase are more nearly equal, but with plagioclase in excess of orthoclase. There is rather less quartz, on the whole, than in granite; the dark minerals are mainly hornblende and biotite. Granodiorite generally has a somewhat higher mafic content than granite (see diagram, p. 145); the rock has the granitic texture and is similar in appearance to granite. It is transitional between granite and diorite.

A good example of granodiorite (often referred to as granite) is the rock from the Mountsorrel Quarries, near Leicester; these are perhaps the largest producers of road-stone among granite quarries in England. Micromeasurements on thin sections of the Mountsorrel rock give the following mineral composition: Quartz, 22.6; orthoclase, 19.7; plagioclase, 46.8; biotite, 5.8; hornblende and magnetite, 5.1 per cent. The rock has grey and reddish varieties, a medium to coarse texture, and a moderately high crushing-strength.

Another granodiorite is worked at the Hollybush Quarry at the southern end of the Malvern Hills, where it occurs with a number of other igneous types. It has a mineralogy similar to that of the one just quoted.

Many granodiorites (formerly called granites) are found in Scotland, e.g. the Moor of Rannoch intrusion and the Criffel-Dalbeattie mass (see p. 252). The latter is quarried at Dalbeattie and is a grey rock which has been much used in dock construction (e.g. at Liverpool and Swansea), and for the King George V and other bridges at Glasgow.

Apart from the relatively small British occurrences quoted here, granodiorite also occurs in extremely large bodies in fold-mountain belts, including the Coast Range batholith of British Columbia which has a length of outcrop of about 2000 km (1250 miles) and a width of 200 km (Note 17). In a great igneous mass of this extent there is considerable variation in the composition of the rock in different parts of it.

Special names are used for particular varieties of granodiorite, such as *tonalite* (from the Tonali Pass) and *adamellite* (Mt. Adamello, N. Italy).

QUARTZ-PORPHYRY AND ACID LAVAS

Quartz-Porphyry

This rock is the dyke equivalent of granite, and has a similar mineral composition. It has the porphyry texture (Fig. 4.5) with porphyritic quartz and orthoclase in a microcrystalline base composed of quartz and feldspar; small crystals of mica are often present (Plate 5, *middle*). Dykes and sills of quartz-porphyry occur in most granite areas; when porphyritic crystals are absent the rock is called *felsite*. A West of England quarrymen's term for quartz-porphyry is *elvan*.

Acid lavas, which include *rhyolite* and *dacite*, have a restricted occurrence and their amount is small by comparison with basic rocks.

Rhyolite

This rock (named from the Greek, *rheo*, = flow) characteristically shows *flow-structure*, i.e. a banding in the rock developed by relative movement in the layers of the viscous lava as it flowed during extrusion. It may be glassy or cryptocrystalline, and often contains porphyritic crystals of quartz and orthoclase. It frequently shows *spherulitic* structures, which are minute spheres of crystallization formed of quartz and feldspar crystals radiating from a centre. In course of time the original glassy rock may become cryptocrystalline, when it is said to be *devitrified;* any flow-structure originally present is preserved during devitrification.

Obsidian, a black glassy-looking rock which breaks with a conchoidal fracture, is a variety of rhyolite, almost entirely devoid of crystals. Obsidian Cliff in Yellowstone Park, U.S.A., is a classic locality for this type. *Pitchstone* is another glassy lava, generally with a greenish tint and pitch-like lustre, which approximates to rhyolite in composition but may contain up to 10 per cent of water. The pitchstone flow which forms the Sgurr of Eigg in the Hebrides, contains porphyritic quartz and sanidine. Small curved contraction cracks, formed during cooling and known as *perlitic* structure, are sometimes shown by these glassy rocks.

Dacite is the acid lava which approximates to granodiorite in composition, and is distinguished from rhyolite by having porphyritic plagioclase instead of orthoclase. It is found associated with andesites and is distinguished from them by its content of quartz crystals.

Pumice is a highly vesicular 'lava froth,' formed by escaping gases and making the rock so light as to float on water. Pumice may have the composition of rhyolite or it may be of a more basic character. Of recent years, pumice has been used in light concrete slabs for interior partitions, with sound insulating properties.

SYENITE, PORPHYRY, AND TRACHYTE

This group of igneous rocks, of which syenite (named after Syene, Egypt) is the plutonic representative, is treated separately here because its members do not form part of the diorite—granodiorite—granite series already described. Syenites and their related rocks contain a higher proportion of alkalies than is found in rocks of similar silica percentage such as granodiorite; this is demonstrated for syenite by the curves in Fig. 4.7. In terms of minerals the same fact is expressed by the large content of alkali-feldspars which is common in syenites, and sometimes by the presence of feldspathoids. Syenite, with the incoming of quartz, grades on the one hand into quartz-syenite and thence into an alkaline variety of granite; on the other hand, with decrease of orthoclase and increase in the proportion of plagioclase and mafic minerals, syenite grades into an alkaline type of gabbro. (Note 18).

Rocks of the syenite family, taken as a whole, are not very abundant by comparison with the great bulk of the world's granites. Where, however, they are locally well developed they may be quarried and used in engineering construction; examples of this are provided by the syenites and alkaline gabbroic rocks which are extensively worked for road-metal and concrete aggregate in the neighbourhood of Montreal, Canada; and the laurvikites of Southern Norway, which are attractive decorative stones and are exported on a considerable scale.

Types which are undersaturated, such as nepheline-syenite, are numerous and provide many varieties which are of petrological interest, but their description is beyond the scope of this book and only the chief saturated types will be dealt with.

Porphyry (or feldspar-porphyry, i.e. without quartz) is the dyke equivalent, and trachyte the extrusive equivalent, of syenite.

Syenite

Minerals

Orthoclase (or some other variety of alkali-feldspar) is the chief constituent, forming well over half the rock, with a smaller amount of plagioclase (oligoclase); mafic minerals are frequently hornblende

Plate 5

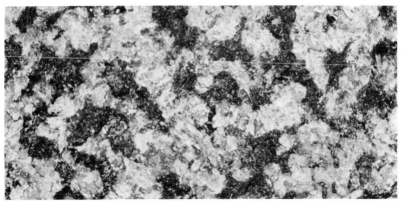

(a) Gabbro of ring-dyke, Ardnamurchan, Scotland (x 1).

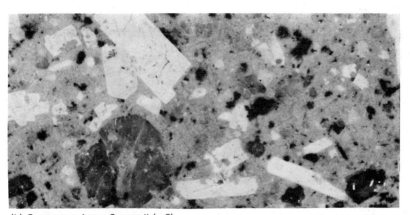

(b) Quartz-porphyry, Cornwall (x 3).

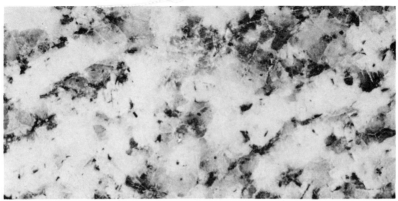

(c) Biotite-granite, Merrivale, Dartmoor (x 1).

Plate 6

(a) Biotite-gneiss, Sutherland, showing foliation and augen-structure.

(b) Boudins in sandstone and slate, in cliff face near Trebarwith Strand,
North Cornwall. (*Courtesy of the Institute of Geological Sciences, London.
Crown copyright reserved*)

and biotite, sometimes augite; accessories include iron oxides, apatite, sphene, zircon. A small amount of quartz may be present, filling interstices between the other minerals. Feldspathoids occur in undersaturated syenites, when their name is used as an adjective to the rock name, e.g. nepheline-syenite.

Texture

Coarse-grained holocrystalline, sometimes porphyritic. Hand specimens appear rather lighter in colour than most diorites, but this depends on the proportion of the dark minerals in the rock. In thin section the texture is granitic, with interlocking crystal plates.

Varieties

These are named by prefixing the name of the chief mafic constituent, and include: hornblende-syenite; augite-syenite; biotite-syenite; also quartz-syenite, when quartz is present in small amounts and insufficient for the rock to be called a granite.

The type hornblende-syenite comes from Plauen, Saxony, and is a rock with a light purplish tint. There are few occurrences of syenite in Britain, but great developments of these rocks occur in Canada and Norway, as already mentioned, and in the Kola peninsula, Russia, where they are associated with important ore bodies.

The soda-rich syenite (with average $Na_2O = 6$ per cent) called *laurvikite*, from Laurvik, Norway, is important because of its use as a decorative stone for the facing of buildings. It consists mainly of the feldspar anorthoclase with some titaniferous augite, mica, and iron oxides. The anorthoclase shows beautiful blue and green schiller effects which give the rock its value for decorative purposes; it also takes a good polish. Various differently coloured laurvikites are marketed and are known to the trade by names such as 'light pearl,' 'blue granite.'

Porphyry

Porphyry, sometimes called syenite-porphyry, is the dyke equivalent of syenite, and contains porphyritic crystals of orthoclase in a microcrystalline base made of feldspar with hornblende or biotite. *Rhomb-porphyry* is the dyke rock corresponding to laurvikite; boulders of it have been found in East Anglia and owe their transport from Norway to the agency of ice.

Trachyte

Minerals and texture

Trachyte, the volcanic equivalent of syenite, is typically a pale-coloured, rough-looking lava (Greek: *trachys*, rough). It has porphyritic crystals of orthoclase in a groundmass composed mainly of feldspar microliths (orthoclase and plagioclase), with some biotite or hornblende as a mafic constituent. A glassy (non-crystalline) base is sometimes present.

The *trachytic texture* has a characteristic appearance under the microscope, the microliths of the groundmass having a sub-parallel arrangement, and showing lines of flow around porphyritic crystals.

Sanidine-trachyte contains porphyritic crystals of sanidine, a glassy variety of orthoclase formed at a high temperature.

Varieties

Trachytes are found at a number of localities in the Central Valley of Scotland, e.g. in the Braid, Garleton, and Eildon Hills. A feldspathoidal trachyte called *phonolite* (Gk. *phone*, a sound) in which nepheline takes the place of some of the feldspar, forms the hill of Traprain Law, Haddington; the Wolf Rock off the coast of Cornwall is composed of nosean-phonolite. The Eifel district of Germany provides many leucite-bearing trachytic lavas and tuffs, and the Auvergne district of Central France is noted for its trachyte domes, which are extrusions of viscous lava that solidified around the vents as dome-like masses.

Pegmatites and aplites   *Pegmatites* are very coarse-grained rocks which occur as dykes and veins in the outer parts of an intrusive mass and in the surrounding country-rocks; they are composed of minerals similar to those of the parent igneous body, and represent residual portions of the magma. Thus, granite-pegmatites, containing quartz, microcline, and mica, are common in many granite areas. The mica (e.g. muscovite or phlogopite) used in industry is obtained from pegmatites, where individual crystals may be many centimetres in diameter, yielding large plates of the mineral; the United States, Canada, and India produce mica from such sources. Some minerals grow to very large sizes in pegmatite veins: crystals 12.8 m long of the rare pyroxene spodumene, which contains lithium, have been measured in a granite-pegmatite in Dakota. Syenite-pegmatites in Norway contain large crystals of apatite and rare earth minerals.

These coarsely crystalline growths are the products of residual magmatic fluids which are rich in volatile constituents, and are late injections in the cooling history of an igneous mass. The volatiles, which are largely aqueous, act as fluxes and lower the crystallization temperatures of minerals, which can continue to grow as big crystals in the mobile medium. Rare constituents (such as lithium, tungsten, cerium, thorium) may become concentrated in these residual fluids, and the resulting pegmatites are sometimes worked as ores from which the metals are extracted.

Aplites are fine-grained, 'sugary' textured rocks, found as small dykes and veins in and around granites and other intrusives. They are composed chiefly of light (felsic) minerals, such as quartz and feldspar, with few or no dark minerals. Their fine texture points to a derivation from more viscous and less aqueous fluids than in the case of pegmatites; but they are commonly associated with the latter, and streaks of aplite often occur within a pegmatite vein. Aplites also contain fewer of the rare elements that are often,found in pegmatites.

Igneous rock series   A series of rocks such as the granite—grano-diorite—diorite suite, in which there are gradations between the different members, can be illustrated by means of a *variation diagram*. This may be constructed in several ways from the chemical analyses of the rocks; in Fig. 4.7 the percentages of the different constituents are plotted as ordinates against silica percentage, giving a series of curves. The values used in the figure are taken from the following table of average compositions of plutonic rocks.

	Granite %	Grano-diorite %	Diorite %	Gabbro %	Syenite %
SiO_2	70.2	65.0	56.8	48.2	60.2
Al_2O_3	14.5	15.9	16.7	17.9	16.3
Fe_2O_3	1.6	1.7	3.2	3.2	2.7
FeO	1.8	2.7	4.4	6.0	3.3
MgO	0.9	1.9	4.2	7.5	2.5
CaO	2.0	4.4	6.7	11.0	4.3
Na_2O	3.5	3.7	3.4	2.6	4.0
K_2O	4.1	2.8	2.1	0.9	4.5
H_2O	0.8	1.0	1.4	1.5	1.2
rest	0.7	0.8	1.2	1.4	1.1

It will be seen that the points representing syenite do not fall on the smooth curves given by the other four rock types, indicating that syenite does not belong to the same magma series as the rest. The curves also show the increases in magnesia, lime, and iron towards the basic end of the series. Variation diagrams can be drawn for other rock series and comparisons made.

Origin of igneous rocks   While there are many different sorts of igneous rock, it is thought that there are only a few, perhaps two, kinds of primary magma from which they have been derived. The science of Petrology is concerned with the explanation of the origin of magmas and the derivation of different rocks from them. Some of the processes which are believed to have contributed to the formation of igneous rocks are briefly outlined below.

The splitting of a magma into fractions of contrasted composition is called *differentiation*. It may be brought about, for instance, by the separation of crystals from a melt during crystallization if the crystals sink and accumulate at a lower level. The early formed crystals in a

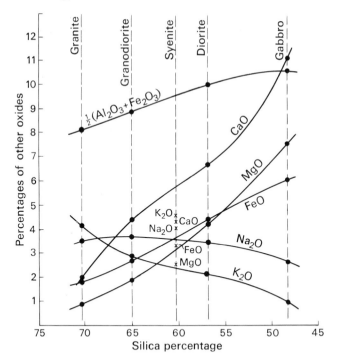

Fig. 4.7   Variation diagram for plutonic igneous rocks.

basic magma, for example, are heavier than the fluid in which they grow. Olivine as it crystallizes would sink because of its higher relative density, and also pyroxene. Many instances of intrusive masses having an olivine-rich layer near their base are on record, as in the Palisades sill (p. 138), and the Lugar sill (p. 150). As cooling continues, with the separation of pyroxene and olivine, the specific gravity of the remaining liquid is continuously being reduced; but if viscosity increases, the separation may be less complete than before. The crystallization of mafic minerals, which use up Mg, Fe, Ca, and some Si and Al, results in the remaining fluid becoming relatively richer in silica, alkalies, water (steam) and other volatiles, and also in rare constituents originally present in the magma as traces. Thus at successive levels in the cooling mass, if crystallization is undisturbed, there might be formed zones containing olivine, olivine and pyroxene, pyroxene and plagioclase, and so on. These layers would consolidate as peridotite, olivine-gabbro, and gabbro, grading into one another.

The residual fluid, now much smaller in quantity, has a composition completely different from that of the original magma, because of the abstraction of mafic material and some silica by crystal growth; were it to be separated from the crystal phase at this stage and injected (by the operation of crustal stresses), it would consolidate possibly as diorite. Supposing it to remain in place and the process of differentiation by crystallization to continue, with the separation of further crystals such as feldspar and hornblende, the final silica-rich residue, relatively small in amount, would consolidate as a quartz-feldspar rock (granite or microgranite). Thus a series of igneous rocks could be produced from one basic magma; and the intrusion of magma might take place at any stage during differentiation, if earth-movements intervened. It should be noted, however, that any granite so produced must be in relatively small quantities, which would be quite insufficient to account for the vast bulk of granitic rocks which exist in every continent. But the theory helps to explain the many basic and ultrabasic rocks, and some small occurrences of granite.

Other theories of the origin of granite have been put forward. Some geologists have postulated the existence of primary granitic magma. Others have suggested processes such as the melting at depth of sedimentary and other rocks of high silica content. During orogenies (p. 248), compression has produced intense folding in certain belts of the earth's crust, and the lower parts of the folds must have been pushed down to depths where temperatures were high enough to melt the rocks. Thus magma would be generated by the melting, and it would be of acidic composition because of the mainly siliceous nature of the rocks involved. In this way a likely origin for granite magma in bulk is suggested. Its formation is associated with mountain-building, and it is a fact that granite batholiths are found in the cores of the fold-belts of the earth (p. 140).

Another process which could result in the formation of granitic rocks is called granitization, which involves the permeation (or soaking) of rocks of suitable composition within the crust by igneous fluids, particularly by alkaline silicates. The original rock is made over, as it stands, into a rock of granitic composition and appearance. The nature of the 'igneous' mass thus formed is therefore dependent not only on the character of the permeating fluids, but also on the composition of the pre-existing rocks which have been granitized: some, such as shales and sandstones, are more readily transformed than others. The hot permeating fluids might also transform solid rocks at depth into a mobile mass, which could then move upwards and become intrusive at higher levels in the crust.

There may thus be more than one way in which granite has been formed, and granites at different levels in the earth's crust have probably been emplaced by different mechanisms. These ideas lead to the conception of a 'granite series'. (Note 19).

During the time that a hot igneous mass is in contact with the rocks into which it is injected, reactions take place between the magma and its walls (see also p. 207). Some of the country rock may become incorporated in the magma, a process known as *assimilation*; and magmatic juices may permeate the zone of country rock next to the contact, thereby producing a variety of new rocks immediately around the igneous body. The assimilation of pre-existing rock by an intrusive magma, if the mixing of the two is complete, will yield a resulting rock different from that which would have been produced by the uncontaminated magma. Incorporation of shale, for example, would render an intrusive granite less acid, i.e. the granite might be changed locally to granodiorite or even to diorite. Field observations have shown that some granite masses have margins of dioritic composition which merge gradually into granite at some distance inwards from the contact; this is attributable to assimilation. Assimilation of country-rock by basic magma may also take place. The partial incorporation of sediments by an igneous body is often observed near a contact, where the process has been frozen solid by the cooling of the mass. In general, therefore, assimilation tends to produce contaminated rocks, different from the normal type which the magma alone would have evolved.

Ores of igneous origin   *Mineral deposits* are local accumulations or concentrations of useful minerals; many of these deposits are *ores*, i.e. minerals or rocks from which a *metal* or metals may be profitably extracted. As well as ores, mineral deposits include coals, oil, and non-metalliferous minerals such as barytes, gypsum, and sulphur.

Metalliferous deposits associated with igneous rocks are here discussed briefly, and can be grouped as follows:

(i) those formed by direct segregation from magma.

(ii) those formed as veins or lodes, whose material is largely derived from magma. These may be sub-divided into (*a*) *pneumatolytic deposits*, due to the action of gaseous emanations at a high temperature (500°C or over), including water in the gaseous state; and (*b*) *hydrothermal deposits*, due to the operation of hot aqueous fluids, at temperatures from about 500°C downwards. It is difficult to draw a sharp distinction between the higher temperature hydrothermal minerals and those of pneumatolytic origin.

(iii) certain deposits are sometimes formed in the aureole of contact metamorphism which surrounds an intrusive mass such as granite; these *pyrometasomatic deposits*, as they can be called, result from the replacement of the wall-rocks by material from the magma, limestone in particular being often replaced in this way (Fig. 4.8). They are not further discussed here. As examples, the copper deposits of Clifton, Arizona, and the magnetite deposits of Iron Springs, Utah, may be cited.

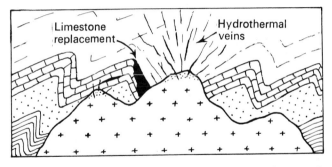

Fig. 4.8   Veins filling fractures near a granite intrusion (crosses), shown in section.

(i) *Magmatic segregations.*   During the solidification of a magma, early formed heavy minerals may sink and become concentrated at or near the base of the mass. Such minerals include magnetite (Fe_3O_4) and chromite ($FeCr_2O_4$). The large magnetite deposits of Kiruna, Sweden, which have yielded millions of tons of high grade ore, were formed by the concentration of the mineral in this way, after crystallization from a magma which consolidated mainly as syenite. Deposits of chromite occur, for example, as segregations at the base of the Bushveld norite of the Transvaal, South Africa. The metals chromium, nickel, and platinum are generally associated with basic rocks.

(ii) *Veins and lodes* consist of the infillings of fissures and fractures developed in the outer part of an intrusive body or in the surrounding roof and wall-rocks. Veins which contain metalliferous minerals are termed lodes. Magmas of acid composition (such as granite) are frequently rich in volatile constituents, and may also contain small quantities of many metals. As crystallization proceeds, with the formation of minerals like feldspar and quartz which form the bulk of the resulting rock, the metals—which were originally disseminated throughout the magma and were not incorporated in the feldspar and other crystals—become concentrated in the residual fluids. These also contain the volatile constituents, and are thus able to remain fluid to lower temperatures than would be possible without the fluxing effect of the volatiles. If, then, fissures are formed in the outer (first solidified) part of the granitic mass and its surrounding rocks, by the operation of stresses, they become channels into which the residual fluids migrate, there to crystallize as lodes and veins. (See also pegmatites, p. 160.)

Metals which are commonly associated in this way with acid rocks include copper, lead, zinc, arsenic, tin, tungsten, gold, and silver. The volatile constituents, consisting chiefly of water vapour but often also including fluorine, chlorine, boron, and other gases, act as carriers for the metals into the area of mineralization around the intrusion.

As they pass outwards into zones where lower temperatures and suitable pressures prevail, they deposit e.g. tin as cassiterite, SnO_2; tungsten as wolfram $(Fe, Mn)WO_4$; and copper as chalcopyrite, $CuFeS_2$. Thus tin and tungsten lodes may be formed in and around granite masses, as in Devon and Cornwall; they may be regarded as of pneumatolytic origin. Minerals such as quartz, pyrite, topaz, and tourmaline are commonly associated with the metalliferous minerals in such lodes. Iron, lead, and zinc may also be carried outwards and deposited as haematite, galena, and blende in joints and fractures in cooler rocks at somewhat greater distances from the igneous source.

Veins of this type grade into the group of *hydrothermal deposits*, which are formed largely through the agency of hot aqueous solutions. These fall into three sub-groups, according to the temperature conditions under which they were formed: (*a*) those deposited between about 500° and 300°C are the high-temperature hydrothermal deposits; (*b*) those at 300° to 200°C are the intermediate deposits; and (*c*) those from 200° down to about 50°C the low-temperature deposits.

(*a*) Minerals generally formed under high-temperature hydrothermal conditions are mainly the sulphides of iron, copper, lead, and zinc (see p. 127). They occur in lodes, as described above, in association with non-metalliferous minerals such as quartz, fluorite, calcite, or dolomite. The non-metalliferous minerals are referred to as the *gangue*, so that a lode consists of gangue and ore.

(*b*) Among the deposits formed at intermediate temperatures are some lead and zinc veins, certain gold-bearing quartz veins, and some copper and pyrite deposits. Lead and zinc commonly occur as the sulphides, galena and blende, but sometimes as compounds with arsenic and antimony, often in association with pyrite, and quartz, calcite, fluorite, or barytes. In the formation of such veins, mineral-forming materials are carried by the hydrothermal solutions, which may penetrate country-rocks along any available channels for considerable distances from the igneous mass, at a late stage in its cooling history. As the solutions enter cooler regions the minerals are deposited in fissures and cavities. Some replacement of soluble rocks like limestone may be effected (Fig. 4.8), as in the Carboniferous Limestone of Derbyshire and Cumberland. Many veins containing galena, blende, and calcite are found in these limestones; other British occurrences are in Cornwall, Durham, and Cardiganshire.

(*c*) Deposits formed under low-temperature hydrothermal conditions occur at shallow depths and are often associated with andesites. They include certain gold occurrences, in which the gold (together with some silver) is combined as a telluride. Gold telluride ores are worked at Kalgoorlie, Australia, and were formerly worked at Cripple Creek, Colorado. Mercury, combined as cinnabar (HgS), and antimony, as stibnite (Sb_2S_3), are two other examples of low-temperature hydrothermal minerals. They are sometimes associated in veins with minerals of the zeolite group, indicating a temperature of formation not greater than 200°C.

Secondary enrichment. When ore-deposits undergo weathering and decomposition at their outcrops, secondary alterations take place which are often of economic importance. The weathered upper part of the deposit is known as gossan. Sulphide ores, in particular, are affected in this way. Above the water-table (*q.v.*) there is a downward movement of water percolating from the surface, which carries solutions from the weathered rocks. Below the water-table cementation and reaction take place, leading to the concentration there of metals derived from higher levels, and the process is termed secondary enrichment. Soluble sulphates thus become precipitated as sulphides near the water-table. Copper provides important examples: the sulphides covellite (CuS) and chalcocite (Cu_2S) are formed in the way described from the primary copper minerals of the ore, giving rise to local enrichment of the deposit.

For greater detail than is given in the above outline reference should be made to works on Mineral Deposits.

NOTES AND REFERENCES

1. Bodvarssen, G. and Walker, G. P. L. 1964. Crustal Drift in Iceland, *Geophys. Jour. Roy. Astr. Soc.*, **8**, 285.
2. Materials ejected from a volcanic vent may be classified according to their size as follows:

    volcanic bombs   .   .   .   greater than 32 mm
    lapilli .   .   .   .   .   between 32 and 4 mm
    coarse ash .   .   .   .   from 4 to 0.5 mm
    fine ash   .   .   .   .   from 0.5 to 0.05 mm
    dust   .   .   .   .   .   less than 0.05 mm

    Tuffs are described as coarse or fine according to the size of the fragments they contain. The fragments may be *lithic* (i.e. rock fragments), *crystal*, or *vitric* (glassy, from exploded magma).
3. The research carried out by the Environment and Space Sciences Administration, Boulder, Colorado, 1969, on Old Faithful geyser is described in *Journ. Geophys. Research*, **74** 566.
4. Solid substances which form incrustations around fumaroles may in some cases be *sublimates*, i.e. deposited from gaseous emanations in a cooler environment. It has been suggested that in such a situation, siliceous sinter could be formed by decomposition of silicon tetrafluoride (a gas).
5. See The Breiddalur Volcano, by Walker, G. P. L. 1963, *Q. Jour. Geol. Soc.*, **119**, 50.

6. For a general account of the geology of the area see, for example, The Geology of South Africa, by du Toit, A. L. 3rd. Edition, 1954. (Oliver & Boyd, Edinburgh).

7. The occurrences in Ardnamurchan, and the ring-dykes and other features of Mull, are described in *British Regional Geology*, Scotland: the Tertiary Volcanic Districts, by Richey, J. E. 3rd. edition, 1961.

    Cone-sheets are also discussed by Durrance, E. M. 1967, in *Proc. Geol. Assoc.*, **78**, 289.

8. Described by Stillman, C. J. 1970, in Structure and evolution of the Northern Ring Complex, Nuanetsi Igneous Province, Rhodesia. *Mechanism of Igneous Intrusion*, (*Ed.* Newall and Rast), *Geol. Journ.* Special Issue No. 2, p. 33.

9. It is believed that the formation of the ring structures was related to vertical (epeirogenic) uplift in an unfolded continental area, where large pressures operated, the pressures originating in the sub-crystal region. The rocks and processes are described by Jacobson, R. R. E., Macleod, W. N. and Black, R. 1958, in Ring-complexes in the Younger Granite Province of Northern Nigeria, *Geol. Soc. London*, Memoir No. 1.

10. The term 'stope' has long been used in mining for the opening up of a vertical or inclined shaft by the piecemeal removal of the rock. Stoping as applied to magma was discussed by Billings, M. P. 1925, in a paper On the mechanics of dyke emplacement, *Journ. Geol.* **33**, 140; and by others subsequently.

    In a recent review, Pitcher, W. S. discusses stoping and other aspects of granite emplacement, and quotes examples from the Coastal Batholith of Peru. (In Mechanism of Igneous Intrusion, 1970, *Ed.* Newall and Rast; Liverpool Univ. Symp., published as *Geological Journal* (Liverpool) Special Issue No. 2).

11. Details are given in The Geology of the northern end of the Leinster Granite, by Brindley, J. C.; *Proc. Roy. Irish Acad.* 56B, 159, and 58B, 23, 1954–56.

12. See The Main Donegal Granite, by Pitcher, W. S., Read, H. H. *et al.* 1959, *Quart. Jour. Geol. Soc.*, **114**, 259–300; and references in this paper.

13. The borehole at Rookhope, in the northern Pennines 24 miles west-south-west of Newcastle-upon-Tyne, used continuous coring and passed through 1259 feet (381 m) of Carboniferous sediments with intrusive dolerites, before entering granite at a depth of 1281 feet (389 m). The dolerites penetrated were the Great Whin Sill, 193 feet (58 m) thick, and at a higher level the Little Whin Sill (1.8 m). Boring into the granite, named the Weardale Granite, continued to a depth of 2650 feet (804 m), The granite, of which the top 5 feet (1.5 m) was weathered, carries muscovite and oriented biotites, and contains thin layers of white rock (aplite) and coarse layers (pegmatite). (Data taken from the paper by Dunham, K. C. *et al.*, 1965, in *Q. Jour. Geol. Soc.*, London, **121**, 383.)

14. The igneous rocks of the Lizard, including gabbros, are discussed in A re-study and re-interpretation of the geology of the Lizard peninsula,

Cornwall, by Green, D. H. 1964, *Roy. Geol. Soc. Cornwall*, Present views of some aspects of the geology of Cornwall and Devon, p. 87.

In the above paper the Lizard serpentinite (or preferably, peridotite) is considered to be a plug-like body with a south-easterly extension which was intruded as a sill into flat-lying metamorphic rocks, seen in the coastal exposures on the east side of the Lizard peninsula.

15. The complex is described by Wager, L. R. and Deer, W. A. 1939, in Geological investigations in east Greenland: Part III, the Petrology of the Skaergaard intrusion, Kangerdlugssuaq. *Meddel. om Grφnland*, **105**, no. 4.

16. Along the southern margin of the Insch Gabbro, north-west of Aberdeen, several bodies of serpentinite are emplaced in a zone of strong dislocation, and are described as 'tectonic pods' by Read, H. H. 1956, *Proc. Geol. Assoc.* **67**, 73. That is, they have been brought into their present position, from a deep-seated source, by tectonic movements.

A similar situation exists in connection with the body of serpentinized peridotite which extends for over 4 miles (6.4 km) along the faulted southern boundary of the Cabrach igneous area, west of the Insch gabbro. The emplacement of this ultrabasic mass has been ascribed to vertical movements in a tectonic zone (Blyth, F. G. H. 1969, *Proc. Geol. Assoc.* **80**, 63).

17. In The Coast Range Batholith near Vancouver, British Columbia, by Phemister, T. C. 1945, a discussion is given of a part of this gigantic batholith. (*Q. J. Geol. Soc.*, **101**, 37).

18. Such a series of rocks is called an *alkaline series*, as distinct from *calc-alkaline* rocks, which are typically represented by granite, granodiorite, and diorite.

19. From a study of granites situated in orogenic belts it has been suggested that magma which was generated at depth early in an orogeny (e.g. by melting of sediments in the roots of folds) may, in part, be moved to higher levels in the crust at a later stage, to solidify there as intrusive bodies. Thus there are deep-seated granites that are formed *early*, and intrusive granites which are *late*, in the history of an orogenic belt. The later granites are found as high-level plutons; and between them and the early granites others may have been emplaced in intermediate positions. This idea of a *granite series* is due to H. H. Read; it can be represented diagrammatically as follows:

↑ Last in time:	High-level granites	Mechanically emplaced, permitted intrusions, e.g. cauldron subsidence
Later in time:	Intrusive granites at higher level	Forceful intrusion of magma; metamorphism of surrounding rocks
Early in time:	Deep-seated granites, batholiths	Formed *in situ* by permeation; melting of folded sediments at depth.

See also the discussion in Read and Watson (reference below); and Read, H. H. 1949, A contemplation of Time in Plutonism (Pres. Address), *Q. Jour. Geol. Soc.* (Lond.), **105,** in which an example of the Granite Series is given from the Hercynian fold-belt, extending from the Auvergne of France to Brittany and south-west England.

GENERAL REFERENCES

HATCH, F. H., WELLS, A. K. AND WELLS, M. K. 1961. Petrology of the Igneous Rocks, 12th Edition, Allen and Unwin, London.

HYNDMAN, D. W. 1972. Petrology of Igneous and Metamorphic Rocks. McGraw-Hill, New York.

READ, H. H. AND WATSON, J. Introduction to Geology (Chapters 7, 10 and 11); 2nd edition, reprinted 1971 in Macmillan paperbacks.

STANTON, R. L. 1972. Ore Petrology. McGraw-Hill, New York.

WILLIAMS, H., TURNER, F. J. AND GILBERT, C. M. 1954. Petrography. Freeman, San Francisco.

5 Sedimentary Rocks

Sediments form a relatively thin surface layer of the earth's crust, covering the underlying igneous and metamorphic rocks. This sedimentary cover is discontinuous and of varying thickness; it averages about 0.8 km in thickness but locally reaches over 12 km in long narrow belts, the sites of former geosynclines (p. 173). It has been estimated that sediments constitute only about 5 per cent of the crustal rocks (to a depth of 16 km), in which the proportions of the three main types are approximately: shales and clays, 4 per cent; sandstones, 0.75 per cent; limestones, 0.25 per cent. Sedimentary rocks also include varieties which are composed of the remains of organisms, such as certain limestones and coals, and others which are formed by chemical deposition.

The rocks with which this chapter is concerned were, when formed, not in the condition in which they are found today. Accumulations of loose sand, for example, derived from the breakdown of older rocks in ways described earlier, and brought together and sorted by water and wind, have become hardened rocks such as sandstone and quartzite. Pore spaces in the original sands have been partly or completely filled with mineral matter brought by percolating water and deposited as coatings on the sand grains, thus acting as a cement to bind them together. These processes are known as *cementation*. In muddy sediments, the very small particles of silt and clay of which they are mainly composed have been pressed together by the weight of sediment; interstitial water has been squeezed out and in course of time the mud has become a coherent mass of clay, shale, or mudstone. *Compaction* of this kind affects the muddy sediments to a greater degree than the sands, and during the compaction process much of the pore-contained water in an original mud is pressed out (Note 1). Some of the water, with its dissolved salts, may remain in the sediment after its compaction, and is known as *connate water* (connate = 'born with'). The general term *diagenesis* is used to denote the compaction of a sediment into a sedimentary rock, and includes the processes outlined above and also chemical processes such as re-crystallization and replacement.

When rocks come again into the zone of weathering, after a long history, soluble substances are removed and insoluble matter is released, to begin a new cycle of sedimentation in rivers and the sea. The broad groupings used in the Table of Sedimentary Rocks (p. 176) are:

I. *Detrital sediments* (mechanically sorted), e.g. gravels, sandstones, clays and shales.

II. *Chemical, and biochemical (organic)*, e.g. limestones, coals, evaporites.

Environments of deposition The composition, texture, and sedimentary structures of any sediment are dependent on the processes which operated during its formation, and on the length of time over which they acted. These factors are governed, in turn, by the environment in which sedimentation proceeds. In turbulent water, for example, sedimentary particles are well sorted and rounded, especially if the processes go on for a long time. This is seen in beach deposits, formed in a stable environment. When, on the other hand, the duration of the formative processes is shortened by the onset of crustal movements such as uplift or depression, or from other causes, the weathering of the sediments is incomplete, the transport is shorter, and the sedimentary particles are deposited more rapidly. There is a contrast, therefore, between *stable* and *unstable* environments, which is expressed in the characters of the sediments accumulated.

(i) *Stable environments.* On the continental shelf, in the marginal part of a sea, deposits of pebbles and sand (of various grades) are formed, together with muds and limestones. The rough water caused by wave action along a coast results in the rounding of rock particles (to give pebbles), and mineral particles (sand) in which quartz is the main constituent. Minerals which are less hard than quartz do not persist for so long; nor do minerals with cleavages, such as feldspar. Still finer particles (silt) also come to consist mainly of quartz. Current-bedding and ripple-marks are common and may be preserved in the resulting sandy and silty rocks. Thin shale beds are formed from muds deposited in the somewhat deeper and less turbulent zone at a distance from a shore, at depths below the limit of wave-action (p. 53). Limestone-forming materials derived from calcareous skeletal remains (e.g. algae, crinoids, shell debris) are commonly associated with such sands and muds. A sequence of fossiliferous sedimentary layers is thus formed, in a marine area, and they may lie uncomformably on older rocks in which the platform of deposition was eroded (p. 248). Because of its content of fossil shells and shell fragments, such a sedimentary series is often referred to as a *shelly facies*, or *shelf facies*. It is frequently of relatively small thickness (hundreds of metres rather than thousands); but if gentle subsidence has affected the area of deposition during the formation of the deposits, a much greater thickness of these shallow-water sediments may have accumulated. Under these circumstances successive pebbly layers of the fore-shore, for example, have been put down farther and farther inland from the original coast, as in time the water slowly deepened over the area of sedimentation while the fore-shore zone migrated inland (Fig. 2.19).

Other relatively stable environments of deposition include the land areas on the continental margins where *desert*, *piedmont*, and *lacustrine* deposits are accumulated. Desert deposits such as eolian sandstone and loess have been discussed earlier (p. 33). Piedmont deposits, which are formed during rapid weathering of mountains at the end of an orogenic upheaval and lie at the foot of steep slopes which are undergoing denudation, may include *arkoses* (feldspathic sandstones, *q.v.*). In lakes of still water lacustrine clays are slowly deposited; and where water is impounded in glacial lakes (p. 70), seasonal melting of ice leads to the formation of *varved clays*, with alternations of coarser (silty) and finer (muddy) layers.

(ii) *Unstable environments.* The elongated basins of sedimentation called *geosynclines* (Fig. 5.1) are at first filled with sediment during

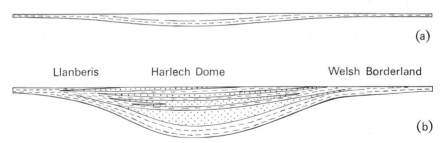

(a)

Llanberis   Harlech Dome   Welsh Borderland

(b)

Fig. 5.1   Sections through a geosyncline. (*a*) the trough in an early stage; (*b*) the Welsh geosyncline, filled with thick Cambrian deposits (see p. 251) during deepening; dots show sediment probably brought by currents along axis of trough (after J. L. Knill).

slow subsidence extending over a long period of time. In the later part of their development the geosynclines and their adjacent land areas undergo great disturbances. Much coarse detritus and other sediment is then poured into the deepening trough, filling it with elongated lenses of deposit derived from nearby land areas by rapid weathering, often with contributions from volcanic eruptions. The sediment is accumulated rapidly, without having undergone lengthy processes of sorting and rounding. Slumping affects the finer muddy sediment built up on the steepening submarine slopes, and mud-flows run their course to lower levels. Thicknesses of many thousands of metres of sediment are piled up in the over-deep trough.

Typical *greywackes* are formed under these conditions; they are badly sorted muddy sediments with much coarse clastic material. With them are associated breccias, and lenses of poorly graded conglomerates in which partly rounded pebbles are set in a matrix of angular mineral grains, the minerals being of more than one variety (in contrast to the predominant quartz of the pebbly beach deposits referred to earlier). The textural features shown by greywackes and associated rocks (p. 183) indicate rapid accumulation with little transport and sorting; shales which accompany them often show slump structures; limestones if present are generally thin and represent temporary conditions of greater stability which prevailed for a time, especially when the trough of sediment was nearly filled.

When subsequently all these sediments have undergone compression, their folded and faulted condition at the present day reveals something of the movements which affected them as they became elevated to form fold-mountains on the site of the former geosyncline (p. 252).

(iii) *Deep sea.* The muds and oozes of the deep sea, some of which are being formed to-day, are less easily investigated than the more accessible sediments, but are interesting because of their wide extent and distinctive characters: they usually contain no large fragments, no remains of shallow-water animals, and no features due to current or wave action. They are spread in horizontal layers over great distances and connected with shallower water sediments by a gradual transition. Our knowledge of these deposits in the past was due to the *Challenger* Expedition of 1872–6, which obtained many hundreds of samples dredged from the ocean floors of the world. More recently, new methods of core-sampling in the ocean floors have been developed, and the results of the second *Challenger* expedition (1952) and many later explorations are yielding much new information, as indicated on p. 2.

The land-derived *muds* which lie on the continental slopes, and continue down to the deep seas, are composed of very finely divided material and are blue, red, or green in colour. The blue muds extend to depths of about 2.75 km (1500 fathoms). At still greater depths are found *oozes* made of the skeletons of small floating organisms, such as foraminifera and coccoliths, which sink and accumulate on the bottom at a slow rate. The *Globigerina* ooze is a calcareous deposit, composed largely of minute globular shells (tests) of the foraminifer *Globigerina;* it covers some 130 million km^2 (50 m. sq. miles) in the Atlantic and Pacific.

Siliceous organisms persist at greater depths than those reached by calcareous particles, which are slowly dissolved as they sink through deep water. The *Radiolaria* form a siliceous ooze which covers an area of some 7 million km^2, chiefly in the South Pacific.

From about 4000 m (2200 fathoms) down to the greatest known depths, the deposit named the *Red Clay* covers great areas of submarine surface. This soft, plastic material is derived from fine volcanic dust which has fallen from the atmosphere and slowly settled through great depths of water. Hard parts of fishes, such as sharks' teeth and bones of whales, are found embedded in samples of the Red Clay which have been dredged up. Spherules of meteoric iron were found in the deposit at depths of 5.5 km (3000 fathoms) in the North Atlantic; and many manganese nodules ('sea potatoes') lie in places on the ocean floors (Note 2).

Composition of sea water   The chemical properties of sea water are due to matter in solution, brought by rivers and from volcanic eruptions, and to the presence of sediments and marine organisms. Modern studies have been made of the concentration of many constituents, and the main ions from dissolved salts are listed below.

Concentration of the Main Components
of Sea Water for a Salinity of 35%
(after Turekian, see Note 2)

Component	Grammes per kg
Chloride	19.35
Sodium	10.76
Sulphate	2.71
Magnesium	1.29
Calcium	0.41
Potassium	0.39
Bicarbonate	0.14

Many other metal and non-metal ions are present in sea water, including Br, Sr, B, Si, Rb, F, P, I, Cu, and Ba (in decreasing order of abundance).

The density of sea water varies between 1.025 and 1.028 depending on the salinity and temperature.

THE SEDIMENTARY ROCKS

I. DETRITAL SEDIMENTS, mechanically sorted

This group is subdivided according to the size of the particles into (*a*) *pebbly* (or psephitic), (*b*) *sandy* (or psammitic), and (*c*) *muddy* (or pelitic) sediments, as in Table 5.1. Convenient limits are adopted for the different particle sizes, or *grades;* two scales are given below, the Wentworth scale which is in common use among geologists, and the

Table 5.1   Classification

Dominant Character	Raw Materials	Rocks (Consolidated)
	I. DETRITAL (terrigenous, mechanically sorted)	
Pebbly (psephitic)	Pebbles, gravel (over 2 mm) Scree (talus) Boulder clay	Conglomerate Breccia Tillite
Sandy (psammitic)	Sand ⎰very coarse, coarse, medium, fine, very fine⎱ between 2 mm and 0.06 mm Silt (0.06 to 0.002 mm)	Greywacke, Arkose Sandstones: Varieties according to cements ⎰siliceous (orthoquartzite), calcareous, ferruginous, clayey (argillaceous)⎱ Micaceous, feldspathic, glauconitic and other varieties according to constituents Ganister, brickearth, loess
Muddy (pelitic)	Clay, clay minerals (less than 0.002 mm)	Shale, Mudstone, Clay Fireclay, Fuller's Earth, Ball-clay
	II. CHEMICAL and BIOCHEMICAL (organic)	
Calcareous	Shells, calcareous algae, corals, crinoids, foraminifera Coccolith fragments $CaCO_3$ precipitated from solution $CaMg(CO_3)_2$ of precipitation or replacement origin	Limestones: shelly, algal, crinoidal, reef (bioherm), and other varieties Chalk Oolitic limestone, pisolite, travertine, stalactite, tufa Dolomite Dolomitic limestone
Siliceous	Silica gel, radiolaria, diatoms	Flint, chert, jasper Radiolarite, diatomite
Saline	Salt lake precipitates marine precipitates	Evaporites: gypsum, anhydrite, rock-salt, potash salts
Carbonaceous	Peat Drifted vegetation Liquid hydrocarbons	Coal Series: lignite, (brown coal), humic ('bituminous') coals, anthracites Cannel coal Oil-shale, bituminous shale
Ferruginous	Ferrous carbonate Ferrous silicate Colloidal ferric hydroxide	Siderite ores Chamosite ores Bog-iron ores

Atterberg (or M.I.T.) scale, frequently employed in soil mechanics. (Other scales have been proposed.)

Wentworth		Atterberg	
Grade	*Size*	*Grade*	*Size*
Pebbles	over 2 mm	Gravel	over 2 mm
Sand { very course	2–1	Sand { coarse	2–0.6
coarse	$1-\frac{1}{2}$	medium	0.6–0.2
medium	$\frac{1}{2}-\frac{1}{4}$	fine	0.2–0.06
fine	$\frac{1}{4}-\frac{1}{8}$	Silt { coarse	0.06–0.02
very fine	$\frac{1}{8}$–0.06	medium	0.02–0.006
Silt	0.06–0.002	fine	0.006–0.002
Clay	below 0.002	Clay	below 0.002

A particular sediment generally contains particles of several different grades and not all of one size, e.g. a 'clay' may contain a large proportion of the silt grade and even some fine sand, as in the London Clay. The determination of the grades or sizes of particles present in a sample of sediment, known as the *mechanical analysis*, is carried out by seiving and other methods, as discussed in Chapter 10. The results of such an analysis can be represented graphically, grade size being plotted on a logarithmic scale against the percentage weights of the fractions obtained (Fig. 10.2).

Unconsolidated sands, silts, and clays are of considerable importance in civil engineering and are known to the engineer as *soils*. Mechanical analysis may give valuable information about the physical properties of a deposit. The proportion of clay present in a sediment, for instance, affects the engineering properties of the material. A good moulding sand should contain a small proportion of particles of the clay grade; a good sand for glass manufacture would be composed mainly of one grade of clean grains. It is to be noted that in general sand and silt particles are physically and chemically inert, but clays (0.002 mm and less) are often highly reactive.

In the descriptions which follow, the rocks will be treated under the general headings of (1) Pebbly Deposits, (2) Sands, (3) Sandstones, (4) Clays and Shales; and in Group II, (5) Limestones, (6) Siliceous Deposits, (7) Evaporites, (8) Coals and Ironstones, (9) Residual Deposits. The general relations of raw materials to the consolidated rocks are shown in Table 5.1.

(1) Pebbly deposits

Pebbles are generally pieces of *rock*, e.g. flint, granite, and by definition, are greater than 2 mm in diameter; most pebbles are much larger than this. Sand grains, on the other hand, usually consist of *mineral* particles, of which quartz, because of its hardness and resistance to weathering, forms the greater part of most sands.

Marine pebbly deposits are formed at the foot of cliffs from the break-up of falls of rock and from material drifting along the coast. Sorting by wave action may result in a graded beach, as at Chesil Beach, Dorset. The pebbles consist of the harder parts of the rocks which are undergoing denudation, wastage of the softer material being more rapid; thus, a gravel derived from flint-bearing chalk is composed mainly of rounded flints. Pebbles derived from the denudation of boulder clay deposits, as on the Yorkshire coast, are generally of many kinds (e.g. granite, gneiss, vein quartz), and may have travelled far from their source.

River gravels are laid down chiefly in the upper reaches of streams, after being carried over varying distances, and often contain an admixture of sand; at a later date a river may lower its bed and leave the gravels as terraces on the sides of its valley, marking its former course. In the Thames valley area such gravels are mainly composed of one constituent, flint; 'Thames Ballast' contains as much as 98 per cent flint. In areas where many different rocks have contributed fragments, gravels laid down by the rivers have a more varied composition; e.g. deposits of the River Tay, Scotland, contain pebbles of granite, gneiss, quartzite, etc. (see p. 509).

Some gravels are accumulated by the action of sudden storms which wash down much detritus. Lens-shaped beds of sand are often intercalated in deposits of river gravels, the variation in size of material being related to the currents which carried it.

Pebbly deposits or gravels when cemented are known as *conglomerate*; the spaces between the pebbles are often largely filled with sand. Two groups of conglomerates may be distinguished: (1) the beach deposits, in which rounded pebbles are well sorted, and commonly derived from one kind of rock (such as flint, quartzite). These often occur at the base of a formation which follows above an uncomformity, e.g. the orthoquartzite at the base of the Ordovician in South Shropshire. 'Pudding-stone' is a flint-conglomerate formed by the cementing of pebble beds in the Lower Tertiary sands, and is found locally as loose blocks lying on the surface of the Chalk north-east of London. (2) those conglomerates in which the pebbles are less well rounded and are derived from several different sources. Pebbly layers of this kind are often found within thick beds of sandstone, as in the Old Red Sandstone of Herefordshire, Scotland, and other areas. Such deposits represent rapidly accumulated coarse detritus washed into place by storms. *Banket* is the gold-bearing conglomerate, composed of pebbles of vein quartz in a silicous matrix, from the Witwatersrand System, South Africa; alluvial gold was washed into stream gravels which were later cemented into a hard rock.

The name *breccia* is given to cemented deposits of coarse angular fragments, such as consolidated scree. Such angular fragments, after limited transport, have accumulated at the foot of slopes and have not been rounded by water action, in contrast to the rounded pebbles which are distinctive of conglomerates. Breccias composed of limestone fragments in a red sandy mixture are found in Cumberland, e.g. near Appleby, and are called *brockrams;* they represent the screes which collected in Permian times from the denudation of hills of Carboniferous Limestone (see p. 256). The formation of scree or talus is specially characteristic of semi-arid or desert conditions.

The Dolomitic Conglomerate of the Mendips is a deposit of Triassic age, containing partly rounded fragments of dolomitic limestone of Carboniferous age and much angular material contributed by rapid weathering; it is a consolidated scree.

Breccias may also be formed by the crushing of rocks, as along a fault zone, the fragments there being cemented by mineral matter deposited from percolating solutions after movement along the fault has ceased. These are distinguished as *fault-breccias.* The explosive action of volcanoes also results in the shattering of rocks at a volcanic vent, and the accumulation of their angular fragments to form *volcanic breccias.*

(2) Sands

Mineral content. Most sand-grains, as stated above, are composed of quartz; they may be rounded, sub-angular, or angular, according to the degree of transport and attrition to which they have been subjected. Windblown grains, in addition to being well rounded often show a frosted surface. Other minerals which occur in sands are feldspar, mica (particularly white mica), apatite, garnet, zircon, tourmaline, and magnetite; they are commonly present in small amounts in many sands, but occasionally may be a prominent constituent. They have been derived from igneous rocks which by their denudation have contributed grains to the sediment. Ore mineral grains of high specific gravity are present in stream sands in certain localities, e.g. alluvial tin or gold.

Porosity and packing. Sands in their natural condition have a wide range of porosity because of differences in the packing of their component grains (partly arising during deposition) and in the sizes of the grains. Porosity, n, is defined as the percentage of void space in a deposit (or in a rock), i.e.

$$n = \frac{\text{volume of voids}}{\text{total volume}} \times 100.$$

The ratio of void space to solid matter, by volume, is called the *voids ratio*, e, or $e =$ (volume of voids/volume of solids) $\times$ 100. The relationship between these two quantities is given by $n = e/(1 + e)$, and $e = n/(1 - n)$. And if ρ is the density of the dry solids, then the moisture content $w = e/\rho$ when all the voids are filled with water.

The following list is given as a guide to the range of porosity for different materials, but values vary widely even for similar kinds of rock:

	Porosity (%)
Soil	50 and over
Clay	40 or more
Sand and gravel	20 to 47
Cemented sand	5 to 25

Soil, which forms the uppermost layers of the ground, is a very variable and porous material and can hold much water. Clays, though highly porous, are composed of very small particles (p. 185) and are therefore microporous.

Sands and gravels: the porosity of these deposits depends on several factors, including: (i) The grade sizes of the grains (p. 177); a deposit containing mainly grains of one grade will possess a higher porosity than one consisting of a mixture of grades, since in the latter case the smaller grains will partly fill interstices between the larger and so reduce the pore space (Fig. 5.2c). This applies also to sandstones.

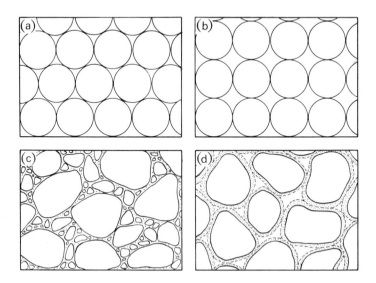

Fig. 5.2   Porosity and texture. (*a*) Close packing of spheres, porosity = 27 per cent. (*b*) Open packing of spheres, porosity = 47 per cent. (*c*) Poorly graded sand, porosity reduced by presence of small grains. (*d*) Well-graded sand (i.e. grains mainly on one size), porosity reduced by cement (stippled) between the grains.

(ii) The amount and kind of packing which the grains have acquired. Figure 5.2(*a*) and (*b*) shows two kinds of packing for theoretical spherical grains, the pore spaces being inter-connected though irregularly shaped. The arrangement in (*a*) is known as hexagonal close packing; in a layer overlying the one shown, each sphere would lie above a group of three below it, and succeeding pairs of layers repeat the arrangement. In (*b*), each grain of the next layer rests directly on the one below it, and the packing is more open, with larger interstices; an unstable structure not found in nature.

(iii) The amount of cement present. Interstices may be filled wholly or partly with mineral matter which binds the grains together, and in this way porosity may be considerably reduced (Fig. 5.2*d*).

It will be seen that many different possibilities arise from a combination of the three factors indicated above (Note 3).

(3) Sandstones

A sandy sediment, after natural compaction and cementation has gone on, is converted into a relatively hard rock called *sandstone*. It has a texture of the kind shown in Fig. 5.3; there is no interlocking of the component grains, as in an igneous rock, and the mineral composition is often simple. Bedding planes and joints develop in the consolidated rock, and form surfaces of division; these often break the mass into roughly rectangular blocks and are useful in quarrying (see Chapter 17). Widely spaced bedding-planes indicate long intervals of quiet deposition and a constant supply of sediment. Sands laid down in shallow water are subject to rapid changes of eddies and currents, which may form current-bedded layers between the main bedding planes, and this current-bedding is frequently seen in sandstones (see Fig. 2.12).

Cementation occurs in two main ways: (1) By the enlargement of existing particles; rounded quartz grains become enlarged by the growth of 'jackets' of additional quartz, derived from silica-bearing solutions, the new growth being in optical continuity with the original crystal structure of the grains (a fact which can be tested by observing the extinction between crossed polars). This is seen in the quartz of the Penrith Sandstone (Fig. 5.3*a*), or of the Stiperstones Quartzite of south Shropshire. (For another use of the term *quartzite* see p. 205). (2) By the deposition of interstitial cementing matter from percolating waters. The three chief kinds of cement, in the order of their importance, are:

(i)   silica, in the form of quartz, opal, chalcedony, etc.
(ii) iron oxides, e.g. haematite, limonite.
(iii) carbonates, e.g. calcite, siderite, magnesite, witherite (p. 126).

There are also rarer sandstones cemented by sulphates (e.g. gypsum, barytes), sulphides (pyrites), and phosphates. Sometimes there is a clay bond between the grains. Sandstones are named according to their cementing material or according to constituents other than quartz (Table 5.1).

Siliceous sandstone, with a cement of quartz or cryptocrystalline silica between the grains of quartz (Fig. 5.3a), is generally a very hard rock because of its high content of silica, and is therefore resistant to weathering (Note 4).

The term *orthoquartzite* is now used for nearly pure, evenly graded quartz-sands (Note 5) which have been cemented to form a sandstone. Examples are the Cambrian quartzite of Hartshill, Warwickshire, and the Stiperstones quartzite (p. 181) at the base of the Ordovician.

Ferruginous sandstone is red or brown in colour, the cement of iron oxide forming a thin coating to each grain as in many rocks of Old Red Sandstone age and in the Triassic of Britain (Note 6). Permian sandstones from Ballochmyle, Ayrshire (Plate 2a) and from Locharbriggs, near Dumfries, are two red structural stones of this type.

Calcareous sandstone has a cement of calcite, a relatively weak material, easily weathered by acids in rain-water. The 'Calcareous Grit' of Yorkshire, of Corallian age (see Fig. 8.8), and the Calciferous Sandstones (Lower Carboniferous) of Scotland are examples, as also the Sandgate Beds in Surrey (Lower Greensand). In some rocks, the Downton Castle sandstone among others, calcite crystals have grown around the quartz grains, which are thus enclosed in them. Reflection of light from cleavage surfaces in the calcite reveals the crystals, when the rock is handled; the appearance is described as 'lustre mottling.'

Argillaceous sandstone has a content of clay, which acts as a cement; it is relatively weak and such rocks have a low strength. Sandstones of this kind occur in the Carboniferous of Scotland, and some are crushed to make moulding sands for use in the steel industry.

Sandstones cemented with barytes ($BaSO_4$) are occasionally found, as at Alderley Edge, Cheshire, and at Bidston; this very hard cement has resulted in the preservation of isolated stacks of rock, such as the Hemlock Stone, near Nottingham, on account of its resistance to weathering. Sandstones cemented by gypsum are found in arid regions like the Sahara; the gypsum may grow and enclose sand grains, then forming 'sand-gypsum crystals.'

Sandstones which are named after constituents include *micaceous sandstone*, in which mica flakes (generally muscovite) are present, dispersed throughout the rock. Other sandstones have parallel layers of mica flakes, spaced at intervals of a few centimetres; along these layers the rock splits easily, and it is called a *flagstone*. The structure arises when a mixture of mica and quartz grains is gently sedimented

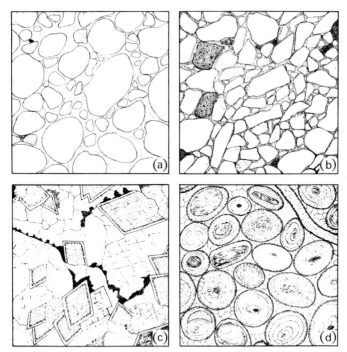

Fig. 5.3 Sedimentary rocks in thin section. (*a*) Penrith sandstone, showing enlargement of grains by new growth of quartz. (*b*) Feldspathic sandstone (arkose). (*c*) Dolomitic limestone. (*d*) Oolitic limestone. (Magnification × 15.)

in water; the micas settle through the water more slowly than quartz because of their platy shape, and are thus separated from the quartz grains and form a layer above them. This may be repeated many times. Flagstones are used in slabs with the bedding vertical for facing steel-framed buildings; e.g. the Moher Flags from the coast of Co. Clare, west Ireland.

Feldspathic sandstone has a small but noticeable content of feldspar. *Glauconitic sandstone* contains the green clay-mineral glauconite (a hydrated silicate of iron and potassium, of marine origin), for example the Greensands of the Lower Cretaceous. The glauconite may occur as rounded grains and as infillings to the cavities of small fossils. Some of the Upper Eocene marine sands are rich in glauconite, as in parts of the Bracklesham Beds above the London Clay.

Greywacke (German 'grauwacke') is a dark grey coloured, badly sorted rock in which many coarse angular grains of quartz and feldspar are present, together with mica and small rock-fragments (e.g. slate) and fine matrix material. Greywackes were formed in an unstable environment (p. 173) during the infilling of a geosyncline, when much coarse detritus was being washed into the trough of deposition. As a result of the eventual compression of the contents of the trough, typical greywackes are now found in areas of sharply folded strata, for example among the Lower Palaeozoic sediments of central Wales and the Southern Uplands of Scotland. Many greywackes show *graded-bedding*, in which the sediment passes from coarser to finer particles from the bottom of a bed upwards; this structure is produced by the settling of a mixture of sand and mud in water, after movement over the sea floor as a turbidity current. (Turbidity currents give rise to a mass of disturbed sediment which on later consolidation forms the rocks known as *turbidites*.) Elongated projections called *flow-casts* or *sole-markings* are frequently found on the undersides of greywacke beds; they show the direction of currents which operated at the time of their formation.

The term *arkose* is used to denote typically pale-coloured sandstones, coarse in texture, composed mainly of quartz and feldspar in angular or partly rounded grains, usually with some mica; the minerals were derived from acid igneous rocks such as granites, or from orthogneisses, which were being rapidly denuded at the end of an orogenic upheaval. The feldspar may amount to a third of the whole rock in some instances; the constituents are bound together by ferruginous or calcareous cements. Much arkose is found in the Torridon Sandstone of Ross-shire (Fig. 5.3 *b*).

Ganister is a fine-grained siliceous sandstone or siltstone which is found underlying coal seams in the Coal Measures; it is an important source of raw material for the manufacture of silica bricks and other refractories. *Loam* is a deposit containing roughly equal proportions of sand, silt, and clay.

The term *freestone* is used for sandstones *or* limestones which have few joints and can be worked easily in any direction, yielding good building material; e.g. the Clipsham Freestone of Rutland (Note 7).

(4) Shales and clays

Shales are compacted muds, and possess a finely laminated structure by virtue of which they are fissile and break easily into parallel-sided fragments. This lamination is parallel to bedding-planes and is analogous to the leaves in each book of a pile of books. Starting from a deposit of very fluid mud with a high water content, water has been slowly squeezed out from the sediment as a result of the pressure of superincumbent deposits, until the mud has passed into the condition where the water content is perhaps 10–15 per cent or more. With further compaction and loss of more water, the deposit has ultimately taken on the typical shaley parting parallel to bedding; there is thus a gradation from a mud to a shale. In some cases the shaley parting is not developed and the rock is called *mudstone*.

Muddy or argillaceous rocks are of various colours; when they contain finely divided carbon or iron sulphide they appear grey to black. The presence of iron oxide gives a red, brown, or greenish colour, according to the state of oxidation of the iron; thus, greenish-blue London Clay turns brown on exposure to the atmosphere owing to the change from ferrous to ferric oxide.

Muds, like clays, are composed of minute particles of the order of 0.002 mm diameter or less, which were deposited very slowly in still water on the continental shelf or in lakes. Some of the particles are of colloidal size, and in some cases colloids form a high proportion of the clay. The term 'fat clay' is used in some engineering literature for a clay which has a high content of colloidal particles, and is greasy to the touch and very plastic; a 'lean clay', in contrast, has a small colloidal content. Soon after the deposition of a mass of muddy sediment, sliding or slumping may occur on submarine slopes, and the incipient bedding structures are destroyed or disturbed; considerable thicknesses of disturbed shales are seen intercalated between normal beds at many localities, and are attributable to slumping contemporaneous with deposition. Some silt particles are commonly present, and by lateral variation, with increase in the proportion of the silt, shales pass gradually into fine-grained sandy deposits.

Compaction and consolidation. Compaction processes have been defined (p. 171), and in argillaceous sediments they result in the progressive change from a soft mud or clay, through to a mudstone or shale. The main process involved is the squeezing out of pore water under increasing weight of overburden, called *consolidation;* but other processes during compaction include cementation and the development of bonding between particles.

A freshly deposited mud on the sea floor has a high porosity (and high void ratio, p. 179), represented in Fig. 5.4 by the point (*a*). As deposition continues, the overburden pressure increases and the sediment compacts, e.g. to point (*b*) or further to (*c*). The curve *abc* in the figure is called the sedimentation compression curve for that particular clay. The clay is said to be *normally-consolidated* if it has never been under a pressure greater than the existing overburden load. If, however, the clay has in the past been subjected to a greater overburden pressure, due to the deposition of more sediment which has subsequently been eroded, it is said to be *over-consolidated;* this is represented by point (*d*) in Fig. 5.4. Thus two clays, at points (*b*) and (*d*), which are under the same effective pressure and are identical except for their consolidation history, have very different void ratios. The construction of curves such as those shown in the figure has been discussed by Skempton (1970) (Note 8). The London Clay is an example of an over-consolidated clay; its consolidation took place under

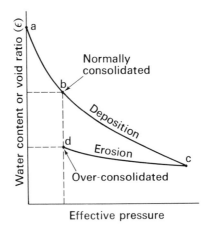

Fig. 5.4   Normally-consolidated and over-consolidated clay (after Skempton).

the weight of the overlying Upper Eocene beds, the estimated load in the central London area being about 35 kg/cm². Uplift, and the erosion which resulted from it, reduced the overburden to its present amount.

Other over-consolidated clays include certain ball-clays from Devon, in which over-consolidation is thought to have resulted mainly from desiccation of the individual beds of clay when they were exposed to the atmosphere at an early stage in their geological history (see p. 188 and Note 11).

Sensitive clays.   When the strength of an undisturbed sample of clay or silt is compared with the strength of the material after it has been re-moulded, the ratio of the undisturbed strength to the re-moulded strength is called the *sensitivity* of the material. Re-moulding breaks down the structure of the clay, although its natural water content remains the same. Values of sensitivity up to 35 or more have been recorded; but many clays have sensitivities of about 10, i.e. they lose 90 per cent of their strength on being re-moulded. Apart from laboratory tests, such *sensitive clays* when disturbed during engineering construction may undergo great loss of strength, and it is therefore important that their character should be understood when excavations in them are to be made.

Clays which are highly sensitive are also known as 'quick' clays. Certain Norwegian quick clays, of late-glacial age (deposited in lakes) or post-glacial (marine), have been investigated, as well as others from the St. Lawrence valley, Ontario. Many of the post-glacial marine clays have been leached by the passage of fresh water through them after the isostatic uplift of the area in which they were deposited, and this has resulted in the removal of some of the dissolved salts in their original pore-water. The connection between reduced salt content of the pore-contained water and increased sensitivity of the clay was demonstrated experimentally for a quick clay from Horten, Norway, by Skempton and Northey in 1952 (Note 9).

Mineral content. Clays and shales have a complex mineral composition, which is the more difficult to investigate on account of the very small size of the particles involved. The *clay minerals* have been described in Chapter 3; they are hydrous aluminium silicates, sometimes with magnesium or iron replacing part of the aluminium and with small amounts of alkalies. They form minute flaky or rod-like crystals, which build up the greater part of most clays and are important in determining their properties. Clays may also contain a variable proportion of other minerals, such as finely divided micaceous and chloritic material, together with colloidal silica, iron oxide, carbon, etc., and a small proportion of harder mineral grains (e.g. finely divided quartz). Organic matter may also be present. In addition, water is an important constituent and on it the plasticity of the clay depends; the water forms thin films around the very small mineral particles and fills the minute pore spaces. These films of water which separate the mineral flakes act as a lubricant between them and endow the clay with plastic properties. The water-absorption capacity of clay in turn depends on the nature of the clay minerals.

During the expulsion of water from clay in the process of compaction, mineral re-constitution takes place and the resulting *shale* consists essentially of sericite (cf. illite), chlorite, and quartz; small crystals grow out of the raw materials available and form part of the shale, which, however, is not completely crystalline. Chemically, shale is characterized by a high content of alumina and is also generally rich in potash.

Some particular kinds of clay and shale are now described.

Marl is a calcareous shale or clay. In the Keuper Marl (p. 259) the mineral dolomite is present, in addition to clay minerals (mainly illite), chlorite, and a little quartz. The carbonate content ranges from about 7 per cent to over 20 per cent (Note 10).

Fuller's earth, a clay largely composed of montmorillonite, has low plasticity and disintegrates in water. It readily absorbs grease and is used for the bleaching and filtration of oils, for cleaning cloth and taking grease out of wool, and for medical purposes. Deposits of fuller's earth are worked in the Sandgate Beds (part of the Lower Greensand) at Nutfield, Surrey, and in Jurassic rocks at Combe Hay near Bath and at Woburn. Supplies of commercial bentonite (below) are obtained by processing these deposits.

Bentonite is a clay derived from the alteration of volcanic dust and ash deposits, and is mainly composed of montmorillonite. Owing to the capacity of this mineral to absorb water within the crystal lattice (p. 121), as well as acquiring a film of water around each particle, bentonite clays swell enormously on the addition of water, forming a viscous mass. This property renders the material useful for various purposes, such as the thickening of drilling mud in sinking oil wells; it is also used in America as an ingredient of moulding sands for foundries, and as an absorbent in many processes. Commercial deposits are worked in the Black Hills of Wyoming and South Dakota. In civil engineering bentonite is employed as a sealing layer in trenches and cofferdams to prevent the percolation of water, and as a slurry pumped into sands or gravels to fill the voids and render the mass impervious. It was used, for example, in the construction of the Hyde Park Corner underpass, London, to help support gravels in the walls of the excavations.

Fire-clays are rich in kaolinite, and commonly contain small amounts of quartz and hydromica. They occur beneath coal seams in the Coal Measures, and characteristically have a very low content of alkalies. They can be exposed to high temperatures, 1500 or 1600 degrees C, without melting or disintegrating and are used in the manufacture of refractories.

Bituminous shale or *oil-shale* is black or dark brown in colour and contains natural hydrocarbons from which crude petroleum can be obtained by distillation. It gives a brown streak and has a 'leathery' appearance, with a tendency to curl when cut; parting planes often look smooth and polished. The oil-shales of the Lothians of Scotland provide raw material for the production of crude oil and ammonia.

Alum-shale is a shale impregnated with alum (alkaline aluminium sulphate): the alum is produced by the oxidation and hydration of pyrites, yielding sulphuric acid which acts on the sericite of the shale.

Ball-clays are plastic, kaolinitic clays formed from transported sediment and generally deposited in lacustrine basins. The clays are composed of secondary kaolinite together with some micaceous material (e.g. illite), and may contain a proportion of silt or sand. The kaolinite, which forms 60 per cent or more of the ball-clay, is probably derived from argillaceous rocks such as shales, during their weathering and transport especially under warm temperature conditions; it is less well crystallized than kaolinite derived by hydrothermal decomposition of granite (p. 208), the minute crystals being commonly less than 1 micrometre in size [Note 8, Chapter 3].

Ball-clays are used in the ceramic industry, and when fired at temperatures of 1150 to 1200°C, they have a white or near white colour. In England they are worked in the Bovey Basin near Newton Abbot, Devon, where deposits of Oligocene age are associated with lignites; and in the Wareham district, Dorset. The geotechnical properties of the Bovey deposits have recently been described (Note 11).

Certain argillaceous deposits are formed on land; these include *loess*, a wind-blown deposit of fine particles of the clay and silt grades, which is widely distributed in Central Europe and Asia (see p. 33); and *brick-earths*, brown and red silty clays which are thought to have been formed in glacial waters in front of the ice-sheets during the Pleistocene period (*q.v.*), and were formerly used in the making of bricks.

For a description of the glacial *varved clays*, see Chapter 2; the Red Clay and muds of deep-sea origin are briefly discussed on p. 174.

Apart from uses mentioned above, clays are employed in industry in many ways such as for bricks, tiles, terra-cotta, tile drains, earthenware and porcelain, firebricks, crucibles and other refractories, puddle for engineering structures (e.g. the cores of earth dams), and in cement manufacture. The suitability of a deposit for any particular purpose depends on the presence of different clay minerals and other constituents, and their relative proportions in the deposit.

II.   CHEMICAL AND BIOCHEMICAL (ORGANIC) SEDIMENTS

(5) Limestones (carbonate rocks).

Limestones consist essentially of calcium carbonate, with which there is generally some magnesium carbonate, and siliceous matter such as quartz grains. The average of over 300 chemical analyses of limestones showed 92 per cent of $CaCO_3$ and $MgCO_3$ together, and 5 per cent of SiO_2; the proportion of magnesium carbonate is small except in dolomite and dolomitic limestones. Limestones are bedded rocks often containing many fossils; they are readily scratched with a knife, and effervesce on the addition of cold dilute hydrochloric acid (except dolomite). The distance between bedding-planes in limestones is commonly 30 to 60 cm, but varies from a couple of centimetres or less in thin-bedded rocks (such as the Stonesfield 'Slate' p. 519) to over 6 m in some limestones (e.g. the Clipsham Stone, p. 184).

Calcium carbonate is present in the form of crystals of calcite or aragonite, as amorphous calcium carbonate, and also as the hard parts of organisms (fossils) such as shells and calcareous skeletons, or their broken fragments. Thus, a consolidated shell-sand is a limestone by virtue of the calcium carbonate of which the shells are made. On the other hand, chemically deposited calcium carbonate builds limestones (e.g. oolites) under conditions where water of high alkalinity has a restricted circulation, as in a shallow sea or lake. Non-calcareous constituents commonly present in limestones include clay, silica in colloidal form or as quartz grains or as parts of siliceous organisms, and other hard detrital grains. Though usually grey or white in colour, the rock may be tinted, e.g. by iron compounds or finely divided carbon, or by bitumen. The types listed in the table are now described.

Shelly Limestone is a rock in which fossil shells, such as brachipods and lamellibranchs (Fig. 8.9), form a large part of its bulk. It may be a consolidated shell-sand (cf. the 'crags' of East Anglia). *Shell banks*, in which the shells are laid by currents and the spaces between them filled by milky calcite (during diagenesis), are also known as 'conquinites' and are biochemical deposits. ·

Coral, Algal, and Crinoidal Limestones take their names from prominent fossil constituents. *Algae* are aquatic plants allied to sea-weeds; some kinds secrete lime and their remains may build up large parts of some limestones. Some fresh-water algae precipitate calcium carbonate and others silica. Crinoids (or 'sea-lilies.' Fig. 8.5*f*) have a calcareous casing supported on a stem made of innumerable small discs; these disintegrate when the animal dies and help to build up deposits on the sea floor. Corals (Fig. 8.5) are marine animals of very simple structure, with radial symmetry, often living in colonies which build reefs in warm, shallow seas, as in the South Pacific at the present day. Parts of the Carboniferous Limestone contain many corals, and the remains of ancient coral reefs are found in the Wenlock Limestone. The *reef-limestones* are mounds of unbedded rock (reef-knolls or bioherms), rich in fossils, which are found adjacent to bedded lime-stone with fewer fossils, as in the Carboniferous Limestone of Clitheroe, Lancs. Reef-knolls may be several hundred feet long; they are often very porous, and may be reservoirs for oil which has accumulated in them under a cover of shale. Reef facies are abundant at the present day.

Chalk is a soft white limestone largely made of finely divided calcium carbonate, much of which has been shown to consist of minute plates, 1 or 2 microns in diameter. These plates are derived from the external skeletons of calcareous algae, and are known as coccoliths. The Chalk also contains many foraminifera, which differ in kind and abundance in different parts of the formation; and other fossils, such as the shells of brachiopods and sea-urchins (Fig. 8.9). The *foraminifera* are minute, very primitive jelly-like organisms (protozoa) with a hard globular covering of carbonate of lime; they float at the surface of the sea during life, and then sink and accumulate on the sea floor (p. 174). *Radiolaria* are similar organisms which have siliceous frameworks, often of a complicated and beautiful pattern; these too are found in Chalk but are not so numerous as the foramini-fera. Parts of the rock contain about 98 per cent $CaCO_3$ and it is thus almost a pure carbonate rock. It was probably formed at moderate depths (round about 180 m) in clear water on the continental shelf.

The English Chalk is a bedded and jointed rock, the upper parts of which contain layers of flints (p. 193) along some bedding planes and as concretions in joints. Chalk is used for lime-burning and, mixed with clay, is calcined and ground for Portland Cement.

Limestones which contain a noticeable amount of substances other than carbonate are named after those substances, thus: *siliceous limestone, argillaceous limestone, ferruginous limestone, bituminous limestone;* these terms are self-explanatory. *Cement-stone* (or hydraulic limestone) is an argillaceous limestone in which the proportions of clay and calcium carbonate are such that the rock can be burned for cement without the addition of other material, e.g. the cement-stones of the Lower Lias in England (p. 526).

Among the limestones which have originated mainly by chemical deposition two important types are oolite and dolomite.

Oolitic Limestone, or *oolite*, has a texture resembling the hard roe of a fish (Greek, *oon* = egg). It is made up of rounded grains formed by the deposition of successive coats of calcium carbonate around particles such as a grain of sand or piece of shell, as a 'seed'. These particles are rolled to and fro between tide marks in limy water near a limestone coast, and become coated with calcium carbonate; they are called 'ooliths,' and their concentric structure can be seen in thin section (Fig. 5.3*d*). Sometimes the grains are larger, about the size of a pea, when the rock made of them is called *pisolite*. Deposition of $CaCO_3$ may also take place through the agency of algae; each particle, or seed, which is rolled about by currents acquires a coating of calcareous mud, which is surrounded by a layer of blue-green algae. The latter can survive the rolling for a long time. Devonian oolites when carefully dissolved have been shown to contain these concentric organic coatings. Ooliths are forming to-day in the shallow seas off the coasts of Florida and the Bahamas.

In course of time the ooliths, often mixed with fossil shells, become cemented, and then form a layer of oolite. When few shells are present the rock has an even texture and is easily worked, and makes a first-class building stone. The Jurassic oolites of the Bath district, and the Portland oolitic limestone, Dorset, are two English examples (Note 12).

Dolomitic Limestone and *Dolomite* are rocks which contain the double carbonate $MgCO_3 \cdot CaCO_3$, dolomite (the name is used for the mineral as well as for the rock); the mineral occurs as rhomb-shaped crystals (Fig. 5.3c). Dolomite is made entirely of the mineral dolomite, but dolomitic limestone has both dolomite and calcite, the full ratio of magnesium to calcium needed for the double carbonate not being present. For a test for dolomite see page 124.

Magnesian limestone is the term used for a limestone with a small content of magnesium carbonate, which is usually not present as the double carbonate but is held in solid solution in calcite crystals.

The mineral dolomite often replaces original calcite in a rock, the magnesium being derived from sea-water and introduced into the limestone by solutions passing through it. The replacement process is called dolomitization. The change from calcite to dolomite involves a volume contraction of 12.3 per cent and hence often results in a porous rock. Magnesium carbonate may also be precipitated direct from solution. Borings in the coral atoll of Funafuti, in the South Pacific, showed progressive dolomitization of the calcite of the limestone reef with depth, indicating that the change takes place in shallow water.

Beds of dolomite are found in the Carboniferous Limestone, e.g. in the Mendips, and at Mitcheldean, Forest of Dean; in some cases the process of dolomitization has been selective, the non-crystalline matrix of a limestone being dolomitized before the harder calcite fossils which it contains are affected. The Magnesian Limestone of Permian age (p. 257) is mainly a dolomitic limestone.

The general term for replacement processes, including that outlined above, is *metasomatism* (= change of substance, Greek); original minerals are changed atom by atom into new mineral substances by the agency of percolating solutions, but the outlines of original structures in the rock are frequently preserved. The dolomitization of a limestone thus involves the replacement of part of the calcium by magnesium. Another metasomatic change which takes place in limestones is the replacement of calcium carbonate by silica; in this way silicified limestones and *cherts* are formed, as in parts of the Carboniferous Limestone. Silica from organisms such as sponges and radiolaria is dissolved by water containing potassium carbonate, and re-deposited as chert, which is a form of cryptocrystalline silica. The Hythe beds of the Lower Greensand in Kent have beds of chert of considerable thickness. Replacement of calcite by siderite ($FeCO_3$) is another important metasomatic change; it is seen in some Jurassic limestones such as the Cornbrash of Yorkshire and the Middle Lias marlstone of the Midlands. It was formerly thought that the Cleveland Ironstone of Yorkshire originated in this way, but the siderite of this deposit, which is oolitic, is now believed to be a primary precipitate on the sea-floor.

Travertine, *stalactite*, and *calc-sinter* are made of chemically deposited calcium carbonate from saturated solutions. Calc-sinter (or *tufa*) is a deposit formed around a calcareous spring. Stalagmites and stalactites are columnar deposits built by dripping water in caverns and on the joint planes of rocks (see Chapter 2). Travertine is a variety of calc-sinter, cream or buff-coloured, with a cellular texture due to deposition from springs as a coating to vegetable matter.

The latter rots away, leaving cavities in the rock. An important economic deposit of travertine occurs in Italy, on the banks of the River Anio near Tivoli. It is about 140 m thick, is soft when quarried, but hardens rapidly on exposure. Travertine has been much used in slabs for non-slip flooring, as in many London buildings, and for interior and exterior panelling, with or without a filler in the cavities.

(6) Siliceous deposits

Under this head are included the organically and chemically formed deposits mentioned in Table 5.1. The deep-sea siliceous oozes are briefly described on p. 174. *Diatomite* is the consolidated equivalent of *diatom-earth*, a deposit formed of the siliceous algae called diatoms, which accumulate principally in fresh-water lakes. It is found at numerous Scottish localities and at Kentmere in the Lake District, where there is a deposit of Pleistocene age. The United States, Denmark, and Canada have large deposits from which considerable exports are made. Diatom-earth is also known as *kieselguhr* or as *Tripoli-powder*, and was originally used as an absorbent in the manufacture of dynamite. It is now employed in various chemical processes as an absorbent inert substance, in filtering processes, as a filler for paints and rubber products, and for high temperature insulation. Deposits of *sponge spicules*, which are fragments of the skeletons of siliceous sponges, are found to-day on parts of the ocean floor; older deposits of this kind provided the raw material for the cherts which occur in certain limestones.

Flint is the name given to the irregularly shaped siliceous nodules found in the Chalk. It is a brittle substance and breaks with a conchoidal fracture; in thin section it is seen to be cryptocrystalline. The weathered surface of a flint takes on a white or pale brown appearance. In England, flints occur most abundantly in the Upper Chalk where they may lie along bedding planes or fill joints in the form of tabular vertical layers. The silica of flint may have been derived from organisms such as sponge spicules and radiolaria in the Chalk, which were slowly dissolved and the silica re-deposited probably as silica gel by solutions percolating downwards through the Upper and Middle Chalk. The Lower Chalk generally contains no flints. Other modes of origin for flint have been suggested. For *chert* see p. 192.

Jasper is a red variety of cryptocrystalline silica, allied to chert, the colour being due to disseminated Fe-oxide; it is found in Pre-Cambrian and Palaeozoic rocks.

(7) Evaporites (salt lake deposits)

When a body of salt water has become isolated its salts crystallize out as the water evaporates. The Dead Sea is a well-known example;

it has no outlet and its salinity constantly increases. Another instance is that of the shallow gulf of Karabugas in the Caspian Sea, into which salt water flows at high tide but from which there is no outflow, so that salt deposits are formed, in what is virtually a large evaporating basin which is continually replenished. In past geological times great thicknesses of such deposits have been built up in this way in various districts. The first salt to be deposited is *gypsum* ($CaSO_4 2H_2O$), beginning when 37 per cent of the water is evaporated. *Anhydrite* ($CaSO_4$, orthorhombic) comes next, followed by *rock salt* (NaCl). Pseudomorphs of rock-salt cubes are found in the Keuper Marl (see p. 259 for description of the conditions under which these rocks were deposited). Lastly are formed magnesium and potassium salts such as *polyhalite* ($K_2SO_4 \cdot MgSO_4 \cdot 2CaSO_4 \cdot 2H_2O$), *kieserite* ($MgSO_4 \cdot H_2O$), *carnallite* ($KCl \cdot MgCl_2 \cdot 6H_2O$), and *sylvite* (KCl). All these are found in the great salt deposits of Stassfurt, Germany, which reach a thickness of over 1200 m, and have been preserved from solution through percolating water by an overlying layer of clay (*loess*).

Gypsum and rock salt are found in the Triassic rocks of Britain, e.g. in Cheshire, Worcestershire, and near Middlesbrough, but nothing beyond the rock-salt stage was known in Britain until deep borings were made at Aislaby and Robin Hood Bay, north Yorkshire. These penetrated important potash salt deposits lying above rock salt and covered by marl, at depths of about 1250 m. The development of such very soluble deposits is rendered difficult owing to the depth at which they occur.

Gypsum is also mined at Netherfield, Sussex, where a deposit in the Purbeck Beds is found between 30 and 60 m below the surface. Apart from its use in building industry, gypsum is used as a filler for various materials (e.g. rubber), and for a variety of other purposes (p. 126). Rock salt, apart from domestic uses, is employed in many chemical manufacturing processes; potash salts are used as fertilizers, as a source of potassium, and in the manufacture of explosives.

Sodium Salts.—*Soda nitre* (or 'Chile salt-petre,' $NaNO_3$) deposits are found in the Atacama Desert of Chile, in sandy beds called *caliche*, of which it forms from 14 to 25 per cent. It is exported largely as a source of nitrates, iodine (from sodium iodate) being an important by-product. Other sodium and magnesium salts were deposited from the waters of alkali lakes, e.g. in arid regions in the western U.S.A., and the Great Salt Lake of Utah. Sodium sulphate ('salt cake') is important in the chemical industry and in paper and glass manufacture. It is produced chiefly in the U.S.S.R., Canada, and the U.S.A.

The evaporites form an economically important group, of which the foregoing is a brief outline; for fuller treatment the student is referred to books on economic minerals. (See also Salt Domes, p. 227.)

(8) Coals and ironstones

The formation of coal through the burial of ancient vegetation is described in Chapter 8. The average compositions of several varieties of coal are given in the table below, and the carbon–hydrogen proportions in the accompanying graph (Fig. 5.5), which shows the

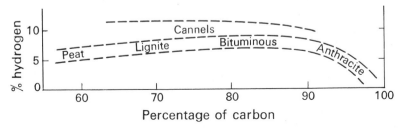

Fig. 5.5   Range of composition of coals (after Raistrick and Marshall). Points plotted for numerous coal analyses fall within the narrow zone between the two lower dotted lines (the 'coal belt'). Cannel coals lie outside this belt.

variation from low-grade to high-grade fuel. The coal seams had their origin in extensive peat-bogs, which supported dense vegetation for a time and were then covered by silts and muds when subsidence and inundation took place. *Peat* is thus the first stage in the formation of coal. Upland peat of the present day, derived from compressed mosses and plants such as sphagnum, heather, and cotton-grass, is cut and burned for domestic use; it has a high ash content and burns with a smoky flame. Peat-fired generating stations for electricity are in use in Ireland, as at Portarlington, south-west of Dublin.

	Average composition of fuels (percentages)				Proportions recalculated with Carbon as 100			
	C	H	N	O	C	H	N	O
Wood	49.65	6.23	0.92	43.20	100	12.5	1.8	87.0
Peat	55.44	6.28	1.72	35.56	100	11.3	3.1	64.1
Lignite	72.95	5.24	1.31	20.50	100	7.2	1.8	28.1
Bituminous Coal .	84.24	5.55	1.52	8.69	100	6.6	1.8	10.3
Anthracite . . .	93.50	2.81	0.97	2.72	100	3.0	1.3	2.9

In limited areas of the peat-bogs, slight erosion before deposition of the overlying silts created hollows in which pockets of drifted plant detritus accumulated; these now form the impersistent seams of *cannel coal* which often accompany the more extensive coals. Cannel is a dense coal of dull grey to black appearance which ignites in a candle flame (hence the name) and burns with a smoky flame. It contains some 70 to 80 per cent carbon, 10 to 12 per cent hydrogen, and much ash, and has a calorific value of 12 000 to 16 000 B.T.U. The ash is attributable to an admixture of sediment which was originally deposited with the drifted vegetable debris.

Lignite (or brown coal) is more compact than peat; it has lost some moisture and oxygen, and the carbon content is consequently higher (but less than 75 per cent). Its texture is rather like woody peat, and when dried on exposure to the atmosphere it tends to crumble to powder. There are small lignite deposits at Bovey Tracey, Devon; extensive beds are found in Germany (Cologne), Canada, and the southern United States.

Bituminous coal includes the ordinary household varieties; it has a carbon content of 80–90 per cent and a calorific value of 14 000 to 16 000 B.T.U. (as against lignite's 7000 to 11 000). Between it and lignite there are coals called *sub-bituminous* (or 'black lignites') with less moisture than lignite and between 70 and 80 per cent carbon. High grade, hard steam coals, sometimes termed *semi-bituminous*, are transitional between bituminous coal and anthracite; they have the greatest heating value of any, and burn with a smokeless flame.

Bituminous coal has a typical banded structure in which bright and dull bands alternate. The three main kinds of material which are present are known as (i) *fusain*, a soft black powdery substance ('mineral charcoal') which readily soils the fingers, occurs in thin bands, and is composed of broken spores and cellular tissue; (ii) *vitrain*, the bands or streaks of bright black coal, several millimetres or more in thickness, representing solidified gel in which the cellular structure of original plants has been detected; and (iii) *durain*, the bands of dull hard coal, containing spore cases and resins. Blocks of coals break easily along the fusain layers, and are bounded by joints ('cleat' and 'end') which are perpendicular to the layers; a thin film of mineral matter (e.g. pyrite) often coats the joint surfaces. 'Bright' coals have a high content of vitrain, which is the best source of coal tar by-products.

Anthracite contains from 90 to 95 per cent carbon, with low oxygen and hydrogen, and ignites only at a high temperature; its calorific value is about 15 000 B.T.U. It is black with sub-metallic lustre, conchoidal fracture, and banded structure, and does not mark the fingers when handled. Anthracite appears to have been formed when coal-bearing beds have been subjected to pressure or to increased temperatures, during gentle metamorphism; transitions from anthracite to bituminous coals have been traced, e.g. in the South Wales coalfield, where also shearing structures are present in the anthracites.

Associated with coals are beds of ganister and fireclay (Fig. 5.6), which lie below the seams and represent the seat-earth in which the Coal Measure plants grew. Overlying a coal seam is a series of shales,

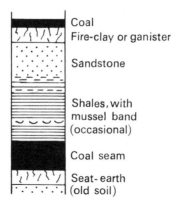

Coal
Fire-clay or ganister

Sandstone

Shales, with
mussel band
(occasional)

Coal seam

Seat-earth
(old soil)

Fig. 5.6   Unit of coal measure rhythm.

of varying thickness, which grade upwards into sandy shales and then sandstones. These are followed by the next coal seam, with its seat-earth and its overlying shales and sandstones. The repetition of this sequence of rocks over and over again, with minor variations, is referred to as the Coal Measure 'rhythm,' each unit of which records the submergence of the vegetation of a coal swamp, and the eventual emergence (as the water shallowed) of sandy shoals on which the swamps grew again.

Blackband ironstone and clay-ironstone are ferruginous deposits associated with the Coal Measures. In Staffordshire the blackband ironstones contain from 10 to 20 per cent of coal, the iron being present as carbonate, and are found replacing the upper parts of coal seams; they are economical to smelt on account of their coal content. Clay-ironstones also consist chiefly of chemically deposited siderite ($FeCO_3$); they occur as nodular masses in the argillaceous rocks of the Coal Measures, and also in the Wealden Beds, where they are of fresh-water origin.

Bog-iron Ores are impure limonitic deposits which form in shallow lakes and marshes, as in Finland and Sweden to-day. The deposition of the iron may be due to the action of bacteria or algae. The ores were much used in the early days of iron-smelting.

Jurassic ironstones. Important bedded iron ores of this age contain the minerals siderite and chamosite (hydrated iron silicate); some of the ironstones are oolitic and others have small crystals of siderite in a matrix of mudstone. The Cleveland ironstone of the English Middle Lias (p. 259) is an oolitic rock containing both the above minerals. A rock of the same age, the marlstone of Lincolnshire and Leicestershire, has a large amount of calcite as well as chamosite and siderite. The Northamptonshire ironstones are partly oolitic rocks with chamosite as the chief constituent, and partly mudstones with siderite and limonite.

Haematite and limonite ores. In some cases iron is present as haematite in a shelly limestone, as at Rhiwbina, South Wales, or as limonite, e.g. the Frodingham ironstone. The haematite deposits of Cumberland are replacements of the Carboniferous Limestone by irregular masses of the iron oxide. The Lake Superior region of North America possesses very extensive haematite deposits; the rocks here are Pre-Cambrian sediments in which iron of sedimentary origin has probably been concentrated as the oxide after alteration from its original silicate or carbonate form.

Magnetite in notable amount occurs in a few sediments, but magnetic iron ores, which are of such importance, for example, in Sweden, are generally of igneous origin and are referred to on p. 165.

(9) Residual deposits

In hot, semi-arid regions, where evaporation from the ground is rapid and nearly equal to the rainfall, and where there is little frost action, chemical decomposition of the rocks proceeds to great lengths and a hard, superficial crust is formed by the deposition of mineral matter just below the soil. The water from the occasional rains carries dissolved salts only a short distance below the surface, where they are retained by capillarity, with the result that as evaporation proceeds a mineral deposit is built up. If solutions are saturated with calcium carbonate the deposit will be a calcareous one, like *kankar*, which covers large areas in India. If the solutions are ferruginous, such as would result from the decomposition of basic igneous rocks, a red concretionary deposit called *laterite* may be formed, as in many parts of Africa. Laterite is hydrated ferric oxide, generally with some alumina and silica; its composition varies according to the nature of the underlying rock and the amount of chemical breakdown that has gone on. 'Iron pan' is a very hard variety of laterite. This material has been used as rough building stone in tropical countries where it is plentiful.

Bauxite, which consists essentially of hydrated alumina, forms residual deposits resulting from the weathering, under tropical conditions, of igneous and other rocks containing aluminium. It is the raw material from which the metal aluminium is produced, by reduction of the oxide by electrolysis. Important deposits of bauxite occur in Guyana, in the southern United States and elsewhere.

NOTES AND REFERENCES

1. Compaction can also occur beneath heavy foundations, and so the loads placed on sedimentary rocks must be carefully considered.

Igneous and metamorphic rocks are less susceptible to such changes.

2. An illustrated description of the deep-sea sediments and ocean basins, and an account of the history of the oceans, is given by Turekian, K. K. 1968, in Oceans (Prentice-Hall, Inc., Foundations of Earth Science Series, paperback).

3. The porosities of sands in relation to the packing of their component particles are discussed by Kolbuszewski, J. 1948, in *Proc. 2nd. Int. Conf. Soil Mech. and Found. Eng.*, Rotterdam, **I**, 158, and **VII**, p. 47.

4. The Craigleith Sandstone, near Edinburgh, is an example of a siliceous sandstone, some varieties of the rock having 98 per cent silica. Darley Dale sandstone, Derbyshire, has over 96 per cent silica and contains a little feldspar and mica.

5. The term 'graded' is used in two senses: in engineering literature a 'well graded' deposit is one containing a range of different sized particles; whereas in geology the same phrase means well-sorted, i.e. having a preponderance of particles of one grade (or size) as shown in Fig. 5.2d.

6. Among the red sandstones of the Trias, the Bunter Sandstone from Woolton, Liverpool, has been used in the building of Liverpool Cathedral. A few structural stones from the Old Red Sandstone have been used in the past, as in Herefordshire buildings.

7. Freestone has also been defined as 'any fine-grained sandstone or limestone that can be sawn easily' (Arkell and Tomkieff, 1953, *Rock Terms*).

8. For a fuller discussion of sedimentation compression curves and examples from clays of different geological ages, see The consolidation of clays by gravitational compaction, by Skempton, A. W. 1970, *Q. J. Geol. Soc. Lond.*, **125**, 373–411.

9. See Skempton, A. W. and Northey, R. D. 1952, in *Géotechnique*, **3**, 30. Quick clays are also discussed by Kenney, 1964, in Sea level movements and the geologic histories of post-glacial marine soils at Boston, Ottawa, and Oslo, in *Géotechnique*, **14**, 203.

   The properties of a clay or silt are affected by the proportion of clay mineral particles smaller than 0.002 mm that it contains. The term *activity* is used to denote the ratio

plasticity index/clay fraction

where clay fraction is taken as the percentage (by weight) of the dry weight of the material which is composed of particles smaller than 2 micrometres (0.002 mm). For plasticity index see p. 313. The value of 'activity' for clays rich in kaolinite (p. 120) is about 0.45; and for those mainly composed of illite, 0.9.

10. Details of the composition and properties of Keuper Marl are given in Keuper Marl Research, by Kolbuszewski, J., Birch, N., and Shojobi, J. O. 1965. *Proc. 6th. Int. Conf. Soil Mechanics and Fndn. Eng.* (Canada), 1/13, p. 59.

    See also Chandler, R. J. 1967, The strength of a stiff silty clay, *Proc. Geotech. Conf.*, Oslo. And Kolbuszewski, J. Feb. 13, 1965, Chemical analysis of Keuper Marl, *The Surveyor*, London.

11.   Ball-clays in Devonshire, in particular the Bovey deposits, are described and illustrated by Best, R. and Fookes, P. G. 1970, in Some geotechnical and sedimentary aspects of ball-clays from Devon, *Q. Jour. Eng. Geol.* (London), **3**, p. 207–239.

See also Chapter 3, Note 8, for their formation.

12.   Portland Stone has been used in many London buildings and other structures, where it weathers evenly over a long period. The pale cream coloured stone is quarried from several beds of the Portland Series at Portland, Dorset, the best material coming from the Whit Bed and the Base Bed. The Roach is a coarser, shelly layer used for rough walling. For an account of the Portland Series see *British Regional Geology:* The Hampshire Basin and adjoining areas, by Chatwin, C. P. H.M.S.O. London.

Much limestone and sandstone waste at quarries is now used in the manufacture of 'artificial stone', i.e. fine-grained concretes made to resemble particular natural stones, with coloured cement and natural stone aggregates.

GENERAL REFERENCES

GILLOTT, J. E. 1968. Clay in Engineering Geology. Elsevier Publication Co., Amsterdam.

GREENSMITH, J. T. 1965. Petrology of the Sedimentary Rocks, Murby, London. (A revision of earlier volume by M. Black).

PETTIJOHN, F. J. 1957. Sedimentary Rocks, Harper & Brothers, New York.

TRUEMAN, SIR. A. 1954. The Coalfields of Great Britain, Arnold, London.

6 Metamorphic Rocks

Metamorphism is the term used to denote the transformation of rocks into new types by the recrystallization of their constituents (from Greek *meta*, after (denoting a change) and *morphe*, shape). Weathering processes are not included. The changes which occur in metamorphism result from the addition of heat or the operation of pressure, i.e. energy is put into the rocks. New minerals grow out of the old, in a solid environment, and the rock undergoes a transformation or metamorphism. The original rock, which may be igneous, sedimentary, *or metamorphic*, is subjected to new physical conditions, and the resulting metamorphic product is in equilibrium with the new conditions. Textures entirely different from those of the original rocks are produced. The changes may be aided by the presence of solvents, such as pore-contained fluids, which gradually work through the fabric of a rock and promote new mineral growth. Variations in chemical composition may also play a part in the process, as when the concentration of particular substances is brought about.

Two broad classes of metamorphism, depending on the controls exercised by temperature or pressure, may be distinguished:

1. *Thermal* (including *Contact*) *Metamorphism*, in which rise of temperature is the dominant factor. Thermal effects may be brought about when sediments are down-folded into hotter regions in the crust; or in contact zones adjacent to igneous intrusions.

2. *Regional Metamorphism*, in which both pressure and temperature have operated, and which commonly affects rocks over a large area. Where stress is locally dominant, as in belts of dislocation, the term *dislocation metamorphism* may be used.

1. *Contact metamorphism.* By the intrusion of a hot igneous mass, such as granite or gabbro, an increase in temperature in the neighbouring rocks is produced. The general effect of this increase is to promote the *recrystallization* of some or all of the components of the rocks affected, the severest changes occurring nearest the contact with the igneous body. When there are no external stresses acting upon a rock, but only heat, new minerals grow haphazardly in all directions, and the metamorphosed rock acquires a granular fabric, which is known as the *hornfels* texture (Fig. 6.3).

In addition, transfer of material sometimes takes place at such a contact, and hot gases or fluids from the igneous mass penetrate the country-rocks; this process is known as *pneumatolysis* (p. 207). During contact metamorphism the country-rock is not melted, but is often subjected to the action of emanations (such as water, carbonic acid, and volatile compounds of boron and fluorine) which percolate through it and result in the growth of new minerals; temperatures probably do not much exceed 500°C.

2. *Regional metamorphism.*   In this class, the operation of stress results in the recrystallization of the rocks affected, with the formation of new crystals which grow with their length or platy surfaces at right angles to the direction of the maximum compressive stress. High temperatures and strong stresses are produced in active (or orogenic) belts of the crust; the rocks formed by regional metamorphism are thus found in the great fold-belts, where they are frequently associated with granites (p. 140). The mineral components of the rocks have acquired a largely parallel orientation, and in consequence the rocks develop oriented or banded textures; the oriented texture produced by platy or columnar minerals is known as *schistosity* (Fig. 6.1), while an

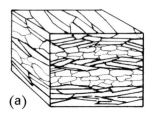

(a)

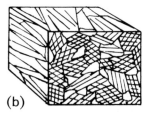

(b)

Fig. 6.1   Diagrams to illustrate schistosity. *Upper:* parallel orientation of mica crystals in mica-schist; *lower:* oriented hornblende prisms in hornblende-schist.

alternation of schistose layers with others less schistose produces the banded texture called *foliation* (Plate 6a). Argillaceous rocks under the influence of moderate stresses without marked rise of temperature develop *slaty cleavage*.

Where stress is the principal control and temperatures are low, as in zones affected by strong shearing movements, the rocks undergo a *dislocation metamorphism* with the production of cataclastic textures. These result from the mechanical breakdown of rocks under stress (p. 209), e.g. shearing, brecciation.

The crystallization form of a mineral partly determines the ease of its growth under metamorphic conditions, e.g. micas and chlorites, with one cleavage, grow easily in thin plates oriented with their surfaces at right angles to the maximum stress; amphiboles (such as hornblende) grow in prismatic forms with their length at right angles to the maximum stress. Some minerals which have a high crystallization strength (e.g. garnet, andalusite) grow to a relatively large size in metamorphic rocks, and are then called *porphyroblasts* (in contrast to the porphyritic crystals of an igneous rock, which have a different origin). Feldspars and quartz have low and nearly equal crystallization strengths, and thus metamorphic rocks composed of quartz and feldspar show typically a granular texture (p. 212).

CONTACT METAMORPHISM

The contact metamorphism of the main sedimentary rock types (shale, sandstone, limestone) is now discussed, assuming temperature to be the controlling factor, stress being entirely subordinate. Consider the contact zone bordering an igneous mass, say a granite, intruded into sedimentary rocks; the latter are metamorphosed for some distance from the contact, and the area over which this metamorphism occurs is called the *contact aureole* (Fig. 6.2). Within the aureole metamorphic zones of increasing severity are traceable as the contact is approached; they are distinguished by the development of certain minerals (see below). The size of an aureole depends on the amount of heat transferred, i.e. on the size and temperature of the igneous body. The width of the aureole indicates the steepness or otherwise of the contact (cf. east and west sides of the aureole in the figure).

1. *Contact metamorphism of a shale (or clay).* An argillaceous rock like shale is largely made up of very minute particles; some are flaky (the clay minerals, p. 119) and are essentially hydrated aluminium silicates; together with them are particles of sericite (secondary white mica) and chlorite, and smaller amounts of colloidal silica, colloidal iron oxide, carbon, and other substances. The two dominant oxides in a clay or shale, as would be seen in a chemical analysis, are thus SiO_2 and Al_2O_3, and when the shale is subjected to heat over a long period the aluminium silicate *andalusite*, or its variety *chiastolite*, is formed (p. 114). *Cordierite* is another mineral frequently formed at the same time; it grows, with andalusite, as porphyroblasts in the metamorphosed shale.

In the outermost zone of the aureole (Fig. 6.2, zone 1) small but

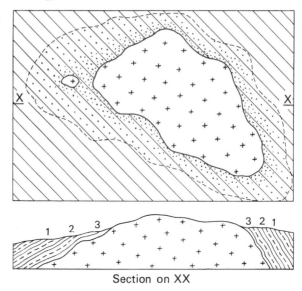

Section on XX

Fig. 6.2   Metamorphic aureole around an igneous intrusion. Aureole dotted. 1, 2, and 3 are zones referred to in the text.

noticeable changes are produced; the shales are somewhat hardened, specks of opaque magnetite are formed, and incipient crystals of chiastolite or cordierite appear as small dark spots; the name *spotted rock* is used to describe this low-grade metamorphic product. Nearer the contact (zone 2), chiastolite or cordierite (or both) are better developed, and give very distinct spots, and small flakes of biotite have formed from the chlorite and sericite in the shale. This recrystallization is continued in the next stage (3) still nearer the contact, together with the growth of muscovite flakes (from any sericite remaining) and of quartz; magnetite (from original Fe) and rutile (TiO_2) may also be present, depending on the original composition of the sediment. The shale has now been completely recrystallized, and is called a *hornfels;* it is hard, with a 'horny' fracture, and has a typical fine granular texture (Fig. 6.3). Nearest to the igneous contact a further mineral change may occasionally take place: some muscovite of the hornfels may give rise to crystals of orthoclase, the excess alumina then entering into combination with silica to form needles of *sillimanite* (Al_2SiO_5, acicular). The shale has now become a biotite-cordierite-sillimanite-quartz-hornfels (Note 1).

(It is assumed in the above discussion that chemically there is neither addition nor subtraction of material during the metamorphism. If chemical transfer does take place across the contact, a variety of entirely new conditions will arise.)

As the igneous rock is approached from outside its aureole the

metamorphic sequence passed over is, therefore: hardened shale →
spotted rock → cordierite-bearing hornfels. This may be seen, among
British occurrences, at places around the granites of Devon and
Cornwall and in Scottish aureoles, e.g. around the Ben Nevis granite
or the Insch (Aberdeen) gabbro. An interesting case is provided by the
Skiddaw granite (Cumberland), which is only exposed over three small
areas, each less than a mile across, in the valleys of streams flowing
from the north-east side of the mountain; over a considerable area of
adjacent ground, however, there are exposures of chiastolite-bearing
rock, indicating the existence of the granite at no great depth below
the present surface, which marks a stage in the process of unroofing
by denudation.

2. *Contact metamorphism of a sandstone.*   A siliceous rock like
sandstone is converted into a metamorphic *quartzite.* The original
quartz grains and siliceous cement are recrystallized into an interlocking mosaic of quartz crystals. Partial fusion of the mass may occur
under special circumstances, but rarely. Constituents other than
quartz in the sandstone or the cement between the grains may give
rise to new minerals, depending on their composition, e.g. a little
biotite (from clay), magnetite (from iron oxide cement). The bulk of
the rock consists essentially of quartz.

3. *Contact metamorphism of a limestone.*   Limestones are mainly
made up of calcium carbonate, generally together with some $MgCO_3$
and silica. (i) In the ideal case of a pure calcium carbonate rock, the
metamorphic product is a *marble* composed entirely of grains of
calcite. The coarseness of texture of the rock depends on the degree of
heating to which it has been subjected, larger crystals growing if the
metamorphism has been prolonged. Dissociation of the carbonate into
CaO and CO_2 is prevented by the operation of pressure. Examples
of such rocks, with a high degree of purity, are provided by the
statuary marbles. (ii) In the metamorphism of a limestone in which a
proportion of silica is present (e.g. as quartz grains or in colloidal
form), the following reaction occurs in addition to the crystallization
of calcite described above:

$$CaCO_3 + SiO_2 \rightarrow CaSiO_3 \text{ (wollastonite[1])} + CO_2.$$

The resulting product is a *wollastonite-marble,* as in parts of the
Carboniferous Limestone of Carlingford, Ireland, where the rock has
been metamorphosed by gabbro intrusions. (iii) During the contact
metamorphism of a magnesian or dolomitic limestone the dolomite is
dissociated, thus:

$$CaCO_3 \cdot MgCO_3 \rightarrow CaCO_3 \text{ (calcite)} + MgO \text{ (periclase)} + CO_2.$$

[1] The mineral *wollastonite,* which occurs under conditions such as these, forms white or grey
tabular, translucent crystals.

The periclase readily becomes hydrated, forming *brucite*, $Mg(OH)_2$, in colourless tabular hexagonal crystals, and the product is a *brucite-marble*. Blocks of this rock have been ejected from Vesuvius; other occurrences of brucite-marble are in Skye and in the North-West Highlands.

When silica is present, the magnesium silicate *forsterite* (p. 96) is formed together with crystalline calcite, CO_2 being lost in the reaction, thus:

$$2CaCO_3 \cdot MgCO_3 + SiO_2$$
$$\rightarrow 2CaCO_3 \text{ (calcite)} + Mg_2SiO_4 \text{ (forsterite)} + 2CO_2.$$

The metamorphic product is a *forsterite-marble* (Fig. 6.3); a British locality for this rock is Kilchrist, Skye. If the further change from

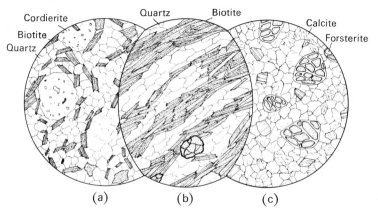

Fig. 6.3 Metamorphic rocks in thin section. (*a*) Biotite-cordierite-quartz-hornfels. (*b*) Mica-schist with garnets. (*c*) Forsterite-marble.

forsterite to serpentine takes place (by hydration), a *serpentine-marble* is formed (also called *ophicalcite*); the white calcite, streaked with green serpentine, can make a beautiful decorative stone, e.g. the Connemara Marble of Ireland.

When some clay is present as well as silica in the original rock, minerals such as tremolite and diopside (Ca-Mg-silicates) and lime garnets are formed by contact metamorphism, in addition to calcite, and the resulting rock is called a *calc-silicate-hornfels*. Its mineral composition varies according to the proportions of different substances present in the original limestone; the texture of these rocks tends to be coarse, because of the fluxing action of dissociated $MgCO_3$. (See Note 2 on the influence of hard calc-silicate-hornfels in tunnelling.)

The above discussion of some varieties of marble does not exhaust all the possibilities, and other types may arise, some beautifully coloured by traces of impurities in the original sediment. Decorative stones of this kind are exported from Italy and Greece. The rock is cut into thin slabs, with one surface polished, and is often displayed as panels for internal decoration; advantage may be taken of any pattern due to veining or zones of brecciation, by arranging the slabs symmetrically in groups of two or four (Note 3).

4. *Contact metamorphism of igneous rocks.* The effects here are not so striking as they are for the sedimentary rocks, because the minerals of an igneous rock were formed at relatively high temperatures and are less affected by reheating; but some degree of recrystallization is often evident. A basic rock like dolerite or diabase may be converted into one containing hornblende and biotite (from the original augite and chlorite), the plagioclase being recrystallized. Secondary minerals occupying vesicles, as in amygdaloidal basalt, yield new minerals such as lime-feldspar (after zeolite), and amphibole (after chlorite and epidote).

Igneous rocks which were much weathered before contact metamorphism yield calc-silicate minerals (e.g. lime garnet) from the calcite of the original rock, and hornblende from original chlorite. Andesites and andesitic tuffs on contact metamorphism develop many small flakes of brown mica and crystals of magnetite; these changes are well seen near the contact of the Shap Granite, Westmorland.

Basic *granulites* (rocks of equi-granular texture, p. 212) are produced by the complete recrystallization of a basalt or gabbro, during prolonged high-grade metamorphism; many instances are on record among the Tertiary igneous rocks of Scotland, e.g. the granulitic patches enclosed in the gabbros of Skye and Mull.

PNEUMATOLYSIS

In the foregoing discussion it has been assumed that no transfer of material from the igneous mass has taken place across the contact. It frequently happens, however, that volatile substances, accumulated in the upper part of a body of magma during its crystallization, pass eventually into the country-rocks, which are impregnated with the vapours at a moderately high temperature stage in the cooling history of the igneous mass. This gas action is called pneumatolysis; it probably takes place at temperatures around 500°C (p. 164). The vapours include compounds of boron, fluorine, carbon dioxide, sulphur dioxide, etc.; characteristic minerals which are produced near the contact under these conditions include *tourmaline, topaz, axinite, fluorspar,* and *kaolin.*

Tourmaline (p. 113) is formed by the action of boron-bearing vapours, with some fluorine. It is a mineral with a high content of alumina, and hence tends to be formed in clay bands in the country-rocks adjacent to the igneous mass. The biotite in a granite may also be changed to tourmaline by the addition of boron and fluorine, the granite itself often being locally reddened. The name *luxullianite* (from Luxullian, Cornwall) is given to a variety of tourmalized granite in which the tourmaline occurs as radiating clusters of slender crystals (the *schorl* variety) embedded in quartz. Quartz-tourmaline veins, often carrying cassiterite, are common in some granite areas as discussed on p. 166.

Axinite is a calcium-boron-silicate which is formed where transfer of boron-bearing vapours into limestone has occurred, at an igneous contact. Axinite crystals are generally flat and acute-edged, brown, and transparent, with a glassy lustre (Fig. 3.1).

Kaolinization. The action of steam together with some carbon dioxide on the orthoclase of a granite results in the breakdown of the feldspar into kaolin; potassium is removed and silica is liberated:

$$4KAlSi_3O_8 + 2CO_2 + 4H_2O$$

Orthoclase

$$= 2K_2CO_3 + Al_4Si_4O_{10}(OH)_8 + 8SiO_2.$$

Kaolinite

Large parts of the granites of Devon and Cornwall, as at St. Austell, have been decomposed in this way into a soft mass of quartz, kaolin, and mica, which crumbles at the touch. In the quarries the kaolinized rock is washed down by means of jets of water; the milky-looking fluid is then pumped from a sump and allowed to gravitate down through a series of tanks, the mica and quartz being first removed in the process, and the kaolin finally being allowed to settle out of suspension. This kaolin or 'china clay' is an important economic product; it is used as a paper filler, in pottery manufacture, and for numerous other purposes.

China-stone is a granitic rock which represents an arrested stage in the kaolinization of a granite; in addition to quartz and kaolinized feldspar, it frequently contains topaz $(Al_2SiO_4F_2)$ and fluorspar, both of which minerals point to the action of fluorine during pneumatolysis.

Greisen is a rock composed essentially of quartz, white mica, and topaz, and is formed in small amounts from granite under certain conditions of pneumatolysis; where, for example, the formation of K_2CO_3 is inhibited (see above equation), the feldspar of the granite gives rise to white mica (secondary muscovite) and the resulting rock is called greisen.

REGIONAL METAMORPHISM

The operation of stresses on rocks on a regional scale has the general effect of inducing growth of new minerals with a parallel orientation, as indicated on p. 202, their length or platy surfaces being developed at right angles to the direction of maximum pressure. Increasing temperature may be associated with the stress conditions. Thus, strong shearing stress and low temperatures arise commonly in the outer part of the crust (epizone); moderate shearing stress and moderate temperatures occur at lower levels (mesozone); and low stress with higher temperature at still deeper levels (katazone), where also hydrostatic pressure reaches high values. Under these different controls (or energy levels), sedimentary rocks of argillaceous composition give rise on metamorphism to *slates*, or *schists*, or certain *gneisses*, respectively. Typical minerals which are formed during such regional metamorphism include chlorite, sericite, epidote, albite (under epizone conditions); and in the mesozone, muscovite, biotite, hornblende, garnet, and kyanite (Al_2SiO_5, in pale blue blade-like crystals, triclinic).

The operation of stresses without much rise of temperature may be locally concentrated in narrow belts of crushing or shearing, where severe mechanical effects are produced in the rocks affected. Under this dislocation metamorphism (p. 203) rocks are badly deformed, and internal rearrangements of a mechanical kind take place among their minerals. Signs of strain which may be observed in individual crystals, in thin sections, include the strain polarization of quartz (a patchy extinction due to the distortion of the crystal structure); granulation (breakdown into smaller crystals); the bending of mica flakes and of the twin lamellae of plagioclases; and the production of strain-slip cleavage (small movements or 'slips' along a series of closely spaced shear planes).

An extreme example of the effect of stress may be seen in the rocks adjacent to a big thrust-plane such as the Moine Thrust (p. 237), where the rocks have been subjected to shearing, together with a considerable vertical load, and are locally transformed into *mylonites* (a name which refers to the 'milling' or rolling out of the original mineral grains).

Slate   Under the influence of stresses which may be of regional extent, argillaceous sediments such as shales take on the property of being easily split along parallel planes which are independent of the original bedding (Fig. 6.4). This property is known as *slaty cleavage;* it is often very regular and perfect, yielding thin, smooth-sided plates which are lithologically similar. The cleavage is related to the new texture which is developed in the rock under the influence of the metamorphic stresses; minute crystals of flaky (layer-lattice) minerals like chlorite and sericite grow with their platy surfaces at right angles to the direction of maximum compressive stress. Some original minerals such as small quartz grains may become reoriented with their length at right angles to the maximum stress. The rock thus develops a preferential split-direction or slaty cleavage (to be distinguished from *mineral* cleavage), which is parallel to the flat surfaces of the oriented crystals. Fossils in the rocks become distorted or broken by the stresses.

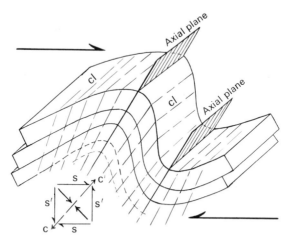

Fig. 6.4   Slaty cleavage in relation to folds. Heavy arrows show deforming forces. Cleavage is developed in the direction of easiest relief, which is perpendicular to the principal compressive stress (see small diagram of stresses, where *cc* = direction of cleavage); it is parallel to the axial planes of folds. Note that the angle between cleavage and bedding varies, and is less on limbs of fold than at the axis. *cl* = trace of cleavage on bedding surfaces; *s, s'* = shearing couples.

Slate is a low-grade metamorphic product, and is essentially a very fine-textured aggregate of chlorite, sericite, and quartz; it is not always completely crystalline. With the action of continued stresses, and especially if some rise of temperature ensues, the above minerals continue to grow, with the production of larger crystals (muscovite and chlorite flakes), and a lustrous, finely crystalline, micaceous rock called *phyllite* results. With continued metamorphism the size of constituent minerals increases further, leading to the formation of rocks known as *mica-schists*. There is thus a gradation, shale → slate → phyllite → mica-schist, the several metamorphic types being produced by metamorphism under different energy conditions, from similar material which was originally muddy sediment.

While most slates are derived from fine-grained sediments like shales, the green Cumberland and Westmorland slates are an example of fine-grained volcanic tuffs (andesitic in composition) which have developed slaty cleavage during regional metamorphism.

The commercial value of slate from any particular locality depends on several factors: (i) The perfection of the cleavage; good roofing slate must be cleavable into thin plates, while poorer slates split less easily and may be useful only for rough walls and similar purposes. (ii) For high-quality slate the cleavage surfaces should be perfectly smooth. This property is found, for example, in the slates of Cambrian age from the Bethesda and Llanberis quarries, North Wales. Some slates have a rougher and somewhat corrugated surface (called 'puckering'), which limits the thinness of the sheets and hence increases their weight. (iii) A good slate should be homogeneous in mineral composition and texture. This affects its water-tightness: its low porosity is due to close packing of the mineral constituents. Colour banding, if due only to iron oxide staining, is not a serious matter; but if due to differences in composition or texture is open to objection. (iv) Accessory minerals such as calcite, epidote, pyrite, should be very few or, preferably, entirely absent. Calcite is an objectionable constituent if there is much of it, on account of its low resistance to weathering, especially to rain-water acids in city atmospheres. A small content of epidote is not harmful. The iron sulphides pyrite and marcasite are objectionable; they weather rapidly, producing ugly iron staining, and if developed in large crystals (porphyroblasts) may leave holes of corresponding size in the slate.

The best slates are those in which the stresses which have induced slaty cleavage have operated in a constant direction. Where two sets of stresses have been active at successive periods, two cleavage directions may be superimposed and the smoothness and regularity of a single cleavage destroyed.

British slate localities, in addition to those mentioned above, include Penrhyn and Bangor in North Wales (Note 4); Ballachulish, Scotland (black slates, some with pyrites); Skiddaw, Cumberland (rough grey slates of Ordovician age); Borrowdale, Cumberland (green slates formed from the ash deposits of Ordovician volcanoes); and Delabole, Cornwall (grey slates of Upper Devonian age).

Schists These crystalline rocks of schistose texture are moderately coarse-grained, so that their main mineral components can be distinguished by eye. They have been formed from sedimentary or igneous material by the operation of moderately high temperatures and pressures.

Schists are largely composed of flaky minerals such as micas, chlorite, talc, or of prismatic minerals such as amphiboles, which have the predominantly parallel arrangement called *schistosity* (p. 202); the flakes or *foliae* into which the rock breaks all have a similar mineral composition and their surfaces are lustrous. (The word 'schist' (Gk. *schistos*, divided) was originally used to denote the property of splitting into foliae.) Some quartz is present, in addition to the flaky minerals, and the quartz grains under the influence of stress have often become elongated (Fig. 6.5) and lie with their length in the surfaces of schistosity. Taking first the schists derived from sedimentary material, a *mica-schist* as already noted is produced by the

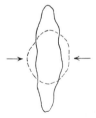

Fig. 6.5   Change of shape of a crystalline grain (e.g. quartz) under stress. Original grain shown dotted. Stress lowers the melting point and is especially prominent where mineral grains touch. Solution proceeds at points of high stress and re-deposition takes place in directions at right angles to the stress direction.

metamorphism of an original clay or shale. It is composed largely of mica (biotite and muscovite), the flakes having the parallel orientation described above, with a varying proportion of elongated quartz grains (Fig. 6.3(b)). Garnet may occur as porphyroblasts in the schist, and the schistosity is sometimes disturbed by the growth of this mineral during metamorphism, the garnets pushing apart the micaceous layers. Rocks rich in quartz and with less mica provide the *quartz-schists*, derived from sandy sediment with a small clay content. Certain coarsely foliated rocks of the same derivation, and formed during the higher grades of regional metamorphism, are the mica-gneisses (p. 214). The regional metamorphism of feldspathic sandstones yields siliceous *granulites*, i.e. rocks composed of granular quartz and feldspar in equi-dimensional, interlocking grains (the granulitic texture). Schistosity is not pronounced in these rocks because platy minerals are scanty, though a little mica is usually present. When the rock is composed almost entirely of granular quartz, the term *quartzite* is used (= quartz-granulite; cf. another use, p. 205).

Marbles result from the regional metamorphism of limestones (as well as by contact metamorphism); they are composed mainly of interlocking grains of calcite (Fig. 6.3(c)). If the original rock was a pure limestone, schistosity is absent in the marble, because the calcite occurs as equi-dimensional grains. If the original rock had other constituents as well as $CaCO_3$, other minerals arise and the metamorphic product may show a directional texture with streaks of different mineral composition.

The metamorphism of basic igneous rocks such as basalt or dolerite gives rise to chlorite-schists (under epizone conditions) or hornblende-schists (mesozone). *Chlorite-schist* consists dominantly of chlorite flakes, in parallel orientation, often with a little quartz and some porphyroblasts of magnetite or garnet; it is formed under conditions of high stress and low temperature. (By high stress is meant a wide difference between maximum and minimum principal stresses, implying strong shearing stress.)

Hornblende-schist is composed essentially of hornblende and quartz, and is derived from similar basic rocks but under a higher grade of metamorphism, with moderate stress and moderate temperature (Note 5). British examples of hornblende-schist are found at Landewednack, S. Cornwall; Start Point, S. Devon; and more abundantly in the Dalradian Series of the Scottish Highlands, as in the neighbourhood of Loch Tay.

Eclogite, a rock composed of pyroxene, garnet, and quartz, is derived from similar original material under conditions of low stress, high hydrostatic pressure and high temperature. (Stress approaching hydrostatic conditions, with small stress differences, exists at a depth of a few kilometres in the earth's crust.) This series (*viz*. chlorite-schist, hornblende-schist, eclogite) illustrates the dependence of the metamorphic product on the conditions of metamorphism.

Derivatives of ultrabasic rocks such as peridotite or dunite include some *serpentine-schists* and *talc-schists*. The latter are composed of talc, a very soft mineral, together with some mica. Soft rocks such as chlorite- and talc-schists, which are somewhat greasy to the touch, and sometimes decomposed mica-schists, may be a source of weakness in engineering construction and fail easily by shearing along the foliae. Instances are on record where underground works have been hindered when rocks of this kind were encountered, in tunnelling and boring. The schist, when left unsupported in a working face, has begun to move rapidly (or 'flow') under the pressure exerted by the surrounding rocks (see Chapter 15, p. 465).

Fabric The term fabric (German, *gefuge*) is now commonly used to denote the arrangement of the mineral constituents and textural elements of a rock, in three dimensions. The term is particularly useful in connection with metamorphic rocks, as it enables the preferred orientation of their minerals, when present, to be described with reference to broader structures. Rocks are either *isotropic* when there is no orderly arrangement of their components (as in hornfels); or *anisotropic* (as in schists) when there is a parallel orientation of minerals, often well developed. Relict textures from an original sediment, such as banding due to variation in composition, may also be preserved in the fabric of a meta-sediment, as in some gneisses (see below).

Gneisses A gneiss has the banded texture called foliation (p. 202), in which light and dark bands alternate; the light bands are composed essentially of feldspar and quartz, the dark bands or streaks of one or more mafic minerals such as biotite, hornblende, or in some cases pyroxene (Plate 6a). Garnets are common accessory constituents; biotite is often accompanied by muscovite. A gneiss breaks less readily than a schist, and commonly splits across the foliation instead of along it; it is often coarser in texture than many schists, though some gneisses are relatively fine-grained.

A gneiss when derived by regional metamorphism from an igneous rock, such as granite, is called an *orthogneiss;* if it results from the metamorphism of a sedimentary rock, the term *paragneiss* is used. An example of the latter is *biotite-gneiss*, the high grade metamorphic derivative of an argillaceous rock such as a shale. It is composed of pale bands made up of quartz and feldspar, which alternate with dark, mica-rich bands or streaks. This banding or foliation may result from variations of composition in the original sediments. Biotite-gneiss thus corresponds to mica-schist (lower grade of metamorphism) in the series: phyllite → mica-schist → biotite-gneiss.

Some gneisses have the composition of a granite or granodiorite, and result from the metamorphism of such rocks (orthogneiss). They are foliated rocks, bands of quartz-feldspar composition being interspersed with streaks of biotite, or biotite and hornblende, in which the mafic minerals are oriented in parallel positions. Generally the feldspar is evenly distributed in the quartz-feldspar bands; but occasionally large feldspars (or aggregates of feldspar crystals) occur in lenticular or eye-shaped areas, to which the name 'augen' is given. Such a rock is an *augen-gneiss;* the foliation is often deflected around the eye-shaped areas. Augen-gneisses arise by regional metamorphism of granitic rocks, or by injection processes. An example of the former is the metamorphosed granite of Inchbae, Rosshire; the porphyritic feldspars of the original granite are now the augen of the gneiss. The granite was intrusive into argillaceous rocks, in which an aureole with andalusite-hornfelses was formed by contact metamorphism. When the mass underwent a second (regional) metamorphism, the hornfelses were converted into kyanite-schists and the granite into augen-gneiss. Several metamorphisms may affect an area, as in the island of Unst, Shetlands, where two successive regional metamorphisms followed by a phase of dislocation metamorphism have been demonstrated (Note 6).

Certain beautifully striped gneisses result from the injection of thin sheets of quartz-feldspar material of igneous origin along parallel planes in the parent rock; these are *injection gneisses*. The introduction of igneous material into country-rocks of various kinds produces mixed rocks which are called *migmatites*. In some migmatites the mixing is mechanical, the introduced material veining or striping the original country-rock. In other migmatites the mixing is chemical, and arises from the permeation or soaking of the country-rocks by the invading fluids (see p. 163, granitization). Migmatites of many kinds are found in the Pre-Cambrian rocks of Scandinavia and Finland (the Baltic Shield, p. 251), and in areas of high grade metamorphism in many orogenic belts. Migmatites in Central Sutherland have been described in detail (Note 6).

The Scottish Highlands are largely built up of metamorphic rocks, gneisses, and schists occurring in great variety in the Lewisian and Dalradian groups (see Fig. 8.7). The Lewisian rocks, which form part of a great belt of regional metamorphism in north-west Scotland, include many high-grade gneisses which were derived from earlier igneous rocks, and possess a foliation interpreted as due to recrystallization from a plastic condition. They often contain veins and patches of granitic material and are then described as migmatites. The Dalradian group comprises a series of metamorphosed sediments, with muddy (pelitic), sandy (psammitic), and limestone bands which have given rise to mica-schists and phyllites, quartzites, and marbles respectively. Dolerite sills intruded into the sediments before their metamorphism were converted into hornblende-schists and related rocks (p. 213).

In the Alps are found much younger schists and gneisses, the metamorphic equivalents of Jurassic, Cretaceous and Tertiary sediments, which owe their transformation to the great orogenic movements by which the Alps were raised. Other large areas of regional metamorphic rocks are found similarly in fold-belts in other continents.

NOTES AND REFERENCES

1. Hornfelses are often suitable for use as road-stone and ballast. Large supplies for railway ballast have been obtained at the Meldon Quarry, Okehampton, in the aureole of the Dartmoor granite. Structures in the Lower Carboniferous rocks revealed by this quarrying are described by Dearman, W. A. 1959, in The Structure of the Culm Measures at Meldon, near Okehampton, North Devon, *Q. Jour. Geol. Soc. Lond.*, **115**, 65–106.

2. A zone of calc-silicate-hornfels (metamorphosed Ballachulish Limestone) was passed through by the Lochaber water-power tunnel at Ben Nevis; this extremely hard rock, together with baked schists at a granite contact, reduced the rate of progress of the tunnel excavation to half the average rate. (Reference: Halcrow, W. T. 1930–31, in *Proc. Inst. Civ. Eng.*, **231**, 54).

3. The term 'marble' is also used as a trade name for any soft rock which will take a polish easily, and includes many limestones which make attractive decorative stones on account of their colouring or content of

fossils. British examples include the Ashburton 'marble' (a Devonian coral limestone with stromatoporoids), the Hopton Wood stone (a Carboniferous crinoidal limestone from Derbyshire), and the Purbeck 'marble' (an Upper Jurassic limestone from Dorset, containing the fossil shell *Paludina*, Fig. 8.9, *j*).

4. See The Slates of Wales, by F. J. North. 5th Edition. Published by the National Museum of Wales.

5. The term *amphibolite* is used as a group name for metamorphic rocks in which hornblende, quartz, and plagioclase are the main minerals; hornblende-schist is one member of this group.

6. The Unst rocks are described by Read, H. H. 1934, in *Q. Jour. Geol. Soc. Lond.*, **90,** and in Metamorphic correlation in the polymetamorphic rocks of the Valla Field block, Unst, Shetlands, *Trans. Roy. Soc. Edinburgh*, 1937, **59,** 195–221.

    A summary and map of Unst are given in The Geology of Scotland (Ed. G. Y. Craig), 1967, Oliver & Boyd, Edinburgh.

    For migmatites, see discussion in Read and Watson, 1970, Chapter 10.

GENERAL REFERENCES

HARKER, A. 1950. Metamorphism. 3rd. Edition, Methuen. A classic text on this subject.

READ, H. H. AND WATSON, J. 1970. Introduction to Geology, chapter 9, (paperback edition). Macmillan, London.

TURNER, F. J. AND VERHOOGEN, J. 1960. Igneous and Metamorphic Petrology, 2nd Edition, McGraw-Hill Book Co., New York.

7 Geological Structures: Folding and Fracture

Sedimentary rocks, which have been discussed in an earlier chapter, occupy the greater part of the earth's land surface and are found in beds or *strata;* a series of such layers constitutes a *succession* of strata. A single stratum may be of any thickness from a few millimetres to a metre or more, and each stratum is bounded by surfaces known as bedding planes (p. 50). We are here concerned with the various arrangements of strata as structural units which build the upper part of the crust; the ways in which these structures appear on geological maps is discussed in Chapter 11. This chapter deals first with horizontal and simply dipping beds, then with folding, faulting, and joints.

Dip and strike Consider a flat uniform bed of rock which is tilted out of the horizontal (Fig. 7.1*a*). On the sloping surface of this stratum there is one direction in which a horizontal line can be drawn: this is called the *strike* of the bed. It is a direction which can be measured and recorded as a compass bearing when part of the surface

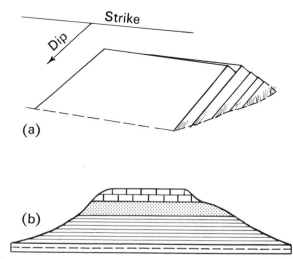

(a)

(b)

Fig. 7.1 Dip and strike. (*a*) dipping strata in an exposure of rock at the surface of the ground; (*b*) section through a hill formed in horizontal Leds (dip zero), limestone at top, sandstone below (dots), and shales (lined).

of the stratum is exposed to view. At right angles to the strike is the direction of maximum slope or dip. The *dip* (or true dip) is the angle of inclination which a line on the stratum in the dip direction makes with the horizontal. It is measured with a clinometer and recorded to the nearest degree; for example, a dip of 25 degrees in a direction whose bearing is N. 140° would be written *25° at 140*. (The bearing may be taken from true or from magnetic North or, in the southern hemisphere, from South). The inclination to the horizontal of a line on the sloping rock-surface in any other direction is less than the true dip, and is called an *apparent dip*. Apparent dips are often seen in steep quarry faces where these faces do not run in the true dip direction (p. 319), but intersect the dipping beds in some other direction.

Dip and strike are two fundamental conceptions in structural geology, and represent the geologist's method of defining the attitude of a sloping stratum at any point. The information is placed on a map in the form of a short arrow with its point at the place of observation and with an adjacent figure, indicating the direction and amount of dip (see Chapter 11). For horizontal beds the symbol + is used, i.e. when the dip is zero.

Horizontal strata   A section through a hill built of horizontal beds is shown in Fig. 7.1*b*. The uppermost strata form the flat top of the hill, and the area over which they are present at the surface is their *outcrop*. Successively lower strata lie at lower levels and crop out on the slopes of the hill. A large flat-topped hill is called a *mesa*, or tableland; it may sometimes be capped by an igneous rock such as a lava flow, or by a resistant sedimentary layer.

A series of dipping beds will have outcrops of varying shape and width, according to their thickness and to the form of the ground surface which they intersect. The appearance of such outcrops on a geological map is discussed in Chapter 11.

Escarpment, outlier, and inlier   A hill or ridge which is formed by hard beds overlying softer rocks, often with a gentle dip, has a steeper slope on one side than on the other and is called an *escarpment* (Fig. 7.2). The length of such a ridge follows the strike direction of the dipping beds; the gentler slope in the dip direction is called the *dip-slope*. Wenlock Edge (Fig. 2.9) and the North or South Downs (Fig. 8.12) are examples of escarpments in England. When beds dip more steeply, a ridge is formed (after denudation) whose dip-slope may have the same inclination as the scarp slope, and the ridge is then called a *hog-back* (e.g. the narrow Chalk outcrop west of Guildford, Fig. 8.10). Steep-sided ridges are also formed in vertical beds, i.e. when the dip is 90°.

Outlier and inlier.   An outcrop of rock which is completely surrounded in plan by older rocks is called an *outlier*. Thus, the beds

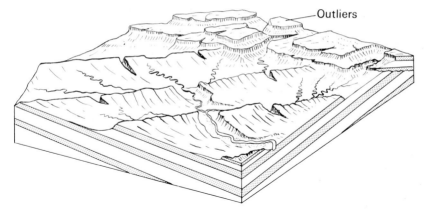

Fig. 7.2   Escarpments and outliers. Hard beds (dotted), gently dipping towards the observer, form a series of escarpments running from left to right. Intervening valleys have been carved out of softer beds. On the extreme right a younger series rests unconformably on the older rocks and is dissected into outliers.

forming the hills shown in the background of Fig. 7.2 are detached from their main outcrop, and constitute outliers.

The converse of this is an *inlier*, *viz.*, an area of older rocks surrounded by newer. Inliers are often developed in valleys, where streams have cut down and exposed locally the rocks which are below (older than) those forming the sides of the valley.

Unconformity   One series of strata is sometimes seen to lie upon an older series with a more or less regular surface separating the two. A junction of this kind is called an *unconformity* (Plate 9b and Fig. 7.2, right); there is frequently an angular discordance between the older beds and the newer. Again, Fig. 7.3 shows an upper series of beds, now gently dipping to the north-west, which has been deposited unconformably upon a lower and older series. The latter were folded by horizontal compression, and then denuded during a period of elevation of the sea floor, after which the area was again submerged

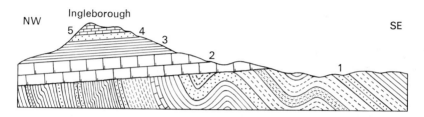

Fig. 7.3   Section through Ingleborough, West Yorkshire, to show unconformity and escarpment (after D. A. Wray, H.M. Geol. Survey). (1) Silurian and older rocks, below the unconformity. (2) Carboniferous Limestone. (3) Yoredale Series, with sandstones (4). (5) Millstone Grit. Length of Section about 5 miles.

and the upper beds were laid down. The unconformity thus marks a break in the process of deposition, and represents an interval of time during which no sediments were laid down in the area. A later uplift with tilting brought the unconformity to its present position. Such a sequence of events is referred to again in Chapter 8.

Another term used in connection with unconformity, i.e. *overlap*, may be noted here. As the sea advances over a land during a sub-mergence, and successive layers of sediment are deposited over wider and wider areas, each layer will cover those below it and overlap further on to the land. An unconformity with overlap is thus formed; the Upper Cretaceous rocks in Britain (p. 263) are an example of sediments deposited over successively wider areas in this way.

FOLDS

It is frequently seen that strata forming parts of the earth's crust have been bent or buckled into *folds;* and dipping beds, mentioned above, are often parts of such structures. An arched fold in which the two sides dip away from one another is called an *anticline* (Fig. 7.5a); the rocks forming its central part (or core) are older than the outer strata. A fold in which the limbs dip towards one another is a *syncline*, and the strata forming the core are younger than the adjacent layers. If the relative ages are not known the terms *antiform* and *synform* (p. 222) can be used.

The term *monocline* is given to a flexure that has two parallel gently dipping limbs with a steeper middle part between them (Fig. 7.4);

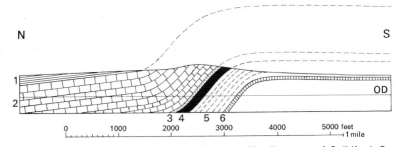

Fig. 7.4 Monoclinal fold at the Hog's Back, about 4½ miles west of Guildford, Surrey (after H.M. Geol. Survey). (1) Tertiary Beds. (2) Chalk. (3) Upper Greensand. (4) Gault. (5, 6) Lower Greensand.

it is in effect a local steepening of the dip in dipping, or sometimes horizontal, beds. It differs from an anticline in that the dip is in one direction only. Large monoclines have sometimes been developed in sedimentary rocks overlying a basement when the latter has been subject to fault movement.

In many cases folding has been due to the operation of forces tangential to the earth's surface; the rocks have responded to crustal compression by bending or buckling, and often form a fold-system whose pattern is related to the controlling forces. Such folding is termed *diastrophic;* the term diastrophism (= distortion, Greek) is used to denote processes connected with major deformations of the earth's crust. Other kinds of folds, such as those formed by movement under gravity or by flowage of weak materials, are called *non-diastrophic*. We can therefore classify folds as follows:

I. Diastrophic (due to tectonic compression)
- compressional folds
- parasitic (or minor) folds
- shear folds

II. Non-diastrophic (due to other causes, e.g. gravity)
- gravity folds
- valley bottom folds
- salt domes, etc.

In many structures (both folds and faults) the strata have behaved as brittle materials; but some folds have shapes which indicate that the rocks have deformed by ductile flow. The former are the more resistant or *competent* strata; the latter are weaker, more ductile, and are called *incompetent* strata. The scale of the folding ranges from the smallest puckers seen only in thin sections of a rock with a microscope, to large folds which may make up single outcrops or a small mountain, with wavelengths measured in tens or hundreds of kilometres.

I. DIASTROPHIC FOLDS

Some *compressional folds* are shown in Fig. 7.5, and we now consider their geometry. In the cross-section of an upright fold (*a*) the highest point on the anticline is the *crest* and the lowest point of the syncline the *trough;* the length of the fold extends parallel to the strike of the beds, i.e. in a direction at right angles to the section.

The line along a particular bed where the curvature is greatest is called the *hinge* or *hinge-line* of the fold (Fig. 7.5*b*); and the part of a folded surface between one hinge and the next is a fold *limb*.

In some folds the surface which bisects the angle between the fold limbs is a plane (as in Fig. 7.5*a*, where it is vertical), and can be called the *axial plane*. More accurately the term *axial surface* is used, and is defined as the locus of the hinges of all beds forming the fold (see Fig. 7.5*c*). This definition allows for the curvature which is frequently found in an axial surface (Note 1).

The intersection of an axial surface with the surface of the ground can be called the *axial trace* of the fold; in some instances it is marked on a geological map as 'axis of folding' (see Chapter 11).

Two other terms for describing folds, especially useful where dissimilar rocks are involved, are *core* and *envelope;* Fig. 7.6*d*

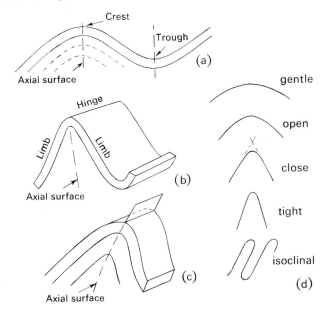

Fig. 7.5 (*a*) Anticline and syncline in upright open folding. (*b*) Diagram to illustrate hinge and limb. (*c*) Asymmetrical fold with inclined axial surface and two limbs dipping in opposite directions. (*d*) Degrees of acuteness in folding: gentle, open, close, tight, and isoclinal, can be defined in terms of the inter-limb angle; e.g. close folds between 70 and 30 degrees, tight folds less than 30. Isoclinal folds have parallel limbs, and the axial surface is parallel to the limbs.

shows that the core is the inner part of the fold and the envelope the outer part.

A number of other fold profiles are shown in Fig. 7.6. In an *overturned fold* the axial surface and both limbs dip in the same direction; this also applies to a *recumbent fold*, but here the axial surface is horizontal or nearly so. Some recumbent folds have developed into flat-lying thrusts by movement of the upper part of the fold relative to the lower. After their formation, folds may be tilted through any angle, or even inverted, by later structural deformations affecting an area. A recumbent fold does not, from its shape alone, provide a distinction between an anticlinal or synclinal structure (Fig. 7.6*d*). But if the rocks of the core of the fold are *younger* than the envelope rock it is a recumbent *syncline;* if they are older it is a recumbent anticline. The terms *synform* and *antiform* can be used to describe such folds when the age relation of the rocks involved is not known.

Figure 7.6*c* and *e* illustrate zig-zag and chevron folds. *Zig-zag folds* have straight limbs and angular hinges, the limbs having unequal lengths (Plate 9a). *Chevron folds* have limbs of equal length.

Parallel folds (Fig. 7.6*g*) may be formed in competent beds by slip

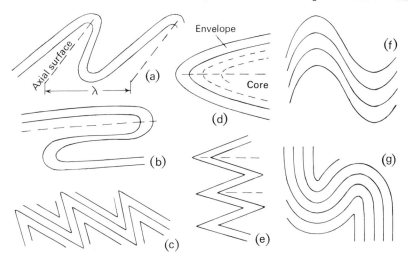

Fig. 7.6 (*a*) Overturned fold in profile, limbs dip in same direction, $\lambda =$ wave-length. (*b*) Recumbent fold with nearly horizontal axial surface. (*c*) Inclined zig-zag-folds, limbs unequal. (*d*) Synform or antiform, recumbent, rocks of core dotted. (*e*) Recumbent chevron folds. (*f*) Similar folds, and (*g*) parallel folds.

between the layers, probably under small confining pressure, and individual beds retain their thickness. When the folded material is more ductile the individual layers tend to become thickened towards the hinge and thinned on the limbs, by internal flowage, giving rise to *similar folds* as shown in the figure. Such folds maintain their identity in the axial direction, whereas parallel folds die out along the axial surface. Descriptions of folds, such as knee-shaped and box folds, not included here, may be found in publications listed at the end of this Chapter (and see Note 1).

Plunge. The attitudes of folds in three dimensions are described in terms of the inclination and direction of the hinge, as well as the style of folding. In most instances the fold hinge is inclined to the horizontal, and is then said to plunge (Fig. 7.7, see right hand side). Thus the level of an anticlinal crest falls in the direction of plunge, and in some cases the anticline diminishes in amplitude when traced along its length in that direction, and may eventually merge into unfolded beds. In a plunging syncline (Fig. 7.8*a*) the trough shallows in one direction along the length of the fold and deepens in the opposite direction, which is the direction of plunge. The angle of plunge (or inclination of the hinge to the horizontal) may be small, as in the figure, or at a steeper angle (Fig. 7.8*c*), or vertical. A *dome* is an upfold having dips radially outwards from a centre.

It should be noted that when a plunging anticlinal fold (Plate 7) is seen intersected by a ground surface, i.e. in outcrop, the strikes of the

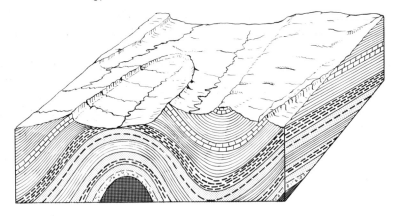

Fig. 7.7   Plunging folds, with hard beds forming topographical features on a denuded surface.

two limbs converge in the direction of plunge until, at the closure of the fold, the strike is normal to the axial trace (Fig. 7.7). Examples of this are sometimes clearly displayed on a foreshore which has been cut by wave action across folded rocks. The representation of such structures on geological maps often gives characteristic outcrop shapes (e.g. Fig. 7.8*a*; and Fig. 11.7, p. 326).

The Wealden anticline of south-east England is a large east-west fold which plunges to the east in the eastern part of its outcrop, and to the west at its western end (Fig. 8.9, p. 262). (The structure can also be thought of as an elongated dome of which the highest point is called the culmination.) The limbs of the fold dip north and south respectively on either side of the main anticlinal axis, and denudation has removed the Chalk and some underlying formations in the crestal area, revealing beds of the Wealden Series in the core of the structure. The Chalk, which once extended across the present Weald, now

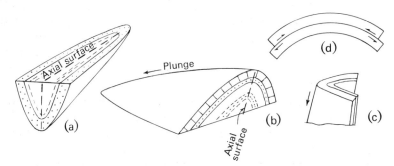

Fig. 7.8   Diagrams to illustrate plunge: (*a*) upright plunging syncline; (*b*) inclined plunging anticline; (*c*) fold with nearly vertical plunge; (*d*) relative slip between competent strata due to folding.

outcrops along the North Downs and South Downs, ending in the coastal cliffs at Dover and Beachy Head respectively; its continuation is seen in the cliffs of Calais and along part of the French coast.

Parasitic (or minor) folds. When folds are developed in competent strata there is necessarily some slip of the beds relative to one another (Fig. 7.8*d*), and this bedding-plane slip is most marked where the curvature is greatest, e.g. near the crest of an anticline. Where competent and incompetent strata are interbedded, such as sandstone and shale, the slip is taken up more by the incompetent beds, and this may cause small secondary folds to develop in them. They are called *parasitic folds* (or drag folds, see Note 2), and are illustrated in Fig. 7.9*a*. In the anticline shown, an incompetent layer (black) has been

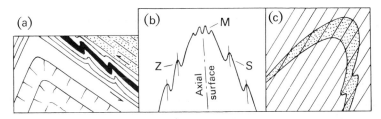

Fig. 7.9   (*a*) Parasitic (or drag) folds on limb of anticline; (*b*) S-, Z-, and M-folds nomenclature; (*c*) Shear-folding; in this drawing the thickness of the dotted layer is the same, in the shear direction, between successive shears.

deformed by relative movement between overlying and underlying competent beds. The hinge-lines of the minor folds are parallel to the hinge of the main fold, and their axial surfaces are in general parallel to that of the main fold. (Exceptions to this rule arise when more than one direction of movement has affected the rocks.) See also Plate 8a.

It will be seen from Fig. 7.9*b* that the minor folds on one limb of an anticline mirror those on the other limb, and for convenience they can be called Z- and S-folds respectively. Near the crest of such a structure minor folds may be present and are called M-folds. Z and S folds can often be useful in the field, for example where they are seen in steeply dipping or vertical strata which are part of a larger structure; they are an indication of the relative movement of the competent beds, and therefore of the direction in which a neighbouring anticlinal or synclinal axis will be present (Fig. 7.9*a*).

Shear folds. This type of fold is produced by movement on successive shear planes which cut across bedded layers (Fig. 7.9*c*); the layers of different material act as passive markers. When the shear planes are extremely close together and not necessarily visible, the layers appear as curved sheets; if the interval between shear planes is greater, the layers will show successive offsets.

Folds formed by buckling of a competent layer within more ductile material, during compression, may approximate in form to parallel folds, as in some *ptygmatic veins*. They have been produced experimentally with materials which have differing viscosities.

Other small scale structures associated with folds, such as fracture-cleavage, strain-slip cleavage, and tension gashes, are referred to on a later page.

II. NON-DIASTROPHIC FOLDS

Gravity folds, which may develop in a comparatively short space of time, are due to the sliding of rock masses down a slope under the influence of gravity. Examples of the masses of sediment which move over the sea floor and give rise to slump structures, on a relatively small scale, have been mentioned in Chapter 5 (p. 173). Submarine slumping often takes place on slopes greater than 10 or 12 degrees but may also occur on flatter slopes; it is frequently related to submarine fault scarps. Sliding may develop on an inclined stratification surface within a sedimentary series, and the partly compacted sediment is thrown into complex folds. This can occur on low-angle faults (Note 3).

Larger structures arise where the vertical uplift of an area covered with sedimentary layers has caused the latter to slide off down-dip soon after the uplift developed. An example of such *gravity collapse* structures is the series of folded limestones and marls which lie on the flanks of eroded anticlinal folds in south-west Iran (Fig. 7.10*a*) (Note 4). The drag effects of the passage of an ice sheet over soft sediments can produce shallow irregular folds, as in certain Tertiary lignites (brown coals) in Germany, or in the upper part of the Chalk outcrop in some English localities.

The disturbances known as *valley bulges* are formed in clays or shales which are interbedded with more competent strata, and are exposed in the bottoms of valleys after these have been eroded. The excavation of a river's valley is equivalent to the removal of a large vertical load at the locality. The rocks on either side of the valley exert a downward pressure, which is unbalanced (without lateral restraint) when the valley has been formed; as a result, soft beds in the valley bottom become deformed and squeezed into shallow folds (Fig. 7.10*b*). Residual stress within the sediments often assists in the process (Chapter 13, p. 382). Bulges of this kind, and related structures, were described in Jurassic rocks near Northampton by Hollingworth *et al.* in 1944 (Note 5). Other instances are found in the Pennine area, where interbedded shales and sandstones of the Yoredale Series crop out, as at the site of the Ladybower Dam in the Ashop valley, west of Sheffield, England.

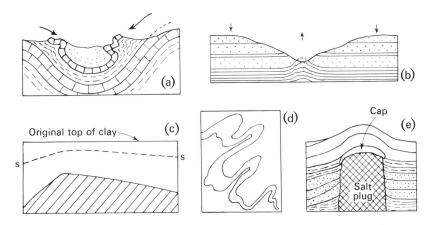

Fig. 7.10   (*a*) Gravity collapse structures in young sediments (after Harrison and Falcon). (*b*) Valley bulging (superfical fold in valley). (*c*) Compaction of clay above a buried hill, *s*- - -*s*, surface after compaction. (*d*) Flow folds in ductile material. (*e*) Doming resulting from intrusive plug of salt (and see Fig. 14.10).

Folds formed by gravitational compaction of sediment are also developed where clays have been deposited upon an uneven surface. There is more compaction in the greater thickness of the clay adjacent to a buried hill, for example, than in the smaller thickness of clay above it (Fig. 7.10*c*). The originally horizontal beds of clay acquire dips away from the buried mass. The presence of a coral reef in argillaceous sediments (as in the Carboniferous Reef-Limestones of the north of England, see Note 6), or a lens of sand, either of which undergoes less compaction than the adjacent clay, leads to the formation of flexures in the rocks. Differential compaction of this kind may also affect peat beds containing sand-filled channels (washouts).

Flow-folding arises in beds which offer little resistance to deformation, such as salt deposits (evaporites), or rocks which become ductile when buried at considerable depth in the crust where high temperatures prevail (Fig. 7.10*d*), as for some gneisses.

Salt domes are formed where strata are upturned by a plug of salt moving upwards under pressure. A layer of rock-salt is more easily deformed than other rocks with which it is associated, and under pressure can rise as an intrusive plug, penetrating and lifting overlying strata. The doming thus formed (Fig. 7.10*e*) is often nearly circular in plan. When the salt outcrops at the surface it makes an abrupt change of slope. Many salt domes are found in the Gulf Coast area of Texas, where they are associated with accumulations of petroleum; and in Iran, Germany, and elsewhere. A sandy, porous cap-rock acts as a reservoir rock for petroleum, and deposits of sulphur may be associated. Recent geophysical exploration of the North Sea basin has revealed the presence of dome-like uplifts which may be due to intrusive bodies of salt (Note 7).

FAULTING

The state of stress in the outer part of the earth's crust is complex, and in the first part of this Chapter we have seen how some rocks obtain relief from compression by folding and buckling. Commonly also, however, fractures are formed in relief of stress which has accumulated in rocks, either independently of folds or associated with folding. The fractures include faults and joints: *faults* are fractures on which relative displacement of the two sides of the break has taken place; *joints* are those where no such displacement has occurred. Groups of faults and sets of joints may both form patterns which can be significant in indicating the orientation of the stresses which resulted in the fracturing, though a clear indication of this is not always forthcoming. In this account our discussion is limited to the geometrical description of fractures and some of the structural patterns to which they give rise.

Stresses in the earth's crust   Theoretically, within a relatively small distance of the surface, the stress due to the weight of rock at a particular depth is given by

$$\sigma_v = \rho z$$

where ρ is the density of the rock and z the depth. (The term rock here includes sediments or 'soils' as well as more coherent and harder materials.) The quantity σ_v is called the vertical *geostatic stress* or *geostatic pressure*, and if the rock is uniform and the ground surface horizontal, it has a linear increase with depth, as shown in Fig. 7.11 (Note 8). When the density of the rock or soil is not constant—for instance, it may increase with depth because of a change of character or because of compression, a value for the vertical stress can be obtained by integrating, thus:

$$\sigma_v = \int_0^z \rho \, dz,$$

or by a summation for different layers.

It is easily calculated that at a depth of say 1 km (0.62 mile) the pressure exerted by the self-weight of rock having a density of 2 g/cm^3 (or 125 lb/cu. ft) is 19.57×10^6 N/m^2 (1.27 tons/in.^{2}). Extrapolation of this kind to further depths, from near-surface conditions, is likely to be inaccurate because of unknown changes in density and the presence of other stresses.

The horizontal geostatic stress, σ_h, which arises because lateral

Plate 7

Plunging anticline,
deeply eroded, Iran.
(*Photograph by
Aerofilms Ltd.*)

Plate 8

(b) Steeply dipping sandstone, Eggesford, North Devon, with cleavage oblique to bedding in narrow (8 cm) shale band. (*Courtesy of the Institute of Geological Sciences, London. Crown copyright reserved*)

(a) Large drag-folds on downthrow side of normal fault (surface on right), in slate and siltstone, Tintagel. (*Courtesy of the Institute of Geological Sciences, London. Crown copyright reserved*)

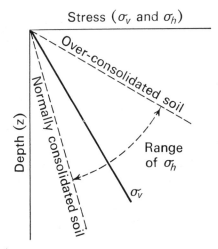

Stress (σ_v and σ_h)

Fig. 7.11 Diagram of geostatic stresses for uniform conditions and a horizontal ground surface.

expansion is restricted by the pressure of surrounding rock, increases as σ_v increases as shown in the figure; where also a distinction is made between normally consolidated and over-consolidated deposits (clays). The ratio of horizontal to vertical stress is called the *lateral stress ratio*, K, i.e. $\sigma_h = K\sigma_v$, and this definition applies whether the stresses are geostatic or otherwise. The latter condition (i.e. not geostatic) would arise, for instance, in greatly over-consolidated sediments; or in stress conditions in rocks that have high lateral stresses which are a residual from earlier tectonic compression that affected them (Note 9). This has been found in some deep underground excavations, and is discussed in Chapter 15.

Brittle fracture When a brittle material is broken in compression, failure occurs along shear planes inclined to the direction of loading; for example, a prism of concrete will, on breaking, yield a rough cone or pyramid. The angle at the apex of such a cone is theoretically 90°, since the maximum shear stresses should occur on planes inclined at 45° to the direction of maximum principal stress. In practice this angle is generally found to be less than 45°, owing to the operation of frictional forces in the material at the moment of failure. Thus an acute-angled cone is formed; or, in two dimensions, two complementary shear fractures intersect at an acute angle which faces the direction of maximum compression. Many rocks when tested break in this way, as brittle materials, with an acute angle between the shear directions which is often in the neighbourhood of 60°.

Tests have been made on cylinders of rock, with lateral confining pressure as well as an axial compression. In these tests brittle failure occurs at low confining pressures, and ductile failure at high confining pressures; a transitional type of failure which may be called semi-brittle can occur at intermediate pressures, with a stress-strain curve lying between the ductile and brittle curves. Applying these results to natural conditions in the earth's crust: for a particular rock different types of failure could arise from different conditions, such as the depth at which failure takes place, and the temperature prevailing at that depth. Granite would behave as a brittle material at depths up to several kilometres if surface temperatures prevailed; but there is an average temperature gradient in the outer crust of 30°C per km (see p. 3), and at a depth of 30 km temperature would be about 500°C. At still greater depths and higher temperature the same rock would behave as a ductile material (Note 10).

Faults   Near the earth's surface, hard rocks which have undergone compression have failed in shear, with the production of single faults, or groups of faults forming a fault pattern. (Under some conditions rocks have obtained relief from stress by folding.) Three main kinds of fault are formed, namely thrust faults, normal faults, and wrench faults. Where the dominant compression was horizontal and the vertical load small, the shear fractures formed intersect as shown in Fig. 7.12(*a*), the acute angle between them facing the maximum principal stress; faults having this kind of orientation are the *thrusts* or thrust faults. Where the greatest stress was vertical (Fig. 7.12*b*) the shear planes are steeply inclined to the horizontal, and faults formed under such stress conditions are *normal faults*. Thirdly, when both the maximum and minimum stresses were horizontal and the intermediate stress vertical (Fig. 7.12*c*), the resulting fractures are vertical surfaces and correspond to *wrench* faults. The two wrench faults of a pair are often inclined to one another at an angle between 50 and 70° (Note 11).

Normal and reverse faults   Here the relative movement on the fault surface is mainly in a vertical direction; and the vertical component of the displacement between two originally adjacent points is called the *throw* of the fault.

Two common types of fault are illustrated in Fig. 7.13, which shows a *normal fault* on left, where a bed originally continuous at *a* has been broken and the side *A* moved down relative to *B*, or *B* moved up relative to *A*. The side *B* is called the 'footwall,' and *A* the 'hanging wall.' In a normal fault, therefore, the hanging wall is displaced downwards relative to the footwall. A *reverse fault* is shown in Fig. 7.13 (right). In this case the hanging wall *C* is displaced upwards relative to the footwall *D*. Notice that the effect of the reverse faulting is to *reduce* the original horizontal extent of the broken bed; the two displaced ends of a stratum overlap in plan. On the other hand, the effect of a normal fault is to *increase* the original horizontal distance

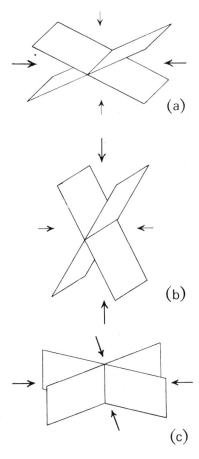

Fig. 7.12   Relationship of faults (shear planes) to axes of principal stress; (*a*) thrusts, (*b*) normal faults, (*c*) wrench faults. Maximum stress, long arrows; minimum stress, short arrows. Intermediate stress omitted, but its direction is that of the line along which the shear planes intersect.

between two points on either side of the fracture.

The custom of describing faults as 'normal' or 'reverse' originated in English coal-mining practice. Faults which were inclined towards the downthrow side were met most commonly. When a seam which was being worked ran into such a fault, it was necessary to continue the heading a short way (as at *a* in the figure), and then sink a shaft to recover the seam; this was the usual or 'normal' practice. When a fault was encountered which had moved the opposite way, that was the reverse of the usual conditions, and the fault was called 'reversed.'

Fault groups.   Several normal faults throwing down in the same direction are spoken of as 'step faults' (Fig. 7.14); two normal faults

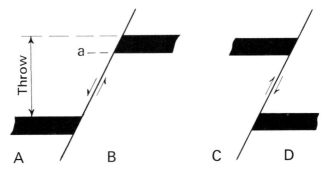

Fig. 7.13   Normal and reverse faults. Cross-sections showing the displacement of a horizontal stratum (in black). *Left:* Normal fault; *A* = downthrow side (hanging wall), *B* = upcast side (footwall). *Right:* Reverse fault; *C* = upcast side (hanging wall), *D* = downthrow side (footwall).

hading towards one another produce 'trough faulting,' and hading apart form a pair of 'ridge faults.' Where a stratum approaches a fault it is often bent backwards a little, away from the direction of movement along the fault plane, as shown in the figure. A sunken block, bounded on all sides by faults, is called a *graben;* the Rhine Valley and the Midland Valley of Scotland are examples of this structure on a large scale. A *horst* is a fault-bounded ridge-block, the converse of graben; an example is the ridge of the Malvern Hills, Herefordshire.

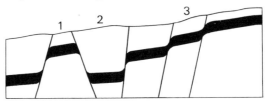

Fig. 7.14   (1) Ridge, (2) trough, and (3) step faults. The ends of the black bed are shown slightly dragged round as they approach a fault.

Fault components   The movement on a fault may be of any amount and in any direction on the fault surface. The complete displacement along a fault plane between two originally adjacent points can be described by means of three components measured in directions at right angles to one another. The vertical component, or *throw*, has already been mentioned; the two horizontal components are the *heave* (*bc* in Fig. 7.15), measured in a vertical plane at right angles to the fault plane; and the *strike-slip* (*cd*), measured parallel to the strike of the fault plane. The total displacement, *ad*, is called the *slip*, or *oblique-slip*. The other component on the fault-plane is the *dip-slip* (*ac* in the figure).

The *hade* of a fault plane is its angle of inclination to the vertical, the angle *bac* in Fig. 7.15. It is often more convenient to speak of the

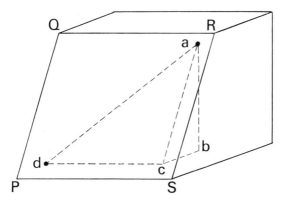

Fig. 7.15 Components of fault displacement. *PQRS* = fault surface. *ad* = total slip. *ab* = throw. *bc* = heave. *cd* = shift. The points *a, c, d* lie on the fault surface.

dip of a fault, i.e. its inclination to the horizontal, than of its hade. Hade + dip = 90°. When a fault is vertical the hade and heave are zero.

Faults are referred to above as 'planes' only for ease of description; they are more often zones of crushed rock which has been broken by the movement, and range in width from a few millimetres to several metres. Sometimes the crushed rock along the fault surface has been ground very small, and become mixed with water penetrating along the line of the fault, producing a clay or *gouge*. Fault surfaces are often plane only for a small extent; curved and irregular fractures are common.

Strike and dip faults Faults are also described from the direction of their outcrops on the ground, with reference to the strata which they displace. *Strike faults* outcrop parallel to the strike of the strata; *dip faults* run in the direction of the dip of the beds; and *oblique faults* are those which approximate neither to the dip nor strike direction. These cases are illustrated below.

Effect of faulting on outcrop on dipping strata The different examples of normal faulting considered here are:

(i) a dip fault;
(ii) a strike fault, dipping in the same direction as the dip of the strata;
(iii) a strike fault, dipping in the opposite direction to the dip of the strata.

(i) The general effect of a dip fault is to displace the outcrops of corresponding beds on either side of the fault line. In Fig. 7.16, the side *AB* of the block is parallel to the dip direction, and the side *BC* shows a strike section. The normal fault (*FF*) throws down to the

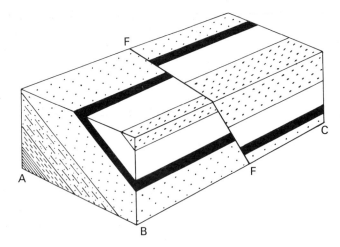

Fig. 7.16   Effect of normal dip fault (*FF*) on dipping strata.

right. The outcrops of the beds (seen on the top of the block) are offset on either side of the fault trace; upper (younger) beds on the downthrow side are brought opposite lower (older) beds on the upcast side.

(ii) Strike faults either *repeat* or *cut out* the outcrop of some parts of the faulted strata. In Fig. 7.17*a*, a series of beds is broken by a normal strike fault (*FF*) which dips steeply in the same direction as the dip of the beds. The result of this is that one bed (striped) does not appear in outcrop at the surface (the upper surface of the block); it is cut out by the faulting. The case where the dip of such a fault is the same as the dip of the strata is referred to as a *slide*, i.e. the fault movement has taken place along a bedding-plane. Where the dip of the fault plane is *less* than the dip of the beds, and in the same direction, there is *repetition* of some part of the outcrop of the beds.

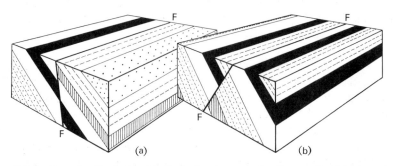

Fig. 7.17   Normal strike faults. (*a*) Fault dips in same direction as dip of beds; outcrop of striped bed is concealed. (*b*) Fault dips in opposite direction to dip of beds, and outcrops of some beds are repeated.

(iii) Figure 7.17*b* shows a series of beds broken by a normal strike fault (*FF*) which dips in the opposite direction to the dip of the beds. The effect is to repeat the outcrop of some beds (the black bed, for instance), as seen on the upper surface of the block.

The above diagrams show outcrops on a plane surface; irregular topography will modify this simple conception, as discussed in Chapter 11. When a fault is vertical it has a straight-line trace, whatever the topography.

Oblique faults to some extent share the features shown by both dip and strike faults.

In some instances the effects of reverse faulting on outcrop are not always distinguishable from those of normal faulting. For example, a reverse dip-fault will still bring older beds on the upcast side against newer beds on the downthrow side (cf. normal faulting, above). On the other hand, a reverse strike-fault will have the opposite effect to a normal fault in the same direction; in the cases illustrated in Fig. 7.17, a little consideration will show that *repetition* of outcrop would be produced by a reversed fault dipping in the same direction as the dip of the strata, and elimination of outcrop in the other case.

If a fault is not vertical, the shape of its trace or outcrop on a map gives an indication of its hade, which may show whether it is normal or reverse. But, in general, it is difficult to distinguish between normal and reverse faults solely by their relation to outcrops on a map. Field evidence is more definite; for example, the ends of the fractured beds may be seen in an exposure to be dragged round so as to indicate the direction of relative movement along the fault plane (Fig. 7.14).

Wrench faults   Faults in which horizontal movement (or strike-slip) predominates, the other components being small or nil, are called *wrench faults*. The term *transcurrent fault* is also used (and formerly 'tear fault'). Wrench faults are commonly vertical or nearly so (Fig. 7.12*c*) and the relative movement is described as follows: for an observer facing the wrench fault, if the rock on the far side of the fault has been displaced to the right the movement is called *right-handed* or *dextral;* if in the opposite direction, i.e. to the observer's left, it is termed *sinistral* or *left-handed.*

A fault of this type, in which the displacement is essentially horizontal, may shift the *outcrops* of dipping beds in an apparently similar way to a normal fault movement. In Fig. 7.16 for example, the staggering of the outcrops shown could have come about by a horizontal movement of one part of the block past the other, along the fault (*FF*). It is important therefore not to make rigid assumptions about the nature of a fault solely from the displacement of outcrops, without looking for other evidence.

A wrench fault in the direction of strike will not change the order of the strata in outcrop unless there is some vertical throw as well as the horizontal movement.

Criteria for recognizing wrench faulting are often a matter of field observation, and three may be noted here: (i) Parallel grooves have in some cases been cut on a fault surface by projecting irregularities on one block as it moved past the other. These grooves are known as *slickensides* and give valuable information about the direction of the fault movement. If they are horizontal or nearly so, it can be inferred that the last movement on the fault had a mainly wrench component. (ii) When vertical beds are faulted and displaced laterally in outcrop, a wrench fault is indicated. (iii) The horizontal shift of a vertical axial plane of a fold is illustrated in Fig. 7.18.

Thrust planes   Most of the fractures so far described have a steep inclination to the horizontal; in contrast to this is the group of faults produced by dominantly horizontal compression and known as

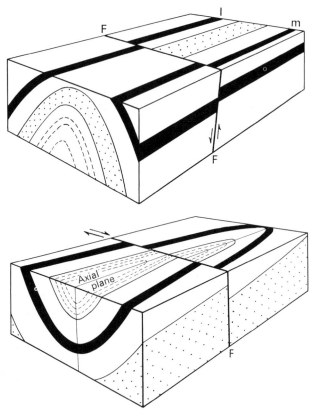

Fig. 7.18   Folded strata broken by faults. *Above:* Faulted anticline; outcrops of limbs are moved apart on upcast side of a normal fault. *Below:* Plunging syncline cut by a wrench fault. Note the horizontal shift of the axial plane.

thrusts or thrust planes (Fig. 7.12*a*). They are inclined to the horizontal at angles up to 45 degrees. (If the inclination of the fault surface is greater than 45 degrees it is called a reverse fault.) The development of a thrust from an overturned fold is illustrated in Fig. 7.19: with

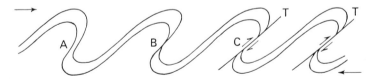

Fig. 7.19   Development of thrusts in folded strata. *T* = thrust.

continuing compression the middle limb of the fold becomes attenuated (*A*), then fracture occurs (*B*), and further compression results in movement along the thrust (or shear) plane so formed, as at *C*. The effect of this kind of fracture is thus to shorten the horizontal extent of folded beds, in the direction of compression, by increasing their vertical extent. Other very large thrusts, which are virtually horizontal, are called *overthrusts;* they do not fit into the classification of faults given above. They are surfaces of large extent, with a small angle of dip, on which large masses of rock have been moved for considerable distances. The Moine Thrust (p. 240), for example, is an overthrust with a displacement of over 10 miles (16 km) (Note 14).

Examples of thrusts have been described from many localities, as in the cliff sections at Tunnel Beach, Ilfracombe, North Devon, where on a small scale thin bands of folded limestone in cleaved shales are broken by repeated thrusts. The thrusting has left isolated S-shaped (sigmoid) sections of the limestone, similar to that shown in Fig. 7.19 between the thrusts *T, T.*

A structure involving both major and minor thrusts, known as the *imbricate structure* (Latin, *imbrex*, a tile; the minor thrust masses overlap like tiles on a roof) was first described from the district of Assynt, in Sutherlandshire. In that region, rocks have been piled up by a series of minor thrusts, as described above, and these have been truncated by flatter major thrusts, or overthrusts, along which large heavy masses of rock have been pushed forward so as to overlap the folded and broken rocks beneath. Figure 7.20 shows the type of structure in section (see also p. 252). The outcrops of the nearly horizontal major thrust planes can be traced for many kilometres across the area, and appear on the map as irregular, winding lines. The rocks immediately below these surfaces of dislocation have been crushed and sheared by the over-riding of the heavy masses, and converted into mylonites.

Nappes   In intensely folded mountain regions, such as the Alps,

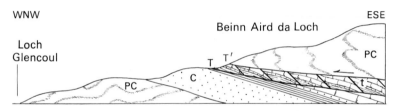

Fig. 7.20 Section at Loch Glencoul, Sutherland; imbricate structure on right, with minor thrusts (*t*) between flatter major thrusts (*T*, *T'*). *PC* = Pre-Cambrian, *C* = Cambrian.

overthrusts occur associated with recumbent folds on a large scale. A recumbent fold driven forward on a thrust plane is called a *nappe*, or thrust-sheet, and structures of this kind form the basis of one interpretation of the complex geology of the Alps (see Fig. 7.21).

Fold-mountains   There is evidence that belts of the earth's crust have passed through periods of intense lateral or tangential compression, and these regions have been severely folded and ridged up into mountain ranges. Mountain-building or orogenic movements of this kind have operated over the sites of long-continued sedimentation in sinking areas. Such areas of deposition or geosynclines (p. 173) are elongated, and from the great thicknesses of sediment which were laid down in them mountain masses have subsequently been elevated by folding and faulting, often with associated intrusive igneous activity. The Alps (Fig. 7.21) where, as already mentioned, the rocks have been thrown into great overfolds and recumbent folds, and often moved for large distances along nearly horizontal thrust planes, have resulted from the compression of a thick pile of geosynclinal sediments. The horizontal compression caused a crustal shortening across the fold-belt of many kilometres. The Appalachian folds in the eastern United States, formed at a different time, and fold-mountain chains such as the Rockies and the Himalayas show similar features. The Caledonian fold system which trends across the Southern Uplands of Scotland in a north-east to south-west direction represents an ancient mountain-building epoch which strongly affected the British Isles; its place in geological time is described in the next chapter.

It is beyond the scope of this book to discuss the many geological problems involved in such large-scale structures of the earth's crust, and the theories of their formation; the subject is one of great interest, and leads to a consideration of the earth's internal mechanism. The student can fill in the details for himself, if he wishes, by further reading; suggestions are given in Note 12.

In summary, the cycle of events that is inferred from a study of the rocks is—first, the accumulation of thick deposits of sediments over sinking areas (the geosynclines); second, the crumpling of the

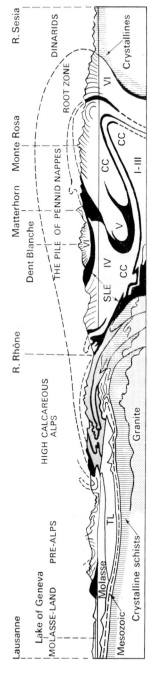

Fig. 7.21 Geological section across the Alps from N.W.–S.E. (after Argand), distance about 140 km. The *molasse* (Oligocene sandstones and conglomerates) is formed of debris eroded from the rising Alps and is overthrust by Mesozoic sediments (centre of section). The Pennid Nappes contain big recumbent folds and are named after localities: Simplon (I–III), Great St. Bernard (IV), Monte Rosa (V), Dent Blanche (VI). Much structural detail was elucidated from deep railway tunnels in the region. *TL* = Triassic lubricant (rock-salt), *CC* = crystalline core, *SLE* = *schistes lustrés* envelope, surrounding the core.

sediments by lateral compressive forces in the crust, accompanied by balancing movements of sub-crustal material, resulting finally in the elevation of a mountain chain. This then becomes subject to break-down by the agents of denudation, with the production of new sedimentary deposits. Isostatic forces are continually tending to produce crustal equilibrium. Into this cycle at various stages, but particularly at the time of mountain formation, may come igneous activity with the injection of magma and the extrusion of lavas.

Minor structures associated with folds The structures to be considered here are *fracture-cleavage*, *tension gashes*, and *boudinage*. Slaty cleavage (or flow cleavage), which results from the new growth of oriented minerals, has been discussed in Chapter 6 (p. 209).

When a series of rocks has been subjected to shearing stresses, as in the neighbourhood of a thrust, groups of closely spaced parallel fractures are frequently developed in them. The structure is known as *fracture-cleavage*, and is dominantly mechanical in its mode of origin; it is to be distinguished from slaty cleavage on the one hand and from jointing on the other. The spacing of the shear fractures depends on the nature of the material; incompetent rocks like shales show fracture-cleavage planes more closely spaced than is the case with harder (competent) rocks.

When beds are folded, shear stresses incidental to the folding are set up between the layers (Fig. 7.22*a*) and may lead to slip along

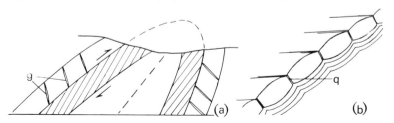

Fig. 7.22   (*a*) Fracture-cleavage in shales folded between harder beds. Tension gashes in the hard bed are marked *g*. (*b*) Boudinage structure formed by extension in a competent bed; quartz veins, *q*, may fill the gapes between boudins.

the bedding planes, especially if the fold is not broken at the crest; the slip is greatest on the limbs of the fold. Under these conditions an incompetent layer lying between competent beds, as in the figure, may develop a fracture-cleavage pattern which crosses the layer obliquely. Observation of fracture-cleavage in folded rocks is useful in the interpretation of structure where only a part of the fold is open to inspection; e.g. the inclination of fracture-cleavage on one limb of a denuded anticline will indicate on which side of the limb the axis of the fold is located. On an overturned fold limb (e.g. the steeper limb in Fig. 7.22*a*) the fracture-cleavage dips at a *lower* angle than the dip of the strata; whereas in the flatter limb it dips at a greater angle (Plate 8b).

Tension gashes are formed during the deformation of brittle rocks. They may be opened as radial fractures by stretching in the crestal part of a fold; or on the limbs of folds and in fault zones they may be developed in relief of local tension (Fig. 7.22a). Tension cracks are also developed in clay which is being subjected to shearing under experimental conditions (Note 13).

Conditions arise in which the earth's crust is put locally into a state of tension, resulting in the formation of fractures which may appear as normal faults, or may tap sources of hot igneous material which rises and fills them to form dykes. One way in which tension is produced is from the stressing of an area by horizontal shearing forces (couples, *ss* in Fig. 7.23). The tension breaks are aligned parallel to

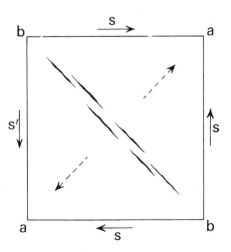

Fig. 7.23   Tension fractures produced by shearing stresses. *aa* = direction of tension. *bb* = direction of compression.

one diagonal of the rectangular horizontal area, as illustrated, the other diagonal being in compression. (See also Fig. 6.4.)

Tension fractures may also arise from torsional forces, for example due to unequal settlement in a horizontal slab of rock. If opposite corners of such a slab (e.g. *aa* in Fig. 7.23, neglecting the shearing couples) are raised or lowered relative to the other pair of corners, torsional stresses are then acting on the slab and a fracture pattern may be developed. Tension zones produced as a result of under-ground mining are discussed in Chapter 15.

Since brittle materials are, on the whole, weakest in tension, it might be expected that tension breaks would be the most common among rock fractures; the very numerous cases of shear failures, however, may be attributed to the predominance of compressive stresses in the rocks of the continental masses.

Boudinage. When brittle beds in a more ductile matrix are subjected to an extension in the plane of the beds, the deformation of the brittle layers results in fracturing across the layers (Fig. 7.22*b*), while the more ductile material yields by flowage and packs into spaces formed during the elongation. Boudins ('sausages') are formed in sandstones or limestone interbedded with shales in fold limbs (Plate 6b), and also in metamorphic rocks such as gneisses and schists.

JOINTS

Parting-planes known as *joints* are ubiquitous in almost all kinds of rocks, and have been referred to earlier in this book (pp. 24, 57). They are fractures on which there has been no movement, or no discernible movement, of one side (or wall) relative to the other. In this way they differ from faults (p. 230). Groups of parallel joints are called *joint sets*, and for two or more sets which intersect the term *joint system* is used. Commonly two sets of joints intersecting at right angles or nearly so are found in many sedimentary rocks, and are often perpendicular to the bedding planes (Fig. 7.24, *s* and *d*).

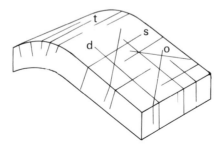

Fig. 7.24   Jointing in a folded stratum. *t* = tension joints at crest of fold, *s* = strike joints, *d* = dip joints, *o* = oblique joints (shear joints).

The formation of joints is a matter that is still imperfectly understood. Many are developed in the relief of tensional or shearing stresses acting on a rock mass. The cause of the stresses has been variously ascribed to shrinkage or contraction, compression, unequal uplift or subsidence, and other phenomena; all are a relief of *in situ* stress.

Shrinkage-joints develop from the drying and resultant shrinkage of sedimentary deposits (*cf.* sun-cracks in newly formed estuarine muds). The contraction of a mass of sediment, if evenly distributed, was held by some to be a cause of joint formation; but objections have been put forward to such a process operating in thick sandstones and limestones. Some recently deposited clays found in the beds of lakes, and other wet sediments have, however, been observed to contain joint sets which intersect at right angles, formed relatively early in the history of the sediment. Jointing in shales is also well known.

The joints in sedimentary rocks sometimes end at a bedding-plane, i.e. they extend from one bedding-plane to the next, but others may cross many bedding-planes and are called *master-joints*. If contraction is excluded as a process in joint formation in some sediments, other processes such as tectonic compression (see below) must be invoked to explain the regular joint patterns found in them.

In dipping or folded sediments the direction of one set of joints frequently corresponds to the strike of the beds, and the other set to the dip direction; they are therefore referred to as *strike-joints* and *dip-joints* respectively (Fig. 7.24).

In igneous rocks *contraction-joints* are formed as a hot mass cools and contracts. A lava flow often develops a hexagonal joint pattern by contraction around many centres, equally spaced from one another, the contraction being taken up by the opening of tension joints. This gives rise to a columnar structure in the lava sheet, the hexagonal columns running from top to bottom of the mass; the columnar basalts of Antrim and of the Western Isles of Scotland (e.g. Staffa) are familiar examples. In an intrusive igneous body such as a granite pluton, joint systems develop during the cooling of the mass after its emplacement (Note 15). Some of these joints act as channels for the passage of residual fluids, which consolidate as veins of aplite or pegmatite; or for the passage of hot gases which affect the walls of open joints (pneumatolysis, p. 207), or fluids which coat them with hydrothermal minerals. All these joints are of early origin. But many granites show other joint systems which are probably tectonic, and often consist of two sets of steep or vertical joints which intersect at 90 degrees or less, and are sometimes accompanied by diagonal joints. (The terminology of such jointing is outlined in Note 16.) *Sheet-jointing*, which crosses such a joint-system and is especially developed near the roof of the igneous body, is probably caused by tensile stresses generated by the unloading of the mass during denudation, as described on p. 24.

Tectonic joints are those which are developed during tectonic folding or thrusting, and show a recognizable relationship to the directions of the resulting folds or thrusts. Such joints, which follow the directions of maximum shearing stress arising from the compression (of sedimentary or igneous rocks), may have been formed when the shear components of the stress system were just enough to overcome the shearing strength of the material before the stresses diminished or ceased to operate. Or again, the tectonic joints may be the visible expression of residual stresses remaining in the rocks after deformation ceased. The fractures generally form two sets which make an acute angle with one another, the acute angle facing the direction of the maximum compressive stress (*cf*. Fig. 7.12c). One of the two directions of shearing may be emphasized, showing many joints, and the other largely suppressed.

It is often a matter of observation that, near a visible fault, the rocks are traversed by joints parallel to the fault surface. This suggests that they were formed in response to the same stress system that resulted in the faulting; their frequency diminishes with increasing distance from the fault.

Where shear fractures intersect at a small angle, 20 degrees or less, and one fracture is a fault, the small fractures which lie at the acute angle with it are called pinnate or *feather-joints*. They can sometimes be used to indicate the direction of shearing movement along the main fracture; the acute angle points away from the direction of movement of that side where the pinnate structures occur.

Size and spacing of joints.   There is a wide range of size or extent shown by joint planes; master-joints in sediments have been mentioned above and may extend for 100 metres or more. Less extensive but well defined joint sets can be called *major joints*, in distinction from smaller breaks or *minor joints*. Minute joints on a microscopic scale are seen in some thin sections of rocks. The spacing or frequency of joints (= number of planes of a particular set in a given direction measured at right angles to the joint surfaces) also varies considerably. These factors are of great importance in quarrying; some rocks such as sandstones and limestones, in which the joints may be widely spaced, yield large blocks of stone which are suitable for masonry (Chapter 17). Whereas other rocks may be so closely jointed as to break up into small pieces—which, however, may be suitable for road-metal or other purposes. The ease of quarrying, excavating, or tunnelling in hard rocks largely depends on the regular or irregular nature of the joints and their direction and spacing. Joints are also important in connection with water supply (Chapter 13), and their presence is a main factor in promoting rock weathering (Chapter 2). (Note 16).

NOTES AND REFERENCES

1. For a full discussion of the nomenclature of folds and their various features see The description of Folds, by Fleuty, M. J. 1964, in *Proc. Geol. Assoc.* **75**, 461–492. The definitions given there are drawn on in this chapter.

    The term *fold axis* which was formerly used is omitted here as the preferred term *hinge* is available.

    Some structural geology text-books are listed on p. 247.

2. Minor folds on the limbs of a major fold are commonly called *drag folds*, but many are thought to have resulted from causes other than frictional drag. The term *parasitic fold* does not carry any implications as to origin and is used in preference to drag fold.

3. Large scale sliding of Tertiary sediments on fault surfaces, with the development of folds, has been described from Peru by Baldry, R. A. 1938, in *Q. Jour. Geol. Soc.* (Lond.), **94**, 347; and from Ecuador by Brown, B. B. *ibid*. 359.

4. See Harrison, J. V. and Falcon, N. L. 1936, Gravity collapse structures and mountain ranges as exemplified in south-western Iran. *Q. Jour. Geol. Soc.* (Lond.), **92**, 91.

5. Hollingworth, S. E., Taylor, J. H., and Kellaway, G. A. 1944, Large-scale superficial structures in the Northamptonshire Ironstone field, *Q. Jour. Geol. Soc.* (Lond.), **100**, 1.

6. Parkinson, D. 1943, The Origin and Structure of the Lower Visean reef-knolls of the Clitheroe District, Lancashire. *Q. Jour. Geol. Soc.* (Lond.), **99**, 155.

7. The term *diapir* (= through-piercing) is used for the structure produced by a rock such as salt, or a mobile granite, which has moved upwards and pierced through the crest of an anticline. For a discussion on diapir structures see Braunstein, J. (*Ed.*), Memoir 8, *Amer. Assoc. Ptrol. Geol.*, 1968.

8. For sedimentary deposits such as sands and clays in which pore-water is present, the geostatic pressure is reduced by pore-water pressure. This is discussed in Chapter 13, p. 376.

9. The term *tectonic* is used to denote the deformation of crustal rocks considered in relation to the forces and movements which caused it. (Greek, *tekton*, a builder). The phrase 'tectonic compression' therefore signifies an episode (or episodes) of compression which have brought about particular structural deformations in the rocks.

10. Early experiments on marble, carried out by Adams, F. D. and Bancroft, J. A., showed that under high confining pressures a normally hard rock would deform by ductile flow (= plastic flow). See Internal friction during deformation and relative plasticity of rocks, 1917, *Jour. Geol.*, **25**, 597.

11. A fuller discussion of the orientation of faults with regard to the principal stress directions is given by E. M. Anderson, The Dynamics of Faulting, 1951 (Oliver & Boyd, Edinburgh).
    The opening up of faults may begin at minute flaws or cracks in the rock material, according to Griffiths, A. A. 1925, The Theory of Rupture, *1st. Int. Congr. Appl. Mech.*, Delft, p. 55. Local concentrations of stress may arise around cracks and initiate failure, although the *average* stress throughout a rock mass is much lower.

12. For a short account of the structure of the Alps see Holmes, A. 1965, Principles of Physical Geology, Nelson, Edinburgh.
    Also Wells, M. K. 1948, *Proc. Geol. Assoc.* **59**, 181.

13. In the experiments a slab of clay is placed over two boards with their long edges in contact, which are then moved slowly one past the other, underneath the clay layer, thus producing a zone of shear deformation

in the clay. The movement of the boards relative to one another is horizontal. (Riedel, W. 1929, in *Centralb. f. Mineral. Geol. u. Pal.*, *B*, p. 354).

This model simulates the formation of a belt of shearing in rocks under natural conditions; tension cracks followed by *en echelon* shear fractures are formed.

See Blyth, F. G. H. 1950, The sheared porphyrite dykes of South Galloway, *Q. Jour. Geol. Soc. Lond.*, **105**, 393; and Skempton, A. W. 1966, Some observations on tectonic shear zones, *Proc. 1st. Congr. Int. Soc. Rock Mechanics*, Lisbon, 1, 329.

14.   A theory of the mechanics of movement on an overthrust, taking account of the pore-water pressure in the very large masses of rock involved, was put forward by Hubbert, M. K. and Rubey, W. W. 1959 (*Bull. Geol. Soc. Amer.* **70**); a summary of it is given by Price. 1966, reference below).

15.   The cooling of an intrusive igneous mass such as a granite takes effect first near its roof and walls; movement of still liquid or plastic material at a lower level may then give rise to fracturing in the outer, more solidified, part of the intrusion. Lines of oriented inclusions such as elongated crystals, in the igneous rock, are called *flow-lines;* they have been formed by viscous flow near and often parallel to the roof of the mass (Fig. 7.25).

Early formed joints include *cross-joints* (or 'Q-joints'), which are steep or vertical and lie perpendicular to the flow-lines. These cross-joints are tension fractures opened by the drag of viscous material past the outer partly rigid shell. They are often filled with aplite.

Longitudinal-joints (or 'S-joints') are steep surfaces which lie parallel to the flow-structures and perpendicular to the cross-joints, and are later in formation.

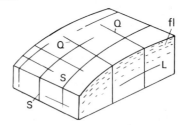

Fig. 7.25   *Q-, S-, and L-* joints. *fl*, flow lines.

Flat-lying joints (or 'L-joints') of primary origin form a third set of fractures at right angles to the Q- and S- joints. Later formed flat-lying joints, or sheet-jointing (p. 243), are developed during denudation and unloading.

The Q, S, and L terminology was due to Hans Cloos (1923). A good summary of Cloos' work is given by Balk, R. 1937, in Structural behaviour of igneous rocks, *Geol. Soc. Amer.* Memoir 5.

16.   For a general discussion of the frequency of joints and related matters see Price (1966; reference below). The relationship between joint

Plate 9

(a) Recumbent zig-zag folds in Carboniferous rocks, Millook, North Cornwall.

(b) Unconformity: Carboniferous Limestone on steeply dipping Silurian flagstones, Horton-in-Ribblesdale, Yorkshire.

Plate 10

Ice-eroded surface in Lewisian gneisses, Sutherland; in distance a Torridon Sandstone hill, Suilven, rising above the Lewisian floor. The ice has eroded along belts of weaker rock more deeply than in stronger rock. (*Photograph by Aerofilms Ltd.*)

frequency and the thickness of strata is also discussed by Focardi, P. *et al.*, 1970, in Frequency of joints in turbidite sandstones, *Proc. 2nd. Congr. Int. Soc. Rock Mechanics*, Belgrade.

GENERAL REFERENCES

ANDERSON, E. M. 1951. Dynamics of Faulting (2nd Edition). Oliver & Boyd.

HILLS, E. SHERBON 1963. Elements of Structural Geology, Methuen, Paperback edition, 1970, J. Wiley & Sons, New York.

PRICE, N. J. 1968. Fault and Joint Development in Brittle and Semi-brittle Rock. Pergamon Press Ltd., Oxford.

RAMSAY, J. G. 1967. Folding and Fracture. McGraw Hill, New York.

8 Historical Geology of Britain

The succession of sedimentary rocks and their fossil content, in Britain, was largely elucidated during the century which followed the publication in 1795 of James Hutton's 'Theory of the Earth'—a work which marked the beginning of modern geological studies. In a pile of sedimentary layers (each stratum of which was formed during a particular interval of geological time) the succession of strata represents a series of events in the geological history of the area. The main divisions of geological time, or Periods, corresponding to the Systems into which rocks are grouped, are given in Table 8.1 below (and see Chapter 1, p. 19). From modern studies of radio-active mineral dating it has been possible to find the absolute ages of many rocks in millions of years. Some of these datings are given in Table 8.1, and they illustrate the enormous length of time during which geological processes have been in operation at the earth's surface. The major mountain-building episodes, called *orogenies*, are stated in the last column.

Breaks in deposition, when no sediments were formed in a given area, are marked by unconformities (p. 219); these represent intervals of time when not only was there no deposition, but when denudation took place, the sea floors with their sediments being raised and becoming subject to denuding agencies. Thus there were periods of quiet sedimentation, when the seas covered the lands, and intervening episodes of disturbance when uplift and folding took place. This broad pattern of events—the transgression of the sea over a land, the regression of the sea, followed by orogenic upheaval—has been repeated many times throughout geological history.

Unconformities between younger and older rocks are often marked by beds of pebbles and sand at the base of the younger series, the beach deposits of an incoming sea (p. 178). Examples of this are the pebbly quartzites at the base of the Cambrian in N.W. Scotland and the wave-rounded flints at the base of the Eocene deposits in south-east England; both mark the oncoming of marine conditions. Boulder-beds and screes formed on an old land surface during erosion, may also be preserved as the lowest members of a newer series of rocks resting unconformably on older rocks; as for instance the boulders and coarse sands which mark the base of the Torridonian in N.W.

Table 8.1

Geological Period (or Group)			Age in Millions of Years	Major Orogenies
QUATERNARY	{ Recent { Pleistocene			
TERTIARY	{ Neogene { { Palaeogene	{ Pliocene { Miocene { Oligocene { Eocene	 70	Alpine orogeny
MESOZOIC	{ Cretaceous { Jurassic { Triassic		136 195 225	
NEWER PALAEOZOIC	{ Permian { Carboniferous { Devonian (and Old { Red Sandstone)		280 345 410	Hercynian orogeny Caledonian
OLDER PALAEOZOIC	{ Silurian { Ordovician { Cambrian		440 530 570	orogeny
PRE-CAMBRIAN	Dalradian { Moinian { Torridonian { Lewisian		 700 900 3500+	Pre-Cambrian orogenies

Scotland, and lie unconformably on the old land surface carved in the Lewisian rocks (Fig. 8.1). The breccias which are found at the base of the Permian similarly represent scree detritus formed on land by the denudation of uplifted areas of folded Carboniferous sediments.

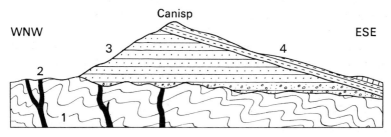

Fig. 8.1 Section through Canisp, Sutherland, to show Lewisian (1) cut by basic dykes (2); Torridon Sandstone and basal conglomerate (3); and Cambrian quartzite (4). Two types of unconformity are shown: the old land surface at the base of the Torridonian, and the tilted plane of marine denudation on which the Cambrian rests.

In a succession of sedimentary layers, tests of their order of formation include: (i) the *order of superposition* (younger beds lie upon older, in an undisturbed sequence); (ii) *included fragments:* inclusions such as pebbles, in a deposit, have been derived from some older formation; (iii) *current-bedding:* the *tops* of current-bedded layers are truncated by *younger* layers (p. 45); and (iv) *graded-bedding:* the size of grains in graded layers range from coarse at the bottom to fine at the top of a bed. (v) Fossils contained in beds of sediment also provide a means of relative age determination, by comparison with a known fossil sequence. The 'way-up' of a series is found from such tests.

In the following account the rock groups and the main events in the geological history of the British area are outlined.

PRE-CAMBRIAN

The ancient rocks collectively called the Pre-Cambrian, which represent a very long span of time, are sub-divided into several groups in Great Britain:

1. *Lewisian* (after the Isle of Lewis): metamorphic rocks, mainly gneisses of igneous origin, with some schists, and cut by basic and ultrabasic dykes. In N.W. Scotland the Lewisian forms a basement, which was eroded into a land surface on which sediments of the next group were deposited (Plate 10).

2. *Torridonian* (from Loch Torridon, the type locality): a thick sedimentary series of brown sandstones and arkoses, with some shales. Boulder beds at the base fill hollows in the old Lewisian topography, well seen for example on the south-west slopes of Slioch (Loch Maree). Torridonian rocks lie unconformably on the Lewisian and outcrop with it along the coastal belt west of the Moine Thrust (Fig. 8.7).

The above two groups are in places covered by Cambrian strata (Fig. 8.1), and are therefore demonstrably pre-Cambrian. The great thrusts which affected the rocks were formed during the Caledonian orogeny (see p. 252).

3. *Moinian* (after the Moine district of Sutherland): a thick group of metamorphosed sandstones and shales, now granulites and schists, which occupy a large area, from the line of the Moine Thrust southwards to where they dip below the Dalradian (Fig. 8.7). Radiometric dating has shown that the Moine rocks are upper Pre-Cambrian in age (see Table 8.1).

4. *Dalradian* (named after the old Irish kingdom of Dalriada): a series of metamorphosed sediments, now strongly folded, which occupy the area between the Moine outcrop and the Highland Boundary Fault in Scotland (Fig. 8.7); younger than the Moinian, the

Dalradian schists range in age from upper Pre-Cambrian to middle Cambrian, and include marble (metamorphosed limestone) as well as mica-schists, quartzites and hornblende-schists. Dalradian rocks also occupy large areas in west and north-west Ireland.

Apart from the Scottish occurrences, Pre-Cambrian gneisses and schists are found in Anglesey, the Malvern Hills, and Charnwood Forest (Leicestershire). In Shropshire lavas and tuffs called the Uriconian Series form fault-bounded ridges near Church Stretton; and the Longmynd is built of steeply dipping Pre-Cambrian sediments. At the Lizard, Cornwall, and Start Point, Devon, hornblende-schists and some mica-schists form the headlands.

Larger areas of Pre-Cambrian rocks exist in Scandinavia (the Baltic Shield), Greenland, eastern Canada (the Canadian Shield), north-east Russia, western Australia, peninsular India, and Africa south of the Sahara. The large areas called 'shields' are old, stable areas of the crust. The Scottish, Greenland, and Canadian areas probably formed part of a single Pre-Cambrian mass before the opening of the North Atlantic rift in Mesozoic times (p. 17).

OLDER PALAEOZOIC

The big group of rocks known as the *Older Palaeozoic* (= 'ancient life') comprises the Cambrian, Ordovician, and Silurian systems. At the end of the Pre-Cambrian a long period of marine deposition began and a large part of the British area was submerged. An elongated trough of deposition, or *geosyncline* (p. 173), extended across it from south-west to north-east, and across the North Sea into Norway. The north-western margin of the trough was at first in Scotland and its south-eastern margin ran from South Wales across England to the Humber estuary. In this trough were laid down shallow-water deposits of great thickness, mainly sands and muds, a slow subsidence of the sea-floor keeping pace with their accumulation. The trough persisted through Ordovician times, with some changes in its boundaries, and into the Silurian (Note 1), when much of the sea became shallower, at times forming a *shelf sea* (p. 172) the sediments of which contain shallow-water fossils. In the deepest part of the main trough, over Wales, a total thickness of some 12 000 metres (40 000 feet) of sediment was accumulated through Older Palaeozoic time; towards the margins of the trough the thickness was less (Fig. 5.1). Farther to the north-east (i.e. along strike) the thickness of the deposits decreases, until in the south of Sweden they amount to only about 1350 metres.

Near the end of Silurian time, mountain-building or orogenic movements, of which a phase had operated at the end of the Ordovician, developed and became intense. The sediments of the geosyncline were compressed in a north-west and south-east direction and ridged up into fold-mountains known as the Caledonian Chain, of which the denuded stumps now remain in Wales, Ireland, and southern Scotland. The general direction of these folds in Scotland (i.e. the strike of the folded sediments) is north-east and south-west; their continuation in south-west Ireland is more nearly east-west (Note 2). During the orogeny slaty cleavage was impressed on the Palaeozoic sediments in the Welsh area and in the Lake District (Skiddaw Slates), and the isoclinal folds of the Southern Uplands were formed.

The Older Palaeozoic rocks at the present day, whose outcrops are shown in Fig. 8.7, are mainly marine sandstones and shales in which the chief fossils are trilobites, brachiopods, and graptolites (Fig. 8.2). Volcanic activity during parts of Ordovician time gave rise to extensive pyroclastic deposits of tuff and agglomerate, as in North Wales and the Lake District. Limestone-forming conditions developed in the shallower seas of the Silurian period; the Wenlock Limestone and Aymestry Limestone ridges of Shropshire (Fig. 2.8) and elsewhere in the Welsh Borderland at the present day are representative of these limestones. In Fig. 8.3 these rocks are shown in relation to the Pre-Cambrian of the Malvern Hills, Herefordshire.

The prolonged Caledonian orogeny at the end of Silurian time continued into the Devonian. In a late phase of the movements the great overthrusts in the N.W. Highlands of Scotland, such as the Moine Thrust (with a horizontal movement of some 24 kilometres), and others in Scandinavia, were formed, over-riding the margins of the old geosyncline; they are associated with north-westerly wrench faults. Extrusion and intrusion of igneous material took place on a large scale: the Lower Old Red Sandstone of Scotland contains many beds of tuff and agglomerate, with lava flows, formed at this time; and the great granite masses of the Cairngorms and of N.E. Scotland also came into position, as did the granites of Galloway, Cheviot, and Shap further south, and in Ireland the Newry and Leinster granites. These together are called the Newer Granites in distinction from the Older Granites of Pre-Cambrian age. The large gabbro masses of Aberdeenshire had also been emplaced (Note 3).

NEWER PALAEOZOIC

Rocks of three Periods go to make up the Newer Palaeozoic: the Devonian (named after Devon), the Carboniferous (so called because of its coal seams), and the Permian (after the old Russian province of Perm). As a result of the Caledonian movements a large land area had been formed, built of the folded older rocks and extending across

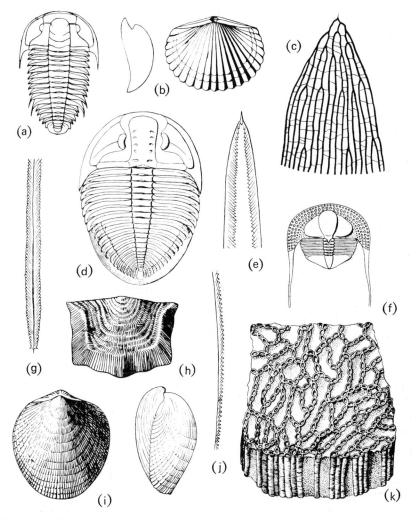

Fig. 8.2   Older Palaeozoic fossils. Trilobites: (*a*) *Olenus* (Cam.), (*d*) *Ogygia* (Ord.), (*f*) *Trinucleus* (Sil.). Brachiopods: (*b*) *Orthis* (Cam.), (*h*) *Leptaena* (Sil.), (*i*) *Atrypa* (Sil.). Graptolites: (*c*) *Dictyonema* (Cam.), (*e*) *Didymograptus* (Ord.), (*g*) *Diplograptus* (Ord.), (*j*) *Monograptus* (Sil.). Coral: (*k*) *Halysites* (Sil., 'chain coral'). All ×1.

northern Europe; it is shown in Fig. 8.4 as the 'Old Red Sandstone Land', and it began to undergo denudation in Devonian times. Britain, except in the south, was part of this land area, with shallow inland seas and gulfs in which non-marine sandy deposits were laid down. The south-west of England, however, was covered by a sea in which marine sediments were formed, containing remains of marine life of the time. Thus during the first part of Newer Palaeozoic time two distinct kinds of deposit were being accumulated: (i) the marine

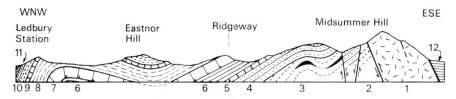

WNW   ESE

Ledbury   Eastnor   Ridgeway   Midsummer Hill
Station   Hill

11   12

10 9 8 7   6   6   5   4   3   2   1

Fig. 8.3   Section from Ledbury to the Malvern Hills, Herefordshire. Distance about 3½ miles. Vertical scale nearly twice the horizontal. (1) Pre-Cambrian. (2) Hollybush Sandstone (Cambrian). (3) Bronsil Shales (Cambrian) with basalt sills. (4–10) Silurian: (4) Mayhill Sandstone, (5) Woolhope Limestone, (6) Wenlock Shales, (7) Wenlock Limestone, (8) Lower Ludlow Shales, (9) Aymestry Limestone, (10) Upper Ludlow Mudstones. (11) Ledbury Shales (O.R.S.). (12) Keuper Marl (Triassic).

Devonian muds and sands, which are now found as slates (e.g. the Delabole Slates of N. Cornwall), shales, sandstones, and some limestones as at Torquay and Plymouth, with fossils of marine shells; and (ii) the *Old Red Sandstone* deposits, of freshwater type, laid down in shallow gulfs, deltas, and transient lakes on the land area. Torrential floods spread the debris brought from mountain tracts, to accumulate as successive sheets of sediment and build up a great thickness of freshwater deposits. In these deposits, now seen in Orkney, Cheviot, Herefordshire, and Kerry, the remains of primitive freshwater fishes have been preserved (Fig. 8.5, *a* and *b*).

Both the Devonian rocks and the contemporaneous Old Red

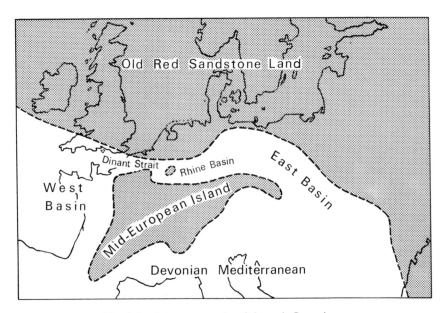

Fig. 8.4   Palaeogeography of the early Devonian.

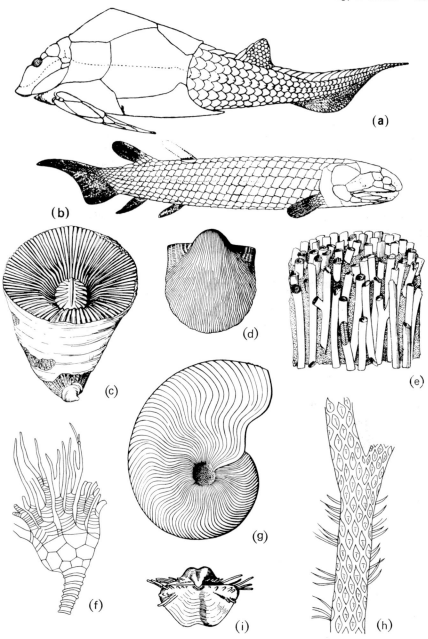

Fig. 8.5 Newer Palaeozoic fossils. Fishes: (*a*) *Pterichthys*, (*b*) *Thursius* (Old Red Sandstone; after Watson). Corals: (*c*) *Dibunophyllum* (Carb.), (*e*) *Lithostrotion* (Carb.), both ×1. Brachiopods: (*d*) *Productus* (Carb.), (*i*) *Productus horridus* (Perm.). (*g*) A goniatite (Carb.), ×2. Plant: (*h*) *Lepidodendron* (Coal Measures), ×⅜. (*f*) A crinoid (Carb.).

Sandstone were later to be folded during the mountain-building movements of the Hercynian (or Armorican) orogeny, now often regarded as a later phase of the Caledonian movements.

At the beginning of *Carboniferous* times the sea invaded a large part of the British area, and in it distinctive deposits of limestone and shales were built up in clear and not very deep water. These deposits are the Carboniferous Limestone of the present day; they form massive escarpments in the Mendips, Derbyshire and the Pennines, and many caverns have been eroded in the limestone by streams of water flowing underground. They are followed by shallow-water sands and deltaic deposits, the Millstone Grit of West Yorkshire and other areas, which represent an uplift of the sea floor at the end of Carboniferous Limestone times with the formation of deltas on the margins of the receding sea.

The deltaic conditions gave place gradually to swamps, and in an increasingly warm climate the swamps supported a dense vegetation. From time to time, as submergence occurred, the growth of vegetation or swamp-forest was buried by incoming sediment; the swamp-forest then grew again at a higher level. This was repeated many times, and the successive layers of vegetation became compressed under the weight of overlying sediment, to be preserved as the coal seams of the present day. In these *Coal Measures*, remains of trees and smaller plants, their spore cases, and other details, give evidence of the vegetable origin of the coals. Beneath each seam a layer of fireclay or ganister (siltstone) represents the ancient 'seat-earth' in which the Coal Measure plants grew (p. 196).

Towards the end of the Carboniferous the Hercynian orogeny (Note 4) began to affect a large part of N.W. Europe, and the rocks of south-west England were folded about east-west axes. Structures having this trend are found in the Mendips of Somerset, in Devon and Cornwall, where cleavage was impressed on rocks of argillaceous composition. North of the 'front' of the Hercynian folds, which lay approximately east and west along the Bristol Channel, other folds having a north and south trend were also formed (perhaps a little earlier than the main folding), as in the Malvern Hills area and in Derbyshire. The Coal Measures and underlying strata were preserved in broad synclines flanking the uplifts; these are the sites of present day coalfields. The Pennine chain in northern England was elevated mainly by faulting (Fig. 8.6).

Lakes became established west of the Pennine uplift and over east Devon, and in them deposits of *Permian* age were formed: breccias, sands, and red marls. On the east of the Pennine uplift lay a land-locked sea in which marine deposits were laid down, now the white Magnesian Limestone of Nottingham and Co. Durham, and red

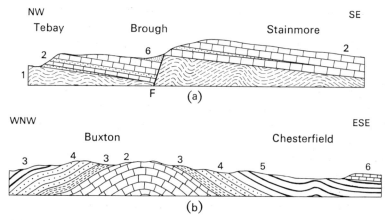

NW   SE
Tebay   Brough   Stainmore

F   (a)

WNW   ESE
Buxton   Chesterfield

(b)

Fig. 8.6 Sections across the Pennine Uplift. (*a*) Northern Pennines (after D. A. Wray, H.M. Geol. Surv.), *F* = Dent Fault. (*b*) Southern Pennines: the Lancashire and Yorkshire coalfields flanking the Derbyshire dome. *1*, Silurian and older rocks; *2*, Carboniferous Limestone Series, with basement conglomerate; *3*, Yoredale Series; *4*, Millstone Grit; *5*, Coal Measures; *6*, Permian (Magnesian Limestone).

marls overlying it (Fig. 8.6*b*, no. 6). Small areas of Permian sandstone around Dumfries and in Ayrshire show coarse dune-bedding and wind-rounded sand grains (Plate 2a) and illustrate the desert conditions of the time. Lavas of Lower Carboniferous age form e.g. the Campsie Fells near Glasgow, and the volcanic rocks of Arthur's Seat at Edinburgh are of this age, as are the numerous dykes and sills of quartz-dolerite intruded into the Carboniferous rocks of the Midland Valley of Scotland. In the north of England the Whin Sill, a quartz-dolerite intruded into the Yoredales with an outcrop some 280 kilometres in length, and the Hett dyke of Durham and others, date from near the end of the Carboniferous period. These dolerites yield good road-stone from many quarries. Igneous activity early in the orogeny resulted in the intrusion of the granites of south-west England, as well as similar granitic masses in Brittany; together they are called the Hercynian granites.

MESOZOIC

Three groups comprise the Mesozoic rocks, the Triassic, the Jurassic, and the Cretaceous. There is a transition from the Permian to the Triassic in Britain, where the succession is complete, and the two groups of rocks, predominantly red in colour, are often referred to as the Permo-Trias or 'New Red Sandstone'. While this is convenient in Britain it is usual to separate the two Systems in Europe.

The earliest *Triassic* rocks, the Bunter Sandstones, are mainly coarse sands which accumulated in inland basins of deposition (or

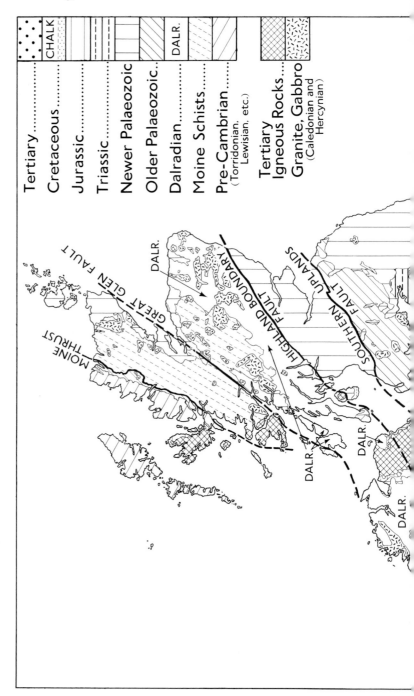

Fig. 8.7 Geological Map of the British Isles. (Scale approximately 1 : 5 000 000).

French, *cuvettes*), the sediment being contributed by the denudation of surrounding mountains in an arid climate. These deposits were moved by torrents, with pebbly layers and sheets of wind-borne sand being formed at times. One area of deposition, in south-west England, is now represented by the red Triassic rocks of south and east Devon; another lay north and west of Birmingham, where a thickness of over 900 metres of Bunter Sandstone is present. Overlying the Bunter are lacustrine sands and red marls of the Keuper Series, formed in a shallow lake covering the Midlands, with extensions at times to the south, and north-west towards southern Scotland and Ireland. The lowest member of this series, the Keuper Sandstone, is succeeded by the Keuper Marl, 760 metres thick in Cheshire (Note 5). Salt lakes and pans came into existence, in which rock-salt, gypsum, and other evaporites were precipitated (p. 194).

At the end of the Keuper a sea invaded and flooded the existing land area, with its salt lakes. This is demonstrated by an abrupt change to marine sediments, which lie on the uppermost marls of the Keuper. The marine beds, known as the Rhaetic (after the Rhaetian Alps), consist of about 18 metres of black shales with thin limestones and sandstones, and have a persistent outcrop from Dorset to the Yorkshire coast. They form a transition from the Triassic marls to the marine clays and limestones of the Jurassic.

Changing conditions of deposition gave rise to alternations of argillaceous and limey sediment. The sequence of *Jurassic* formations is shown in the form of a cross-section in Fig. 8.11. The grey clays and shales of the Lower Jurassic, or Lias, are worked extensively for brick-making near Cheltenham and elsewhere in the outcrop; in places they contain impure limestones ('cement-stones') which render them useful for the manufacture of cement, as at Rugby. Important bedded iron ores are worked in the Middle Lias from Cleveland, Yorkshire, at intervals southwards to the Banbury area; much of the iron occurs as the carbonate (siderite) and silicate (chamosite). Above the Lias clays lie oolitic limestones, which give rise to the main escarpment of the Cotswolds, where for many years they have been worked for building material known as Bath Stone. Some characteristic fossils from the rocks are illustrated in Fig. 8.8. A thick clay formation, the Oxford Clay, lies above the oolitic limestones and forms tracts of rich pasture land; it is extensively worked for brick clays.

Towards the top of the Jurassic sequence lie the Portland Sands and Portland Limestone; the latter yields a cream coloured building-stone from quarries at Portland, Dorset. They are succeeded by the Purbeck Series at the top of the succession; in the Purbeck Limestone freshwater shells (*Paludina*, Fig. 8.8) are plentiful and indicate a change from marine to lacustrine conditions. By Portland and Purbeck times

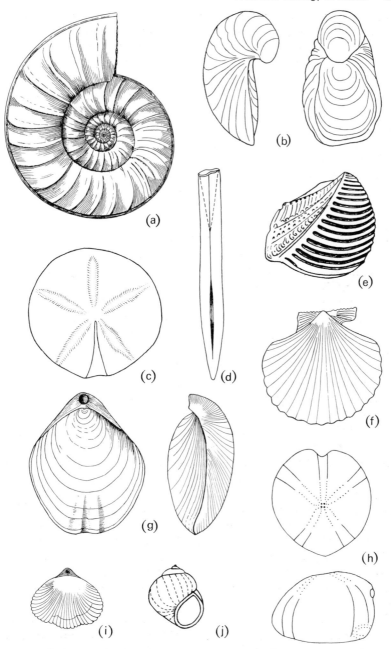

Fig. 8.8   Mesozoic fossils. Ammonite: (*a*) *Asteroceras* (Jur.). Brachiopods: (*g*) *Terebratula* (Chalk), (*i*) *Rhynchonella* (Jur.). Lamellibranchs: (*b*) *Gryphaea* (Jur.), (*e*) *Trigonia* (Jur.), (*f*) *Pecten* (Jur.). Sea-urchins: (*c*) *Clypeus* (Jur.), (*h*) *Micraster* (Chalk). (*d*) A belemnite (Cret.). (*j*) Gastropod: *Paludina* (L. Cret.). All $\times \frac{1}{2}$ except (*a*).

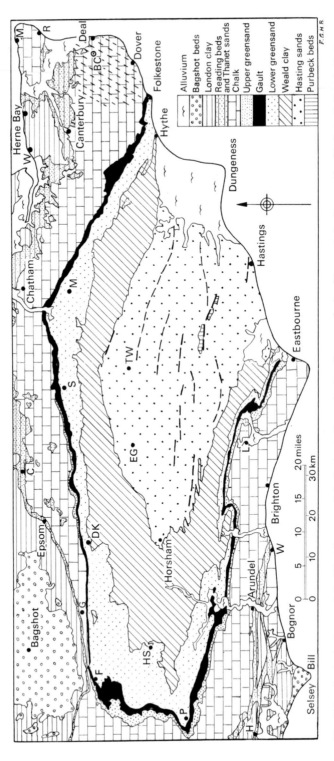

Fig. 8.9 Geological map of the Wealden area, S.E. England. Heavy broken lines are faults. Diagonal shading on the Chalk in E. Kent indicates approximate extent of the concealed Kent Coalfield; *BC* = Betteshanger Colliery. Croydon, *C*; Dorking, *DK*; East Grinstead, *DK*; Farnham, *F*; Guildford, *G*; Havant, *H*; Haslemere, *HS*; Lewes, *L*; Maidstone, *M*; Petersfield, *P*; Ramsgate, *R*; Tunbridge Wells, *TW*; Whitstable, *W*; Worthing.

Alluvium
Bagshot beds
London clay
Reading beds
and Thanet sands
Chalk
Upper greensand
Gault
Lower greensand
Weald clay
Hasting sands
Purbeck beds

the Jurassic sea had contracted to a small area in the south of England and in it sedimentation continued into the Cretaceous period.

The uppermost beds of the Purbeck pass upwards without break into the lowest *Cretaceous* deposits, the Wealden. These were laid down in a large lake or restricted estuary, with deltas, extending over the present Weald of Sussex and Kent, and across the English Channel. The open sea lay further to the south-east. Into the Wealden 'lake' sediments were brought by rivers from the west and north, and these freshwater deposits now constitute the clays and sands of the Wealden Series (Fig. 8.9). The climate was moderately warm, and the life of the time included large reptiles such as the dinosaurs, whose skeletons are found in the Weald Clay. Subsidence began during the formation of the upper clays of the Wealden, which are estuarine, and the sea advanced westwards and northwards. In this widening Cretaceous sea, marine sands (the Lower Greensand) and muds (the Gault clay) were deposited, and eventually, at its greatest extent, the calcareous muds which became compacted to form the Chalk. The Cretaceous succession extends from the Hastings Sands to the top of the Chalk.

The Chalk is a white, fine-grained limestone mainly made of algal fragments, and at its base is the Chalk Marl; it has a large outcrop in south-east England (Fig. 8.7) and, as a result of folding in early Tertiary times, now forms the escarpments of the Chiltern Hills and the North and South Downs (Fig. 8.12). Some fossils from the Chalk, such as *Micraster*, are illustrated in Fig. 8.8. A compact and somewhat harder band, known as the Chalk Rock, is taken as the dividing line between Upper and Middle Chalk, and can be a useful marker in drilling operations. A greater thickness of Upper Chalk is found in the south of England, as in the Isle of Wight where deposition continued longer, than elsewhere in the outcrop of the formation. Many flints (siliceous concretions, p. 193) occur in the Upper Chalk, fewer in the Middle Chalk, and very few below it.

TERTIARY

At the end of Cretaceous times gentle earth-movements resulted in the contraction of the sea in which the Chalk had been formed until it occupied only the North Sea area and adjoining marginal land. In this restricted sea the early sands and clays of the Eocene were deposited. In south-east England they lie on a surface eroded in the Upper Chalk; as a result of a small tilt of the area the inclined layers of Chalk were bevelled off by the sea, and the lowest Eocene beds contain rounded flint pebbles derived from flints in the Chalk. The beginning of Tertiary time dates back to about 70 m.y. Its sub-divisions are named Eocene, Oligocene, Miocene, and Pliocene in

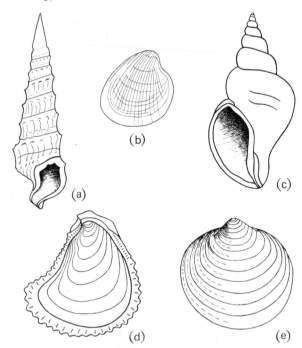

Fig. **8.10**   Tertiary fossils. Gastropods: (*a*) *Cerithium* (Eoc.), (*c*) *Neptunea* (Plio.). Lamellibranchs: (*b*) *Cardita* (Eoc.), (*d*) *Ostrea* (an oyster) (Eoc.), (*e*) *Pectunculus* (Eoc.). All × ½. (After A. Morley Davies.)

ascending order, now grouped as Palaeogene and Neogene (Table 8.1). As well as the shells of marine animals, and at times shells of estuarine type (Fig. 8.10), the fossil remains of many mammals are found in the Tertiary rocks and indicate a great development of mammalian life at this time.

The Palaeogene (Eocene and Oligocene) sediments are preserved in two synclines in England, the London Basin and the Hampshire Basin, and correspondingly in the Paris Basin. In these areas the thick London Clay lies above the earliest Eocene deposits and marks a marine transgression; it is followed by further sands and clays deposited in shallower water, which in the Hampshire Basin extend up into the Oligocene (Fig. 8.13).

Also during the Eocene period, intense igneous activity began in the north of the British area; volcanoes erupted in Mull, Skye, and Arran, and in Northern Ireland the great lava flows of Antrim were poured out. Similar basaltic lavas were being formed in Greenland and Iceland (Note 6). Plutonic igneous activity resulted in the intrusion of the ring complexes of Ardnamurchan and Mull, the gabbros and granites of Skye, Rhum and other islands of the Hebrides, and the

Fig. 8.11 Section across the Jurassic rocks of the Cotswolds to the Chalk near Swindon. *1*, Keuper; *2*, Rhaetic; *3*, Lias; *4*, Inferior Oolite; *5*, Fuller's Earth; *6*, Great Oolite; *7*, Oxford Clay; *8*, Corallian; *9*, Kimmeridge Clay; *10*, Lower Greensand; *11*, Gault; *12*, Chalk. Length of section about 37 km.

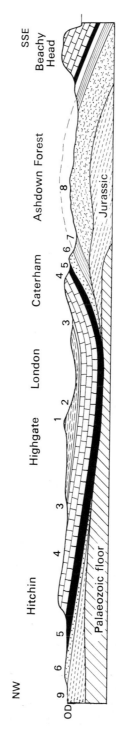

Fig. 8.12 Section across the London Basin and the Weald. Showing the Palaeozoic floor beneath London. Length of section about 160 km; vertical scale exaggerated about 20 times. (1) Bagshot Sands. (2) London Clay. (3) Woolwich and Reading Beds and Thanet Sand. (4) Chalk. (5) Gault and Upper Greensand (the latter is not continuous under London). (6) Lower Greensand. (7) Weald Clay. (8) Hastings Sands. (9) Jurassic rocks.

granites of Mourne and Carlingford. The Tertiary dyke-swarms of these areas came into existence as fractures which had been opened by the operation of crustal stresses, to be filled by rising basic magma.

Orogenic movements which formed the fold-mountains of the Alps Pyrenees, Carpathians, Himalayas, and the Cordillera of North America began in the Eocene and reached their climax in the Miocene. The sediments of a large east-west trough (called *Tethys*), which during the Mesozoic lay broadly in the region of the present Mediterranean, were folded to form the Alpine ranges. The 'outer ripples' of this Alpine orogeny are seen in the south of England as the folds of the London Basin, the Weald, and the Isle of Wight (Fig. 8.13). There are

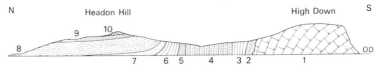

Fig. 8.13   Section across the Isle of Wight near Alum Bay (after Bristow), length about 2 km. (1) Chalk. (2) Woolwich and Reading Beds. (3) London Clay. (4) Lower Bagshot Sands and Clays. (5) Bracklesham Beds (clays with gypsum). (6) Barton Clays. (7) Upper Bagshot Sands. (8) Headon Beds. (9) Osborne and Bembridge Beds. (10) Plateau Gravel (Quaternary).

no Miocene deposits in Britain owing to the elevation of the area above the sea of that time, when the Alpine orogeny was at its maximum. During the Pliocene the climate gradually became colder, leading up to the glaciation of the Pleistocene period.

QUATERNARY

Deposits younger than the Tertiary are called Quaternary, and include the glacial deposits of the *Pleistocene*, together with the *Recent* sediments such as river alluvium and gravel, blown sand, and peat. These lie at the surface over much of the land area of Britain, masking the solid rocks which they cover, and are important from an engineering point of view because much excavation is carried out in them. They are shown on the Drift Edition of the 1:63 360 geological maps (p. 324).

With the very cold climate which spread south over Britain and northern Europe during the Ice Age of the Pleistocene, ice accumulated in the mountain areas of Scotland and other high ground, and ice sheets moved out over the lower ground and into the North Sea and Irish Sea. Ireland also was covered by ice largely of local origin. Glaciers formed on high ground in the English Lake District and the Pennines, and in North Wales; the ice eventually spread about as far south as London (Fig. 8.14). Under the ice-sheets were formed the

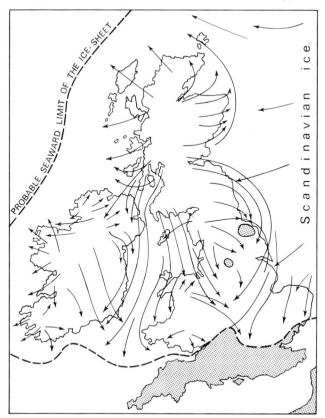

Fig. 8.14   Extent of glaciation in the British Isles and lines of ice-flow. Unglaciated areas (stippled) were subject to periglacial conditions.

boulder clays which cover a large part of northern, central, and eastern England, and the Midland Valley of Scotland. In the latter, three phases of glaciation have been distinguished, represented by three types of boulder clay which succeed one another and indicate advances of the ice. The clays are separated by sandy and peaty deposits which point to milder, 'interglacial' periods when the ice was temporarily being melted. Some of these glacial sands and gravels occur as kames and eskers (p. 69) as at Carstairs.

In northern England boulder clay covers much of Northumberland and Durham. West of the Pennines three advances of the ice-front can again be recognized, with intervals of retreat when water from melting ice deposited sands, gravels, and laminated clays (*varves*). In the Midlands south of the Pennines, the drift of the Cheshire plain extends as far as Birmingham and contains much material derived from Triassic rocks, and also erratics from Scotland and the Lake

District (see arrows, Fig. 8.14). Water-laid deposits again mark the retreat stages of the glaciation. Eastern England was invaded by ice from the North Sea and also from Scandinavia, the latter bringing distinctive erratics. The Chalky Boulder Clay, which contains much ground-up chalk derived from the rocks over which the ice moved, is spread over large areas of Norfolk and Suffolk, and is one of four boulder clays recognized in the area (Note 7).

Changes in the surface drainage were produced by the glaciation, a good example being the diversion of the River Severn into its present course below Shrewsbury. In its upper reaches the river flows north-east from the Welsh hills, and formerly drained into the Irish Sea by way of the Dee estuary. But during the retreat of the Irish Sea ice from the Cheshire plain, melt-water was impounded in front of the ice, forming a glacial lake (Lake Lapworth) whose outlet was by a col at Ironbridge, where a gorge was cut by the overflow. Later, the Severn was prevented by accumulations of drift from following its pre-glacial course, and became established in its present channel through the Ironbridge Gorge.

The old valleys of many rivers, including the Clyde, Forth, Tyne, and Mersey, are filled with glacial deposits and buried beneath the drift. The River Devon in Scotland, a tributary of the Forth, in its pre-Glacial course flowed in a valley which is now filled with late-Glacial clays (i.e. deposited towards the end of the Glacial epoch), with overlying sands and silts; a few miles east of Stirling, borings to Coal Measures below the drift proved the old valley floor at a depth of 107 metres at one point, and a geophysical survey helped to delineate its course (Note 8).

The *Recent* deposits have been formed since the ice retreated from the British area about 10 000 years ago, as the climate ameliorated. They include alluvial deposits along the courses of rivers, consisting of sand, silt, and clay, with occasional seams of gravel; in places this alluvium may attain a thickness of 9 or 10 metres. Terrace gravels which were formed by rivers in an earlier stage are now found at higher levels than the river's present course; the Thames valley gravels are illustrated in Fig. 8.15. Accumulations of blown sand occur at

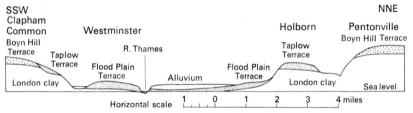

Fig. 8.15 Section across the Lower Thames Valley, to show river terraces. (Gravels stippled.) **Length about 16 km.**

Fig. 8.16 Raised marine platform after isostatic uplift; formed in late-Glacial times and covered by gravels of '100-foot' beach, Islay. The '25-foot' terrace is in foreground, and on right the present storm-beach of quartzite boulders. (From a Geological Survey photo.)

many localities around coasts, as in East Anglia and Kent. The fenlands around the Wash, with their Recent silts, clays and peats, have been discussed on p. 48.

With the melting of the ice-sheets great quantities of water were gradually released, and a world-wide rise of sea-level resulted; this *eustatic* change in sea-level is estimated to have amounted to about 91 metres since the end of the Pleistocene glaciation. It probably continues at a slow rate, about 12 cm per century, at the present day as the remaining ice-caps slowly melt. Also, *isostatic* compensation began to take effect as the weight of the ice was removed, locally re-elevating land areas which had been depressed by the ice load. Where this isostatic rise was greater than the eustatic change the land has emerged above a rising sea-level. Thus the Scottish '100-foot' platform (Fig. 8.16) and other raised beaches mark stages in the process, when the two effects were nearly equal for a long enough time for a platform to be formed by marine erosion.

Of recent years, tide-gauge records taken around the coasts indicate a slow submergence in the south of England, where sea-level continues to rise relative to the land at about 2 mm per year. This effect has led to the construction of higher coastal defence works on the east and south coasts of England, and in other countries bordering the North Sea (see p. 61).

NOTES AND REFERENCES

1. For a recent discussion, with map, see Geosynclinal development of the British Isles during the Silurian Period, by A. M. Ziegler. *Journ. Geol.* **78,** July 1970, p. 445–479.

2. Corresponding structures with a similar trend are found in Newfoundland. If continental drift is assumed, they are the continuation of the Caledonian structures in Britain.

3. The basic intrusions of north-east Scotland, and also the granites of the south-west Highlands (Cruachan, Ben Nevis, Glen Coe, and others), are described in British Regional Geology: The Grampian Highlands, by G. S. Johnstone (3rd edition, H.M.S.O., Edinburgh). A full list of the Regional Guides is given below.

   The main intrusions of Ben Nevis consist of two granites, the earlier forming an outer zone around the younger. A nearly circular down-faulted block of andesite lavas resting on schists lies within the younger granite, which is thought to have welled up around a cylindrical fracture as this block subsided, the lavas being part of the original roof. The mechanism is called *cauldron subsidence;* it was first described from Glen Coe, where andesitic lavas within a ring fault are surrounded by the Cruachan granite (Clough, Maufe, and Bailey, 1909, *Q.J.G.S.* lxv, p. 611). Confirmation of the structure of Ben Nevis was obtained when the 15-mile long tunnel of the Lochaber Power Scheme was driven through the side of the mountain, to bring water from Loch Treig to the generating station at Fort William. New geological evidence was also brought to light in the tunnel exposures, and the instance provides an illustration of the way in which geological knowledge may benefit from information made available in engineering construction.

4. The duration of a major orogenic episode was probably of the order of 10 or 20 m years. The terms Armorican and Variscan are used for fold-systems in Europe formed during the Hercynian orogeny.

5. Engineering properties of the widespread Keuper Marls have been extensively studied in recent years. See, for example, the account by Kolbuszewski, Birch, and Shojobi, 1965, in *Proc. 6th. Int. Conf. Soil Mech. and Fndn. Eng.*, Canada, on Keuper Marl Research.

6. This eruptive activity was probably one episode in the opening of the North Atlantic Rift, which is believed to have begun during Jurassic times, with the gradual separation of the American and Euro-Asian continents (p. 17).

7. See West, R. G. and Donner, J. J. 1956. The Glaciations of East Anglia and the East Midlands, *Quart. Jour. Geol. Soc.* **112**, 69.

8. Parthasarathy, A. and Blyth, F. G. H. 1959. The Superficial deposits of the buried valley of the River Devon near Alva, Clackmannan, Scotland. *Proc. Geol. Assoc.* **70**, 33.

GENERAL REFERENCES

BENNISON, G. M. and WRIGHT, A. E. 1969. The Geological History of the British Isles. Edward Arnold, London. (paperback).
British Regional Geology, a series of 18 small volumes published by H.M. Stationery Office, and comprising:
   Scotland: Northern Highlands. 3rd edition, 1965.
   The Grampian Highlands. 3rd edition, 1966.

The Tertiary Volcanic Districts, Scotland. 3rd edition, 1961.
The Midland Valley of Scotland. 2nd edition, 1948.
The South of Scotland. 2nd edition, 1948.
Northern England. 3rd edition, 1963.
The Pennines and Adjacent Areas. 3rd edition, 1954.
East Yorkshire and Lincolnshire. 1948.
North Wales. 3rd edition, 1961.
South Wales. 2nd edition, 1948.
The Welsh Borderland. 2nd edition, 1948.
Bristol and Gloucester District. 2nd edition, 1948.
Central England. 3rd edition, 1969.
East Anglia and Adjoining Areas. 4th edition, 1961.
London and the Thames Valley. 3rd edition, 1960.
The Wealden District. 4th edition, 1965.
The Hampshire Basin and Adjoining Areas. 3rd edition, 1960.
South-West England. 3rd edition, 1969.

Each booklet gives a description, with maps and sections, of the district covered, and a list of the 1-inch Geological Survey Maps dealing with the district, together with a Bibliography. Available at Inst. Geol. Sciences.

British Association publication: A view of Ireland: (2), Geology. Dublin, 1957.

CHARLESWORTH, J. K. 1953. The Geology of Ireland. Oliver & Boyd, Edinburgh.

CRAIG, G. Y. (Ed.) 1965. The Geology of Scotland, (reprinted 1970). Oliver & Boyd, Edinburgh.

FLINT, R. F. 1971. Glacial and Pleistocene Geology. Wiley & Sons, New York.

READ, H. H. 1949. Geology. Oxford University Press. (Home Univ. Library).

RAYNER, D. H. 1967. The Stratigraphy of the British Isles. Cambridge University Press.

Analytical Techniques

9 *In Situ* Investigations

Sampling, for whatever purpose, is a process that is fraught with difficulties, the most common being to ensure that a representative sample is obtained. This usually means answering the following questions:

(i) Has a large enough sample been collected?
(ii) Has it been collected from the best location?
(iii) Has its character been changed by the techniques used in its collection?

It has long been recognized that one of the ways of overcoming some of the difficulties is to avoid sampling and test the subject in place; hence the techniques for *in situ* testing and investigation were developed. These, however, by no means solve the problem of sampling, for most *in situ* tests are still conducted on a scale that is much smaller than the volume of ground affected by many engineering structures. The investigator is forced to be selective in choosing the areas to be sampled, and so the question of knowing where to test, i.e. where to take the field sample, again arises. Design requirements and geological homogeneity are factors that usually control such decisions. Economics may control the size of an investigation, and it should be realized that many *in situ* tests are very costly. The investigator often has to balance a programme of expensive *in situ* investigations on large samples against less expensive laboratory investigations on smaller samples. Here the question of geological homogeneity has once more to be considered. Thus the philosophy behind the desirability of *in situ* testing is by no means simple, and is largely controlled by the relative scales of the problems that have to be solved. For example, the proposed route for a tunnel of a given diameter passing through rock which has a fracture spacing much in excess of the tunnel diameter will not require the same type of investigation as for a similar tunnel passing through more intensely broken ground.

Geology is therefore important to the investigation stage of any project as the investigation should be tailored to both the needs of the design and the sub-surface geology. The latter, in areas where the geology is poorly known, makes the ideal investigation difficult to achieve and allowance should then be made for the adoption of either a flexible or a multistage investigation. In this way the programme can be changed as ground conditions are revealed, (e.g. Bjerrum, 1973).

The *in situ* techniques that can be used are well described in the literature (see B.G.S. Symposium, 1973) and will not be the principle subject of this chapter. Instead, the emphasis is placed on an appreciation of the geological conditions that should be considered when planning, conducting, and interpreting the results of *in situ* investigations. The general purpose of these investigations will vary from job to job, but in most cases four fairly well defined objectives can be discerned:

(1) To determine, in whatever detail is required, the character of the ground (including both the solid material and the contents of pores and fissures). This accounts for much of the bore-hole drilling that is requested, together with core logging and hole logging, using the variety of optical, electrical, and mechanical devices now incorporated in bore-hole logging instruments.

(2) To determine the variations in the character of the ground within a particular volume and throughout a given time. The former accounts for the position of investigation areas, e.g. the location of bore-holes, their depth and inclination, the core recovery required, the location of geophysical surveys, and so on. The latter controls the length of time instruments such as water level recorders, load cells, and inclinometers, remain in the ground.

(3) To make a quantitative description of the area in terms that are relevant to the programme of testing and the engineering design. This is often a difficult task because many standard geological descriptions are qualitative in nature. Qualitative information is nevertheless of great value when describing an area in which tests have later to be conducted, as it assists in correlation with similar areas, and so aids the application of test results to other ground that has not been tested. The engineer should not be misled by some apparently quantitative rock classifications that now exist. Weathering classifications are a typical example and one of many is given in Table 9.1. Table 9.2 illustrates a classification for rocks that is becoming increasingly popular: for others see Bieniawski (1973), Barton *et al.* (1974) and Dearman (1974). Despite the numerals, these classifications are essentially qualitative and rely on the user correlating some characters of the rock with a suite of physical properties.

(4) To determine the response of the ground to certain imposed conditions of stress, strain, drainage, saturation, etc. This is where

Table 9.1 A Classification for Weathering (from the Geol. Soc. London. Engng. Gp. Working Party Report 1970)

Class	Symbol	Application
IA	F	*Fresh:* no sign of weathering
B	FW	*Faintly weathered;* weathering limited to the surface of discontinuities.
II	SW	*Slightly weathered;* penetrative weathering developed on open discontinuities but only slight weathering of rock material.
III	MW	*Moderately weathered;* weathering extends throughout the rock mass but the rock material is not friable.
IV	HW	*Highly weathered;* weathering extends throughout the rock mass and the rock material is partly friable.
V	CW	*Completely weathered;* the rock is wholly decomposed and in a friable condition but the rock texture and structure are preserved.
VI	RS	*Residual Soil;* a soil material with the original texture, structure and mineralogy completely destroyed.

Table 9.2 A Classification for Rock Quality (Called Rock Quality Designation or RQD, after Deere *et al.* 1966)

RQD (%)	Description of Rock Quality
100–90	Excellent
90–75	Good
75–50	Fair
50–25	Poor
25–0	Very poor

Note: RQD = summed length of core sticks greater than 10 cm (4 in.) in length, expressed as a percentage of the drilled length.

Example: Drilled length = 127 cm (50 in.).
Total core recovered = 101.6 cm (40 in.)
= 80%.
Summed length of core sticks greater than 10 cm (4 in.) long = 68.58 cm (27 in.).
RQD = 54%.

the majority of the *in situ* tests are used. They abound in variety and almost anything that needs to be measured can be measured, given both the money and the time. However, as stated earlier, the cost of an investigation can be excessive, and the time allowed for it is frequently minimal. In consequence, most investigations fall short of what is desired.

With these points in mind the general method of *in situ* investigation is now reviewed under two headings, *viz*. the desk study and the field study.

Desk study   This colloquially describes the preliminary search through records, maps, and other literature that is relevant to the geology of the area to be investigated. The object of the study is to obtain some indication of the local geology and hence the type of investigation required. The main problem in this exercise is to find the information, which is usually disseminated in various libraries. A guide to sources of geological information is provided in the Appendix at the end of this book; further information contained in the report by Dumbleton and West (1976), who recommend the following procedure:

(i) Locate and (if necessary) acquire any maps, papers, and air photographs relating to the site, and interpret as far as possible the geological conditions shown by these sources. In a complex area attempt an analysis of the geology; the preparation of geological sections or block diagrams may help to indicate areas where further information is needed. A short visit to the site is often desirable to confirm observations and predictions already made.

(ii) At this stage it may be fruitful to seek additional information from institutions such as the Geological Survey, geological societies, local authorities and libraries, universities, and from engineers who may have been involved in projects in the area.

(iii) After these enquiries it may be useful to visit the site again to collate all the data so far obtained, and to identify areas where engineering difficulties may exist and areas where particular investigations are needed.

(iv) Compile as good a report as can be made, recording the geological and geotechnical data, the addresses of useful contacts, and references to literature. This preliminary report will assist the ensuing investigation and provide a basis for the final report.

(v) After this ground-work it is possible to consider the construction requirements of the proposed engineering works at the site. In doing this there is a good chance that the ensuing investigation will be reasonably suited to both the geology of the area and the requirements of the design.

The above procedure is designed to help in the examination of a new area where there is little information about sub-surface conditions. It can be trimmed to suit the state of knowledge at the time, but vigilance is needed even in areas where the ground is known.

Field study This is used to prove the geological and geotechnical character of the ground and to accurately describe existing conditions. All this may seem straightforward in view of the techniques that are now available but it need not be so in ground that is variable. It should be remembered that the rocks in many localities contain a three-dimensional array of surfaces which will probably be discontinuous within the boundary of the site and extend beyond the site itself. Strength properties may vary from place to place and in addition there may also be mobile materials in either a liquid or gas phase. If the available field techniques are seen against this background it is possible to appreciate their limitations. The many techniques now available for field studies can be grouped into those that investigate (i) along a line, (ii) over an area, and (iii) through a volume: for helpful reviews see Glossop (1968), Amer. Soc. Civ. Eng. Report (1972), Bell (1975).

(i) *Linear investigations* include all kinds of borehole work and represent the sampling of a line or column of ground. If the columns are fairly close together a reasonably accurate interpretation of the geology can be made. Most *in situ* studies are of this type, and it is not surprising that sophisticated methods have been developed for looking at and logging the holes, as well as the cores that come from them. The ingenuity of these methods should not blind the investigator to their limitations. An important point to note here is the orientation of the line within a geological structure. Vertical bore-holes, for example, are unlikely to intersect vertical joints; the same applies for horizontal bore-holes and horizontal joints. The more general case is illustrated in Fig. 9.1. Hence, for each bore-hole there is a blind zone which cannot be sampled. An analysis of the size and angular disposition of these zones is described by Terzaghi (1965).

The techniques for drilling and coring are well described in the technical literature and will not be considered. However, in selecting the method to be used attention should be paid to the character of the ground, the type of sample (if any) that is required, and the instrumentation that will be subsequently used in the hole. In general, most problems of hole support and core recovery occur in weak or soft ground and an important point to consider here is the diameter of the core barrel (Note 1). EX, AX, and BX core sizes can give a core recovery of 50 to 70 per cent in weak material such as shale. A better

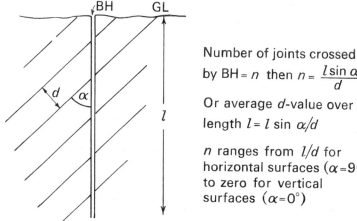

Number of joints crossed
by BH = n then $n = \dfrac{l \sin \alpha}{d}$

Or average d-value over any
length $l = l \sin \alpha / d$

n ranges from l/d for
horizontal surfaces ($\alpha = 90°$)
to zero for vertical
surfaces ($\alpha = 0°$)

Fig. 9.1   Intersection of joints by bore-holes.

overall recovery which can be close to 100 per cent is possible with larger types of barrel, but this is dependent upon the skill of the driller, good equipment (including swivel type double tube core barrels) and a rig that is free from vibration. For important work a barrel of 131 mm ($4\frac{3}{4}$ in.) diameter is often suitable. Drilling lubricants should be selected to suit the ground. Water, formerly the ubiquitous drilling fluid on all sites, is still commonly used. Alternatively air can be employed, and despite certain limitations has led to improved core recovery in friable soil and shattered rock at shallow depth.

Another problem that can arise, in addition to those of hole support and core recovery is sample disturbance, either

(1) as a disturbance to the walls and immediate vicinity of the hole. This normally results from the relief of stresses that follow the removal of a column of ground; it may be associated with processes such as dessication, oxidation, and with the degree of geological stress at depth. The overall effect is to open existing fractures and possibly to generate new ones, so generally loosening the ground; or,

(2) as a disturbance to the core sample collected; this is more readily apparent than (1).

A certain amount of 'bedding in' or 'settling down' can be expected in the initial stages of *in situ* tests carried out in holes that have been affected in this way; care should be taken when the results are interpreted, especially when using visual aids such as bore-hole cameras and periscopes as it is natural to assume that what is actually seen also represents the condition of the unseen ground.

The disturbance of cored samples can vary from skin effects which are restricted to the outer edge of the core to total disarrangement.

Helpful guidelines for sampling delicate ground are suggested by Hvorslev (1948). Rocks can also be disturbed by stress relief, and shales are notorious in this respect; the intensity of fracturing seen in cores of shale is often much greater than that found *in situ*. A useful method for recovering weak and broken rock, called integral sampling, is described by Rocha (1971) and is shown in Fig. 9.2. The procedure is as follows: A hole is drilled to just above the level from which a sample has to be collected. A small hole, coaxial to the first, is then advanced into the sampling area. It is generally recommended that the ratio D/d should be about 3.0. A perforated hollow reinforcing bar of diameter approximately 5 mm less than that of the smaller hole, is next inserted into the smaller hole and bonded to the rock by

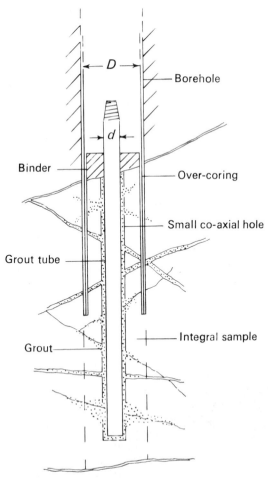

Fig. 9.2   Integral sampling (after Rocha).

the injection of grout through its hollow centre. The sample area is then over-cored, using low drilling pressures, and the sample retrieved. Thin-walled bits are recommended in small diameter holes, and double tube core barrels in weak ground. This method produces beautiful samples because the grout strengthens the weak rock and infills voids, so that the separation between fissure surfaces can be observed.

Having obtained a core it is desirable to have it described and stored. The description is important and can be of great assistance to both the design engineer and the contractor. It is not easily produced by inexperienced personnel and there is much to recommend the adoption of methods set out in the Report on the logging of rock-cores (Geological Society Engineering Group). This Report covers the general requirements of a bore-hole log, the factors that determine logging methods, the handling, labelling and preservation of rock cores, and the information to be recorded on the log, including core recovery, descriptive geology and rock grade classification. (See also Stimpson *et al.*, 1970 and Priest and Hudson, 1976.)

(*ii*) *Areal investigations.* These include all air photograph interpretations, most geophysical reconnaissance techniques (except radiometric and single bore-hole logging techniques), and all geological mapping. Such investigations, which extend over an area, provide a two-dimensional study of the ground and its geological make-up. The areas involved may be exposed on either vertical or horizontal surfaces.

Vertical surfaces can be seen in trenches and pits and are valuable in showing geological variations which occur both laterally and vertically. They are normally a part of any study of near-surface geology, are easily recorded and likely to reveal more information per unit cost than any bore-hole taken to the same depth.

A vertical area of a rather different nature is revealed by certain geophysical surveys, notably those using electrical resistivity and seismic methods. The former measures the potential drop between electrodes, and the latter the time taken by seismic waves (generated by a small explosion) to travel through the ground and return to the surface after reflection or refraction by buried surfaces. Both kinds of survey are conducted along lines laid out at ground level. From the measurements obtained certain deductions can be made concerning the nature of the sub-surface geology and approximate geological sections can be produced. A tabulation of the geophysical methods that can be used is given in Table 9.3. The physical bases for such methods are discussed by F. Dunning (1970) and D. Griffiths (1969).

Geology that is too deep to view in trenches can be seen with the aid of adits, and if these are large enough to accommodate drilling

Table 9.3   Geophysical Methods (after Dunning 1970)

Method	Field Operations	Quantities Measured	Computed Results	Applications
Seismic	Reflection and refraction surveys using, *on land*, several trucks with seismic energy sources, detectors, and recording equipment; *at sea*, one or two ships. Data-processing equipment in central office. Two man refraction team using a sledge-hammer energy source.	Time for seismic waves to return to surface after reflection or refraction by sub-surface formations	Depths to reflecting or refracting formations, speed of seismic waves, seismic contour maps.	Exploration for oil and gas, regional geological studies. Superficial deposit surveys, site investigation for engineering projects.
Gravity	Land surveys using gravity meters; marine surveys gravity or submersible meters.	Variations in strength of Earth's gravity field.	Bougner anomaly and residual gravity maps; depths to rocks of contrasting density.	Reconnaissance for oil and gas; detailed geological studies.
Magnetic	Airborne and marine magnetic surveys, using magnetometers. Ground magnetic surveys.	Variations in strength of Earth's magnetic field.	Aero- or marine magnetic maps or profiles; depth to magnetic minerals	Reconnaissance for oil and gas, search for mineral deposits; geological studies at sites.
Electrical and Electromagnetic	Ground self-potential and resistivity surveys, ground and airborne electromagnetic surveys, induced polarization surveys.	Natural potentials, potential drop between electrodes, induced electromagnetic fields.	Anomaly maps and profiles, position of ore-bodies, depths to rock layers.	Exploration for minerals; site investigations.
Radiometric	Ground and air surveys using scintillation counters and gamma-ray spectrometers; geiger counter ground surveys.	Natural radio-activity levels in rocks and minerals; induced radio-activity.	Iso-rad maps, radiometric anomalies, location of mineral deposits.	Exploration for metals used in atomic energy plant.
Borehole Logging	Seismic gravity, magnetic, electrical and radiometric measurements using special equipment lowered into borehole.	Speed of seismic waves, vertical variations in gravity and magnetic fields; apparent resistivities, self-potentials, etc.	Continuous velocity logs; resistivity of thickness of beds; density; gas and oil, and K, Th, U content.. Salinity of water.	Discovery of oil, gas, and water supplies; regional geological studies by borehole correlation. Applicable to site investigations.

equipment they can be used for making further investigations. Trenches and adits, though valuable, can nevertheless be subject to the effects of stress relief and other disturbances which loosen the ground, and care should be taken when tests are made in them. Attention is also drawn to the support and construction of such excavations (see British Code of Practice, CP 2003: 'Earth works').

Horizontal areas are usually those seen at ground level, either from natural exposures or from sections where the top soil has been removed. They reveal horizontal variations in the ground, and when used in conjunction with trenches and bore-holes can provide a three-dimensional picture of a site. Air photograph interpretation and geological mapping are the techniques most commonly used here, although gravity and magnetic surveys can also be employed (Table 9.3). The interpretation of geology from air photographs, although not strictly an *in situ* method, is a skilled operation that can be very useful in an investigation. The air photograph requirements for geological work have been described by J. Norman (1968) who gives guide-lines for the purchase of photographs and the commissioning of aerial surveys. The handling and interpretation of air photographs are discussed by C. Strandberg (1967), Miller and Miller (1961), and Ray (1960).

The making of a geological map is essentially a job for geologists, but should no geologist be available at a site, any map or record that the engineer can produce will usually be better than no map at all. Geological maps are considered in Chapter 11. A brief review of mapping methods is included here to explain the general nature of the work. The object of a geological map is to record the type and distribution of rocks and structures that can be seen at the surface (Fig. 9.3); for a good example of mapping below ground see Knill and Jones (1965). Normal equipment for mapping includes a topographical map of the area, a map case, soft pencils, compass, clinometer, and hammer. A hammer head weighing about 0.5 kg (or 1 lb) is usually sufficient and is made from forged steel so as not to splinter when breaking hard rock (suitable geological hammers can be purchased). The clinometer is used for measuring the dip of surfaces such as bedding and cleavage (Chapter 7); it may be a separate instrument, as shown in Fig. 9.4, or incorporated in the compass (as in a Brunton compass). In its simplest form it consists of a plummet which hangs vertically against a scale, and can be made by mounting a protractor on to a thin plate of material such as plywood or perspex sheeting, with a plumb-bob hung from the centre of the protractor circle. The compass is needed for measuring directions of dip or strike; many geologists use a prismatic compass because it also allows the accurate sighting of distant objects, as needed for orienting the

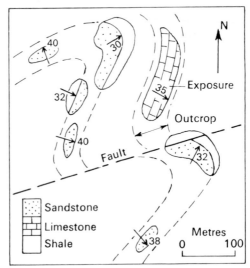

Fig. 9.3 Geological map (large scale) to show exposures and boundaries.

map. Other useful equipment includes a notebook (although some notes can be written on the map), a straight edge or scale, a pocket lens or magnifier (× 5 is a useful magnification for field work) and a haversack. If air photographs are available, especially when the area to be covered is large, they give a comprehensive view of the ground and are a valuable adjunct. But the geology has to be inspected and mapped on the ground itself (Note 5).

Making the map is usually a painstaking operation, and if it is to be detailed the geologist will walk over the whole area and record every exposure, large or small. When enough information has been obtained it is possible to sketch in the position of geological boundaries, as shown in Fig. 9.3. The map not only records the position of the various rock types, but also the dip of any surfaces exposed. The shape

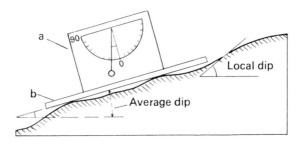

a – Clinometer
b – Base plate for 'average' dips

Fig. 9.4 Simple clinometer.

of exposures seen in the field should be represented as closely as possible on the map. All pencil lines and notes are normally inked in each day with a waterproof ink.

The way in which the geologist covers the ground depends on the geology and on the individual. In a complicated area it is often helpful to walk along predetermined traverse lines, noting all that is seen along them, filling in the gaps later whenever appropriate. Such traverse lines are usually taken at right angles to the general strike, when this can be recognized, so that they cross the geological grain of the area. In simpler situations it may be possible to follow a particular geological feature, e.g. a prominent limestone ridge. The exposures seen in the bed and banks of many streams are useful and stream sections are usually studied early in a survey.

Engineers will not normally be expected to map large areas, as a geologist should be made available for this purpose in the initial stage of a large investigation. But the geologist is not always retained, and on many sites it is usually the engineer who will see geological detail revealed in temporary exposures and other excavations dug during the contract. A map of this detail can be a valuable record if problems of ground engineering are later encountered. Four kinds of information should be recorded:

(i) the rock types; when identification of rocks is difficult it is wise to collect samples of them, which should be labelled and the point of collection shown on the map or sketch (Fig. 9.5);

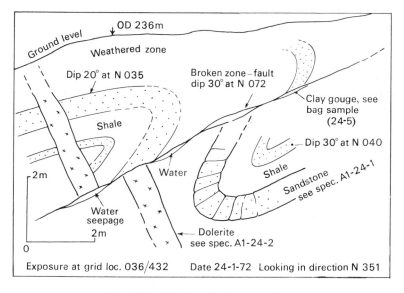

Fig. 9.5   Field sketch of a quarry face.

(ii) the boundaries between exposures; often these present little difficulty, but when indistinct in the field they can be shown on the map by a broken line.

(iii) rock structure; important features to record are the amount and direction of dip of geological surfaces such as bedding planes (p. 217), using the clinometer. When a surface is not plane, and varying values of dip can be obtained, it is useful to mount the clinometer on a thin rigid slat to give it a longer base and obtain an average value (Fig. 9.4). Angles of dip are usually read to 1 degree only; the direction of dip (p. 218) is recorded as a compass bearing, e.g. a surface dipping 60° to the NW would be recorded as a dip of 60° at 315. (And see Chapter 11.)

(iv) water; springs and similar issues from the face of an exposure should be noted. Examples of these observations are shown in Fig. 9.5 (Note 2).

The two geophysical methods mentioned in connection with areal investigations, magnetic and gravity surveys, are usually conducted by geophysicists. Both measure natural field forces (Table 9.3) and are susceptible to such features as buried pipelines, conduits, and the like. In built-up areas these can produce severe background interference and render the survey results difficult to analyse.

(iii) Volumetric investigations. These are primarily concerned with determining the 3-dimensional characters of the geology, thereby differing from investigations which study either a local area, or a large area as described above. Many of the large *in situ* tests fall within this group, e.g. blasting tests, pumping tests, load bearing tests, shear strength tests. The volume of ground involved is an important factor and ideally it should be related to both the geological fabric and the size of the proposed engineering structure.

The ground to be explored can be considered as either a continuous or a discontinuous material, depending on the volume that is tested (the effect of the test volume upon the results obtained, in hydrogeological investigations, is described in Chapter 12). Many authorities consider that the optimum volume of ground to be tested is that above which an increase in size has no appreciable effect on the results obtained. This can sometimes mean testing exceedingly large samples, particularly in rock masses that have a coarse fabric, e.g. widely spaced joints and bedding surfaces. A more practical procedure, which carries testing to the point where no improvement in test results can be expected by increasing the volume of the samples, can sometimes be adopted. Bieniawski (1969) describes how *in situ* compression tests on $1\frac{1}{2}$ metre cubes were adequate for designing mine pillars that were several times bigger.

The size of the test having been decided, its orientation in relation to the engineering structure should then be considered. Most rock masses are anisotropic and their performance will vary according to the direction in which they are tested. This is most marked when the anisotropy is regular, as in a well-bedded or cleaved rock mass. It is usual to apply loads, promote drainage or initiate displacements in the direction which the proposed structure is likely to load, drain, or displace the ground. Finally it should be remembered that tests are best conducted in ground that has not been disturbed (see p. 272).

In situ tests are mostly designed to investigate either the state of stress in the ground or the ground's reaction to certain environments; the latter are usually concerned with measurements of one of the following: deformability, dynamic behaviour, shear strength, and permeability.

Two measurements of *in situ* stress can be made, that of absolute stress and that of relative stress; instruments for the former can usually measure the latter. Absolute stress cannot be determined directly; most methods monitor the strains that result from the relief of stresses, and relate them to the stress-strain character of the rock, or ground, or to the forces required to counteract them (Note 3). The absolute stress method relies upon measuring the strains which occur in a volume of rock when it has been separated from its surrounding *in situ* stresses. This is usually accomplished by inserting a monitoring device in a bore-hole that is later over-cored; the general procedure is illustrated in Fig. 9.6. Instruments of this type are described by Obert *et al.* (1962), Hast (1958), Roberts (1965), and Leeman (1966). The relative stress method relies on counteracting strains with an applied pressure. A slot is cut into the rock which closes a little as a result of the *in situ* stress. An hydraulic jack is inserted into the slot and pressure applied to force the sides of the slot back to their original positions. Calculations of the stress are then made using the values that were required to complete the test. Typical flat-jacks and curved-jacks are described by Hoskins (1966) and Jaeger (1964). Most of the techniques for measuring stress depend on the elastic rebound of the rock tested, and are not suitable in plastic materials where creep is an important mode of failure. Furthermore, the accuracy of the methods is sensitive to the disturbance produced during the cutting of the initial holes or slots. These and other related problems are discussed by Hauzé (1971).

Measurements of deformability. In commonly used methods for measuring deformability a static load is applied to the ground and the resulting deformations obtained. It is customary to interpret the results on the basis of the theory of elasticity and assign values for Young's Modulus and Poisson's ratio to the ground. Tests which

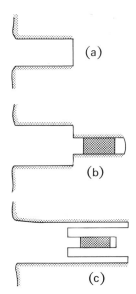

Fig. 9.6   A technique for measuring *in situ* stress. (*a*) An initial hole is drilled to the required location. (*b*) A co-axial hole is used for housing the deformation meter. (*c*) Over-coring releases the stresses around the meter which measures the resulting deformations.

operate within the linear, elastic portion of the stress-strain curve for the ground are those usually chosen for analysis. The static load is normally applied in one of three ways: over the area of a rigid plate, over the area of a tunnel, and over the area of a bore-hole.

In plate bearing tests the load is applied to a flat surface, usually by an hydraulic jack, and the deformations are recorded. General arrangements are illustrated in Fig. 9.7. In theory the size of loaded area should be related to the rock fabric, but in practice an economic compromise is often made and an area 1 m^2 is often used. This is nearly the smallest area desirable for testing most rock masses. The loading pad should be well mated to the test surface.

In pressure-tunnel tests part of a circular tunnel is sealed off and filled with water which is then pressurized. Changes in diameter resulting from this uniformly distributed radial loading are used to define the modulus of elasticity. The volume of rock tested is larger than in a plate bearing test, and more representative values are usually obtained. However, the cost of the test is much greater and difficulties of coping with water losses can be excessive. Both plate bearing and pressure-tunnel methods are described by Rocha (1955); Fox *et al.* (1964) discuss problems of water losses from tunnels.

Because of the cost of such tests smaller, cheaper methods using hydraulic jacks have been developed for radially loading sections of bore-hole. A typical model is described by Rocha *et al.* (1966). Many

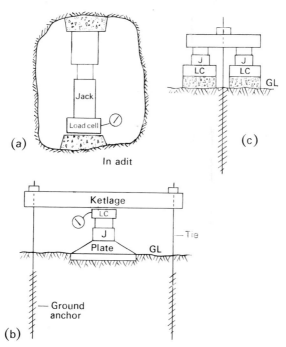

(a) In adit

(c)

(b)

Ketlage — LC — J — Plate — GL — Tie — Ground anchor

Fig. 9.7 Plate bearing tests. (a) Below ground level: (b) and (c) at ground level. LC, Load cell; J, jack. Note; care must be taken with positioning dial gauges for measuring deformations.

investigators believe that these instruments are best suited for measuring rock mass deformability, as the other techniques can induce complicated states of stress and make the analysis of results difficult. All the tests are affected by the disturbance the ground suffers prior to the actual test. For an excellent review see American Society for Testing and Materials (1969).

Values for the elastic modulus can also be obtained by measuring the response of the ground to dynamic loads. These are usually generated at a point either by a falling weight (e.g. a hammer) or by an explosion. Seismic surveying techniques can be used for calculating Young's Modulus from the field velocities. The method is explained by Evinson (1956) and Onodera (1963). Its main advantage is that a large volume of ground can be tested without the need for extensive excavations. A disadvantage is that the Modulus calculated will invariably be greater than that deduced by static loading tests. Explanations have been advanced to explain this discrepancy (see Link, 1964) but none are completely adequate. The problem seems to be related to the nature of the loading systems. Statically applied loads are of long duration and can induce plastic deformation; furthermore the presence of fissures affects the overall deformation. Dynamically applied loads are of much shorter duration and work within the elastic range of the material. Their propagation need not be unduly affected by fissures and pores if these are filled with water. The ratio between the dynamic and static moduli tends to decrease with a decrease in rock quality. Ambraseys (1968) suggests that compressional wave velocities (Note 4) can be used in conjunction with assumed values of 0.27–0.35 for Poisson's ratio, in evaluating a dynamic Young's Modulus for engineering purposes; however, this should only be attempted in competent rock that will transmit at velocities greater than 3000 metres per second (Table 9.4). For less competent rock, especially if saturated, it is advisable to use shear wave velocities. This is because compressional waves can cross water-filled joints and fissures and so follow much shorter paths through the ground than would be possible above the water-table; shear waves travel through neither water nor air.

Measurement of dynamic behaviour. In certain fields such as blast control and earthquake engineering, it is necessary to know the speed with which shock waves are propagated through the ground, and the extent to which they will be reduced or attenuated. Seismic techniques can be employed to investigate this, and such work is normally conducted by geophysicists. A source of waves is provided at a point on or in the ground by a mechanical device such as a hammer or a vibrator, or by an explosion. Geophones are placed at intervals from this source and record the arrival of waves so produced. Examples of *in situ* seismic velocities are given in Table 9.4.

The subject of dynamic behaviour is complicated because many factors can exert an influence, and the factors need not be geological.

Table 9.4   Some Typical Seismic Velocities

Material	Seismic velocity (m per s)	
	Compressional	Shear (Note 3)
Air ⎱ pore and Water ⎰ fissure filling	330 1450	— —
Sand	300–800	100–500
Shale	880–3900	400–2000
Sandstone	1400–4200	700–2100
Glacial moraine	1500–2000	900–1300
Limestone	3500–6500	1800–3800
Evaporites	4000–5500	2000–3200
Granite (massive)	5500–7000	2500–4000
Granite (weathered)	680–3000	250–1200
Low quality rock	<3000	<1500
Fairly good rock	>3000	>1500

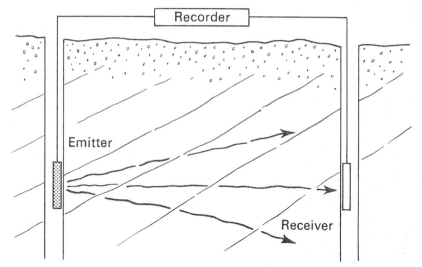

Fig. 9.8   Cross hole test.

For an introduction to the subject see Ambraseys and Hendron (1968); and for a fuller coverage Clarke (1970). The dynamic behaviour of ground can often be related to physical characters which are of interest to other fields of engineering; e.g. its quality can be assessed from cross hole tests, illustrated in Fig. 9.8. Compressional waves are generated and radiate out from their source, at a velocity V_f that is governed by the ground conditions, especially by the amount of fissuring. The effect of discontinuities in the ground can then be estimated by comparing the field velocity (V_f) with the velocity of similar waves through an intact core sample of the same rock (V_c) subjected to an axial stress equal to the overburden load and with a similar moisture content. The ratio of the two velocities, V_f/V_c, will approach unity as the rock approaches an unfissured state. Knill (1969) has used this method for studying the grout-take of dam foundations.

Seismic velocities have also been correlated to the ease with which ground can be excavated, as described in Chapter 15 (see Table 15.3).

Measurements of shear strength.   These are normally obtained from shear box tests. For reasons of scale no completely adequate test has yet been devised for hard rocks; a commonly used test arrangement is illustrated in Fig. 9.9, and others are described by Link (1969). A block of ground is carefully trimmed so as to cause as little disturbance to the sample as possible. Most blocks have an area of about 1 m², though the comments made above on the size of plate bearing tests also apply here. Loading pads, usually of epoxy resin or concrete, are then mated to the sample and hydraulic jacks positioned

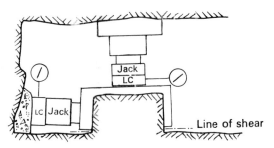

Fig. 9.9 Shear strength test; arrangement for use in adits: (displacement gauges not shown). LC, Load cell.

to supply the normal and shear loads. Measurements recorded during the test include the horizontal and vertical displacements and the applied loads. Values for *in situ* shear strength can also be obtained by pressing a plate into the ground until failure occurs. The method is applicable to soils and seems to apply to soft rocks, but it is doubtful whether it describes the shear strength of hard rocks.

A further method, applicable to weak materials such as soft clay, is the *vane test*. Here a vane consisting of four thin rectangular blades, usually four times as long as they are wide, is pressed into the ground and twisted in the soil at a uniform rate of about 0.1 degree per second. A cylindrical surface of rupture develops at a certain torque, the value of which is measured and used to calculate the shear strength. This is a commonly used method of soil testing and is described more fully in most texts on soil mechanics.

Measurements of permeability. Permeability is the speed with which water will flow through rocks in response to the head provided by a hydraulic gradient. It can be determined in a number of ways, two of the most common in engineering practice being the pumping test and the packer test. In a pumping test a well is sunk into the ground and surrounded by observation holes of smaller diameter, which are spaced along lines radiating from the well. Two observation holes are generally held to be a minimum requirement, as in Fig. 9.10. Pumping from the well lowers the water level in it and in the surrounding ground, so that a *cone of depression* results. By using values for the discharge from the well at given times, and the drawdown measured in the holes at those times, and the known distances of the holes from the centre of the well, the permeability of the ground can be calculated. Values for permeability can similarly be found from the rise in water levels which occurs when pumping stops: this is called the recovery method. A great variety of tests can be conducted, all of which are based on this fundamental procedure; they are described by Kruseman and de Ridder (1970). A more general account is provided by Slater and Ineson (1963).

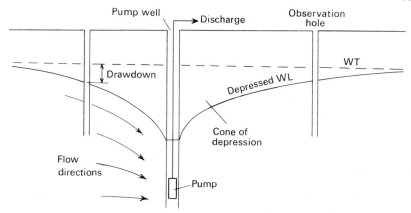

Fig. 9.10 Pumping test with observation holes.

In *packer tests* only one hole is required. Into this is lowered a tube to which one, two, or four inflatable packers are attached, for sealing off lengths of the hole as illustrated in Fig. 9.11. Water is passed into the tube, and from its rate of discharge into the ground, the area of the test section, and the head of water in the system, it is possible to calculate field permeability. Here too there are various tests that can be conducted, all of which are based on this general

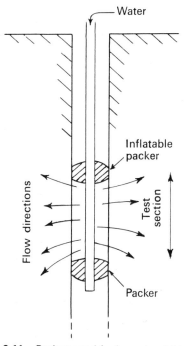

Fig. 9.11 Packers used for 'pumping-in' test.

procedure. The relevant calculations can be found in Cedegren (1967). Single hole tests of this kind are of great value to site investigations because they can be conducted in existing bore-holes and require no additional observation holes.

Care should be exercised in choosing the testing technique for a particular site because the value of permeability obtained will depend on the direction of flow from or to the test hole. Different values for permeability will be found by tests made in different directions in anisotropic ground, and a system and orientation should be chosen to simulate as closely as possible the flow regime that will operate when the final engineering structure is complete.

Water engineers use a standard pumping test because the flow around the well when in production will be similar to that in the test. Ground-water-lowering and de-watering schemes might use such a test if they consist of a series of pumped wells. However, it is unlikely that the permeability derived from these tests would apply to seepage problems beneath large dams or to the flow towards tunnels. In later Chapters the importance of ground-water to the stability of engineering structures is described, and it will be seen that it is important for the engineer to design with the correct values for permeability. A valuable contribution to this field of study has been made by Maini (1971).

This chapter has briefly reviewed the more commonly occurring items of *in situ* investigations. Serious problems can arise if the design of an engineering structure has been based on incorrect field data (e.g. Eide, 1974); care should therefore be taken in sampling and describing the ground and a code of practice followed wherever possible, so that different investigations are comparable. The relevant code in the United Kingdom is CP.2001, 'Site investigations'.

NOTES

1. Equipment dimensions tend to vary from country to country; some that are fairly common throughout the world are listed below.

Diamond Drill Core Casings

Size	Outside Diameter	Approx. Diameter of Bit	Approx. Diameter of Hole Made for Core Barrel	Approx. Diameter of core	
	mm	mm	mm	mm	(in.)
EX	46	47	38	22	(0.825)
AX	57	58.5	46.3	28.6	(1.125)
BX	72	74.5	60.3	41.2	(1.625)
NX	89	90.5	76.2	54.0	(2.125)

2. For further details on field mapping see Himus and Sweeting; Elements of Field Geology (reference on p. 294). If much field work is anticipated, it may be useful to have a polaroid camera for making 'instant pictures' which can be annotated on the spot.
3. Another method is described by Haimson and Fairhurst (1969) which relies on the measurement of the pressure at which hydraulic fracturing occurs. It is not commonly used by engineers but does permit the determination of stress at great depth, i.e. hundreds of metres.
4. See Chapter I for a comparison of compressional and shear waves.
5. Much geological surveying is now done off-shore; see for example Exploring the Geology of Shelf Seas. McQuillin, R. and Ardus, D. 1976. Graham and Trotman Ltd., London.

REFERENCES

AMBRASEYS, N. and HENDRON, A. 1968. Dynamic Behaviour of Rock Masses, in *Rock Mechanics in Engineering Practice*. *Ed*. Stagg and Zienkiewicz. J. Wiley & Sons, New York.

AMER. SOC. FOR TESTING AND MATERIALS. 1969. Determination of the *in situ* modulus of deformation of rock. *ASTM STP* 477.

AMER. SOC. CIVIL ENGINEERS. 1972. Report of Task Committee on Subsurface investigation of foundations of buildings. *Proc. Amer. Soc. Civ. Eng., Jour. Soil Mech. and Foundations Div.*, **98**. Note Appendices A and B.

BARTON, N. *et al*. 1974. Analysis of rock mass quality and support practice in tunnelling, and a guide for estimating support requirements. Internal Report of the Norwegian Geotechnical Institute, Oslo.

BELL, F. 1975. Site investigations in areas of mining subsidence. Newnes-Butterworths, Kent.

BJERRUM, L. 1973. Geotechnical problems involved in foundations of structures in the North Sea. *Géotechnique*, **23**, No. 3.

BIENIAWSKI, Z. 1969. *In situ* large scale testing of coal. *Proc. Conf. on In Situ Investigations in Soils and Rocks*, British Geotech. Soc., London.

BIENIAWSKI, Z. 1973. Engineering classification of jointed rock masses. *The Civil Engineer in South Africa*. December.

BRITISH STANDARD CODE OF PRACTICE, CP 2001. 1957. Site investigation, British Standards Institution.

BRITISH STANDARD CODE OF PRACTICE, CP 2003. 1959. Earthworks, British Standards Institution.

CEDEGREN, H. 1967. Seepage, Drainage and Flow Nets. J. Wiley & Sons, New York.

CLARKE, G. (Ed.). 1970. Dynamic Rock Mechanics. *12th Symp. Rock Mechanics, Univ. Missouri*. Pub. by Amer. Inst. Mining Engineers, New York.

DEARMAN, W. 1974. The characterization of rock for civil engineering practice in Britain. Special publication for the centenary of the Geological Society of Belgium, Liège.

DEERE, D. *et al*. 1966. Design of surface and near surface construction in rock. *8th Symp. Rock Mechanics. Univ. Minnesota*. Pub. Amer. Inst. Mining Eng. 1967.

DUMBLETON, M. and WEST, G. 1976. Preliminary sources of information for site investigations in Britain. Road Research Laboratory Report LR403.

DUNNING, F. 1970. Geophysical Exploration. Institute of Geological Sciences, H.M.S.O.

EIDE, O. 1974. Marine Soil Mechanics. Pub. 103. Norwegian Geotechnical Institute, Oslo.

EVINSON, F. 1956. The seismic determination of Young's Modulus and Poisson's ratio for rocks *in situ*. *Géotechnique*, **6**, No. 1.

FOX, P., MAYER, A. and TALOBRE, J. 1964. Foundations of the Pablavi Dam on Dex River. *8th Int. Congress Large Dams, Edinburgh*.

Geological Society Engineering Group Working Party Report on The logging of rock cores for engineering purposes. *Quart. Journ. Eng. Geol.*, **3**, No. 1, 1970.

GLOSSOP, R. 1968. 8th Rankin Lecture. *Géotechnique*, **28**, No. 2.

GRIFFITHS, D. and KING, R. 1969. Applied Geophysics for Engineers and Geologists. Pergamon Press, Oxford.

HAIMSON, B. and FAIRHURST, C. 1969. *In situ* stress determination at great depth by means of hydraulic fracturing. *Proc. 11th Symp. Rock Mechanics, Univ. California*. Published by Amer. Inst. Mining Metal. and Petrol. Engrs, New York.

HAST, N. 1958. The measurement of rock pressures in mines. *Sveriges Geol. Undersokn. C.52.* No. 3.

HEUZÉ, F. 1971. Sources of Errors in rock mechanics, field measurements and related solutions. *Int. Journ. Rock Mechanics and Mining Sciences*, **8**, No. 4.

HIMUS, G. and SWEETING, G. 1951. The Elements of Field Geology. University Tutorial Press, Ltd.

HOSKINS, E. 1966. An investigation of the flat jack method of measuring stress. *Int. Journ. Rock Mech. and Mining Sci.*, **3**, No. 4.

HVORSLEV, M. 1948. Sub-surface exploration and sampling of soils for civil engineering purposes. *U.S. waterways Experimental Station*, Vicksberg. (Reprinted by Engineering Foundation in 1962 and 1965).

JAEGER, J. 1964. State of Stress in the Earth's Crust, pp. 381–396, Elsevier.

KNILL, J. 1969. The application of seismic methods in the prediction of grout take in rock. *Proc. Conference on In situ investigations in soils and rocks*. Brit. Géotech. Soc., London.

KNILL, J. and JONES, K. 1965. The recording and interpretation of geological conditions in the foundations of the Roseires, Kariba and Latiyan dams. *Géotechnique*, **15**, No. 1.

KRUSEMAN, G. and DE RIDDER, N. 1970. Analyses and Evaluation of pumping test data. Int. Inst. for Land Reclamation and Improvement. Wageningen. The Netherlands. Bull. No. 11.

LEEMAN, E. 1966. The determination of the complete state of stress in a single bore-hole. C.S.I.R. Report 538. Pretoria.

LINK, H. 1964. Evaluation of elasticity moduli of dam foundation rock determined seismically in comparison of those arrived at statically. **1**, *8th Int. Cong. Large Dams, Edinburgh.*

LINK, H. 1969. The sliding stability of dams. *Water Power*, March, 1969.

MAINI, Y. 1971. *In situ* Hydraulic Parameters in Jointed Rock. Their measurement and interpretation. Ph.D. Thesis, London University.

MILLER, V. and MILLER, C. 1961. Photogeology. McGraw-Hill, New York.

NORMAN, J. 1968. Air photograph requirements of geologists. *Photogrammetric Record*, **6**, Oct.

OBERT, L., MERRILL, R. and MORGAN, T. 1962. A bore-hole deformation gauge for determining the stresses in mine rock. *U.S. Bur. Mines.* Rept. Investigation 5978.

ONODERA, T. 1963. Dynamic investigation of foundation rocks *in situ. 5th Symp. of Rock Mechanics, Minnesota.* Pergamon Press, Oxford.

PRIEST, S. and HUDSON, J. 1976. Discontinuity spacings in rock. *Int. Journ. Rock Mech. Min. Sci. and Geomech. Abst.* **13**, 135–148.

RAY, R. 1960. Aerial photographs in Geologic Interpretation and Mapping. *U.S.G.S.* Prof. Paper 373.

ROCHA, M., SERAFIM, J. and DA SILVEIRA, A. 1955. Deformability of foundation rocks. *5th Congress. Large Dams, Paris,* **3**, 531.

ROCHA, M. *et al.* 1966. Determination of the deformability of rock masses along bore-holes. *1st Int. Cong. of Int. Soc. Rock Mechanics, Lisbon,* **1**, Paper 3, 77.

ROCHA, M. 1971. A Method of Integral Sampling of Rock Masses. *Rock Mechanics,* **3**, No. 1.

ROBERTS, A., HAWKES, I. and WILLIAMS, F. 1965. Field applications of the photoelastic stressmeter. *Int. Journ. Rock Mech. and Mining Sci.,* **2**, No. 1.

SLATER, R. and INESON, J. 1963. Applications and limitations of pumping tests. *Journ. Inst. Water Eng.,* **17**, No. 3.

STIMPSON, B., METCALFE, R. and WALTON, G. 1970. A new field technique for sealing and packing rock and soil samples. *Quart. J. Eng. Geol.,* **3**, No. 2.

STRANDBERG, C. 1967. Aerial Discovery Manual. J. Wiley & Sons, New York.

TERZAGHI, R. 1965. Sources of error in joint surveys. *Géotechnique,* **15**, No. 3.

10 Laboratory Investigations

This chapter reviews topics relating to the laboratory testing of rocks and soils for engineering purposes, and is in three sections: samples and sampling, laboratory tests, and classifications based on test results. A general discussion of sampling was given in the previous chapter.

SAMPLES AND SAMPLING

One aspect of sampling that is important to the engineer is the degree to which a sample is disturbed during collection in the field and transport to the laboratory; in fact a distinction is made between undisturbed and disturbed samples. An undisturbed sample is one which has not been changed during collection and storage prior to testing (Note 1). As such it represents the ground in the area from which it was collected; a disturbed sample need not do so. Common sources of sample disturbance are changes in moisture content, losses of material and local over-stressing.

Changes in moisture content, either by wetting or drying can be avoided by sealing the sample with an impermeable material such as paraffin wax or polythene (see Stimpson *et al.*, 1970). This should be done as soon as possible after the specimen has been collected; samples should not be collected from areas where the natural moisture content of the ground has been changed by wetting or drying. Loss of material is a hazard in sands, gravels, and similar deposits which have little or no cohesion. Comparatively undisturbed samples of moist sand may be taken from natural exposures, excavations, and borings, above the water-table, by gently forcing a sampling tube into the ground. Samples are less easily obtained below the watertable, and if they are to be taken from a borehole, the water level in the hole should be kept higher than that in the adjacent ground. The flow of water outward from the borehole will then tend to prevent the loss of finer particles which would otherwise be flushed from its sides by inflow of water from the surrounding ground. Having obtained a sample, loss of material from it should be avoided during its transport to the laboratory. These problems are less severe when dealing with rocks than with weak materials and drilling fines, which should be handled with care.

Local over-stressing often occurs with the trimming and transport of soft materials such as soils. Disturbance from trimming is affected by the sampling and cutting tools that are used, and cutting edges should be sharp so as to cut smoothly and cleanly. They should also be thin so that they cause little displacement when passing through the soil. This is important in the design of tubes that are pushed into the ground to collect a sample; an area ratio of the form $(Da^2 - Db^2/Db^2) \times 100\%$ should be observed, where Da is the outside diameter of the cutting edge, and Db its inside diameter. The above represents the area of ground displaced by the sampler in proportion to the area of sample. Severe disturbance is likely to occur on the margins of a sample if the ratio is greater than 25 %, as illustrated in Fig. 10.1.

Disturbance can also occur during the extrusion of samples either from sample tubes or core barrels, and care should be taken that the friction between the sample and sampler wall is kept as low as possible, and that excessive pressure is avoided during extrusion. Harder materials such as rocks can be sampled by coring, as discussed in Chapter 9 (p. 276). Hand specimens of rocks should be at least $(8 \text{ cm})^3$ or $(3 \text{ in.})^3$ and representative of the formation sampled. Ground water samples can be collected in either glass or polythene screw-cap beakers, the choice depending upon the future analysis that is required.

Table 10.1 indicates the size of samples needed for most of the standard tests made with conventional laboratory apparatus, and Table 10.2 the methods that are normally used for their recovery.

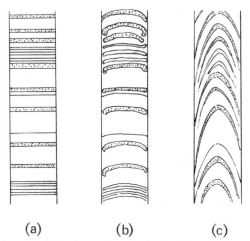

(a)   (b)   (c)

Fig. 10.1 Distortion of bedding by a tube sampler. (*a*) Smooth penetration with correct sampler. (*b*) Jerky penetration with correct sampler. (*c*) Smooth penetration with incorrect sampler.

Table 10.1 A Guide to Sample Requirements

Purpose of Sample	Material	Volume or Weight of Sample Required[a]			
		Bulk or Block Samples		Cored or Tube Samples[b]	
		(Imperial)	(SI)[c]	(Imperial)	(SI)[c]
Chemical composition	Clays and silts	1–2 lb	0.5–1.0 kg	$1\frac{1}{2}$ × 3 in.	38 × 75 mm
	Sands	1–2 lb	0.5–1.0 kg	$1\frac{1}{2}$ × 3 in.	38 × 75 mm
	Gravels	7 lb	3.0 kg	4 × 8 in.	90 × 200 mm
	Rocks	$\frac{3}{4}$–$\frac{1}{4}$ lb	0.5 kg	$1\frac{1}{2}$ × 3 in.	38 × 75 mm
	Ground-water	$\frac{1}{2}$ gal	2.5 l.	—	—
Structural characters including grain size, porosity, etc.	Clay and silts	1–2 lb	0.5–1.0 kg	4 × 4 in.	90 × 90 mm
	Sands	2–6 lb	1.0–2.5 kg	4 × 8 in.	90 × 200 mm
	Gravels ($\frac{1}{4}$ in.–$2\frac{1}{2}$ in.)	10–100 lb	4.5–45 kg	4 × 8 in.	90 × 200 mm
	Rocks (Coarse grained)	(12 in.)³	0.3 m³	4 × 4 in.	90 × 90 mm
	Rocks (Fine grained)	(6 in.)³	0.15 m³	$1\frac{1}{2}$ × 3 in.	38 × 75 mm
Strength characters including elastic moduli, shear strength, consolidation, etc.	Clays and silts	(12 in.)³	(0.3 m)³	$1\frac{1}{2}$ × 3 in.	38 × 75 mm
	Sands	(12 in.)³	(0.3 m)³	$1\frac{1}{2}$ × 3 in.	38 × 75 mm
	Gravels	($\frac{1}{2}$ yd)³	(0.5 in.)³	8 × 12 in.	0.2 × 0.3 m
	Rocks (Weathered)	2 off (12 in.)³	2 off (0.3 m)³	4 × 8 in.	90 × 200 mm
	Rocks (Unweathered)	1 off (12 in.)³	1 off (0.3 m)³	3 × 6 in.	75 × 150 mm
Hydraulic characters including permeability, specific yield, etc.	Clays and silts	(6 in.)³	(0.15 m)³	$1\frac{1}{2}$ × 3 in.	38 × 75 mm
	Sands	(9 in.)³	(0.2 m)³	$1\frac{1}{2}$ × 3 in.	38 × 75 mm
	Gravels ($\frac{1}{4}$ in.–$2\frac{1}{2}$ in.)	($\frac{1}{2}$–$\frac{3}{4}$ yd)³	(0.5–1.0 m)³	8 × 16 in.	0.2 × 0.4 m
	Rocks (Coarse grained)	(12 in.)³	(0.3 m)³	4 × 8 in.	90 × 200 mm
	Rocks (Fine grained)	(6 in.)³	0.15 m³	3 × 6 in.	75 × 150 mm
Comprehensive examination	Clays and silts	50–100 lb	20–45 kg	4 × 8 in.	90 × 200 mm
	Sands	50–100 lb	20–45 kg	4 × 8 in.	90 × 200 mm
	Gravels	100–200 lb	45–90 kg	8 × 16 in.	0.2 × 0.4 m
	Rocks	2 off (12 in.)³	2 off (0.3 m)³	4 × 8 in.	90 × 200 mm
	Ground-water	1–2 gal	4.5–10.0 l.	—	—

a Samples sent to the laboratory should not be much smaller than the dimensions suggested.
b Diameter given first, then length.
c Not equivalents of the Imperial Units.

Table 10.2   A Guide to Sampling Methods

Method	Comments	References
Open drive samplers	Thin walled open tubes from 50 mm (2 in.) to 100 mm (4 in.) diameter. (The U-4 sampler in Britain) which are pressed into the ground. Disturbance is common and accepted. Unsuitable for cohesionless materials, hard clays and soft rocks.	Hvorslev (1948) Lang (1967)
Piston samplers	Thin walled tubes from 50 mm (2 in.) to 100 mm (4 in.) diameter, but can be larger. Contains a piston which is withdrawn when sample is collected. Generally less disturbance caused than with an open drive sampler. Normally used for cohesive soils but can cope with granular materials. Special designs are required for the collection of fine sands (Bishop sampler).	Kallstenius (1963) Bishop (1948)
Foil samplers	Special case of piston sampler that protects the sample in a sheath. Gives little disturbance and collects long samples. Can be used in soft sensitive clays and expansive soils. Cannot be used for coarse granular soils but will collect sand samples from above the water table, if moist. 'Delft sampler' is of similar design and capability.	Kjellman and Kallstenius (1950) Begemann (1961)
Rotary core drilling	For all hard rocks with diameter to suit purpose. Difficult to collect coarse granular soils. Becoming a popular method for sampling stiff clays. Improved sampling obtained using double-tube samplers	Chap. 9 and Note 1 Chap. 9 Wakeling (1970) Earth Manual (1968)
Hand trimming	Least disturbance of all methods in soft and cohesionless materials.	Earth Manual (1968)
Mechanical and Manual excavations	For all bulk sampling, either disturbed or undisturbed.	Earth Manual (1968)

Certain precautions should be observed when handling and storing samples, particularly those of soils and soft rock. Adequate containers should be provided that will protect them from any further disturbance after collection. At least 2 large durable labels giving the location and depth from which the samples were taken, their date of collection and serial number should be written for every sample. One label is placed inside the container, the other attached to the outside of it. The labels should be written in indelible ink and be able to withstand the wear and tear of site work. It is usually desirable to test soil samples within 2 weeks of their collection, during which interval they are best stored in a cool room of controlled humidity.

Samples of hard rock are less delicate but still require reasonably careful handling. Individual samples should be labelled securely. This can be done by painting a serial number on the samples. Those which have to be transported should be individually wrapped (newspaper is ideal) and crated in stout boxes. A list of the serial numbers of the specimens should be included in the crate. Cores are normally stored and transported in core boxes, which should be clearly labelled with the borehole number and site, and the levels from which the cores have come. Further information on the sampling and handling of soils and rocks is given by Code of Practice 2001, 'Site Investigations' (1957), by Hvorslev (1948), and the 'Earth Manual' (1968).

LABORATORY TESTS

The majority of laboratory tests on geological materials can be grouped under the following headings: Tests for (1) composition, (2) structure, (3) strength, and (4) hydraulic properties. These headings will be followed here.

(1) **Tests for composition** are essentially designed to determine the mineral content of the solid phases and the chemistry of the liquid phase that together comprise most geological samples. Gas, although often present, is usually only studied if pertinent to the engineer, as in mining. Three basic types of test can be conducted on the solid phase, *viz.* physical, physico-chemical, and chemical:

(i) Physical analyses normally involve a section or microscopic mount of some kind, as described in Chapter 3. Very small particles can be studied with the aid of either an electron or a stereo-scanning microscope, though this is not common practice in a routine investigation. One routine analysis is the determination of specific gravity of mineral grains; normally either density bottles (for granular materials as sands and clays) or a steelyard apparatus such as a Walker Balance (p. 77) is used. (See Zussman (1967) and British Standard No. 1377.)

(ii) Physico-chemical analyses, usually conducted on a small volume (1 or 2 cm^3) of fine grained powder, obtained in most cases by crushing the sample. One example of this is differential thermal analysis (Mackenzie, 1957). Here the powder is heated so that physico-chemical reactions may occur at certain temperatures. These reactions may either liberate or absorb heat; a clay lattice, for example, may expand, or water may be driven off. These variations in heat absorption which occur at particular temperatures can be used in many cases to identify mineral constituents in the powder. X-ray diffraction analysis (Nuffield, 1966) is another useful method; here the scatter of X-rays as they are diffracted by the atomic structure of the minerals in the powder is recorded. As the scatter from a mineral is unique its pattern serves to identify the mineral itself. These techniques are normally only used when special problems arise, as was the case with the Sasamua clay (Chapter 17, p. 525).

(iii) Chemical analysis, in its simplest sense, is mainly used to determine the organic and sulphate content of a sample. Both wet and dry combustion methods can be used for the former, whilst a water extraction method can be used for the sulphate (other methods exist, see British Standard 1377). The pH value is also important and can be obtained by a routine colorimetric test. Chemical analyses of a special kind are used to estimate the age of a sample, e.g. radio-active dating (Dalrymple, 1969). These methods are occasionally required to confirm the presence of movement, as in unstable slopes, or near potentially dangerous faults.

The tests that are normally conducted on the liquid phase of a sample determine the amount of water present (i.e. moisture content) and its chemistry. Moisture content is found from the weight lost on drying a saturated or partly saturated sample. The chemical characters normally of interest to the construction engineer are the concentration of sulphates and the pH, both of which can be determined by simple chemical tests as already mentioned; pH meters are commonly used. The tests conducted on a gas phase, if present in a sample, determine the amount and type of gas. The sample is crushed in a sealed container and the gas released is analysed. No standardized test appears to exist, but mine laboratories normally follow a set procedure. Samples should be undisturbed, and sealed as well as possible when collected, so that their included gases cannot be lost prior to testing.

The samples required for the majority of the composition tests can be disturbed because it is their composition and not structure which is determined. Samples for gas analyses (above) are an exception to this. All samples should be uncontaminated and care should be taken that those required for determining moisture content are neither wetted nor dried before testing. The standard precautions for sampling for chemical analyses should be observed. Tests for the quality or composition of ground-water for either a potable or industrial supply, must be conducted according to the prescribed standards; two useful references on this subject are Hem (1970) and the World Health Organization (1963).

(2) **Tests for structure**   These are designed to assess fabrics and deal with (i) the whole fabric, solids, and voids, or (ii) the solids only, or (iii) the voids only.

(i) Whole fabric studies require an undisturbed sample which preserves the shape and distribution of all the voids. Thin sections are commonly used for most rocks (see p. 83). Weaker materials, such as clays and other soft sediments can be strengthened by impregnation with a wax or an epoxy resin; delicate fabrics can then be observed (Morgenstern and Tchalenko, 1967). A stereo-scanning microscope can be used for small fabrics. Indirect assessments of a fabric, usually from its porosity, can be obtained from laboratory measurements of the electrical resistivity of samples (Emerson, 1969). Sonic velocities are also used for this purpose (Rinehart *et al.*, 1961). Both methods are affected by the composition of the sample and by the presence of water in its voids. Two further tests which investigate the nature of whole fabrics are those for determining shrinkage and density. The dimensional changes of a sample which occur while it is being dried are measured in shrinkage tests. The procedures used measure either linear shrinkage (British Standard 1377) or volumetric shrinkage (Soil Mechanics for Road Engineers, 1952). In density tests the density of the whole sample is measured. Values of porosity can then be found from the formula:

$$1 - \frac{\text{bulk density}}{\text{average density of solid mineral grains}}$$

Many soil engineers need to know the relationship between moisture content and bulk density. Standard compaction tests are used (British Standard 1377) to find the optimum density to which a fabric can be raised with respect to a known moisture content.

(ii) Tests for the solid part of a fabric usually determine the shape and size of the solids. Shape is not important to most jobs, but is considered in classifications for concrete aggregates and usually applies to the shape of crushed stone rather than to natural aggregates. Grain size is more commonly determined, and for rocks can be obtained from thin sections. In uncemented granular material such as sands, gravels, and clays, it is found either by sieving or by sedimentation methods. In sieving the granular sample is passed through a stack of graded sieves, the largest aperture size being at the top of the stack. The sieves are normally designed to some standard, e.g. British Standard B.S.410 (Test Sieves) or the Institution of Mining and Metallurgy sieves. The weight of sample retained on any sieve is measured, and expressed as a percentage of the whole sample passing that sieve. The weights retained on successive sieves are then plotted as a cumulative curve against the sieve sizes, as shown in Fig. 10.2. This method is employed for grains down to about 0.1 mm diameter, i.e. from cobbles to fine sands. Finer materials such as silt and clay are normally determined by a sedimentation technique: a suspension in water is placed in a cylinder and allowed to settle out, each particle settling at a rate according to its size and specific gravity. The grain size distribution in the suspension can be calculated by taking small samples from the suspension at a given level in the cylinder, over a period of time. Both these techniques, together with a third using an hydrometer, are described in British Standard 1377. An old method of analysing fine-grained samples employs an elutriator (Mills, 1950) but this is no longer a routine test. The shape of the particle size distribution curves can be expressed approximately by a *uniformity coefficient*, defined as the ratio D_{60}/D_{10}, where D_{60} and D_{10} are the particle sizes corresponding to the cumulative percentages 60 and 10 respectively. The uniformity coefficient for the wind blown sand shown in Fig. 10.2 is 1.0, that for the sandy alluvium is 9.0 (Note 2). Samples for grain size analysis need not be undisturbed but they should be uncontaminated and complete, i.e. with no fraction of the deposit missing from the sample.

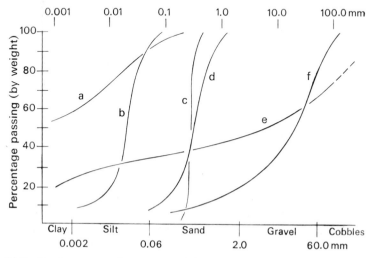

Fig. 10.2 Grading curves for some granular materials. *a* = marine clay; *b* = loess; *c* = wind blown sand; *d* = sandy alluvium; *e* = boulder clay; *f* = gravelly alluvium.

(iii) Tests concerned with the voids in a sample are an attempt to determine the total volume of voids rather than their individual volumes or shape. The latter can be obtained from thin sections as described above (p. 302), but the method is limited to sampling in one plane at a time and is difficult to use for investigating the fine textures found in silts and clays. The volume of voids in rocks is usually assessed from values of bulk and dry density, as described previously, or from the relative weights of the sample in a saturated and a dry state. The results are expressed as a void ratio (see Chapter 5, p. 179) or as a porosity. These tests require undisturbed samples.

(3) **Tests for strength**   The numerous tests for strength can be placed in one of three categories: instant, short term, and long term. 'Instant' tests are of short duration, designed to give 'on the spot' assessments of strength. They are not usually accurate, but their speed enables many tests to be completed and a general assessment of strength to be made. Short term tests are of longer duration but usually not much longer than half a day. Their accuracy depends largely upon the pore pressures within the samples. Long term tests account for the behaviour of pore pressures, in particular their dissipation. In general they are the most accurate of all strength tests, and can take up to a week or more to complete, depending upon the size of the sample. Three points should be remembered concerning the strength of geological materials:

(i) pore pressures: tests which are identical except for pore pressures can give different results.
(ii) fabric: anisotropic rocks can have different strengths in different directions (Fig. 10.3): the same applies to soils.
(iii) size: large samples will generally have a lower unit strength than small samples; Pratt *et al.*, 1972.

The laboratory tests are considered under eight headings: elastic constants, unconfined compressive strength, shear strength, tensile strength, consolidation characters, penetration characters, creep, and index tests relevant to strength. A useful text covering the testing of rocks is Obert and Duval (1967); that for soils is the British Standard 1377. The interested reader is nevertheless advised to refer to recent geotechnical literature as the procedure and analysis for many of these tests is still a subject of research.

Elastic constants can be obtained from statically loading cylinders of rock material and noting the resulting strains in directions normal and parallel to the direction of applied load. The Modulus of Elasticity and Poisson's Ratio can then be determined. The slope of the initial stress-strain curve in many rocks is less than that of curves obtained in subsequent tests; this is generally attributed to the closing of voids and

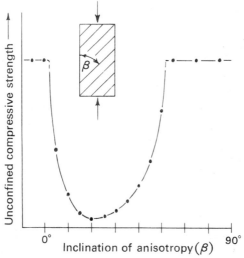

Fig. 10.3   Influence of anisotropy on strength (after Donath).

fissures which have opened during the collection of the sample. It is normal to use the curves from second and subsequent loading cycles for analysis. The elastic constants, under dynamic conditions, can be indirectly determined by measuring the velocity of propagation of compression and shear waves through the material. The relationship between the dynamically and statically determined values is discussed in Chapter 9. Undisturbed samples should be used in all these tests.

Unconfined (or *uniaxial*) *compressive strength* is normally determined by statically loading a cylinder of rock to failure, the load being applied across the upper and lower faces of the sample. The results obtained are in part a function of the length–breadth ratio of the sample and of the rate of loading. The simplicity of the test is somewhat deceptive (Hawkes and Mellor, 1970). Samples should be undisturbed.

Shear strength is measured by the use of either a shear box or a triaxial cell. A shear box houses a normally loaded rectangular sample in such a way that its upper and lower halves can be subject to shear displacements about a centrally placed horizontal plane (Note 3). Figure 10.4 illustrates typical stress-displacement curves, and the form of normal stress—shear stress envelopes that are commonly obtained. Two strengths can be derived for any normal loading, a peak strength which is the maximum shear stress obtained, and a residual strength which is the minimum shear stress supported by the sample once displacement is occurring about a continuous failure surface.

A triaxial cell is a cylinder which can be internally pressurized, usually with water. A cylindrical sample is placed on a pedestal within the vessel and jacketed with an impermeable membrane. This

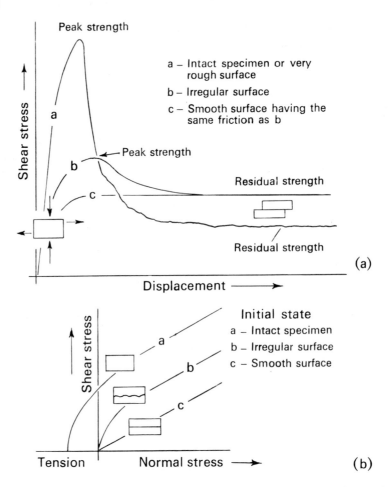

Fig. 10.4 (a) Typical stress-displacement curves for soils and rocks. (b) General form of normal stress—shear stress envelopes in rocks.

isolates the sample from the pressurizing fluid which is to surround it. A loading platten is placed on top of the sample. A ram which passes through the roof of the vessel bears against this platten and transmits an axial load from a loading frame to the sample. The load on the sides of the sample is supplied by the pressurized fluid. Figure 10.5 illustrates a simple triaxial apparatus. Hoek and Franklin (1968) describe such a cell for testing rocks. Bishop and Henkel (1962) describe the equipment used for testing soils. In the test the axial load on the sample is increased until failure occurs. Tests can be conducted with or without control of pore pressures, depending upon the sophistication of the equipment. A vane test, as described in Chapter 9, but on a reduced scale, can also be used for the laboratory determination of shear strength.

In all these tests a static load is applied, however, to measure the shear strength of materials that will be subject to earthquake shocks it is necessary to load the samples dynamically. Tests of this nature are described by Seed (1968).

Tensile strength can be measured in two ways. The simpler method loads test cylinders in tension until failure occurs, and a standard tensile testing rig is used. The second method loads test discs in compression along a diameter, so inducing tensile failure on a diametral surface. This is sometimes referred to as a Brazilian test. The apparent simplicity of these tests is illusive and results can vary with the specimen preparation, test procedure, and equipment used (Mellor and Hawkes, 1971). Undisturbed samples are required.

Consolidation characters. The characters normally required are the *coefficient of compressibility*, which is the change in unit volume that occurs with a change in pressure (used for calculating the magnitude of settlements), and the *coefficient of consolidation*, which is proportional to the ratio of the coefficients of permeability and compressibility (used for calculating the rate of settlement). Theoretical considerations are given in all good texts on soil mechanics. In the two laboratory methods generally used, a sample is compressed

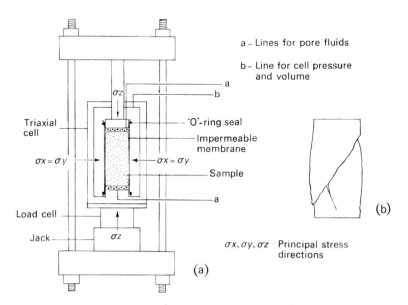

Fig. 10.5 A simple triaxial cell. (*a*) Basic triaxial apparatus: note, some provision is normally made to ensure the alignment of specimen, e.g. ball seating. (*b*) Failed specimen of stiff clay.

with a known load and the resulting changes in its volume with respect to time are measured. The simplest apparatus is an oedometer (Fig. 10.6). In this equipment a sample is axially loaded in a cylindrical container that allows consolidation in one direction only, i.e. vertically. Consolidation in three dimensions can be obtained using triaxial apparatus (Fig. 10.5), where a jacketed cylindrical sample can discharge pore-water from its upper and lower ends. It is subjected to all-round pressures by pressurizing the fluid in the test cell, so that $\sigma x = \sigma y = \sigma z$ in Fig. 10.5. Drainage facilities allow the pore-water to escape from the sample, so that its volume decreases and consolidation occurs. Undisturbed samples are normally required.

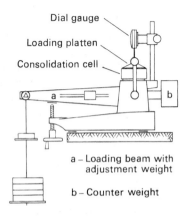

Dial gauge

Loading platten

Consolidation cell

a

b

a – Loading beam with
    adjustment weight

b – Counter weight

Fig. 10.6   An oedometer.

The swelling character of a sample, i.e. the increase in volume which occurs with the release of pressure, can also be determined using an oedometer. The sample is loaded to a pressure equal to that under which it exists in the ground and then progressively unloaded. (Bara and Hill, 1967.)

Penetration characters are commonly assessed for soils by the California Bearing Ratio Test. This empirical test does not measure any basic strength parameter but is normally accepted as a strength test. It consists of pushing a plunger at a constant rate into the top of a compacted sample, housed in a cylindrical mould. The load required to attain a certain penetration is measured. The results are sensitive to the preparation of the sample; standard procedures exist for this test.

Creep characters are usually found by either axially or triaxially loading a cylindrical sample in suitable apparatus, and maintaining the load at a constant level for a long time. The resultant deformations are recorded. Creep tests on rock and clay have been described by Griggs (1939) and Bishop (1966) respectively.

Index tests relating to strength. Many laboratory tests have been designed which do not measure any property such as tensile strength, but instead provide a gauge to the behaviour of material when it is simultaneously subjected to a variety of conditions. This behaviour may be related in a simple way to some strength parameter, and as such be an index to it. Four examples of such index tests are included here, the first of which has long been established in soil mechanics testing.

(i) *Atterberg Limits.* Atterberg designed two tests which reflect the influence of grain size and mineral composition upon the mechanical behaviour of clays and silts: they are the Liquid and Plastic Limit tests. In the Liquid Limit test, the clayey sample is mixed with water to a creamy paste and placed in a shallow brass dish. A V-groove is then cut through the sample and the dish tapped on its base until the sides of the groove just close. A sample of clay is then taken and its water content measured. The remaining clay is remixed to a new water content and the test repeated. This can be done for a number of values of water content. Water content is plotted against the number of blows; the Liquid Limit is the water content at which the groove closed on the 25th blow. This, as defined by Atterberg, is the water content above which the remoulded material behaves as a viscous fluid and below which it acts as a plastic solid. Procedures for the test are standardized (Note 4). The Plastic Limit is the water content below which the remoulded sample ceases to behave as a plastic material and becomes friable and crumbly. Atterberg's test consists of finding the water content at which it is no longer possible to roll the clay into an unbroken thread of about 3 mm ($\frac{1}{8}$ in.) diameter. These two tests have a definite relationship to the mechanical behaviour of a clay sample and are used for classifying soils for engineering purposes (Note 5).

(ii) *Drilling Indices* are an attempt to predict the resistance a rock will offer to the penetration of a drill, and the rate at which a cutting edge will be worn (see Chapter 15). The tests rely on dynamically loading a sample, often with either a pointed or rounded tool, and measuring the resultant damage. None of the indices correlate well with field rates of penetration and wear, but they do provide an indication of drill performance which can be of help. Examples are described by Olsen and Blindheim (1970) and Furby (1964).

(iii) *Blasting Indices* are based on the ratio between maximum and minimum sonic velocities of rocks, as measured in the laboratory. To this ratio are added various factors which account for such items as tensile strength of the rock, and number of joints in the field;

the final figure produced relates to the amount of explosive required to produce a given volume of broken ground. An example of this index is given by Christensen and Olsen (1970).

(iv) *Slaking Index:* This test measures the disintegration of shaley materials; the weight of material which is detached from a block, when wetted, is expressed as a percentage of the original weight of the sample. The index indicates the performance of such materials once exposed in a construction site. (Franklin and Chandra, 1972.)

(4) **Tests for hydraulic properties** The properties most commonly determined are permeability, effective porosity, and specific yield. Permeability tests measure the velocity with which a fluid will flow through a porous sample under the hydraulic head operating within the sample; the sample is housed in a permeameter. That shown in Fig. 10.7 is typical of many used for testing sands; silts and clays require slightly different apparatus. Permeability is normally calculated from the following relationship:

$$K = Q/iA,$$

where Q = the discharge,

  i = the hydraulic gradient = $\Delta h/L$

  A = cross sectional area of the sample measured at 90° to the general direction of flow,

  K = coefficient of permeability.

This is sometimes described as Darcy's Law, although it is not a law in the strict sense of the word. The value of permeability obtained from a test is a function of both the character of the fluid, i.e. its specific weight and viscosity, and the nature of the pore spaces in the solid.

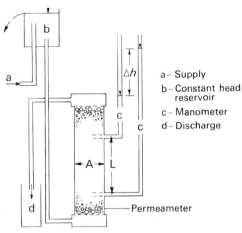

a – Supply
b – Constant head reservoir
c – Manometer
d – Discharge

Fig. 10.7   A constant head permeameter.

Hence the permeability obtained for a sample varies with the character of the fluid used. This limitation can be overcome by considering the fluid characters in a more specific formula and calculating an *intrinsic* permeability; this value is a constant for the solid skeleton of the sample. Once calculated it can be used for finding the relationship between Q and i for any fluid passing through the sample, e.g. oil, seawater, fresh water, industrial effluent, etc. Similar tests can be used to find the flow characters of gases through rocks and soils; gas permeameters are used. A theoretical consideration of permeability is provided by Muskat (1937) and test methods are described by Lovelock (1970). The relationship between permeability and stress in rocks can be studied in the laboratory, using the techniques described by Bernaix (1969). That for soils has been discussed by Bishop and Al-Dhahir (1969).

Effective porosity is a measure of the porosity which can be used for storing fluids. Not all the pores in a porous material will be interconnected and those which are not can neither receive nor release fluid. Effective porosity is therefore a measure of the volume of interconnected pores. It is measured by comparing the weight of a sample in its saturated and oven-dried states, the weight difference being expressed as a volume $(1 \text{ g H}_2\text{O} = (1 \text{ cm})^3 \text{ H}_2\text{O})$, and this in turn as a percentage of the total volume of the sample.

Specific yield is the volume of fluid which can be drained from a porous material and can be expressed as a percentage of the effective porosity. Specific yields of 100% are never obtained because of capillary tension a porous rock will always hold a certain amount of water against the force of gravity. The amount so held is called the specific retention; thus specific yield and specific retention = effective porosity (Note 6). Samples can be drained in various ways depending on the condition to be simulated, e.g. de-watering with a vacuum or by electro-osmosis. Drainage under the force of gravity is the method most applicable in the majority of contracts, but is often time consuming; alternative methods are used; in one method water is forced out of a cylindrical sample by jacketing its walls and blowing air through it from one end. In another method a sample is gently centrifuged so as to increase the gravitational loading on it and hence speed the drainage. Undisturbed samples are normally used and only strong rocks are tested in a centrifuge.

LABORATORY CLASSIFICATIONS

The development of laboratory tests has encouraged the quantitative description of rocks, and allowed other similar materials to be identified on a semi-numerical basis. Many classifications based on

Table 10.3   Unconfined Compressive Strength of Soil and Rock

Very soft soil	less than 0.25 kg/cm²
Soft soil	0.25–0.5
Medium soil	0.50–1.0
Stiff soil	1.0–2.0
Very stiff soil	2.0–4.0
Extremely stiff soil	4.0–6.0
Strong soil	6.0–8.0
Very strong soil or extremely weak rock	8.0–10.0
Very weak rock	below 12.5 kg/cm²
Weak rock	12.5–51
Moderately weak rock	51–128
Moderately strong rock	128–510
Strong rock	510–1024
Very strong rock	1024–2050
Extremely strong rock	over 2050

laboratory tests have been developed, some of which are of interest to the engineer. They are usually either simple, in that they are based on one property such as strength or grading, or composite, in that a number of characteristics are used. Examples of each are briefly described.

Simple classifications   One of the simplest of these describes the character of the material in terms of its unconfined compressive strength. Table 10.3 illustrates a commonly adopted scheme.

Grading is another basis for classifications which are important in earthworks. That adopted by the U.S. Bureau of Soils is shown in Fig. 10.8, where, for example, a sand is defined as having 80–100% sand, 0–20% silt, and 0–20% clay.

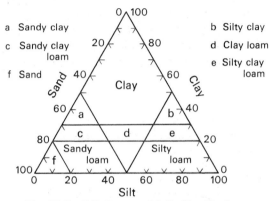

a Sandy clay
c Sandy clay loam
f Sand
b Silty clay
d Clay loam
e Silty clay loam

Fig. 10.8   U.S. Bureau of Soils Classification.

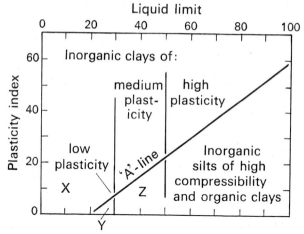

Fig. 10.9 Plasticity Chart (after Casagrande). *X* = cohesionless soil; *Y* = inorganic silts of low compressibility; *Z* = inorganic silts of medium compressibility and organic silts. The *'A'*-Line defines the Plasticity Index = 0.73 (Liquid Limit = 20).

An important classification for soil engineers is based on the Plasticity Index and Liquid Limits, as illustrated in Fig. 10.9. With this scheme it is possible to use the Atterberg limit tests to reveal the more general characters of a soil.

Composite classifications Perhaps the most famous example of this group is that produced by Casagrande (1947) based on grain size and Atterberg Limits. It is applicable to soil mechanics because the classes of soil defined by it have recognizable engineering characters. The classification is widely used, and when desirable can be extended to suit soils of a particular region. (See Soil Mechanics for Road Engineers (1952), and A.S.T.M. (1970).

Laboratory classifications of rock for engineering purposes are not as well advanced. Cottiss *et al.* (1971) have tentatively proposed a scheme that is based on certain quick laboratory tests and could be useful if extended. One of the most comprehensive schemes to date, proposed by Deere and Miller (1966), describes many rock types in terms of their uniaxial compressive strength and Young's Modulus.

NOTES

1. In practice there is no such thing as an undisturbed sample.
    Completely undisturbed samples are rare, and Kallstenius (1958) divides them into three categories:
    1. Simple: not seriously disturbed but collected by unspecialized personnel without skilled supervision.
    2. Routine: only slightly disturbed, as collected by specialist organizations for important jobs.

3. Research: of the highest quality, and quite unlike types 1 and 2; will have been collected 'regardless of cost'.
Simple and Routine methods are those commonly employed in site investigations.

2. Grain size curves of soils having equal uniformity are identical in shape on a semilogarithmic plot.

3. Bishop, et al. (1971) have recently described a method of testing that uses an annular sample which is housed in a ring shear box.

4. A cone penetrometer is now the preferred instrument for measuring the Liquid Limit, it being easier to use than the Casagrande apparatus described on p. 309 (see British Standard 1377;1975).

5. The difference between Liquid Limit and Plastic Limit is the range of water content over which a remoulded clay behaves as a plastic material; this difference (LL − PL) is referred to as the *Plasticity Index*.
The water content of re-moulded clays with respect to the Liquid and Plastic Limits, is an approximate measure of their strength, and can be described by the Liquidity Index, LI, equal to (Water Content − Plastic Limit)/(Liquid Limit − Plastic Limit).
When LI = 0, the water content is equal to the Plastic Limit, and when LI = 1.0 the water content is equal to the Liquid Limit.
Another relationship of significance is defined by the *activity* of the sample, i.e. the ratio Plasticity Index to clay fraction. Activity is related to the mineral composition of a deposit and to the geological conditions of its formation. (See Standard Soil Mechanics texts).

6. Specific yield can also be expressed as a percentage of the total volume of the sample from which it has been recovered.

REFERENCES

AMERICAN SOCIETY FOR TESTING AND MATERIALS. 1970. The sampling of soil and rock. *Special Tech. Pub.* 483.

BARA, A. and HILL, R. 1967. Foundation rebound at Dos Amigos Pumping Plant. *Amer. Soc. Civ. Eng. Jour. Soil Mechanics and Foundations Div.*, **93.**

BEGEMANN, H. 1961. A new method for taking samples of great length. *Proc. 5th Int. Conf. Soil Mechanics and Foundation Engineering.*

BERNAIX, J. 1969. New laboratory methods of studying the mechanical properties of rocks. *Int. Journ. Rock Mechanics and Mining Sciences*, **6,** No 1.

BISHOP, A. and HENKEL, D. 1962. The measurement of soil properties in the triaxial test. Edward Arnold, London.

BISHOP, A. and AL DHAHIR, Z. 1969. Some comparisons between laboratory tests, *in situ* tests, and full scale performance, with special reference to permeability and coefficient of consolidation. in Conf. on *in-situ* investigations in soils and rocks. British Geotechnical Soc. London 1969.

BISHOP, A. et al. 1971. A new ring shear apparatus and its application to the measurement of residual strength. *Géotechnique*, **21,** No 4.

BISHOP, A. 1966. The strength of soils as engineering materials. *Géotechnique*, **16,** No 2.

BISHOP, A. 1948. A new sampling tool for use in cohesionless sands below ground water level. *Géotechnique*, 1, No 2.

BRITISH STANDARD 2001. 1957. Site Investigations.

BRITISH STANDARD 1377. 1967. Methods of Testing Soils for Civil Engineering Purposes.

CASAGRANDE, A. 1947. Classification and Identification of Soils. *Proc. Amer. Soc. Civil Eng.*, 73.

CHRISTENSEN, BERG J. and OLSEN, SELMER R. 1970. On the resistance to blasting in tunnelling. *Proc. 2nd Cong. of Int. Soc. Rock Mech.*, *Belgrade*.

COTTISS, G., DOWELL, R. and FRANKLIN, J. 1971. A rock classification system applied in civil engineering. *Civil Engineering and Public Works Review.* June, 1971.

DALRYMPLE, B. and LAMPHERE, M. 1969. Potassium-argon dating; principles, techniques and applications to geochronology. W. Freeman & Co., San Francisco.

DONARTH, F. 1961. Experimental study of the shear failure of anisotropic rocks. *Geol. Soc. Amer. Bull.*, 72, No 6.

DEERE, D. and MILLER, R. 1966. Engineering classification and index properties for intact rock. *Tech. Report*, AFWL-TR-65-116. Air Force Weapons Lab., Kirtland Air Force Base, New Mexico.

EARTH MANUAL. 1968. Revised Edition. U.S. Dept. of Interior. Bureau of Reclamation.

EMERSON, D. 1969. Laboratory electrical resistivity measurements of rocks. *Australian Inst. Mining and Met. Proc.*, No. 230.

FRANKLIN, J. and CHANDRA, R. 1972. The Slate durability Test. *Int. Journ. Rock Mech. & Min. Sci.*, 9, No 3.

FURBY, J. 1964. Tests for rock drillability. *Mine and Quarry Eng.*, 30. July.

GRIGGS, D. 1939. Creep in rocks. *Journal of Geology*, 47, No 3.

HVORSLEV, M. 1948. Sub-surface exploration and sampling of soils for civil engineering purposes. U.S. Waterways Experimental Station, Vicksburg.

HEM, J. 1970. Study and Interpretation of Chemical Characters of Natural Waters. *U.S. Geol. Surv.* Water Supply Paper 1473.

HAWKES, I. and MELLOR, M. 1970. Uniaxial testing in rock mechanics laboratories. *Engineering Geology*, 4, No 3.

HOEK, E. and FRANKLIN, J. 1968. A simple triaxial cell for field or laboratory testing of rock. *Trans. Inst. Mining & Metall. Sect. A.*77.

KALLSTENIUS, T. 1963. Studies on clay samples taken with the standard piston sampler. *Royal Swed. Inst. Proc.*, No. 21.

KJELLMAN, W. and KALLSTENIUS, T. 1950. Soil sampler with metal foils. *Roy. Swedish Geotech. Inst. Proc.*, No. 1.

LANG, T. 1967. Longitudinal variations of soil disturbance within tube samples. *Proc. 5th Austr. and N. Zeal. Conf. Soil Mechanics.*

LOVELOCK, P. 1970. The laboratory measurement of soil and rock permeability. Tech. Communication No. 2. *Water Supply Papers.* Inst. Geol. Sciences, London.

MACKENZIE, R. 1957. The differential thermal investigation of clays. *Mineral. Soc.* (Clay Minerals Group), London.

MELLOR, M. and HAWKES, I. 1971. Measurement of tensile strength by diametral compression of discs and annuli. *Engineering Geology*, **5**, No. 3.

MILLS, W. 1950. Suggested method for mechanical analysis of soil by elutriation in Procedures for Testing Soils. A.S.T.M.

MORGENSTERN, N. and TCHALENKO, J. 1967. Microstructural observations on shear from slips in natural clays. *Proc. Geotech. Conf., Oslo*, **1**.

MUSKAT, M. 1937. The flow of homogeneous fluids through porous media. McGraw-Hill, Maidenhead.

NUFFIELD, E. 1966. X-ray diffraction methods. J. Wiley & Sons, New York.

OBERT, L. and DUVAL, W. 1967. Rock Mechanics and the design of structures in rock. J. Wiley & Sons, New York.

OLSEN, SELMER R. and BLINDHEIM, O. 1970. On the drillability of rock by percussive drilling. *Proc. 2nd Cong. Int. Soc. Rock Mechanics, Belgrade*, Paper 5.8.

PRATT, H. *et al.* 1972. The effect of specimen size on the mechanical properties of unjointed diorite. *Int. Jour. Rock Mechanics and Mining Sci.*, **9**, No 4.

RINEHART, J., FORTIN, J. and BURGIN, L. 1961. Propagation velocity of longitudinal waves in rocks. Effect of stress, stress level of waves, water content, porosity, temperature, stratification and texture. *4th Symp. Rock Mech., Pennsylvania State Univ.*

SEED, B. 1968. Sand liquification under cyclic loading simple shear conditions. *Amer. Soc. Civ. Eng., Jour. Soil Mechanics and Foundations Div.*, **94**, SM3.

SOIL MECHANICS FOR ROAD ENGINEERS. 1952. Department of Scientific and Industrial Research Road Research Lab. H.M.S.O., London.

STIMPSON, B. *et al.* 1970. A new field technique for sealing and packing rock and soil samples. *Quart. Jour. Eng. Geol.*, **3**, No. 2.

WAKELING, T. 1970. Developments in Soil sampling. *Brit. Geotech. Soc.*, Nov. 1970. (Unpublished synopsis of paper).

WORLD HEALTH ORGANISATION. 1963. International standards for drinking water. Geneva.

ZUSSMAN, J. 1967. Physical methods in determinative Mineralogy. Academic Press, London.

Applications

11 Geological Maps

Maps record a distribution of features as if they had been observed from above. Some of the features shown on geological maps may be visible on the ground and some may be concealed, e.g. by a cover of vegetation or drift. Naturally the maps become less particular in such areas, and this should be borne in mind when reading them; however, it is usually a matter of local detail and the general accuracy of the maps is rarely in question. This is because a distinction is made, on geological maps, between the features which have been seen and those which have been inferred. For example, the known position of a boundary is normally shown by a solid line, whereas a broken line is used for its inferred position. It is therefore important to read the symbols that have been used on a map before studying it; they are usually found in the margin, together with a stratigraphical column and other information concerning the geology of the area represented. References may also be given to an explanatory report or memoir in which geological details noted in the field, such as thickness, anisotropy, grading of materials and so on, may be recorded. Maps should be used in conjunction with these reports, when available, if they are to be fully understood. The colours or symbols shown on standard geological maps, for example, need not necessarily indicate material of uniform character, as such codes are normally used to define stratigraphical boundaries rather than physical characters. Valley fill of one age is often shown by one colour or one symbol, but reference to a memoir may reveal that it contains a variety of materials, e.g. sands, gravels, clays, and organic matter.

Engineers should remember that the amount of information given on a geological map is often controlled by the scale of the map, and that its interpretation is not always a straightforward task.

This chapter discusses the maps that engineers are most likely to encounter and summarizes the methods by which they are made. However, some geometrical constructions that are commonly used,

and which illustrate how the probable positions of concealed geological boundaries can be derived, are first described.

True and apparent dips   The inclination of any point on a geological
surface can be uniquely described by its angle of true dip and the
direction in which this angle was measured, as defined in Chapter 7
(Note 1). The true dip of any surface is its maximum angle of inclination as measured from the horizontal. This angle lies in the direction
of greatest gradient; the dip in any other direction is called an *apparent
dip* (Fig. 11.1). It is necessary to calculate apparent dips when constructing geological sections in directions other than that of true dip
(see p. 331).

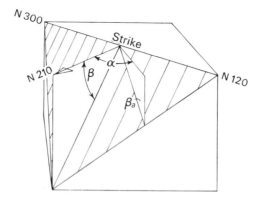

Fig. 11.1   Relationship between true and apparent dips. The true dip of the shaded
surface is β, at N 210°. β_a is an apparent dip, and tan β cos α = tan β_a.

Because of the difficulties involved in measuring the true dip of
many geological surfaces (see p. 284) it is often advisable to use the
larger lengths of surface that may be exposed in the sides of pits,
quarries, and other excavations; a typical example is illustrated in
Fig. 11.2(*a*). In the figure an inclined surface intersects two vertical
quarry faces, and the inclination of its trace on one face differs from
that on the other; the direction of true dip, and hence the angle of true
dip, are unknown. At least one of the dips seen on the quarry face
must be an apparent dip and it is prudent to assume that both are so,
as the angle and direction of true dip can be calculated from two
apparent dips. Consider the situation shown in Fig. 11.2(*a*) when
re-drawn as a geometrical model in Fig. 11.2(*b*). To construct this
model it is only necessary to measure the two apparent dips, 32° and
43° in this example, and the strike of the surfaces on which they have
been measured, *viz.* N 016° and N 094°. Points A' and B' are both a
vertical distance V above some horizontal datum, so making the line
$A'B'$ a contour for the geological surface. Because the true dip lies in

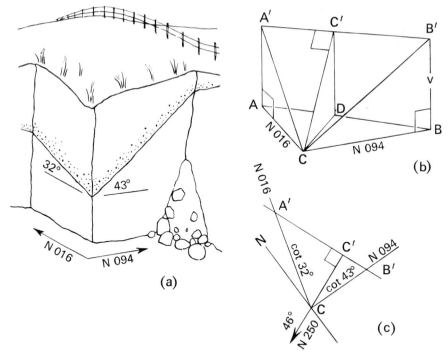

Fig. 11.2   Apparent dips exposed in the walls of an excavation. (b) Solid geometry of the situation shown in (a). (c) Graphical construction used for determining true dip from the apparent dips.

the direction of greatest gradient, lines at 90° to $A'B'$ will define that direction; $C'C$ is one such line. The true dip, or inclination of $C'C$ to the horizontal, is conveniently found by making a two-dimensional plan, as at Fig. 11.2(c). This diagram can be drawn directly from field data using the following steps:

(1) draw a line to represent the direction of magnetic north;
(2) select a point on this line to represent C;
(3) from C draw the two lines CA' and CB' so that they are correctly oriented to North (Note 2). It is then necessary to locate the points A' and B' on these lines.
(4) from Fig. 11.2(b) it can be seen that $AC = AA'$ cot 32°, and $BC = BB'$ cot 43°. Because $AA' = BB' = V$, the actual length of V is immaterial to the construction and can be conveniently taken as unity. Whence the point A' is located a distance equal to cot 32° from C along the line CA', and the point B' a distance equal to cot 43° along the line CB', to some suitable scale.
(5) a line drawn at 90° to $A'B'$, i.e. $C'C$, will lie in the direction of true dip, and this direction can be measured directly from the plan, i.e. N 250° in this example.

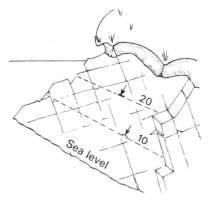

Fig. 11.3   The relationship between strike lines and structural contours illustrated as units above sea level on a costal exposure.

The angle of true dip is $C'CD$ in Fig. 11.2(b). The length $DC = DC' \times \cot C'CD$; because $DC' = V$, which has been set at unity, the equation reduces to $DC = \cot C'CD$, where DC is equal to CC', in plan, in Fig. 11.2(c). Hence the angle of true dip is that angle whose cotangent is equal to the length CC', i.e. 46° in this example.

This graphical method is usually referred to as the cotangent construction, and it can be used to interpret the geology exposed in trial pits, trenches, adits, and similar excavations. Other methods are described by Phillips (1971).

Strike, strike lines, and structural contours   The *strike* of a surface is the direction of a horizontal line drawn at 90° to the direction of true dip. The direction N 120° (or N 300°) is the strike of the surface shown in Fig. 11.1 as is the direction of the line $A'B'$ in Fig. 11.2(b). Because of its horizontality the direction of strike is equivalent, at its point of measurement, to an elevation contour for the surface. Figure 11.3 illustrates how the strike directions of a planar surface can be extended to produce strike lines which can also be used as structural contours for the surface. However, the majority of geological surfaces are not planar and the unlimited extension of strike lines away from the points at which dip and strike are measured can result in incorrect predictions (Fig. 11.4).

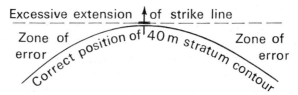

Fig. 11.4   Difference, shown in plan, between the correct position of a stratum contour and its predicted position. The direction of dip and hence strike (the short thick line) was measured in the field at a point 40 m above sea level.

Sampling the dip and strike of rock surfaces is important and care should be taken on site to ensure that the elevation, dip, and strike of concealed surfaces have been sufficiently sampled and that no variations of significance to a contract have been missed. Both engineers and geologists rely largely on bore-holes for finding the depth of underground surfaces. Measurements of dip and strike are not normally made in bore-holes, but the depth of surfaces encountered in them can be used in a 3-point construction for determining their general dip and strike.

3-Point construction   This   construction   requires   the   bore-holes to be grouped into a series of triangles as shown in Fig. 11.5. The

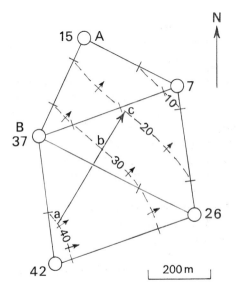

Fig. 11.5   The 3-Point construction. Broken lines are stratum contours constructed between 'known' points of elevation. Small arrows are local directions of true dip. The gradient along the line *a-c* is 1 in 20 (i.e. 2° between *a-b*, steepening to 4° between *b-c*).

level at which a given surface is encountered in each of the bore-holes is then marked on the map. The difference in values at the corners of the triangles is then *uniformly* distributed along their sides. For example, the difference in the elevation of the surface in bore-holes *A* and *B* (Fig. 11.5) is 22 m, and the horizontal distance separating the bore-holes is 170 m, hence a uniform distribution of this difference along the side *AB* would give a 10 m drop in every 77.3 m, as shown. This process is repeated for each side of every triangle. Contours are then drawn between points of similar elevation on the sides of the triangles, so producing a general contour map of the buried surface.

Having obtained these contours it is possible to calculate the general angle of dip in any direction along the surface, as is shown in Fig. 11.5. Note that the angle of true dip lies in the direction of greatest gradient, i.e. at 90° to the structural contours. The bearing of both true dip, and hence strike, can be measured directly from a correctly orientated map. Obviously the accuracy of any prediction based on this construction is closely related to the number of boreholes that are used.

The position of concealed boundaries The constant separation of strike lines on inclined planar surfaces is the basis of a construction that is used for assessing the position of partially concealed boundaries as shown in Fig. 11.6. Here a single exposure revealing a junction

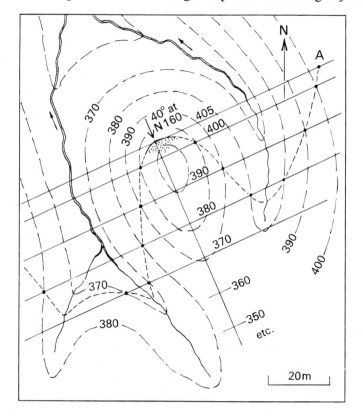

Fig. 11.6   The location of concealed boundaries, using stratum contours.

between sandstone and shale, occurs on the side of a hill. The dip of this junction at the exposure is 40° at N 160°. In order to predict the approximate position of the junction in the area around the exposure the following construction is used.

A line is drawn on the map in the direction of true dip from the point at which the dip was measured; any lines drawn at 90° to this line will be strike lines, although they can also be considered as structural contours for the sandstone-shale junction. The strike line drawn through the point at which the dip was measured will have a value equal to the elevation of the point, in this example 405 m above sea level. By drawing this line it is assumed that the junction between the sandstone and the shale, where present along the line, is 405 m above sea level. It therefore follows that the junction represented by that line will only occur at ground level at points where the strike line intersects the 405 m topographic contour; such a point is shown at point A in Fig. 11.6. It is then necessary to calculate the position of other strike lines and these are usually chosen so that they have the same value as the topographic contours on the base map i.e. in this example, 400 m, 390 m, 380 m etc. The separation of the strike lines is calculated from the angle of true dip and the horizontal scale of the map. In this example a dip of 40° is equivalent to a gradient of 1 in 1.19, i.e. a 1 m vertical drop in the elevation of the junction occurs in every 1.19 m, measured horizontally in the direction of true dip; this would be along the bearing N 160°. Hence a drop of 5 m occurs in 5.95 m. Thus the 400 m strike-line can be drawn as a line parallel to the 405 m strike line but separated from it by a horizontal distance equivalent to 5.95 m on the map. Points of outcrop along the 400 m strike-line can then be located along the 400 m contour. This construction is continued until the position of the boundary in the area in question is completely predicted. Boundaries which have been located in this manner are usually shown on a map by a broken line to distinguish them from those positions where the boundary was actually seen.

This construction assumes any surface to be planar within the area of the map, because the strike lines are shown parallel and the dip is constant. This rarely occurs in practice and therefore the construction is most appropriately used in small areas which occur between exposures. Obviously the predicted position of boundaries should agree with their actual position and hence it is advisable to use all known exposures and dips for the construction.

General geology maps Most maps of general geology are simply two-dimensional displays of the position of geological features as they occur at some surface. For example, mining engineers are interested in the geology at a particular horizontal level below ground level, and hence a map of the subsurface geology as it occurs on a plane at that level, will be used. In contrast, civil engineers are usually more interested in the geology that will be encountered nearer ground level and will often use maps which show the distribution of geological features over the undulating surface of the ground. Geological maps can therefore be made to suit the job in hand, and a great variety of maps can be used.

Published geological maps vary in scale and the most detailed map normally produced has a scale that is somewhere around 1:10 000. In the United Kingdom the nearest maps to this would have a scale of 1:10 560 (six inches to one mile) and be commonly called the six inch maps. These are the base maps that are normally used for detailed field mapping, and the geological maps produced on them should be consulted by engineers whenever possible; many of the smaller scale maps are simply reductions and hence simplifications of these larger scale records. Base maps having a scale that is greater than 1:10 000 are normally only used when it is necessary to map a small area in considerable detail, e.g. a particular foundation or cutting.

The next smaller scale which is commonly used is around 1:50 000 and the nearest to this in the United Kingdom is the 1:63 360 (one inch to one mile). England and Wales are covered by 360 such sheets, called the New Series by the Institute of Geological Sciences (formerly the Geological Survey), Scotland by 131 sheets and Northern Ireland by 72 sheets; not all the sheets have yet been produced. Each sheet is coloured and covers an area of 18 × 12 miles (Scotland 24 × 18 miles). Solid and drift editions are available for many of the sheets (see below). Maps of the 1:50 000 scale are useful for assessing the general geology of a region and should be consulted to appreciate the regional setting of an area; this is important in the fields of mining and groundwater. The maps can also be of great assistance to the location of mineral reserves and construction materials (Note 3).

Maps on scales which are appreciably smaller than 1:50 000 are normally of little use to most engineers. There are three smaller scales which are commonly employed for geological maps, viz., 1:200 000, 1:500 000, and 1:1 000 000. These maps illustrate regional tectonic patterns (of interest to engineering seismology) and general geology. Two well known maps of the United Kingdom which are based on scales close to these are the 1:253 000 sheets (four miles to one inch, commonly called the quarter-inch maps) and the 1:625 000 maps (ten miles to one inch).

This Chapter describes the maps which are most likely to be encountered by engineers; there are many others that can be used; geochemical maps and soil maps are just two (Note 4). However, all these maps no matter what their subject or their scale can be grouped under one of the two headings used in this Chapter, viz., maps of surface and maps of subsurface geology.

Maps of surface geology Geologists recognize two ways of recording surface geology and produce maps which are called *Solid* and *Drift* editions. The Drift edition is more correctly the map of surface geology because it shows the position, and the general character, of all geological materials that occur at ground level. These will include not only the harder rocks which are exposed at ground level, but also such materials as alluvium, glacial drift, mud flows, sand dunes, etc., which conceal the more solid rocks beneath them. It is because these materials have normally been transported, or drifted, to their present position that the maps which record them are called Drift maps. However, not all the drift which occurs at ground level is thick, indeed much of it can be quite thin, and so it is often easy to construct a map which records the geology beneath the drift, as if the drift had been physically removed. No drift is shown on these maps, only the solid geology, hence they have become known as 'Solid' maps. Thus the geology of an area can be studied using two maps, one showing the drift and the solid geology between areas of drift, and one showing only the solid geology as it occurs between, and as it is thought to occur beneath, the areas of drift. It is important to check the title of a map, where the edition either Solid or Drift should be mentioned. There is so little drift in some of the areas that the boundaries of both solid and drift geology can be clearly shown on one map which is then called a 'Solid and Drift' edition.

Rocks are said to be *exposed* when they occur at ground level. An exposure should be distinguished from an outcrop which is the geographical position of a geological unit, regardless of whether or not it is exposed. Figure 9.3 illustrates the two terms and shows that every exposure is part of an outcrop. By studying the geographical distribution of outcrops at ground level, as is shown by a map of surface geology, it is often possible to determine the overall geological structure which exists below ground level. Such a study is part of the wider subject of map interpretation, the fundamentals of which are outlined below. Good references on this subject include Blyth, Simpson, Bennison, and Roberts, titles on p. 340.

Interpreting maps of surface geology   Basically the art of interpreting geological maps lies in the ability to perceive how the two-dimensional pattern of outcrops which occur over a topographical surface which is not a plane, can be explained in terms of three-dimensional geological structures. Certain structural characters are easily distinguished; for example, vertical surfaces such as the sides of the dyke in Fig. 11.7 have an outcrop that is unaffected by topography whereas horizontal surfaces, such as boundary *B* always outcrop parallel to the topographic contours. Vertical and horizontal surfaces can therefore be located by simple inspection. Many geological surfaces are inclined, the boundary shown dipping at 40° in Fig. 11.6 is a typical example. The outcrop pattern of this boundary, as it crosses the valleys, should be noted because it points, or as the

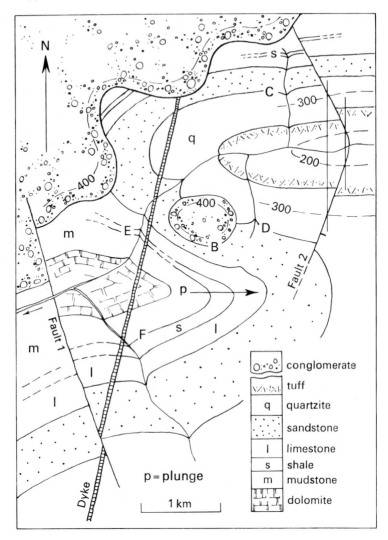

Fig. 11.7   Examples of outcrop patterns commonly seen on maps of surface geology.

geologists say, 'vees', in the direction of dip. (Check this by looking at the strike line values and the direction of dip as shown by the dip arrow.) This V-ing is characteristic of the outcrop of inclined surfaces intersecting topographic depressions. Fault 2 in Fig. 11.7, is another example. Hence the presence of inclined surfaces and the direction of their inclination can also be noted by simple inspection.

Observations of this kind form the basis for any interpretation of geological structure because once the dips are known it is possible to suggest structures that will account for them. For example, at *C*

(Fig. 11.7) the dip of the northern outcrop of sandstone is towards the south, whereas at D, it is to the north. This sandstone outcrop is part of an east–west trending synform. Similarly the boundary to the outcrop of shale dips to the north-north-east at E and to the south-south-east at F, and is part of an antiform which is plunging towards the east. Hence if the directions of dip are marked on a map the presence and shape of fold structures becomes apparent (Note 5).

Outcrop boundaries can also be used for assessing the thickness of strata. Unfortunately the meaning of the word 'thickness' varies with its usage. For example, a mining engineer sinking a vertical shaft through the strata shown in Fig. 11.8(a) would reckon its thickness as t_v, whereas the same engineer driving a horizontal tunnel through the same sandstone would measure its thickness as being t_h. Geologists usually speak only of stratigraphical thickness when dealing with upper and lower strata boundaries and by this they mean the thickness measured at 90° to the bedding surfaces, t_s in Fig. 11.8(a). Knowing the stratigraphic thickness and the dip of the strata it is possible to calculate the 'thickness' in any other direction by using the following formula:

$$t_x = t_s \cdot \sec \alpha \qquad \text{see Fig. 11.8}(b),$$

where t_x = the distance required through the stratigraphical unit in a direction x; this need not be the direction of true dip,

t_s = the stratigraphic thickness of the unit,

α = the angle between the direction in which t_s is measured, and the line t_x.

The vertical thickness (t_v), which is encountered in vertical boreholes

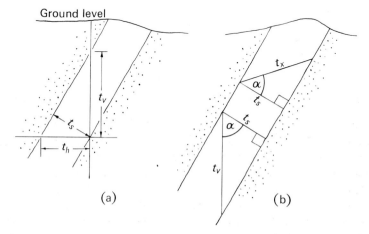

(a)   (b)

Fig. 11.8 (a) Various definitions of thickness. (b) Relationship between stratigraphic and other thicknesses.

and commonly recorded in site investigation reports, is related to the stratigraphical thickness by the formula:

$$t_v = t_s \sec \alpha$$

where α = the angle of dip in the plane containing tv and ts. (See Fig. 11.8(b)).

Width of outcrop. From Fig. 11.9(a) it is apparent that outcrop width is in some measure a function of the dip of the strata, the wider outcrops occurring in the areas of smaller dip, as is the case with the

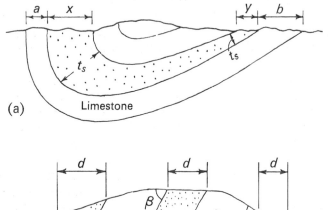

(a) Limestone

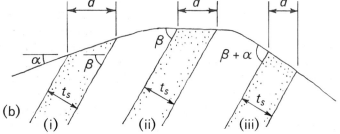

(b)

Fig. 11.9 (*a*) Outcrop width as a function of both dip and stratigraphical thickness. (*b*) Outcrop width (*d*) as a function of geological dip (β) and slope of ground (α). (i) $d = ts \cdot \text{cosec} (\beta - \alpha) \cdot \cos \alpha$, (ii) $d = ts \cdot \text{cosec} \beta$, (iii) $d = ts \cdot \text{cosec} (\beta + \alpha) \cdot \cos \alpha_0$.

limestone. If the stratigraphical thickness of the strata is fairly uniform the symmetry of a fold can often be postulated by comparing the widths of the outcrop on the fold limbs. The width of the sandstone outcrop on the limbs of the synform in Fig. 11.7 is fairly constant and it follows that if the sandstone has a uniform stratigraphic thickness the synform must be upright and nearly symmetrical. In contrast, the outcrop of the sandstone on the southern limb of the antiform is considerably wider than on its northern limb, suggesting that the fold is asymmetrical and that its axial surface dips towards the south.

This is a simple and quick way of assessing the fold structures of an area, but there are two possible sources of error, *viz.*, excessive variations in topography and stratigraphical thickness. Considering topography, it is evident from Fig. 11.9(*a*) that the outcrop of strata of uniform stratigraphical thickness and dip can vary as a function of the shape of the ground surface. Care should therefore be taken when considering the outcrop of geological units whose stratigraphic thickness is appreciably smaller than the variations of topography. Compare this situation with that of the limestone shown in Fig. 11.9(*a*). Here the stratigraphical thickness of the limestone is larger than the amplitude of the topography on its outcrop and outcrop width (*a*) and (*b*) can be used as an indication of relative dips. The sandstone horizon in Fig. 11.9(*a*) illustrates the effect of variations in stratigraphic thickness. Here the greater outcrop (*x*) occurs on the steeper limb of the fold. Significant variations of stratigraphical thickness are usually noted in the stratigraphic column for the map (Note 6).

Faults.   Two faults are shown in Fig. 11.7. Faults are discussed in Chapter 7; they are surfaces or zones about which some displacement has occurred and the outcrops on either side of a fault are therefore usually displaced, as shown in Fig. 11.7 by the sandstone in the synform and the unconformity in the region of the antiform; other examples are illustrated in Chapter 7. Every normal or reverse fault has two characters which should, if possible, be determined, *viz.* its dip and its downthrow side; both can sometimes be assessed from a map. The outcrop of an inclined fault surface will V in the direction of dip when crossing valleys. (Strike lines can be sketched, as in Fig. 11.6, and used to confirm both the direction and the dip of the fault. Two such lines are shown in Fig. 11.7.) The outcrop of vertical faults will be unaffected by topography whereas that of horizontal, or low dipping faults (i.e. thrusts) will run parallel or almost parallel to the topographic contours. Because the movement on a fault displaces rocks there will nearly always be a point on the fault where the rocks of one age are brought adjacent to those of another. Reference to the fault diagrams in Chapter 7 will confirm that the younger rocks are always found on the downthrow side of a fault which cuts uninverted strata and has a vertical component of movement, i.e. a normal or reverse fault.

The fault displacement, e.g. the components of throw and heave and so on, can often be calculated by measuring the distance between points which were coincident prior to faulting. The constructions used for these calculations are beyond the scope of this book, but can be found in most good texts on methods used in structural geology, of which Badgley is one.

Unconformities.   The juxtaposition of outcrop boundaries is a feature which can also be used to detect the presence of unconformities, i.e. surfaces at which there is a break in the continuity of the geology as discussed in Chapter 7. When seen on a map an unconformity often appears as a stratigraphical boundary whose course is usually unrelated to the outcrop of the geological structures beneath it. The conglomerate in Fig. 11.7 lies unconformably on the folded rocks beneath and the limit of the conglomerate marks the outcrop of the unconformable surface at ground level. Because unconformities can be considered as surfaces it is possible to determine whether they are either horizontal or inclined. That shown in Fig. 11.7 is horizontal because its outcrop is parallel to the topographic contours; had it been inclined it would V in the direction of dip when crossing valleys and other topographic depressions, such as road cuttings.

The identification of folds, faults, and unconformities forms the basis of all general interpretations of geological maps. Two other features which are commonly seen on maps and often mentioned in reports are outliers and inliers. An *outlier* is an outcrop which is completely surrounded (in plan) by rocks of a greater age. The conglomerate within the limits of boundary *B* in Fig. 11.7 is an outlier. As noted in Chapter 7, outliers are usually found close to escarpments. An *inlier* is an outcrop which is completely surrounded by younger rocks. As the dolomite in Fig. 11.7 is older than the surrounding mudstone its outcrop forms a faulted inlier in the core of the antiform. Inliers are sometimes developed in valleys where streams have cut down and locally exposed, in the valley floor, rocks which are older than those forming the sides of the valleys.

Age relationships can be deduced from geological maps. The folds in Fig. 11.7 were formed before the intrusion of the dyke, and the formation of faults 1 and 2. The conglomerate was deposited after the period of folding, dyke intrusion and movement about fault 2 as it covers all three. Fault 1 displaces the conglomerate and is the latest geological event which can be deduced from the map.

Although a considerable amount can be inferred from maps of surface geology it should be remembered that they are only two dimensional plans and give no definite information about the geological variations which occur below the surface. One way of partially overcoming this problem is to construct a number of maps to record the geology which exists at different levels below ground level and to stack them one above the other so as to produce a layered model. Such models are commonly used by mining geologists to assess the size and shape of mineral lodes. A less informative, but much easier method of illustrating subsurface geology consists of drawing one or more geological sections.

Geological sections   Geological sections illustrate the geology which exists below ground level, Figs. 7.3 and 7.20 are typical examples.

The first step in drawing a section is to obtain the ground level profile along the line of section; this is done by laying a sheet of section paper along the line of section on the map and marking on its upper edge the separation of the various contours (Fig. 11.10a). These

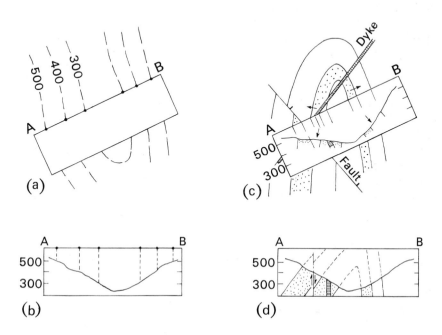

Fig. 11.10 Steps involved in producing a geological section. Topographic contours not shown in (c).

points are then plotted on the section paper against a suitable vertical scale and joined by a smooth line which will represent ground level (Fig. 11.10b). It may be necessary to extrapolate between these points in order to accommodate local variations in topography which may occur between the contour intervals recorded. The section paper is then re-positioned on the map as before (Fig. 11.10c) and the geological boundaries marked on its edge, with due regard being paid to the direction in which the boundaries are dipping. These positions are then transferred onto the topographic profile and the boundaries extrapolated downwards so producing a sketch of the geological structure as it probably occurs below ground level (Fig. 11.10d). The accuracy of such sections can be greatly improved if borehole logs and similar data from shafts and well bores can be incorporated.

Special care should be taken with two features when either constructing or interpreting geological sections: (i) the orientation of the section with respect to the direction of true dip at any point, and (ii) the scale of the section. It is common practice to orientate geological sections so that they illustrate the geology that would be exposed on a vertical planar surface which lies in the direction of true dip. However, it is often necessary to draw sections in other directions and in these cases they will display apparent dips. True and apparent dips which occur on vertical planar surfaces are related to each other according to the formula given in Fig. 11.1. The angles of dip shown on sections will also be a function of the horizontal and vertical scales that are used. It is customary to make the horizontal scale of a section the same as the horizontal scale of the map from which it was constructed; in this way a comparison between the two is facilitated. However, in order to illustrate details such as variations in the thickness of the strata, it is often necessary to make the vertical scale larger than the horizontal. Dips shown on such sections are not those which exist in the ground; these can be calculated by converting the gradient of any surface shown on the section into an equivalent angle. A further consequence of having one scale greater than the other is the convergence or divergence, on the section, of the upper and lower boundaries of geological horizons. This effect varies with the angle the geological horizon makes with the axis of exaggerated scale, as shown in Fig. 11.11.

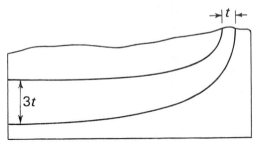

Fig. 11.11   Effect of exaggeration of vertical scale. Note the apparent change in thickness of a bed in passing from the vertical to the horizontal. Vertical scale = 3 × horizontal scale.

Geologists also use sections for illustrating the relative ages of geological events, in particular the relationships between folding, faulting, and unconformities; many examples of this can be seen in the sections which illustrate Chapter 8.

Maps of subsurface geology   These maps illustrate geological features which are not exposed and are often wholly based on information which has been obtained from boreholes, tunnels, well bores, geophysical surveys and the like. Hence the sampling of subsurface geology becomes important and should be considered when producing and interpreting such maps (Koch and Link, 1970). Seven subsurface maps are described below; each is used by geologists and engineers.

Horizontal-plane maps. These record the geology, as it occurs on a horizontal plane at some level below the surface. The outcrops shown on the maps are influenced by dip and thickness alone, and so dips must be recorded as they cannot be deduced from the outcrop pattern, as with maps of surface geology. Horizontal plane maps are useful to engineers involved in underground excavations.

Structural contour maps. These record the shape of structural surfaces such as faults, unconformities and the bedding surfaces of folds, below ground level. Often they are produced with the aid of 3-Point constructions (Fig. 11.5) or a related computer programme. The spacing between the contours is a function of the dip of the surface being illustrated, the contours being closest in those areas where the dip is greatest. The angle of dip can be assessed as the contour spacing divided by the contour interval equals the cotangent of the dip in the direction along which the contour spacing is measured. These maps are normally used when it is necessary to know the position of a definite structural surface below ground level.

Isochore maps. Isochores are lines joining points of equal vertical thickness, so isochore maps record the vertical thickness of geological units. The maps are readily produced using the data obtained from vertical boreholes which have fully penetrated the units being studied. These maps are often used to illustrate such features as the depth of overburden above some deposit, or the areal variations in the vertical thickness of some concealed unit such as a confined aquifer, a mineral deposit or a zone of weak rock.

Isopachyte maps. Isopachytes join points of equal stratigraphical thickness and are used to produce maps which are usually of greater interest to the geologist than the engineer. The maps cannot be interpreted as quickly as those of isochores even though the stratigraphic thickness of an horizon can be related to its vertical thickness by the formula

$$t_s = t_v \cos \alpha \ldots.$$

where α = the true angle of dip. (See Fig. 11.8*b*.)

Obviously vertical thickness will decrease as dip decreases and will eventually become equal to the stratigraphical thickness when the true dip is zero: Isopachytes then become synonymous with Isochores.

Geophysical maps. The geophysical methods described in Chapter 9 are frequently used to assess subsurface geology and the information obtained from them is often well displayed as a map showing either the physical values obtained, e.g. electrical resistivity, gravitational acceleration etc., or an interpretation of these values in terms of real geological features. In order to do this it is necessary to calibrate and confirm the physical values with borehole data, and the 1.50 000 maps produced by the Netherlands Geological Survey (Fig. 11.12)

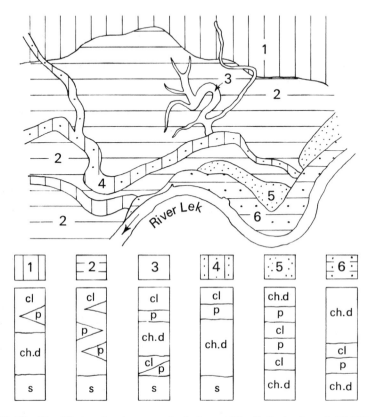

Fig. 11.12   Simplified portion from a geological map of the Netherlands on 1 : 50 000 scale. *cl* = clay; *p* = peat; *ch* = channel deposits; *s* = sand.

are an excellent example of what can be achieved. Geophysical methods, in particular, borehole methods, are also employed to assist in the construction of another group of maps which are of importance to engineering, *viz.* those which record hydrogeological data.

Hydrogeological maps.   Most hydrogeological maps attempt to show the distribution of five basically different subjects;

(i) geological controls, e.g. faults, folds, unconformities, lithologies.

(ii) hydrogeological features, e.g. aquifers, aquicludes, transmissivity, porosity.

(iii) groundwater, e.g. phreatic and piezometric levels, water quality.

(iv) surface water, e.g. rivers, springs, lakes, zones of recharge.

(v) installations, e.g. dams, weirs, wells, adits.

As a result many hydrogeological maps are overcrowded with information (Note 7). One of the difficulties with these maps is that they record features which can change with time, e.g. water levels and water quality. Care should therefore be taken when using these maps as a source of information on these subjects as conditions could have changed since the time of their production. It is generally accepted that these maps are a record of the unchanging hydrogeological characters such as aquifers and anquicludes, and a guide to the more variable characters such as water levels. Water levels themselves have long been of interest to both geologists and engineers and as a result there are many maps simply of water levels, both phreatic and piezometric. The method used for constructing them is identical to that used in producing maps of structural contours, the only difference being the nature of the boundary (i.e. water contours).

Engineering geology maps.   Many geological descriptions of rocks are inappropriate for engineering requirements and make the use of geological maps a difficult task for engineers. To overcome some of the problems arising from this situation it is necessary to map in terms of engineering parameters, so that geological materials that are similar for engineering purposes are grouped under similar headings. It is important to remember that the boundaries so drawn may sometimes bear no relationship to those that would have been recognized by pure geology. For example, a sequence of clays whose stratigraphy would normally cause them to be differentiated on a geological map may be shown, by reason of their similar fissuring and stiffness, as one unit on an engineering geology map. Harder materials such as igneous rocks could be grouped on the basis of their strength parameters instead of their petrography. Some engineers in the United Kingdom already produce maps of new sites purely in terms of engineering parameters, e.g. foundation requirements. These maps may show no geology at all and simply delineate areas to be avoided from areas suitable for stiff rafts, reinforced strips, normal strips, and so on. There are many features which are of interest to the engineer, e.g. depth of weathering, degree of fracturing, variations in strength, presence of landslides, probability of seismic shocks, areas of subsidence, areas of high tectonic stress, etc., and it is easy for these maps to become saturated with information. Certain countries, particularly in Eastern Europe, therefore produce a series of maps or overlays in order to illustrate these features (Note 8).

These maps should only be used as a guide to general conditions, for they purport to be no more; they should never be used as a substitute for a site investigation. Unfortunately they are likely to become dated very quickly. Some geologists believe that as these general maps of engineering geology can be no better than the records from which they come, it would be far better to develop a good filing system where copies of the borehole records of every site investigation can be stored, and made available for reference. It should also be remembered that these maps are only plans and that they do not illustrate the variations which occur with depth unless they use a system similar to that adopted by the Netherlands Geological Survey (Fig. 11.12).

Maps of engineering geology are relatively new to geology and are an interesting indication of things to come. Clearly the environment is going to become an increasingly important factor in engineering decisions which will require the use of hybrid maps showing such features as the availability of deposits, rather than the deposits themselves, or the erodibility of soils rather than the soils themselves. These kind of maps have already been made for specific projects and their production has shown that they require much preparation. It thus appears that they will normally only be produced as specialist maps for clients and not as published maps for general sale, which means that most engineers will have to cope with standard geological maps for many years to come (Note 9).

NOTES

1. This is not always possible in areas where large ore bodies disturb the global gravitational field. The problem in these situations is essentially one of surveying.

2. One of the commonest mistakes made when completing this construction is to draw the lines $A'C$ and $B'C$ on the wrong side of the N–S line. This can easily be done because each direction is the strike of one face of the exposure which can be given either as $x°$ or $(x° + 180°)$ as both readings would be correct. Thus the strike of faces A and B in Fig. 11.2(a) could equally well have been recorded in the field as N 196° and N 274° respectively. If these directions had been plotted directly in plan they would have put the triangle $A'B'C$ (to be constructed) on the west side of the N–S line. The angle of dip calculated would be the same as before, but the direction of dip, if measured in the $C'-C$ direction, would be in error by 180°. It is therefore advisable to think of the situation in three dimensions before commencing with the construction.

3. In fact, it was the quest for 'puddle-clay', i.e. clay which could be trodden to an impermeable state and used for sealing canals, that led to the first geological maps of England and Wales being produced. Strangely enough they were not made by a geologist, but by a civil engineer named

William Smith. Smith collected geological data and published in 1815 a Coloured map of the Strata of England and Wales on a scale of five miles to one inch. The maps are remarkably accurate and are a tribute to Smith's surveying. Private enterprise continued to lead the field, and mapping of the mineral lodes in Cornwall and Devon then commenced under the supervision of Henry de la Beche, who later became the first director of the Geological Survey for the whole of the country.

4. A group of maps which should be mentioned here are those which have fairly recently been produced by geomorphologists, because they are closely related to geology and in particular the processes of denudation. For example, there is a considerable overlap of interests by geomorphologists studying land forms and engineers studying land evaluation; both are concerned with the instability which occurs on slopes. Similarly the engineer in search for sands and gravels will literally cover the same ground as a geomorphologist studying the fluvial history of their transport and deposition. Geomorphologists are now producing what can be described as maps of geomorphological environments; these relate the dynamics of geomorphological processes to the land forms which develop. This is not a completely new idea, for Strahler produced topographical maps in terms of gradients rather than contours as long ago as 1965 (cf. Strahler (1965). *Bull. Geol. Soc. Amer.*, **67**. Quanitative Slope Analyses.); see also Savigear R.A.F. (1965). *Annals. Assoc. Amer. Geographers* **55,** A Technique of Morphological Mapping. These maps are likely to become of increasing importance to engineers. Information on their production in the United Kingdom can be obtained from the British Geomorphological Research Group, c/o Institute of British Geographers, 1 Kensington Gore, London S.W.7.

5. This is rather an over-simplification because the limbs of isoclinal folds dip in the same direction (Fig. 11.13). The repetition of the outcrop as shown in Fig. 11.13 is often an indication of isoclinal folding. This can be confirmed in the field by using 'way-up' structures (Chapter 8) to identify overturned strata.

6. Stratigraphical columns are scale drawings of the succession of strata; the column printed in the margin of a geological map represents the succession within the area of the map. A simple example is given in Fig. 11.7, and a more detailed example in Fig. 11.14a. The divisions in a column are normally based on stratigraphy and lithology, and are arranged so that the oldest strata are always at the base of the column. Fold structures, which have been described as either antiformal or synformal (Chapter 7, and p. 327), can be identified as anticlines or synclines by reference to the relative ages of strata. The thicknesses of lithological formations shown in a column are usually the average stratigraphical thicknesses in the area of the map. Marked local variations are recorded by tapering bands in the column. If there are significant variations in the thickness of formations, over the area of the map, it may be necessary to give more than one stratigraphical column. Stratigraphical columns should not be confused with vertical sections because these are obtained from boreholes, well bores, cliff sections and similar exposures,

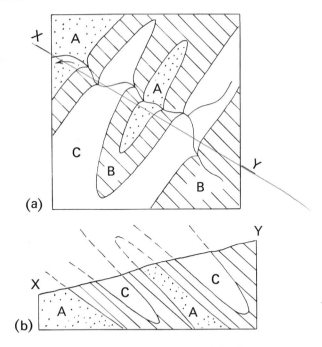

Fig. 11.13   Isoclinal folds. (*a*) Plan, (*b*) Section.

and the thickness of the strata shown on them is the actual thickness of the strata penetrated at a particular location. Another difference is that vertical sections represent the vertical succession as it occurs at a particular location, whereas the stratigraphical column simply represents the stratigraphical position of all the horizons shown on the map. Vertical sections which reveal gaps in the stratigraphical column indicate the presence of either unconformities or faults. A typical vertical section is illustrated in Fig. 11.14*b*. In these sections it is customary to write the vertical thickness of the strata on one side of the column and the depth below ground level on the other. The positions of recognizable unconformities are usually emphasized by wavy lines.

7. The International Association of Scientific Hydrology have agreed upon an international set of symbols for use on these maps. The original reference for the working party report is: Publication No. 60. (1962), A legend for hydrogeological maps. This report has been revised and the best publication to consult is now the joint publication produced in 1970 by UNESCO, the Institution of Geological Sciences, the International Association of Scientific Hydrology and the International Association of Hydrogeologists. The publishers are Cook, Hammond & Knell Ltd. London, and the title of the publication is International Legend for Hydrogeological maps. 1970.

8. It has been suggested that a better approach may be to produce maps which identify zones of engineering hazard, e.g. areas susceptible to

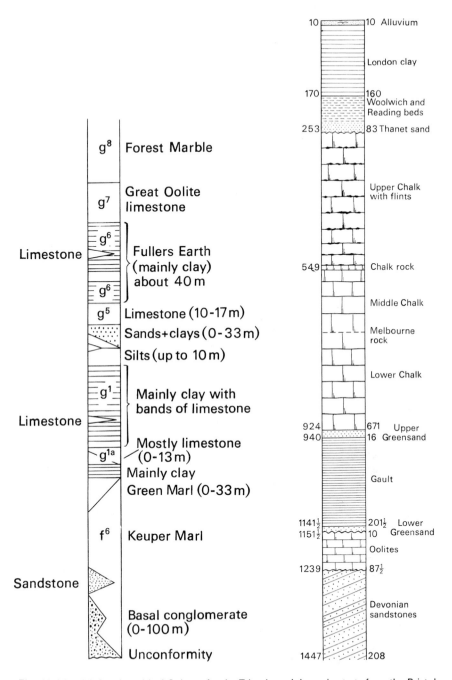

Fig. 11.14 (*a*) Stratigraphical Column for the Triassic and Jurassic strata from the Bristol Sheet. (*b*) Vertical Column from deep boring at Richmond. Thicknesses in feet on right of column, depths from surface on left. Unconformities shown by wavy lines across column.

subsidence rather than the areas of subsidence themselves, or slopes susceptible to natural instability rather than actual landslides. An interesting selection of papers on this subject of mapping in terms of engineering units can be found in Engineering Geological mapping symposium with UNESCO, section 7, First International Congress of International Association of Engineering Geologists, Paris, Sept. 1970.

9. Photographs of the earth, taken from the United States satellites, are now readily available for use by engineers. Much information can be obtained, particularly when the pictures have been obtained by remote sensing techniques. Definition is so good that small-scale maps can be made from the information transmitted back to earth.

REFERENCES

BADGLEY, P. C. 1959. Structural Methods for the Exploration Geologist. Harper's Geoscience Series.

BENNISON, G. M. 1969. Introduction to Geological Structures and Maps. Edward Arnold Ltd., London.

BLYTH, F. G. H. 1965. Geological maps and their Interpretation. Edward Arnold Ltd., London.

KOCH, G. and LINK, R. 1970. Statistical Analysis of Geological Data. J. Wiley and Sons, New York.

PHILLIPS, F. 1971. The use of Stereographic Projection in Structural Geology. 3rd edition. Edward Arnold Ltd., London.

ROBERTS, A. 1958. Geological Structures and Maps. Cleaver Hume Press Ltd.

SIMPSON, B. 1968. Geological Maps. Pergamon Press, Oxford.

References on Selected items are also given in the Chapter Notes 4 & 8.

12 Geology and the Movement of Water

A brief description of the chemical character of water is given in Chapter 17 and the measurement of storage and transmission properties of rocks and rock masses in Chapters 9 and 10. Water supply and pollution are considered in Chapter 16. This chapter considers the movement of water from its point of entry into the ground to its eventual point of exit.

Hydrological cycle The fossil imprints of raindrops on the bedding planes of sedimentary rocks confirm that rain has been a source of water for millions of years. But this was not always the case, for the primeval earth probably had little or no atmosphere and the greater part of the water which is now in the oceans existed as discrete volumes in the evolving magmas. Such water is referred to as being magmatic, or primary. Minute quantities of magmatic water can sometimes be found as very small liquid inclusions in the crystals of igneous rocks. Larger quantities of what may be truly magmatic water can be found associated with active volcanoes and geothermal springs, however, much of the vapour which issues from these sources may be surface water which has become heated at depth.

With the development of an atmosphere came the meteoric environment which promoted the concentration of water vapour and the development of meteoric reserves of water. The quiescence of igneous, particularly volcanic, activity permitted the solar source of energy to dominate the geothermal source in controlling the heat on the surface of the earth, and so permitted a relatively systematic pattern of global movements within the atmosphere. It also meant that the atmosphere became the major transporter of surface water and that the hydrosphere became the major source of supply for the atmosphere. Water therefore began to move in a regular cycle, being evaporated from the seas and carried as vapour into the atmosphere, from which it was precipitated and returned either directly or via some combination of overland and underground routes to the sea. This pattern of movement exists at the present and is called the *hydrological cycle*, the general character of which is shown in Fig. 12.1; for further details see Ward (1967) and Domenco (1972). The supply of magmatic

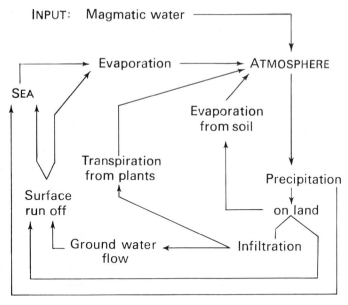

Fig. 12.1   The Hydrological Cycle.

water to the cycle is now so small that hydrologists speak of precipitation as being the 'source' of water.

The volume of water circulating via the atmosphere is generally accepted as being no greater than 2.80 % of all the water represented by the cycle; by far the greater volume of water in the cycle is in the seas. The approximate distribution of water is as follows:

Seas and Oceans	97.20%	
Snow and Ice	2.14	
Groundwater	0.63	Circulating via the
Lakes and Rivers	0.02	atmosphere
Water Vapour	0.01	

Geology significantly affects three stages of the cycle, *viz.* infiltration, percolation, and ground-water flow.

Infiltration   is the movement of surface water into the ground and is closely related to percolation, which is its subsequent movement to a water-table. Infiltration is partially controlled by; (i) the transmissive properties of the surface, (ii) the stability of these properties under changing surface conditions, and (iii) the transmissive and storage properties of the material beneath the surface.

The transmissive properties of a surface are controlled by the character of the pores and fissures which are exposed at the surface. These must not be too small, otherwise their walls will exert a capillary force which will hold most of the water in their mouths and never permit it to move far from its point of entry. Such a surface will accept water but will not permit sustained infiltration.

The rocks and soils which are most likely to permit appreciable infiltration have large pores and fissures; poorly cemented gravels, grits, and sandstones usually provide such surfaces as do soils which have an open crumb structure. Similarly, appreciable infiltration will occur over rocky scree-slopes and colluvial fans (p. 72). It should also be remembered that many impermeable rocks have joints which allow them also to accept surface water; thus the surfaces of many igneous and metamorphic rocks accept water as do those of well cemented and jointed sediments. Even clays can permit infiltration when they are cut by desiccation cracks. Zones of preferred infiltration can be expected in areas of tension and fracture such as the crests of anticlines, the boundaries of faults, the cooling zones of igneous rocks, the sides and floors of stress relieved valleys, the backs of landslides and the tensional areas flanking mining subsidence.

Soils (Chapter 2) have a more delicate structure than rocks and their pores and fissures can easily be closed or blocked by movements, such as shrinking, swelling and aggregation of soil particles associated with the periodic drying and wetting of the surface and the disturbance resulting from the impact of raindrops on to the soil. Here, pores can become blocked and surfaces gradually rendered temporarily impermeable, as commonly occurs in the fine-grained cohesive soils, found for example, on clayey outcrops. Vegetation rather than geology appears to exert a dominant control over this phenomena, and areas not covered are susceptible to changeable infiltration characters. Other disturbances resulting from freezing, compaction by animal traffic, ploughing, etc. will similarly affect infiltration. Careful pedological and geological surveys are thus necessary to delineate most existing and potential infiltration areas.

An appreciation of these characters has assisted attempts either to supplement ground-water reserves, or simply dispose of unwanted water, by recharging the ground from artifically prepared recharge sites. The water to be disposed of must first be filtered so that it is chemically and physically clean and free from any suspended matter which could block the infiltration surface. It is then discharged into the recharge pits, with care being taken to avoid unnecessary turbulence and subsequent disturbance of the surface. There will be invariably some reduction in the rate of infiltration even with the most gentle flooding of sites (Fig. 12.2).

Infiltration is therefore a character which changes with time. Horton (1933) described the maximum rate at which rain can be absorbed by a soil in a given condition as its *infiltration capacity*. If this is exceeded by the rate at which water is being applied to the

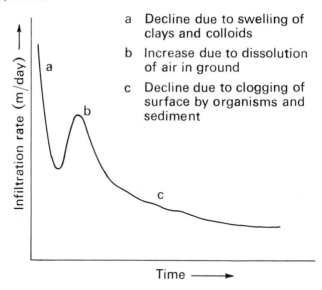

a Decline due to swelling of clays and colloids

b Increase due to dissolution of air in ground

c Decline due to clogging of surface by organisms and sediment

Fig. 12.2   Variation of infiltration rate with time (after Todd).

surface there will be a build-up of surface water. In recharge pits this would mean that the water level in the pits would begin to rise; on natural surfaces it usually means that water will leave the surface as run-off. The relationship between infiltration capacity and rainfall intensity (both in units of length/time) is important, because intense rainfall rarely gives infiltration surfaces the opportunity of accepting much water and most of the rain will leave as run-off. Gentle rainfall will usually be accepted by the land surface provided the transmission and storage properties of the rocks beneath the surface are adequate. Water cannot continue to penetrate a surface if it is not moving away from its point of entry and, in practice, infiltration cannot be separated from percolation.

Percolation is the movement of underground water towards a watertable; it occurs in an essentially vertical direction and is largely controlled by the force of gravity. Percolating water is sometimes called 'gravity water' to distinguish it from those volumes of underground water which are held by the attractive forces of the solid skeleton itself; these are referred to as capillary forces and the water, so held, as capillary water. The strength of these forces depends upon the area and dryness of the surfaces surrounding the voids; of these, dryness is likely to change because the degree of saturation of any soil will usually vary with its distance from ground level, as shown in the ground-water column illustrated in Fig. 12.3.

One factor affecting percolation is the total unsatisfied force, capillary or otherwise, which is available to retain water in the soil. A

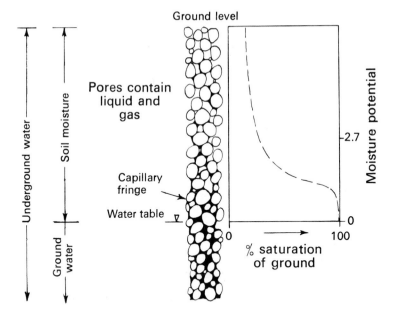

Fig. 12.3   Schematic presentation of liquid occurrence in a porous medium (after Meyboom).

measure of this force is the *moisture potential*, which is expressed as either the vacuum necessary to balance the soil suction, or the logarithm of the head of water (in centimetres) equivalent to the pressure difference between this vacuum and atmospheric pressure; logarithmic values of head are given in Fig. 12.3. The importance of this unsatisfied force can be appreciated by considering the infiltration of water into a soil profile which is dry at the top and saturated at the bottom.

When wetted the upper surface of the soil will first satisfy its own suction requirements before allowing water to enter far into the openings which lead to underground voids. Once in these openings, the water will be attracted by the walls of the voids to satisfy their own requirements; this continues until the surfaces can hold no more water against the pull of gravity. Soil so affected is said to have reached its *field capacity*. At this point the walls of the voids will be covered with a film of water and the moisture potential will have dropped to 2.7. Being unable to hold any more water the surfaces permit subsequent supplies, coming from the infiltration surface, to pass them by as gravitational water and either enter unfilled pore spaces or move on down to satisfy the requirements at lower levels. Gradually this wetting front of gravity water penetrates the full depth of the zone of aeration and reaches the zone of saturation. From that time onwards, water infiltrating through the ground surface will be in a drainage system which carries it to the water-table.

This drainage of gravity water will continue after the cessation of infiltration, until the pull of gravity on the pore water is once again balanced by the surface tension of the film of water around the individual soil surfaces, i.e. until the soil is restored to its field capacity.

The importance of surface area to percolation means that the significant geological factors in the process are essentially grain size and packing. Fine-grained materials such as clays, silts, and silty sands will develop very high potentials when dry, which will only be satisfied by considerable quantities of water, and areas covered by such deposits cannot be expected to allow the percolation of much water. The more open sandy soils will usually require less water to bring them to field capacity and can be expected to provide good recharge areas. The structures which are important in soils are aggregates and fissures. Aggregates are collections of soil matter bound together by either colloidal material or soil moisture. Colloids produce more permanent structures which commonly group together to form larger aggregates, called *crumbs;* soil moisture produces temporary aggregates which owe their existence to cultivation, frost, and root action. Fissures are commonly found in the more cohesive soils and can break the soil into roughly cuboidal, prismatic, columnar, and platy fragments; rootlet holes and animal burrows are also of importance. Structures which assist infiltration into rocks will also assist percolation through them, and mention of these has been made above (see infiltration).

Although grain size and texture are both important, mineral composition should not be overlooked, particularly if the soils and rocks contain minerals which expand when wetted. Clay minerals of the montmorillonite group have this property because water molecules can be taken directly into their lattice (p. 121). The disintegration or slaking which can accompany the wetting of shales and dessicated clays should also be noted.

Water-Table and Capillary Fringe.   Percolating water will eventually reach a zone of saturation where all interconnected voids are full of water; however, a hole drilled to just below this level would not encounter a water level, even though its sides would be moist. This is because the water in the pores is still attracted to the solid skeleton of the rock by capillary forces. In such a situation water would drain out of the sides of the hole and trickle down to its base where it would re-enter the ground. By deepening the hole, a level will be reached where the water coming from the sides will not drain into the bottom of the hole, but begin to pond as a body of water. The moisture potential of the ground at this level is zero and the pore water is at atmospheric pressure (Fig. 12.3). This is the level of the *water-table*. The zone of saturation above the water-table is called the *capillary fringe*, and the height to which this rises above the water-table largely depends on the size of the voids in the ground. Normally it rises no more than 1.5–2.4 m (5 to 8 feet) above the water-table.

The presence of capillary water can be important to the costing of earth-moving contracts, for it may be necessary to use saturated densities when calculating the weight of material above the water-table to be moved. It should also be remembered that capillary water will be recorded by electrical geophysical surveys.

The water-table is the level at which pore water is at atmospheric pressure; its depth below ground level is virtually the same as the depth to water in an uncased borehole. Care should be taken to ensure that the water level recorded in a borehole does represent a water-table and is not some other level (Note 1).

The shape of the water-table usually reflects the topography of the region, attaining its greatest elevation beneath the highest ground (Fig. 12.4). The intersection between topography and water-table is

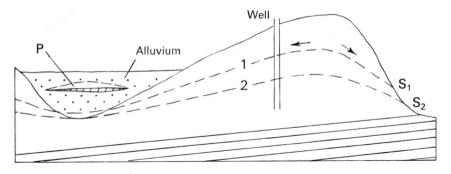

Fig. 12.4   Water tables: *1* = level of the water table after a period of recharge. *2* = a lower level resulting from natural drainage. *p* = perched water table on an impermeable lense. *s* = position of the spring line.

called a *spring-line*, and is an important hydrological boundary because it represents the elevation below which water is discharging from the ground. No infiltration can therefore occur below the spring-line of any geological unit. *Perched water-tables* develop over impersistent horizons of relatively impermeable material such as a seam of clay or silt and are of local, rather than regional, extent. These bodies of water can easily be missed by a site investigation even when penetrated by a borehole, and regular measurements should be made of the level of water in holes which are being drilled, because it is normal for the water level in a drill hole to drop fairly rapidly after the base of the impermeable layer on which water is perched has been pierced. Often it is only by plotting such variations against the bore-hole log that the presence of a perched water-table is discovered.

Water-table movement   This is a function of the natural rate of drainage from the mass. When percolation exceeds this rate the voids in the rock mass begin to fill and the level of saturation within the mass begins to rise. Once the rate of percolation becomes smaller than the rate of drainage the level of saturation starts to fall, and will continue to fall until the hydraulic gradient in the mass is no longer sufficient to permit further drainage. The elevation of the water-table can therefore change with time, varying between an upper limit and a lower limit beyond which ground-water flow can no longer occur (Fig. 12.4). Normally water-tables oscillate within these limits rather than between them, as the length of time required to drain a rock mass to its limit is usually much longer than the time separating periods of recharge. Even so, changes in the elevation of the water-table can be of the order of tens of metres. Figure 12.5 illustrates vertical fluctuations that may occur with time at any one location and is an example of the way in which water level measurements are often presented in reports. The fluctuations measured in one observation hole are rarely those which occur over a large area, as these often vary from place to place and tend to be greatest beneath upland areas. Water level changes should therefore be considered in three dimensions, and one of the easiest ways of doing this is to construct maps which show such changes (Davis and de Weist, 1966).

Areal variations in the magnitude of water level fluctuations are largely controlled by the distribution of infiltration, the proximity of discharge points, and the variation of storage and transmissivity.

(i) The geological characters which affect infiltration usually vary over an outcrop and will almost certainly vary from one outcrop to another. Hence a uniform distribution of rain will rarely result in a uniform distribution of infiltrated water and the consequent rise in water-table will be greater in some places than in others. It is also important to remember that many rainstorms do not distribute their rain evenly over an area and can accentuate the unevenness of a rising water-table.

(ii) Discharge points can be both natural and artificial: springs, sink holes, underground caverns, discharging wells, pumped sumps and drainage galleries are examples. They can be considered as *sinks* within the general system of ground-water flow and their strength can vary with time. As a consequence, water-table fluctuations vary with the proximity and the capacity of local sinks.

(iii) The storage available in pores and fissures varies from rock to rock, and often varies within any one formation. It is this storage which controls the volume of water which can be recharged to a rock

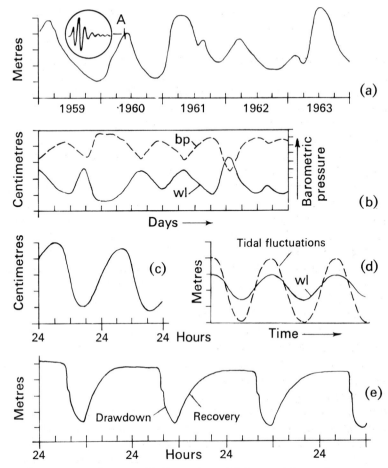

Fig. 12.5 Examples of water-table fluctuations. (*a*) Seasonal fluctuations measured in an unconfined aquifer in England: *A* = short term fluctuations resulting from Chilean earthquake in May 1960. (*b*) Fluctuations in a piezometric level from a confined aquifer reflecting the effects of barometric pressure. (*c*) Daily variations in water table resulting from the transpiration of surrounding vegetation. (*d*) Water levels in a coastal aquifer. (*e*) Periodic fluctuations resulting from the discharge of a nearby well operating between 7 a.m. and noon.

mass for a given rise in water level. Hence the rise of water level which follows a period of uniformly distributed infiltration can easily vary from place to place, and be greatest in areas of least storage. Hence areas which experience a large rise in water levels need not always be those which contain large quantities of water. Pore spaces and fissures also affect the transmissivity of a rock mass; for the diameter, opening, and continuity of such features influence the frictional forces which retard the flow of water. As already mentioned, the water-table rise in response to infiltration is partly controlled by the rate at which water can drain from the area of recharge; hence it will sometimes be greatest in areas of lowest transmissivity.

Each of the above factors has been considered in isolation, but in practice they usually operate together in some degree and contribute to the movement of water levels. The interpretation of these movements can only be satisfactorily completed by studying both the hydrology and the geology of the areas in question. This necessitates an investigation of the characters of the aquifers and aquicludes that are present.

Aquifers and aquicludes   An *aquifer*, by definition, is a 'bearer of water', and rocks which yield water comparatively freely are called aquifers. Unfortunately it has become customary to emphasize only one part of the definition, i.e. 'water', so that saturation has become the criterion used by many for identifying aquifers. This is a potentially deceptive procedure because the degree of saturation can change with time, and rocks which were dry when met in a site investigation may bear considerable quantities of water when actually encountered later in a contract. It would seem better to define a rock mass as an aquifer on the basis of its transmissive properties, because all aquifers are either permeable or pervious and many are both (Note 2).

Some rocks transmit water at a much slower rate than others, and this has encouraged the belief that they do not transmit water at all; such rocks are called *aquicludes*. However, it is now realized that all rock formations will transmit water given sufficient time, and the distinction implied by the term aquiclude is somewhat relative. Materials such as clays and shales are commonly regarded as aquicludes. But it should be remembered that impermeable materials can still be pervious, i.e. transmit water, through fissures, joints and similar openings, and that such leakage may have to be estimated. Special care should be taken with near-surface deposits which crack when dried.

Confined and unconfined aquifers.   Some aquifers have ground level as their upper boundary and are described as being *unconfined*, to distinguish them from those found at depth which are confined by aquicludes. Many large aquifers, such as the Chalk of the London Basin, are unconfined at their outcrops, but become confined down-dip as they pass beneath the outcrops of overlying aquicludes.

A confined aquifer possesses special properties, one of the most important being the ability to store water under pressures that are greater than those which would result from a column of water equal in height to the vertical saturated thickness of the aquifer itself. This occurs wherever the buried upper surface of the confined aquifer is lower than the water level at the aquifer outcrop. The magnitude of the pressure can be measured by drilling into the aquifer and observing the height to which water rises in the hole. The water level in such a hole is called a *piezometric level*, and the pore pressure which supports it a piezometric pressure (Note 3). Sometimes the piezometric level of a confined aquifer is above ground level; holes drilled into such an aquifer would overflow with water and are described as *artesian wells*, after the French province of Artois where the phenomena of overflowing wells was first recorded.

The level of ground-water in the outcrop of a synclinally folded confined aquifer can be high enough to overcome the losses of head associated with flow through its structure, and still maintain a pressure which brings the piezometric level to the surface, so that most wells drilled into the aquifer overflow. Such synclines are described as *artesian basins*. The London Basin (Fig. 12.6*a*) used to be overflowing

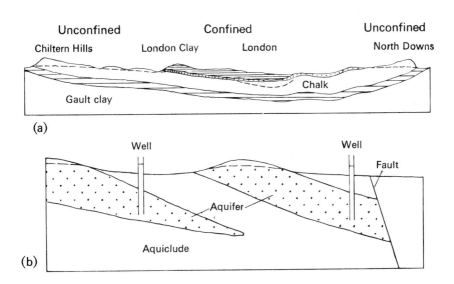

Fig. 12.6   (*a*) Section across the London Basin; length about 85 km. London and Gault clays are the aquicludes and the Chalk is the aquifer. Note the dewatered chalk beneath London. (*b*) Details of artesian slopes. Broken line = water table.

but pumping has now reduced the piezometric level to below ground level, and in some places has actually de-watered part of the Chalk aquifer. Much larger artesian systems are found in other countries and special mention should be made of the Great Artesian Basin of Queensland, Australia. This extends for approximately 600 000 square miles west of its outcrop along the Great Divide, and supplies water, via deep wells, to the arid country of central Queensland. Much smaller structures can also support artesian conditions and examples of two which often occur in sequences of sedimentary rocks are shown in Fig. 12.6*b*. The tapering aquifer is frequently found in drift and shallow water deposits.

Piezometric levels are important to the water engineer because they affect pumping lifts. They are also important to the construction engineer because they indicate the magnitude of the uplift pressures which are acting at the top of an aquifer (see Chapter 14). When water is pumped from a confined aquifer in which it is under pressure, the drainage of water will lower pore pressures and change the magnitude of intergranular stresses within the solid skeleton. This can promote a partial collapse, or consolidation, of the aquifer and settlement at ground level (Davis *et al.*, 1964). The existence of a mechanically strong skeleton does not necessarily preclude the possibility of settlement, because drainage from a confined aquifer may be sufficient to initiate drainage from its confining aquiclude which in turn may consolidate. This can occur when de-watering sands and gravels which lie beneath silts and clays.

Flow below the water-table   The nature of this flow depends on the geology, and can be markedly influenced by certain geological features such as joints, faults, and impermeable boundaries. Two aspects of any aquifer which is to be encountered in engineering work should therefore be appreciated: the geological variability of the aquifer (expressed as anisotropy or heterogeneity), and its flow regime.

Anisotropy and heterogeneity.   The transmissive properties of many soils and rocks are commonly a function of the direction in which they are measured, because most geological materials are not uniform in character. For example, rootlet holes and animal burrows give many soils their greatest transmissivity in the vertical direction. Accumulations of fine material on bedding planes produce zones of low permeability, so that many sediments have their greatest transmissivity in the direction parallel to bedding. The fracturing of relatively impermeable rocks usually results in their greatest transmissivity being parallel to fissure systems. Soils and rocks whose properties differ in this way, i.e. with direction, are described as being *anisotropic*. When the degree of anisotropy varies from place to place within any one aquifer the aquifer is further described as being *heterogeneous*. It is important that not only the small scale features such as porosity, bedding, and other forms of layering which can occur within a rock mass should be noted, but also the larger features such as joints, faults, and folds, which influence the character of regional flow; limestones can contain solution channels and caverns; lavas, lava tunnels; shallow water deposits, e.g. alluvium, varying thicknesses of differing materials. Similarly, variations in the thickness of aquifers affect the volume of storage for a given porosity, as does the extent of the confining cover of an aquifer, particularly if the confining boundaries permit local leakage to or from the aquifer.

Flow regime. Most aquifers have an interconnected system of pores, fissures, and other openings through which water will flow. This means that ground-water can travel by two rather different routes, either through the tortuous tunnels produced by any interconnected system of pores and other small openings, or through the larger and more regular channels produced by open fissures and joints. Fortunately it is usually possible to determine which of the two is dominant in an aquifer by simply inspecting its geology; for the majority of aquifers the flow is either essentially through pores or through fissures. When such a distinction is not easily apparent, e.g. in jointed but well cemented porous sediments, it is necessary to use some borehole test which will measure the flow across jointed and unjointed portions of the aquifer (Note 4). The formulae which are used to assess the velocity of flow and the magnitude of seepage forces associated with flow, use values for water levels yet the formulae themselves differ for each flow regime. Hence, water levels measured in any aquifer can be incorrectly applied if the flow regime which supports them is not known. Similarly, values for water levels, which include piezometric levels, and values for the water pressure in pores and fissures, can be wrongly predicted if the formulae describing flow through a permeable aquifer are applied to an essentially pervious aquifer.

Flow through aquifers: a general assessment   The careful measurement of water levels will define the fluid potentials of a ground-water system and reveal the basic directions of flow. These can be quite variable, particularly in the upper levels of a rock mass where valleys intersect the water-table and exert a dominant influence upon the drainage paths (Fig. 12.7). In such cases it appears that a ground-water system can contain many cells which are distributed about lines of discharge, normally streams. These cells increase in size with depth below the surface (Note 5).

Geology also influences the time dependency of flow, i.e. whether flow is either steady or unsteady. The flow in most natural ground-water systems is unsteady because it is usually the result of discharge from storage. Infiltration increases the saturation of the ground, raises water levels and results in increasing flows. Conversely, the rate at which water drains from storage decreases with time, and results in ever decreasing flows. Hence calculations for determining the discharge, head, and other factors associated with naturally induced flows should normally include functions for time.

Truly steady flow is usually only associated with the artificial conditions induced by engineering works, e.g. de-watering or recharge,

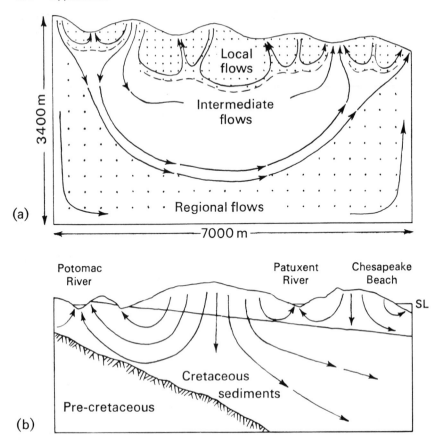

Fig. 12.7   (a) Theoretical distribution of flow (after Toth). (b) Actual example of flow from Maryland U.S.A. (after Back), and see Note 6. Cretaceous sediments are overlain by Tertiary sediments. *S.L.* = sea level.

and then only occurs when there are fixed boundaries of either recharge or discharge within the system which will ensure that there is no net change in storage with time, once the natural system has become adjusted to the new environment.

The procedures for analysing various ground-water conditions are described by Todd (1959), Walton (1970), Heath and Trainer (1968), Kruseman and de Ridder (1970) and others.

Flow through permeable aquifers   Permeable aquifers transmit most of their water through pores and similar openings. Poorly cemented sands and gravels are typical examples of the many geological formations which fall into this category; others include porous limestones, sandstones, conglomerates, fossil soils, and weathering profiles. The majority of permeable aquifers are sedimentary in origin, and though many igneous and metamorphic rocks are porous their pores are usually not continuous, as they are in sedimentary rocks, and their permeability is therefore low.

Flow through permeable aquifers is a diffuse process with the water travelling through the channels made by the pores, and no exact analytical description of it has yet been produced which can be used in practice. Many engineers still rely on Darcy's relationship which expresses the average velocity of flow through such a system in terms of permeability and head; $v = ki$ (Chapter 10). This relationship, known as Darcy's 'Law', was determined experimentally using sand as the permeable material and the engineer has to decide what confidence should be placed in conclusions derived from applying it to field situations. The equation will only apply when the inertial forces of the system, i.e. those produced by the velocity of flow (v) and the density of the fluid (γ) are much smaller than the resistive forces, i.e. those produced by the viscosity of the fluid (μ) and the diameter of channels carrying flow (D). In hydraulic studies these are commonly compared by means of the Reynolds number (Re) defined as:

$$Re = \gamma \frac{vD}{\mu}.$$

As the size of the pore spaces is never known, so applicability of the Reynolds number is restricted (Note 7).

It has been found that the great majority of naturally induced flows through permeable aquifers must have a Reynolds number that is less than one to be described by Darcy's equation. However, flows generated by artifically induced regimes such as those associated with de-watering projects and some field tests can be outside the limits of the equation's applicability. The difficulties involved in this problem are discussed by many authors, including Muskat (1946), Scheidegger (1960), and Mansur and Kaufman (1962).

The velocity with which water will flow through permeable aquifers is, naturally enough, a function of their permeability; this can range from less than 10^{-8} cm/s for materials finer than silty sands to more than 10^1 cm/s for gravels. A list of values is given in Table 12.1. From this it can be seen that porosity has little relationship to permeability, for as far as the velocity of flow is concerned the critical dimension in any pore is not the volume of the pore but the smallest dimension through which water must pass. This is controlled by the shape of the grains, their packing, their size and the degree of cementation between them (Note 8 and see p. 180).

The errors involved in taking porosity as a guide to permeability are well illustrated by comparing clays with sands. Clays, have small and platy shaped minerals which can support an open fabric of high

Table 12.1 Representative Values of Porosity and Permeability

	Permeability (cm/s)	Porosity (%)
Clean gravel	$10^{-2}-10^{0}$	45–40
Sandy gravel	$10^{0}-10^{-3}$	40–25
Fine sands and silts	$10^{-3}-10^{-5}$	45–50
Silty clays and clays	$10^{-5}-10^{-9}$	35–55

porosity, i.e. greater than 40%. But the pores are so small and the surface area of the fabric relatively large that capillary forces in the pores greatly retard any movement of water, and so give the clay a low permeability. Sands tend to have rounder grains which pack together more closely and result in porosities which rarely exceed 30%–40%, i.e. porosities which are normally lower than those of clays. But the grains and so the pores are larger than those of clays so their permeability is greater than that of clays.

The flow of water through any medium occurs when the frictional forces which are retarding flow are exceeded. They are overcome at the expense of the total head in the system, which decreases in the direction of flow. The energy represented by the drop in total head per unit length of flow path, i.e. the hydraulic gradient, is present in the form of a force which acts in the direction of flow on the walls of the pores. It is therefore called a *seepage force*, and its magnitude per unit volume of ground is equal to $i\gamma$ where i = drop in total head per unit length of flow path, γ = weight of unit volume of water.

Seepage forces have the dimension of unit weight, and their line of action at any point in the aquifer is in the direction of flow. Thus, they can significantly reduce the stability of slopes, cuttings, tunnels, and other free faces towards which water is flowing; it is therefore important to know both their magnitude and direction. This is reasonably easy in permeable materials, for although the individual water particles weave their way through a labyrinth of pores, the deviations from a macroscopically smooth path of flow are small. Hence the macroscopic direction of flow is a sufficiently accurate one to consider. This need not be the case in pervious aquifers, where much larger features, e.g. fissures, can impart a markedly erratic character to the paths of flow.

Pervious aquifers: general characters The transmission of water through a pervious aquifer occurs principally along joints and fissures. These are systems of separated planar and semi-planar surfaces which are normally more open than the voids of most permeable aquifers.

The frictional resistance to flow through such fissure systems can be much lower than in a network of pores, and allows the transmission of considerable volumes of water; some of the best of all aquifers are jointed and fissured basalts. Table 12.2 illustrates some differences which have been recorded between the permeability of rock samples

Table 12.2   Values for Pore Permeability (pk) and Joint Permeability (jk) (after Louis 1969)

	pk (cm/s)	jk (cm/s)	Width of Joint (mm)
Limestone	10^{-13}	10^{-4}	0.1
	10^{-9}	10^{1}	6.0
Sandstone	10^{-11}	$10^{-3}-10^{-2}$	0.4
Granite	10^{-10}	10^{1}	2.0

and the joint permeability of the formation from which they were taken. Typical pervious aquifers are well-jointed igneous rocks, well cemented but jointed sedimentary rocks, and jointed metamorphic rocks. Many soils can be pervious, e.g. fissured clays.

The characters responsible for retarding the flow of water through natural channels in rocks are the roughness and separation of the channel walls. Of these, the latter should be noted as it can be affected by load and can markedly alter the hydraulic characters of a rock mass. Theoretical and practical studies have shown that pores tend to be far stronger structures than fissures. Although their shape may change in response to stresses within the granular skeleton their volume will only change if either intergranular movement or solution of the mineral grains occurs, resulting in either consolidation or recrystallization. Fissures are far more susceptible to volume changes because they are usually kept open by discrete point contacts. The stresses resulting from the application of loads at higher levels in a rock mass can cause these contact points to fail; the joints then begin to close and the transmissivity of the rock decreases. This effect can have important consequences in the foundation design of heavy structures such as dams (Chapter 14).

Another character which is fairly typical of pervious aquifers is the decrease in joint 'porosity' (volume of joints per unit volume of rock mass) which occurs with depth; this is usually associated with a decrease in transmissivity with depth. One of the reasons for the decrease is that many joints found in rocks have been formed by the release of residual stresses, the release being made possible by the reduction in vertical pressure that results from the removal of rock by erosion at ground level (Chapter 2, p. 24). The opening of joints by stress relief has important implications in major excavations, especially in large underground works, for rocks which were relatively impervious prior to an excavation can later be made pervious by the release of stresses around it (p. 467). This effect can similarly influence the walls of deep bore-holes, and care should be taken with the interpretation and application of results obtained from packer tests conducted at depth (Chapter 9).

The separation of joint faces can therefore change with time and tends to make the condition of pervious rocks less stable than that of permeable rocks. Investigations of pervious formations can benefit considerably from a study of their structural setting, for many joint fabrics may be related to the thickness and deformation of geological units. For example, tension joints in the crestal parts of folds give a greater perviousness than in the same rocks on the flanks of the folds. Shallow works within 30 m (100 ft) of ground level are usually, though not always, free from the opening-up of joints on a scale that significantly affects the ground-water flow, and the number and position of joints remains reasonably constant with time. Their separation, however, can change with either erosion along them or compression across them. Deeper investigations should take account of the existing state of stress in the ground (Chapter 15).

Flow through pervious aquifers   For most practical purposes it is better to measure rather than to calculate the flow through a complicated system of natural conduits, and to obtain from the data some relationship which describes the flow in parameters that are measurable in the field.

One point to be considered is the difference between the size of the measurable parameters and the size of the rock element being studied. In permeable aquifers the pores and voids are so small that they can be taken to be an integral part of any volume of rock likely to be considered by a construction engineer. The permeability of laboratory specimens from a permeable isotropic aquifer will be the same as that measured in the field, and the aquifer can be considered as a continuum inasmuch as its structure (and hence its permeability) is constant, regardless of the volume of aquifer considered. This is not the case with pervious aquifers, because joints and fissures which transmit water are much more discrete structures that may or may not be an integral part of a particular volume of aquifer. Hence the permeability of a laboratory specimen containing one fissure, of given opening and roughness, will differ from that containing two similar fissures, which in turn will differ from that measured in the field where there are many fissures. In this respect the pervious aquifer can be considered as a discontinuous material, as its properties are a function of the size of the volume studied. However, there must be a volume beyond which, for practical purposes, the pervious aquifer can be treated as a continuum. The calculation of this volume is, as yet, not possible, but research by Maini (1971) suggests that, by the use of borehole packer tests, it should be feasible to determine an 'effective volume' of aquifer beyond which it can be considered as a continuum. The type of graph that will probably be used for this is shown in Fig. 12.8: see also Wittke 1973.

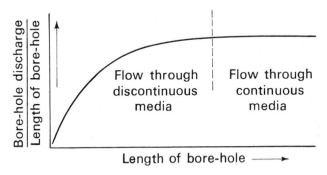

Fig. 12.8   A relationship which reflects the number of fissures encountered in a unit of ground (after Maini, 1971).

Once an element is sufficiently large to behave as a continuum, then flows through it can be described (and hence predicted) with reasonable accuracy, using Darcy's relationship $Q = KiA$, where the quantities Q, K, i, and A have the same meaning as that applied to permeable systems. Below this threshold, however, it is necessary to use other descriptions of flow which will take into account the volume of the conduits involved as distinct from the volume of the whole rock mass. The parameters that are important here are the opening, roughness, spacing, and continuity of fissures and the nature of their inter- sections. Further, because the size of the channels is significant with respect to the unit of aquifer, it is necessary to consider the velocity head as well as the elevation and pressure head. The kind of formulae which describe flow in these situations will probably be determined in practice from field experiments, and an outline of the approach required is given by Maini (1971) and Louis (1969). The mathematical treatment is beyond the scope of this text. It is necessary to observe the fissure fabric of a pervious aquifer in order that the correct tests and analysis can be applied to the unit volume being considered. This applies both to the determination of *in situ* joint permeability and also the distribution of fluid potentials in the region of flow, and hence the assessment of seepage forces and normal stresses that are generated by the presence of water within fissures.

Boundaries   Any geological surface can be considered as being either a permeable or an impermeable boundary. Permeable boundaries allow the passage of water through them, whilst impermeable boundaries do not.

Permeable boundaries. The nature of these boundaries, i.e. recharge or discharge, depends upon the relative head on either side of the interface.

Natural recharge boundaries are the infiltration surfaces of aquifers, e.g. the surface of an unconfined aquifer, the bed of an influent stream (one which is recharging water to the ground), or the surfaces which border a leaky aquifer (i.e. one having a higher head than its neighbours and discharging water to them). Natural discharge boundaries are the seepage surfaces of aquifers, e.g. the surfaces of a leaky aquifer, the bed of an effluent stream (i.e. one to which groundwater is flowing), or the area below a spring-line. Artificially created boundaries can be found associated with the cut-offs beneath dams, with active wells, and in tunnels, shafts, cuttings, and other excavations which extend below the water-table. No permeable boundary will be active if the fluid potentials on either side of it are equal. The ground-water potentials in an aquifer bordering a body of water such as a lake or a river may be equal to those associated with its level, and hence the permeable interface between the aquifer and the free water would be an inactive boundary. Abstraction of ground-water would lower the potentials in the aquifer and water would flow into it from the lake, or river; the boundary would then become one of recharge. Reversing the situation, i.e. recharging water to the aquifer, could raise the potentials in the aquifer so that they exceed those in the body of water; water would then drain from the aquifer and the boundary would become one of discharge. This reversal of boundary characters is important in aquifers bordering tidal waters (see Chapter 16).

The movement of water across a permeable boundary which separates aquifers of differing permeability is accompanied by deflection of flow lines, as shown in Fig. 12.9a. The amount of deflection depends upon the difference in permeability of the adjacent materials and can be calculated (Todd, 1959). It is therefore important to include such boundaries in flow nets, analogues, and other models which are constructed in order to study the movement of groundwater. Their location and recognition is a matter which should be borne in mind when a site investigation is proposed. Care should be taken when shallow-water deposits such as alluvium and glacial debris are examined, for these deposits accumulated in environments which allowed clays to be laid in direct contact with coarse gravels, which themselves could be interbedded with sands and silts (Fig. 11.12). Too often attention is only given to the impermeable boundaries.

Impermeable boundaries. No rock formation is truly impermeable,

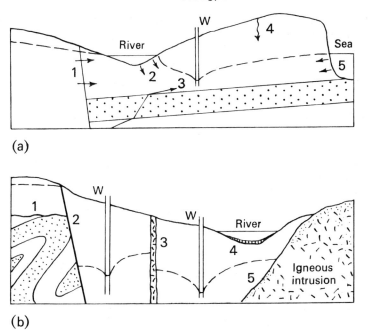

Fig. 12.9   Some hydrogeological boundaries. (*a*) Permeable, (*b*) impermeable.
(*a*) Permeable boundaries: *1*. permeable fault; *2*. influent river channel; *3*. leaky aquifer
boundary; *4*. infiltration surface; *5*. coastal seepage. W—discharging well.
(*b*) Impermeable boundaries: *1*. unconformity with impermeable strata; *2*. impermeable
fault; *3*. dyke; *4*. sealed base of river channel; *5*. igneous boundary; W—discharging well.

but the flow through some deposits is so very slow that for engineering
purposes they can be considered as being impermeable. Flow is
essentially along the boundaries of these materials rather than across
them. Typical examples of natural occurrences of this kind are
illustrated in Figs. 12.6, 12.7, and 12.9*b*. Some impermeable boundaries may behave as underground dams. Dyke intrusions are an
excellent example of this, and pond up water so that a large hydraulic
gradient can exist across them; the unexpected penetration of such a
boundary by tunnels and other underground works can be a serious
matter. Similarly, the de-watering of excavations which are in the
vicinity of relatively thin impermeable layers can create large differences in pressure on either side of a boundary; this may even cause the
ground to deform and the boundary to break, so exposing the excavation to the risk of sudden flooding.

Two other boundaries should be mentioned, which are not really
geological in the sense of those shown in Fig. 12.9, but are in-
separable from the geology; they are the ground level and the water-
table. The intersection of these two marks a line below which the

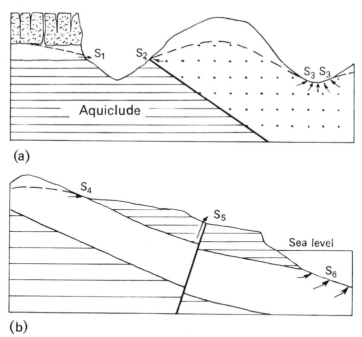

Fig. 12.10 Geological structures commonly associated with springs. S_1 = stratum spring between an aquifer and an aquiclude; S_2 = fault spring; S_3 = valley spring; S_4 = overflow spring; S_5 = artesian spring associated with a fault; S_6 = submarine springs.

discharge of ground-water in the form of springs, and related flows, takes place.

Springs As mentioned earlier, the hydrological cycle involves the water molecule in three fundamental journeys: its movement as vapour into and in the atmosphere; its precipitation back to the surface of the earth; and its subsequent journey, either overland or underground, to the sea. The underground journey begins with the recharge of water to the ground, i.e. infiltration, and ends with its discharge from the ground. The latter occurs at and below the spring-line of any geological unit as either a diffuse flow from many pores and fissures, or as a concentrated flow from pores and fissures in a restricted area. All the discharges are described as some kind of spring flow, but it is only those at which concentrated flows occur that are called *springs* (Note 9). Terminology in this respect is somewhat confusing and care should be taken when terms relating to spring flow are incorporated in contracts. The following terms all describe different aspects of what is essentially the natural discharge of water from the ground.

Spring: first appearance of a concentrated flow derived from ground-water.

Seepage: small discharge of water which only appears at the surface during periods of low evaporation.

Resurgence: re-appearance of an underground stream which has known upper reaches on the surface (these commonly occur in limestone country; see Chapter 2).

The location of springs (i.e. concentrated discharges) is controlled by the position of geological boundaries; Fig. 12.10 illustrates commonly occurring situations. It should be noted that the boundaries need not be impermeable; a difference in permeabilities is often sufficient to promote the ponding of water in the rock of higher permeability, and the formation of springs along the outcrop of the junction (Note 10).

A *spring-line* (p. 347) is the junction between topography and a water-table, and as such need not show the more concentrated issues of water which are described as springs. It is often simply the upper boundary of a zone of diffuse seepage. It is rarely an easy junction to find on the ground, and may sometimes be best located directly from the air or from air photographs. But it should be remembered that spring-lines will move whenever the water-table which supports them moves, and that they should not be thought of as hard and fast boundaries. Spring-lines tend to migrate up a topographical slope in response to periods of recharge accompanying wet weather, and recede back down the slope during the intervening periods of dry weather. The head waters of rivers fed by springs which migrate in this manner thus tend to move up and down the valley. Such streams have been called *bournes* (Note 11).

The intermittent flow of bournes clearly illustrates that the discharge of springs can vary during a season; even those which flow continuously usually vary in their discharge. Some measure of this variation can be obtained by the formula given by Meinzer (1923):

$$\text{Percentage variability} = \frac{(Q_{max} - Q_{min}) \times 100}{Q_{median}}$$

As with other variations (e.g. fluctuations of water level), it is essential to state the period over which the variation was measured. Measurement of spring discharge is often difficult because the boundaries of many springs are not easily defined. It is therefore normal practice to measure the flow of the stream fed by the spring, even though the flow has to be analysed in order to separate the volume of water which has come essentially from overland routes and that from the ground-water system; the latter is referred to as the ground-water component of river discharge.

Ground-water component of river discharge   River run-off is the result of many discharge processes. Consider a storm which precipitates rain at various intensities over a catchment. Those areas which receive rain at a slower rate than that of the infiltration capacity of the surface will act as recharge sites for the water-table. This will rise, if there is sufficient water for percolation, and increase the discharge from springs and seepages, which in turn will increase the flow to streams and rivers. Such a discharge involves ground-water, and the volume contributed by it to rivers is called the *ground-water component* of river flow. However, not all parts of a catchment need be undergoing infiltration; those which receive rain at a faster rate than their infiltration capacity will not transmit the water far below the surface. As a result the surface soon becomes waterlogged and impermeable, so that further rain has to leave as a surface or near-surface flow, which eventually finds its way into the rivers. The volume of river run-off derived from this form of drainage is called the *surface water component* of river flow. The river discharge from a storm which occurs over a catchment containing exposed aquifers can therefore be considered as the sum of the discharges from surface water and groundwater systems, as shown in Fig. 12.11 (Note 12).

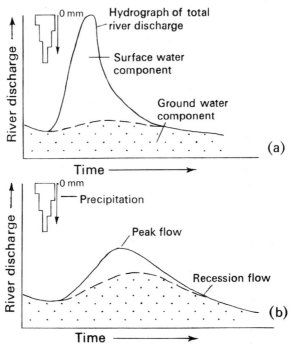

Fig. 12.11   Hydrographs of river flow. (*a*) From a relatively impermeable catchment; rapid runoff with a large component of surface water. (*b*) A permeable catchment: slower response and larger component of ground water.

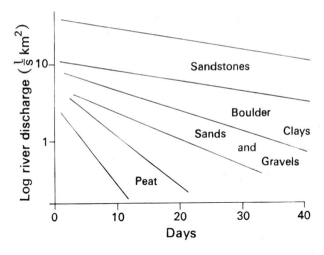

Fig. 12.12 Some recession curves related to geology (after Wright 1970). Discharge expressed in litres per second per kilometre².

The flow of rivers in dry weather must be substantially supported by the discharge from aquifers when they outcrop in the catchment, indeed it must reflect their degree of saturation and their specific yield (Note 13). Figure 12.12 illustrates the dry weather flows which have been identified from some aquifers in Scotland; it can be seen that certain aquifers sustain larger flows for longer periods than do others, possibly because of their size rather than their specific yield. Information about dry weather flows is useful to an engineer for although the flow records require careful interpretation they are, nevertheless, derived from the natural flow systems of the catchment. As such, their study can be a useful adjunct to any survey which attempts to assess the geological factors affecting the storage and movement of water.

NOTES

1. Unlike the water levels in piezometers (Note 3) the water levels in uncased holes are a function of all the water pressures encountered by the hole and not just those at their tip. This means that the water level in uncased holes need not always represent the true elevation of the water-table. Similarly, the water levels in piezometers may not represent the water-table. Examples of situations which occur in practice are shown in Fig. 12.13.
2. From a practical point of view it is often convenient to consider an aquifer as being any rock mass which either can or does transmit quantities of water which are significant to the problem in hand.

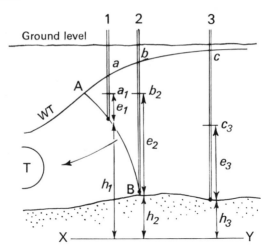

Fig. 12.13 Piezometric levels. e = pressure head; h = elevation head; x—y = datum from which head is measured. A drainage tunnel, T, in clay overlying gravel, depresses the water table, WT. A–B is line of equal potential, i.e. the total hydraulic head at all points along it is equal, velocity head being ignored. Piezometers 1 and 2 register this equipotential. The water levels encountered during drilling would be a and b respectively falling to a_1 and b_2 once the piezometers were installed. Piezometer 3 enters the gravel. Its water level falls from c to c_3, the piezometric level for the gravel, once it is installed. The pressure of water at a piezometer tip is = $e\gamma$ where γ = unit weight of water.

3.   With the development of soil mechanics it became necessary to measure the distribution of pore pressures within natural materials. One of the ways of obtaining this information is to carefully seal tubes, containing permeable tips at their lower end, into the ground so that the tips are located at the positions from which pore pressure measurements are required. The level to which water will rise in these tubes is called a *piezometric level*, and the installations *piezometers*. This terminology is applied to all such installations, which means that piezometric levels are being recorded from aquifers and aquicludes (clays, etc.) and from confined horizons and unconfined horizons. Hence the term 'piezo-metric level' has come to have two meanings, *viz.* (i) any water level measured by a piezometer, and (ii) the water level above the top of a confined aquifer, Fig. 12.13. Piezometric levels can equal but not exceed the maximum static water level for any aquifer, this being the level to which water will rise when the aquifer has reached its maximum saturation and contains no flowing water, as in Fig. 12.6*b*.

4.   The magnitude of flow from fissures can be obtained by comparing the field values for permeability (K_f) with those obtained from laboratory samples (K_l), because field permeability will be a function of both pores and fissures; whereas the size of laboratory samples usually means that their permeability will only be a function of the pores. Hence the relationship ($K_f - K_l)/K_f$ gives some indication of the contribution made by fissures to the field permeability.

5. Care should be taken to ensure that the water level information obtained for such an investigation is sufficiently accurate to describe the systems involved, because the scale of these systems varies with depth. Hence the depth and the spacing of observation holes should also vary so as to be compatible with the scale of the system to be studied and the geological variations that are present. This is important, particularly in limestone terrain where marked changes in the direction of flow can occur because of the presence of solution caverns and similar ducts.

6. Even greater distortions occur in the geometry of flow lines as they cross from material of one transmissivity to that of a differing transmissivity (see Boundaries). It is important to know of such geological variations when attempting to construct either models or flow nets of ground-water systems.

7. In order to obtain a Reynolds number for flow through a permeable aquifer it is necessary to use
   (i) values for apparent velocity, given by Q/A, as in the Darcy equation (p. 310),
   and (ii) values for the diameter of the *grains*; this can be difficult because most aquifers are composed of grains of differing size. Two methods for obtaining an average value for grain diameter, D, have been used with reasonable success:

   (1) $$D = \sqrt[3]{\frac{\sum n_s d_s^3}{\sum n_s}}$$

   where n_s = number of grains of diameter d_s.
   Flow systems which can be described using the Darcy equation are usually associated with Reynolds numbers of less than 1.0, which have been calculated from this value for D.
   (2) $D = d_{10}$, i.e. the effective diameter, as obtained from a standard grading curve. Reynolds numbers calculated with this value for D are usually below 10.0 for flow systems which can be described by the Darcy equation. Limiting values are usually greater than 3.0 and often occur in the range 3.0–10.0.

8. Formulae which attempt to predict the permeability of uncemented materials from grading analysis should be used with caution, for a grading analysis gives no information about packing.

9. Hot mineral springs associated with igneous activity either at the surface or at depth are described in Chapter 4.

10. Many local supplies of water are taken from springs. An example of a large flow in England is the Bedhampton Springs, Hampshire, where water from fissured Chalk is discharged near the coast. The flow varies from 18 to 32 m.g.d. according to the season, and a supply is taken by the Portsmouth Water Company. (See Ineson, J. and Downing, R. 1964. *Jour. I. Water Eng.*, 18).

11. *Bourne* is the old French word for a boundary and its first, albeit restricted, appearance in the English language occurred in the mid 1500's. In 1602 Shakespeare, in his play Hamlet, used the word figuratively to mean the limit or goal of a traveller. Hence the limiting springs of a stream whose head waters moved up and down its valley became appropriately called bournes. To use the word for the stream itself, as is now done, is strictly incorrect; a bourne is a spring, and it provides a bourne-flow, which is a stream. Some bournes only become active after prolonged periods of rainfall, and in England their issue (often in Chalk valleys) is usually associated with bad weather. Flow from such springs was thought to be a measure of the seriousness of the situation for the farmer, and was generally considered to be a bad omen; in farming communities it was often called a woebourne. Many English rivers, valleys, and valley towns have the word 'bourne' incorporated in their name, e.g. Bournemouth in Hampshire lies at the mouth of the River Bourne, Winterbourne is a village in Dorset and Hurstbourne another in Hampshire; there are many others.

12. To consider river run-off as being composed simply of surface run-off and subsurface run-off is to simplify what can be a complicated series of processes. No hydrologist would identify a particular volume of water as having travelled to a stream exclusively as either part of a subsurface or part of a surface drainage system. Water may run off upland areas and be infiltrated in lowland areas at the beginning of a storm, and continue in that way until the infiltration areas themselves begin to receive water at a faster rate than they can accept; run-off then occurs from these areas. In dividing the hydrograph into various units, as shown in Fig. 12.11, it is recognized that water arrives at a river by various routes, either above or below ground, and that some of these routes are faster than others. The quickest routes are generally those which do not involve water in a substantial journey through the groundwater system.

13. There can be no dry weather flow when the water-table is level with the base of a valley. Hence a decrease in dry weather flow, sometimes called 'base flow', can be correlated with a drop in water-table (see water-table movement, p. 348). From this it is possible to assess the specific yield of an aquifer, assuming that:

volume of dry weather flow for a given period
     = volume drained from aquifer storage,
     = drop in water level × effective porosity,
     = specific yield.

Reference to Fig. 12.7 indicates that care is needed when considering the volume of aquifer which contributes to the river flow of any catchment, for not all the rain falling in a catchment need enter the river; clearly, this will affect the calculated values for specific yield. A proper study of dry weather flow should be preceded by the preparation of vertical cross-sections through the drainage basin, as shown in Fig. 12.7.

REFERENCES

BACK, W. 1960. Origin of hydrochemical facies of groundwater in the Atlantic Coastal Plain. *21st Int. Geol. Congr. Pt. 1, Copenhagen.*

DAVIS, S. and WEIST, R. 1966. Hydrogeology. J. Wiley & Sons, New York.

DAVIS, G. *et al.* 1964. Land subsidence related to decline of artesian pressure. In *Engineering Geology Case Histories No. 4.* Pub. Geol. Soc. Amer.

DOMENCO, P. 1972. Concepts and models in groundwater hydrology. McGraw-Hill, Maidenhead.

HEATH, R. and TRAINER, R. 1968. Introduction to ground-water hydrology. J. Wiley & Sons, New York.

HORTON, R. 1933. Role of infiltration in the hydrologic cycle. *Trans. Amer. Geophys. Union,* **14.**

KRUSEMAN, G. and DE RIDDER, N. 1970. Analysis and evaluation of pumping test data. Bull. II. International Institute for Land Reclamation. The Netherlands.

LOUIS, C. 1969. A study of groundwater flow in jointed rock and its influence on the stability of rock masses. A translation in Imperial College Rock Mechanics Research Report No. 10, London SW7.

MAINI, Y. 1971. *In situ* hydraulic parameters in jointed rock—their measurement and interpretation. Ph.D. Thesis. Univ. London, Imperial College.

MANSUR, C. and KAUFMAN, R. 1962. Dewatering in Foundation Engineering. Ed. G. Leonards, McGraw-Hill, Maidenhead.

MEYBOOM, P. 1967. Hydrogeology; in Groundwater in Canada. *Geol. Surv. Canada. Econ. Geol. Rep.,* **24.**

MEINZER, O. 1923. Outline of groundwater hydrology with definitions. *U.S. Geol. Surv.* Water Supply Paper 494.

MUSKAT, M. 1946. Flow of homogeneous fluids through porous media. McGraw-Hill.

SCHEIDEGGER, A. 1960. The physics of flow through porous media. 2nd Ed. Univ. Toronto Press.

TODD, D. 1959. Groundwater hydrology. J. Wiley & Sons, New York.

TOTH, J. 1962. A theoretical analysis of groundwater flow in small drainage basins. *Proc. Hydrol. Symp. No. 3 (Groundwater) Nat. Res. Council of Canada.*

WALTON, W. 1970. Groundwater resource evaluation. McGraw-Hill, Maidenhead.

WARD, R. 1967. Principles of Hydrology. J. Wiley & Sons, New York.

WITTKE, W. 1973. Percolation through fissured rock. *Bull. Int. Assoc. Engng. Geologists. No. 7.*

13 Geology and the Movement of Slopes

Introduction   Slope movements can vary in origin and magnitude, and range from near-surface disturbances of weathered zones to deep-seated displacements of large rock masses. They occur when the strength of the slope is somewhere exceeded by the stresses within it. Thus movements which are restricted to surface layers reflect the presence of disturbing stresses which are controlled by surface or near-surface environments, e.g. precipitation and temperature, whereas movements which originate at depth indicate the presence of adverse stresses at depth. The movements can range from microscopic strains, e.g. creep (p. 51) to the very large displacements of catastrophic slides. Those more commonly responsible for engineering problems are illustrated in Fig. 13.1 and Plate 13; all are essentially a response to the gravity-produced shearing stresses which exist in every slope (Note 1).

This chapter describes some of the geological factors involved in slope movements; they can be grouped as follows, viz: 1. material properties, 2. geological structures, 3. ground-water, 4. seismic disturbances, 5. *in situ* stresses, 6. weathering, 7. palaeo-climates. Many of the phenomena described under these headings are related to each other as cause and effect, and in this respect the above divisions are rather artificial, however, they emphasize the various geological features which should be considered and provide a series of headings around which a geological field investigation can be organized. References relating more to the analysis of slope movement are given in a selected bibliography, p. 404–5.

1. **Material properties**   The properties which are most commonly sought for the analysis of slopes are those associated with the shear strength of the geological materials involved (Note 2). These materials will always contain discontinuities, the scale of which can range from the microscopic networks of mineral boundaries to the megascopic systems of joints and similar partings. Geologists would describe the smaller features as fabrics and the larger as structures, although there is no universally accepted definition of these terms. In this chapter the fabric of an element will be taken to include all the geological discontinuities which affect its strength, regardless of their size. Geological structure will be regarded as the orientation of fabric with respect to the

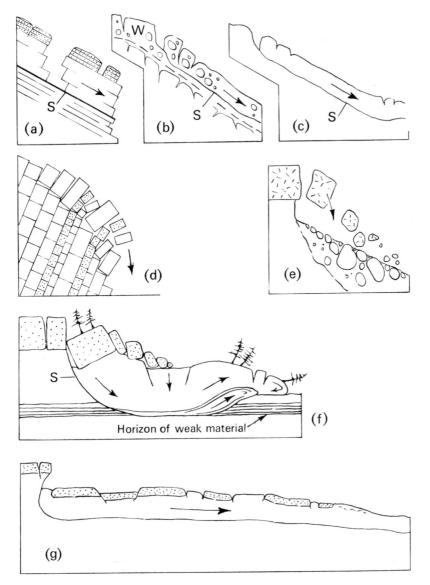

Fig. 13.1 Some common slope movements. S = sliding surface, W = weathered zone. (*a*), (*b*), and (*c*) = slides on planar surfaces in rock, scree or weathered mantle, clays and soft sediments; (*d*) toppling failure; (*e*) rock falls; (*f*) rotational failure; (*g*) flow slide.

force of gravity. In using these definitions it must be remembered that the fabrics themselves are merely networks of surfaces, and that the strength of the mosaic they produce comes from the strength of their surfaces and the material which they bound.

In any discontinuous material the volume of the element considered affects the properties being investigated as an increase in the size of the element is normally accompanied by an increase in the family of surfaces it contains. The smallest fabric that is likely to be of engineering significance is that produced by minerals. A good example of this is provided by the delicate fabrics of the Norwegian quick clays: these are most sensitive to strain and are capable of a rapid collapse which can quickly reduce the shear strength of the clay. Electron microscope photographs of their structure have been obtained by Karlsson and Pusch (1967).

Mineral fabrics can be well developed, schists, slates and laminated clays are typical examples; the shear strength of such fabrics can be markedly anisotropic. A large volume of material having a mineral fabric may also have the additional fabric resulting from joints, cleavage and other partings. These may exert a greater influence on the strength of the element than the mineralogical fabric alone, which would then become a second order fabric. Still larger elements can include fault surfaces and other shear zones, which could make joints and cleavage a second order fabric and relegate mineral orientation to a third order.

Various scales of geological fabric should be noted in the field, and referred to when the data from strength tests is used, as the careless choice of samples can result in testing fabrics which are not of primary importance to the strength of a whole mass as it occurs in a slope. Geological fabrics to note are mineral orientation, stratification, and fracture; the priority to be given to each should be considered on site. The dip, strike and continuity of each fabric should be noted. Often thin zones, only a centimetre or so wide, are crucial to slope stability, because they may be extensive features which can act as surfaces of sliding for large volumes of material.

The shear strength of surfaces within a fabric is a function of their *cohesion* and *friction*. Cohesion is zero when the surfaces are separated, as with open joints, but it can be considerable when such surfaces are jointed by some mineral infilling. Hence the contents of fractures should be examined and particular note made of any soft clay-like material which may be found in joints. Friction too can be a variable character and depends upon the mineralogy, structure and roughness of the surfaces and the load under which they either are, or have been, subjected. Care should therefore be exercised when alotting values for strength to the elements in a slope.

2. **Geological structures** In this chapter geological structures are considered to be the end products of geological processes which result in rock fabrics having a given orientation with respect to the force of gravity. The stability of any slope is a function of its direction

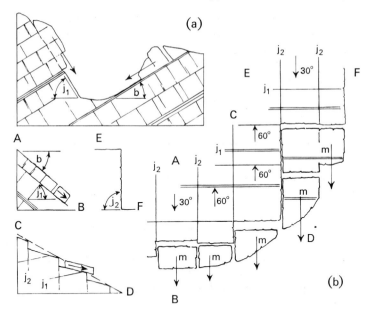

Fig. 13.2   Effect of structure on slope stability. (*a*) In section: *j* = jointing, *b* = bedding; (*b*) in plan: dip of bedding = 30°, of joint 1 = vertical, of joint 2 = 60°: *m* = blocks sliding into the cutting.

within this orientation. For example a cutting excavated in uniformly dipping strata, as shown in Fig. 13.2(*a*) has three sets of orthogonal surfaces, *viz*. bedding planes and joints (two sets j_1 and j_2 of which j_2 is vertical and lies in the plane of the paper). The cutting is parallel to the strike of the inclined surfaces and blocks can move down the dip of bedding and joint surfaces. As a result, the two slopes of the cutting have different angles of stability. If the slopes lay in any other direction they would have failed by the detachment of wedges, and different overall angles of stability would have resulted, as illustrated in Fig. 13.2(*b*); examples of this can be seen in the Cheddar Gorge, Somerset (Note 4). Terzaghi (1962) has commented on similar situations; and Henkel (1961) describes a typical wedge failure that occurred on the east bank of the R. Avon at Bristol. Here orthogonal sets of joints in the limestone are normal to the bedding, and the strike of one set lies at 10° to the direction of the valley, permitting the oblique movement of limestone blocks towards the river.

Another example of the effect of slope orientation on stability is provided by the folded Carboniferous rocks of the North Devon–North Cornwall coast between Bideford and Bude. Here sandstones and shales have been folded with an east–west trend. Two main cliff directions exist, east–west and north–south. The stable slopes on the east–west cliffs are commonly inclined at 45° or less to the horizontal, as their stability is largely controlled by the continuous surfaces of the seaward dipping bedding planes on the limbs of the folds. By contrast, stable slopes in the same rocks exposed on north–south cliffs are often standing at angles of 80–90°, their stability being largely controlled by near-vertical joints which trend at high angles to the strike of the folds.

Folding can have an important effect on the shear strength of bedded rocks. As noted in Chapter 7, certain folds develop by the shearing movement of a layer (or stratum) over the one below it (Fig. 7.9). Morgenstern and Tchalenko (1967) have demonstrated that shearing a soft material such as a clay produces a series of shear surfaces which eventually become oriented parallel to the direction of shear, and that sliding then occurs along well-developed slickensided surfaces. The shear strength of the material has then been reduced to its residual value. The process of folding can generate a series of low shear-strength surfaces. Many of them were encountered in the folded Siwalik Series in India (Henkel, 1966), and particular zones caused considerable problems in the stability of excavations at the Mangla Dam site, W. Pakistan. Similar features can be found in other areas of folded rocks, and their general character has been described by Skempton (1966). The deformation which occurs within fault zones is similar and often identical in end result.

Zones of reduced shear strength have also been observed in Coal Measure rocks (Stimpson and Walton, 1970). They are widespread, and represent an easily overlooked hazard in the stability of open pits and other slopes in coal-bearing sequences.

Structure is largely responsible for those situations which permit the simple detachment of rock in the form of *rockfalls*. These sometimes involve large quantities of material and pose serious engineering problems (Bjerrum, 1968). Various factors contribute to cause rockfalls, e.g. frost shattering, chemical decomposition, temperature variations, water pressures, snow cover, the wedging effects of the roots of plants. Rockfalls of a special type can occur on a large scale in toppling failure (Fig. 13.1*d*). This occurs when the structure, i.e. orientation of blocks, and fabric, i.e. their size, combine to produce a rock mass which contains units whose centre of gravity is beyond their base. Fig. 13.3 illustrates the situation. Toppling is a mechanism which can permit considerable volumes of rock to move down slope (Ashby 1971; de Freitas and Watters 1973).

3. **Ground-water**   In many ways it is incorrect to separate the water contained in pores and fissures from the 'material properties' of a geological mass as its presence has a marked effect upon the strength of the mass. It is treated separately in this chapter to emphasize that it is often a medium whose presence or absence can be

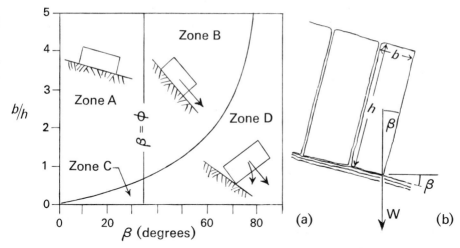

Fig. 13.3   General fields of stability (after Ashby 1971). ϕ = angle of friction; taken as 35° in this figure, β = dip of basal plane. (*a*) Zone *A*: b/h > tan β; $\beta < \phi$: a stable zone. Zone *B*: b/h > tan β; $\beta > \phi$: sliding only. Zone *C*: b/h < tan β; toppling only. Zone *D*: b/h < tan β; $\beta > \phi$: sliding and toppling can occur. (*b*) Relevant dimensions for analysis of toppling.

controlled. The ground-water regime is usually the only natural parameter which can be economically changed on a large scale in order to increase the stability of slopes; see Plate 11a.

Water can reduce the strength of geological materials in five fundamental ways:

(i) By chemically changing the mineral constituents (see examples already given) or simply dissolving them, as in limestones.

(ii) By filling the pores and thus reducing the capillary tension which acts as a force binding mineral grains together.

(iii) By increasing the bulk density of the material, so changing the stresses within the mass. This can be most important in masses containing impermeable barriers which behave as underground dams.

(iv) By producing pore pressures which tend to force the units of the solid skeleton apart, an effect which resembles the application of an all-round, uniform tensile stress upon any element.

(v) By producing seepage forces (see p. 356) which, depending on their orientation, may or may not increase stability.

Slope movements resulting from the presence of water can rarely be attributed to any one of these phenomenon as it is usually the combination of two or more which results in movement and eventual failure.

Three types of movement involving water will be considered: (a) that resulting from excessive pore pressures; (b) that resulting from the adverse orientation of seepage forces; and (c) that resulting in the saturation which promotes earth-flows and similar movements which grade into forms of mass transport.

(a) *Excessive pore pressures.* Water below the water-table exerts a hydrostatic pressure on the solid skeleton of the material; this has the same effect as an equivalent all-round tensile stress. Hence the normal stress (σ_n) at the grain contacts in a granular material or point contacts in joints and fissures, is reduced by an amount equal to the pressure of the water around the contacts (u). The effective stress at the contacts is then $(\sigma_n - u)$. Hence an increase in pore pressure in purely frictional material, where the resistance to shear stress is $(\sigma_n - u) \tan \phi'$ (see Note 2), results in a decrease in shear strength. Similarly, an increase in pore pressure in cohesive soils (where drained shear strength is a function of both cohesion and friction) to the extent that $(\sigma_n - u) = 0$, can make their strength entirely dependent on their cohesive strength. The effect of locally high pore pressures on progressive failure should be considered.

Holm (1961) describes a drainage system that was used to increase the stability of a slope on the west side of Oslofjord, Norway. This was a 20° natural slope some 10 m high and 300 m long, cut into horizontal layers of sand and gravel which overlay 10 to 18 m of Norwegian quick clay. This rested on a blanket of dense sands and gravels with bedrock below. Water within the bedrock supported a piezometric surface 5 m above ground level at the shoreline and water leaked steadily from the lower half of the slope. A retaining wall and other structures were to be built at the top of the slope, and it was decided to increase its stability by decreasing the pore pressures in the basal layers of sand and gravel. A system of vertical sand drains was installed to act as relief wells, and pore pressures fell considerably during their construction (Note 5). A further example of the influence of ground-water is given on p. 396 in connection with the slides at Folkestone Warren (Plate 12b).

Water pressures acting over joint faces and other discontinuities have a markedly directional character. Figure 13.4 illustrates one description of the problem (Muller, 1964a). A set of inclined joints which contain water is oriented either parallel to or at an acute angle to the slope. The unbalanced forces (Δh) acting on each block are normal to the joint surfaces, and thus add to the shear stress resulting from the self-weight of the material. Hence the stability of the rock mass ABC is a function of the normal and shear stresses on the surface BC, which come from the self-weight of the partly saturated block ABC, the uplift water pressure on the surface BC, and the unbalanced water thrust on surface AB. The accurate measurement of total head and mapping of discontinuities are both needed in the investigation of

Plate 11

(a) Excavation in dewatered beach gravel. (*Photograph by Soil Mechanics Ltd.*)

(b) Tunnelling in chemically grouted gravel. Note the stability of the unsupported face in front of the heading. (*Photograph by Soil Mechanics Ltd.*)

Plate 12

(a) Adit in chalk for water collection. (*Photograph by George Stow & Co. Ltd.*)

(b) Folkestone Warren landslip looking east. (*Photograph by Air Ministry; Crown Copyright*)

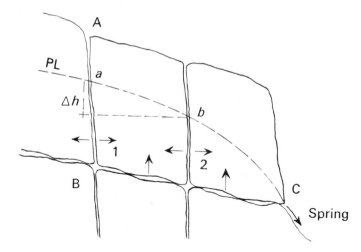

Fig. 13.4   Some water pressures in slopes (after Muller, 1964). *PL* = piezometric level of water in joint *BC* which rises as a water level to points *a* and *b* in joints 1 and 2. The thrust on joint 1 is greater than that opposing it on joint 2 by an amount = $\gamma w \Delta h$, where γw = unit weight of water.

such slopes. Lane (1969) describes an example where both these features were important. A power house was built in 1910 at the base of a 76 m gorge cut into dolomite. Near-vertical joints and fissures which were parallel to the valley sides permitted the drainage of water from a canal at the top of the gorge. The water discharged from drainage holes drilled for the purpose in the back of the power house. The system worked successfully for 45 years, but in 1955 it was decided to stop the seepage into the power house and the holes were grouted. The cliff collapsed shortly afterwards.

(b) *Seepage forces.* These forces can be sufficient to dislodge mineral grains from poorly cemented mineral fabrics, such as loose sands and silts, and wash them out of the ground. The process is called internal erosion, and it reduces the support offered by the layers to overlying materials, which, as a result, commonly collapse as some sort of slide. An example is described by Ward (1948) from the land-slides at Castle Hill, Newhaven, Sussex, Fig. 13.5. Surface water draining from both the topsoil and the road was seeping into the fine yellow sand lying between the clays. The sand became saturated, and groundwater moving through it was washing away its finer fractions at the seaward face, some 9 feet below the level of the drain; failure of the overlying strata followed. Remedial works included sealing off the drain and protecting the seaward exposure of the sand with a suitably graded filter (Note 6).

Internal erosion is not a feature seen in well-cemented materials. Here the seepage forces increase the shearing stresses within the

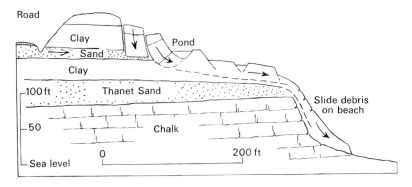

Fig. 13.5   General section across the Castle Hill site, Newhaven.

mass, sometimes to the extent of producing instability. Drainage can control these forces in two ways: (1) by re-orienting the flow lines so that the pressures act in a direction which increases the stability of the mass, and (2) by simple de-watering. Sharp (1972) has commented on the usefulness of drainage galleries in improving the stability of slopes in large excavations (Plate 12a).

(c) *Flow-type disturbances.* These are movements that usually result from the presence of water. They are different in structure from the rotation and slab movements seen in many slopes and have velocities and displacements that are similar to those of a viscous fluid. Many flows in their wetter state become mobile, saturated streams of debris (60–70 per cent solids by weight), which have a considerable erosive potential. The gulleys they cut often become sites for further slips and flows.

Flow-type disturbances have been variously classified as rock-flows, earth-flows, mud-flows, debris flows, and rock avalanches. The majority contain mixtures of poorly sorted debris, often fragments from the weathered mantle, set in a soft clayey matrix. They are commonly found in arid and semi-arid regions where intermittent storms suddenly saturate the ground; they are not, however, restricted to such climates and have been recorded in all latitudes. They are often associated with either unusually high rainfall or sudden thawing of frozen ground. The essential cause of movement is the presence of high pore pressures which initiate failure within and subsequent mobilization of an originally stable mass.

Flow movements which occurred in the Dutch province of Zeeland have been described by Koppejan *et al.* (1948). Tidal streams there have eroded shallow, but steep, slopes in sands which are mainly uniform in grade and below their critical density (Note 7). The result-ing drainage of groundwater stored in these sands is sometimes sufficient, between half and low tide, to initiate movement in banks which have been over-steepened by erosion. Such movements leave a steep unsupported scarp which fails a few minutes later and more sand flows into the stream. This continues, with one flow after another so that the failures progress inland, sometimes at a rate of 50 m per hour. Numerous flow-type movements occur in *sensitive clays*. These suffer a severe reduction in shear strength when re-moulded at a constant water content (see p. 186), and so their strength is reduced when they are re-moulded by deformations and displacements that may occur in slopes. Re-moulding a sensitive clay completely destroys its original mineral fabric and if pore pressures are high it can transform the clay into a thick slurry which flows like a liquid.

The Leda Clay of the St. Lawrence Lowlands in eastern Canada is a deposit which has an open flocculated structure that breaks down under strain to a liquid consistency (Note 3). Crawford and Eden (1967) describe two movements in natural slopes: one, an ancient flow at Green Creek, in which 30 acres of upland ground flowed over some 70 acres of lowland to a depth of 15 to 20 feet. The other, at Breckenridge, occurred during a heavy rainstorm in April 1963, when 25 000 m^3 of the clay were involved, and retrogressive slips occurred intermittently for 3 days following the main flow.

Because of the constant association of flow-movements with saturation it is often assumed that they can only occur in the presence of water. This is not so, for some very large flow-movements have been recorded for materials which were essentially dry at the time of failure. It is thought that the pore pressures which initiated these failures were air pressures, not water pressures; one such slide, in loess at Kansu, is described in the following section, in connection with seismic events.

4. **Seismic disturbances**   The release of energy from earthquakes results in the transmission of seismic waves through the ground (p. 7). These waves locally produce rapid acceleration and generate forces which dynamically load, and often weaken, the ground itself. Dynamic loading commonly affects slopes in two ways:

(1) by augmenting the shear stresses which already exist in the slopes, and

(2) by rapidly decreasing the void ratio of the material of the slope, leading to excess pore pressures.

The first effect can be thought of as providing an external stimulus which increases shear stress, and the second as providing an internal adjustment which decreases shear strength. The end result is equivalent to that which would come from a rapid reduction in the friction of the slope materials.

The geological factors that are significant in any analysis of these conditions are: (i) the magnitude of the accelerations at the slope;

(ii) their duration; (iii) the dynamic strength of the materials affected; and (iv) the dimensions of the slope. (The same points should be considered when dealing with slope stability problems that arise from blasting). Computers have enabled advances to be made in assessing the effects of the above variables (Seed, 1967).

One well documented case-history describes the landside at Turnagain Heights, on the shore-line of the Cook Inlet at Anchorage; this followed the 1964 earthquake. The nearly flat plain, of which the Heights form the seaward edge, is composed of 5 to 20 feet of outwash sands and gravels overlying 100 to 150 feet of the marine Bootlegger Cove clay. The latter is a mixture of stiff clay and a soft sensitive clay which contains seams of silt and sand; the seams range from a few inches in thickness to 30 feet in some lenses. The clay is exposed along the sides of the inlet in 70 feet high bluffs and also forms the floor of the inlet, where it is covered by a thin layer of estuarine silt. The coastal bluffs had withstood earlier earthquakes up to 7.5 intensity (revised Richter Scale). The earthquake of March 27, 1964, was of intensity 8.5 and lasted at least 4 minutes, an unusually long time. The slopes at Turnagain Heights withstood the movement for the first 1.5 minutes and then collapsed and carried 130 acres of ground towards the sea. The seaward edge of the cliffs, in places, moved as much as 2000 feet (600 m) into the bay, and landsliding continued for some time after the 'quake. Subsequent tests made on the sediments revealed that the sand lenses within the clay could liquify rapidly within 45 seconds when loaded by vibrations similar to those arriving during the earthquake; the clay itself failed after 1.5 minutes. It is therefore thought that the sand lenses liquified soon after the earth-quake began, and their loss of strength promoted the eventual failure of the clay (Seed, 1967, b). Fig. 13.6 shows the possible development of successive slides. The sequence of events is thought to be as follows: (i) failure on inland sand lenses within the clay after about the first minute of the 'quake; (ii) failure near the coast by conventional rotations which moved clay onto the estuarine silt, where it further slid into the sea leaving failure surfaces exposed. These continued to move until they merged into the inland zones of moving sand; (iii) change from rotational movement to translation, in a zone that had been weakened by the flowing sand lenses, with failure resulting in the development of ridges and troughs; re-moulding of the sensitive clay during sliding reduced its shear strength and aided the lateral move-ment. Much re-moulded clay in and near the slide zone continued to move after ground motions had ceased.

Turnagain Heights was associated with liquid pore pressures. Air pore pressures, when rapidly increased, can have a similar effect as in the great landslides in Kansu Province, China, in 1922. Thick deposits

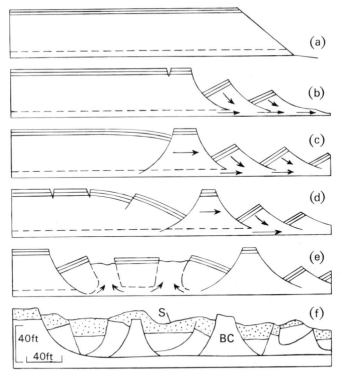

Fig. 13.6 Failure of Turnagain Heights cliffs, Alaska. (*a*) to (*e*), as developed in model tests; (*f*) final situation as observed in the field (after Seed). S = sands and gravel; BC = Bootlegger Clay.

of loess lay on bedrock, and gentle valleys, much terraced by farming, crossed the area. On the evening of December 16 an earthquake began in which ground movements to the north-east were violently jerking back to the south-west, the main movement lasting 0.5 minute. The sides of valleys collapsed and flowed downhill, burying many people, and an area 170 miles by 150 was affected. The loess turned into a powder which apparently flowed like water. Close and McCormick (1922) record that 'the earth that came down had the appearance of having shaken loose clod from clod and grain from grain, and then cascaded like water, forming vortices, swirls, and all the convolutions into which a torrent might shape itself'. It appears that the earthquake initiated a series of enormous dry flow movements, involving a large volume of loess and displacements of up to 1 mile.

5. **In situ stresses**   The presence of stresses in the outer part of the earth's crust has been described in Chapter 7. Joints produced by their relief have been mentioned in Chapter 2, and their presence in a large slide at Vajont is described on p. 387. The occurrence of *in situ* stresses underground is further discussed in Chapter 15.

In general, the *in situ* stresses likely to be encountered, either underground or at the surface, at magnitudes which are significant in engineering works, come from one of three sources: gravitational loading, tectonic activity, and weathering. They can be grouped under two headings, (i) developing stresses, (ii) residual stresses (Note 8).

Developing stresses are those which are generated at the present time, e.g. stresses in materials which are loaded either by natural overburden or by some other loading, stresses that come from tectonic loading, stresses that arise from a change in moisture content that causes materials to either shrink or swell. These stresses are sustained by the resistance of the rock fabric to internal movements.

Residual stresses are those embodied in the materials themselves, remaining there after the external stimulus which generated them has been removed. The abnormally high horizontal stress found in over-consolidated clays is an example (p. 185); the part this plays in initiating landslides has been examined by Bjerrum (1967). The magnitude of such residuals decreases with time, and most of those found in slopes are relics of stress which has been incompletely released by unloading when the slopes were formed. Their magnitude is partly a function of the strength of the geological material, stronger rocks being capable of holding more, and relaxing less, than weaker rocks and thus retaining more energy as residual stress.

The majority of stresses found in most slopes are therefore the sum of the gravitational stress from the self-weight of the material and the residual stresses left over from other phenomena. The geological features to be noted in a survey are therefore the magnitude of the *in situ* stresses, their directions, and their susceptibility to change.

Field (1966) records that lateral movements occurred in excavations at Niagara Falls. A trench 6.1 m (20 feet) wide and 55 m (180 feet) deep was excavated in dolomite that overlay soft grey shale, this formed the lower 6 m (20 feet) of the trench. The walls near the bottom of the trench moved into it and bent the supporting cross-beams. A similar pit on the Canadian side of the Falls experienced 10 cm (4 inches) of lateral movement, together with some upward movement of the floor.

The instability of excavation floors (discussed further in Chapter 15) can rarely, in practice, be separated from movements at the base of the excavation walls, because every removal of material involves the reduction of overburden pressure. As a result, upward and outward movement may take place at the base of a slope. Stress relief of this kind is the basis of the explanation of the valley bulging observed in the Northamptonshire ironstone field (Hollingworth *et al.*, 1944). The area consists of gently dipping Jurassic limestones and ferruginous sandstones overlying clay. The clay is exposed in the floors of some

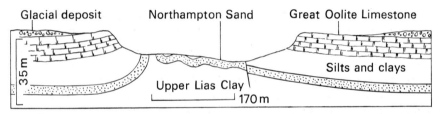

Fig. 13.7 Section across the bulged valley of Slipton in Northamptonshire. (After Hollingworth, 1944.)

valleys and is often found to be domed up there, so that it dips locally towards the valley sides. The general situation is shown in Fig. 13.7. The authors state that '. . . it appears that the clay has reacted . . . to differential unloading resulting from the erosion of overlying rocks. As the down-cutting of the streams proceeded, the excess load on either side of the valley could cause a squeezing out of the clay towards the area of minimum load, with consequent forcing up of the rocks in the valley bottom.' Valley bulging has since been recorded at many other localities.

The 'squeezing out' described by Hollingworth might have been a form of *creep*, a phenomena which is important to the stability of some slopes. Although variously defined, creep in slopes can be considered as the slow process which causes largely irreversible deformations to occur at stresses which are below the maximum shear stress. A well documented record of progressive failure that probably involved creep exists for a slope at Kensal Green in north London (Skempton, 1964). The slope and its retaining wall, in brown fissured London Clay, were completed in 1912. Movements of the wall were recorded from 1929, and were at first small and at a fairly constant rate; they gradually increased and eventually culminated in failure in 1941. The slope had taken 29 years to 'fail' and illustrates that long-term strength can be much lower than instantaneous strength, a point to remember when designing on values obtained from 'quick' tests. Creep tests, performed by Price (1966) on a variety of rocks, have shown that the long-term strength of some geological materials can be as low as 20 per cent of their instantaneous strength. Creep is also known to occur at deep levels in large rock slopes and to be responsible for the movement of great volumes of material. Zischinsky (1966) observed such deformations on the high slopes of the Austrian Tyrol, and Beaujoint and Martin (1966) have noted similar movements in the French Alps. The phenomenon is also thought to have been associated with the catastrophic failures at Turtle Mountain and Vajont (see p. 387).

The creep in near-surface movements found in the zone of weathering is shown by the downslope movement of the soil mantle, weathered outcrops, and surface rocks. The former is most active in regions having wide seasonal variations in climate, and is linked to the freezing and thawing, and wetting and drying, of the soil. Movement increases with an increase in the angle of the slope, or the concentration of colloids in the soil; it is greatest at the surface and diminishes with depth. Lucid accounts of this process have been given by many authors, e.g. Chandler (1971,a) and Barr (1970). The movement of weathered outcrops (Fig. 2.14) is responsible for aligning the fabric so that it is weakest in the downslope direction. This has important implications for the stability of tips placed on slopes (Chapter 15). The downslope movement of surface rocks occurs mainly where bedding in jointed rocks dips downhill. Creep *on* bedding surfaces, particularly if they contain argillaceous material, permits the slow downward migration of slabs from higher levels. An example of the gradually accelerating creep of a large rock slab in the Goldau area, Switzerland, is described by Terzaghi (1950). Creep is one way in which the strength of geological material can decrease with time; weathering is another.

6. **Weathering**   The rate at which chemical changes can occur is variable and can range from a few days to many years. This means that both short term and long term stability of slopes can be affected. Materials most readily involved in these changes are usually soft sediments (the soils of Soil Mechanics, e.g. clays, marls, etc.), and soft fillings found in rock fractures, such as the gouge in some fault zones. Hard rocks, and those which are composed of chemically stable minerals such as quartz, are less susceptible to change. However, as pointed out by Struillon (1966), certain silicates such as feldspars can alter quite readily.

Chandler (1969) records the overall effects of weathering on the mechanical properties of a marl. The liquid limit and natural moisture content of the marl increased with increasing weathering; whereas the bulk density, permeability, c', ϕ', and ϕ' residual (Note 2), all decreased. The rate at which such a deterioration would occur is as much a function of the climate as it is of the original mineral content.

Specific examples of slope failure resulting from particular chemical processes have been recorded; one is described by Drouhin *et al.* (1948) from Algeria, where calcareous rocks overlie a glauconitic marl. The calcium from ground-water seeping from the calcareous strata was taken, by base exchange, into the structure of the glauconite which in turn released potassium. The loss of calcium increased the pH of the water, which caused the marl to deflocculate to a colloidal gel (Note 3). The water content of the marl increased and its mechanical properties changed from those of a solid to those of a viscous liquid. Failure occurs if this process continues until the weight of rocks overlying the marl can no longer be sustained, and blocks of the overlying limestone then move down the slope. It is interesting to note that glauconite has been found in the lower levels of other large landslide systems, e.g. at Dunedin, New Zealand (Benson, 1946), and at Folkestone Warren, England (see p. 395 and Plate 12b).

Another case history involving exchangeable calcium has been described by Matsuo (1957). A railway cutting at Kashio, Japan, failed after being stable for 10 years. Sliding occurred in a series of clayey sands, clays, and gravels. Ground-water issuing from the toe of the slide, at rates between 1.3 and 50 cm^3/s, was found to have a far greater free carbonate content than the rainwater of the area (2.44 mg/1 and 0.039 mg/1 respectively), and it was evident that calcium was being removed from the ground. Laboratory tests showed that the removal of calcium had decreased the strength of the slope and promoted failure. One of the suggested remedial measures was to recharge calcium into the ground in the form of an aqueous solution of calcium salts.

Chemical changes which can occur rapidly in geological materials are the expansion of anhydrite on hydration, and the decay of iron sulphite (marcasite or pyrite) or oxidation (p. 126). Both changes can occur over periods of weeks and can drastically reduce the stability of slopes. The addition of water may cause swelling and compressive stress; and its subtraction, shrinkage and tensile stress. The zones most susceptible to such changes normally lie between ground level and the capillary fringe (p. 345). Coefficients of swelling and shrinkage are used for assessing the behaviour of geological materials; some are described by Kowalski (1970). Both coefficients use a ratio which divides the change in volume on wetting or drying by the original volume, either dry or wet. As would be expected, the greatest shrinking or swelling in sediments generally occurs normal to the bedding. Murayma (1966) described the swelling properties of an over-consolidated clay and a partly decomposed shale; both showed swelling strains, normal to bedding, that were approximately twice those in a direction parallel to bedding. He also noted that swelling in a given direction is not always uniform. This indicates that swelling not only creates additional stresses but can also irregularly mobilize the shear strength of the expanding fabric. The extent to which such physical changes affect slope movement is difficult to assess, but it appears that they are more common in near-surface movements than at depth. Control of swelling and shrinking phenomena may be necessary in the maintenance of permanent slope surfaces.

The periodic expansion of ground subject to freezing and thawing is largely restricted at the present day to about 20 per cent of the total land area, in circum-polar regions (p. 27). Periglacial phenomena, including freezing and thawing, were far more widespread during the Pleistocene period and produced many changes which still affect the stability of sloping ground and an appreciation of periglacial conditions has helped the understanding of some present day slope movements which have resulted from the effects of past climatic conditions.

7. **Palaeo-climates** A number of engineering problems have arisen, concerning the stability of excavated slopes that were cut into apparently landslide-free materials on the sides of gently dipping valleys. Trial pits dug to investigate the ground revealed the presence of ancient mud-flows, like shallow landslides whose surfaces of movement were roughly parallel to the ground surface. Soil testing showed that the residual angle of shearing resistance, ϕ_r', for these surfaces was usually greater than the present angle of slope of the topography. Table 13.1 shows examples from over-consolidated, fissured, Cretaceous and Eocene clays in southern England, given by Weeks (1969):

Table 13.1   Strength Values for Ancient Slip Surfaces (Compiled from Data by Weeks 1969)

Location	Approx. Angle of Topographic Slope (Degrees)	ϕ_r' (Degrees)	Depth of Surface below Existing Ground Level (Feet)
Sevenoaks by-pass	7	15	13 (1)
Sevenoaks by-pass	4	15.6	6
Tonbridge by-pass	4	16	6 and 10 (2)
Tonbridge (Quarry Hill)	7	12.4	$5\frac{1}{2}$ and 15 (3)
Ditton by-pass	3	12.7	4 to $7\frac{1}{2}$
Boughton by-pass	5	14 (4)	$4\frac{1}{2}$
M40, Tetsworth	$3\frac{1}{2}$	14	$5\frac{1}{2}$

(1) Dated (by ^{14}C) at approximately 10 200 B.C. Trial pits revealed that in some localities a series of mud-flows had moved one over the other, the oldest being at the base. Some flows were associated with soil horizons which were later buried by succeeding flows. The soil horizons contain carbonaceous material from which absolute dates can be obtained.
(2) Results from the 6 ft surface.
(3) Results from the $5\frac{1}{2}$ ft surface.
(4) Results from a similar surface at near-by location.

Now the work of Hutchinson and others (1968–9, and see Weeks, 1969) has confirmed that the ultimate angle of slope against landsliding in fissured clay, as determined in the field, is approximately $\phi_r'/2$, but as can be seen from Table 13.1, some slopes have much smaller topographical angles than this, yet still contain surfaces on which sliding has occurred. One explanation for this is that pore-water pressures have at some time reduced the normal load so that sliding could take place. As such conditions do not now exist they must have occurred in the past, and the most probable period when sufficiently high pore pressures could have been present was during periglacial conditions in the Pleistocene.

The Quarry Hill topographical slope (7°) is a little greater than $\phi'_r/2$. At this locality a bed of sand is believed to be underdraining a clay lying above it, and so preventing unstabilizing pore pressures from developing. However, it is thought that the sand may have been sealed off by permafrost during periglacial conditions, thus allowing higher pore pressures to develop in the surface layers in times of thaw. Similar features have been found at other localities; for example, a study of slips at an escarpment in Northamptonshire was made by Chandler (1971,b), who concluded that some of the slips required wetter, colder climates than exist now to encourage a rise in ground-water levels. He suggests that active degradation of the slopes occurred in the last-Glacial period.

Present day examples of instability are easy to identify, but much movement that occurred under past periglacial conditions is not immediately apparent at ground level and should be carefully sought in a site investigation (Note 9). The likelihood of such structures being present in an area can be assessed from a consideration of the local Pleistocene and Recent deposits.

Climate still affects many slopes especially in cold regions. Bjerrum (1968) describes the occurrence of rock-falls in Norway, Fig. 13.8(a), and shows, Fig. 13.8(b), how the rock-falls are related to spring thaw in April and maximum precipitation in October. Most of the major slides occur in April, when the near surface joints still contain ice, so hindering the drainage of slopes that are being progressively saturated by melt waters. During very wet years, unstable slopes are cleaned off by rock-falls and slides, and it may be many years before sufficient material is weakened for further large slides to develop. This situation is shown in Fig. 13.8(c); the period 1720–60 was exceptionally cold and wet and the incidence of landsliding then was ten times greater than in the following 50 years.

The above discussion of climates concludes the list of items (p. 370) which were chosen to illustrate geological factors that can be involved in slope movements. It should be emphasized, however, that particular geological processes rarely work in isolation, and that most large slope movements have resulted from several processes *collectively* producing an overall condition of critical stability. To illustrate this point three case histories are now described: the Vajont slide at a reservoir; the Turtle Mountain slide, above a mine; and the Folkestone Warren landslide, along a natural coastline.

The Vajoint slide   October 9, 1963. (References: Muller, 1964b; Kiersch, 1964; Jaeger, 1969).

The river Vajont flows in a steep gorge which cuts through the

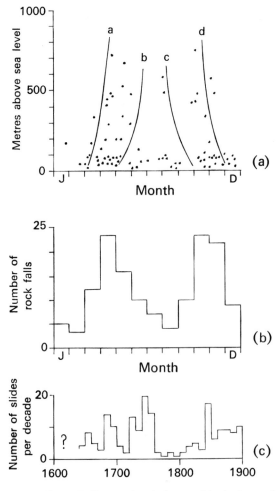

Fig. 13.8 (*a*) Rockfalls in E. Norway in relation to altitude (metres), time of fall and temperature. *a* = average date when the mean daily temperature passes freezing point; *b* = averaqe last day of frost; *c* = average first day of frost. (*b*) Number of rockfalls in the years 1951–55 and their monthly distribution. (*c*) Number of rockfalls and rock slides in fiords.

Alpine folds of north Italy. The broad structure of the valley is a syncline, thought to have been formed in Tertiary times (Fig. 13.9,*a*). The rocks involved in the folding are essentially calcareous sediments of Jurassic and Cretaceous age. The Lias, Dogger, and Malm formations (Lower, Middle, and Upper Jurassic respectively) constitute a thick series of stratified limestones, some pure and others shaley, which contain thin bedded layers of clay. The Lower Cretaceous rocks are also well-bedded limestones which pass up into the marls and limestones of the Upper Cretaceous. It is quite possible that the

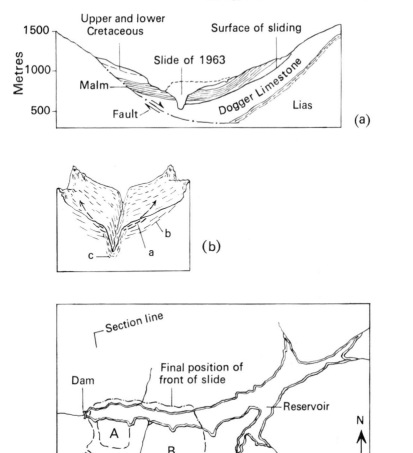

Fig. 13.9 The Vajont Slide (after Kiersch and Muller). (*a*) General section showing the elevation of the geology in metres above sea level. (*b*) Sketch of the gorge. *a* = joints associated with old glacial valley; *b* = opening associated with bedding planes; *c* = joints associated with younger river valley. (*c*) Map of Vajont reservoir. *A* = 1960 slide; *B* = 1963 slide. Both slides moved North, into the reservoir.

folding of the sediments mobilized the shear strength of the weaker clay seams, so that they were at their residual strength when folding ended.

Pleistocene glaciation removed a considerable volume of rock and scoured out the glacial valley along the axis of the syncline (Fig. 13.9,*b*). This unloading of the valley sides promoted the development of stress relief features parallel to the valley itself; in some places new

joints were developed, in others conveniently oriented bedding surfaces were opened. Landslides occurred and may have dammed the
valley until they were overtopped and eroded away. One slide had
actually crossed the valley and its leading edge lay on the glacial sands
and gravels which locally cap the bedrock. By the time the glaciers had
retreated the valley was sufficiently elevated above sea level for its
river to cut a gorge some 195 m to 300 m deep in the valley floor.
This down-cutting initiated a further cycle of stress release which is
active at the present day. Ground-water flow has produced many
solution features in the limestones such as sink holes and cavities
on joints and bedding planes.

A hydro-electric scheme for the valley had been planned, and by
September 1961 a thin arch concrete dam was constructed. Geological
investigations of the valley sides (as distinct from the dam foundations)
had been in progress at intervals since 1928 but were intensified in
October 1960, when accelerated movement into the reservoir area
occurred on the south slope of the valley within 390 m of the dam.
The movement was accompanied by a large M-shaped tension gash
which extended along the south slope (Fig. 13.9,c '1960 scar'). Surveys
revealed that the slide contained approximately 200×10^6 m^3 of
rock (Fig. 13.10,a), and was moving on a zone of sliding-surfaces

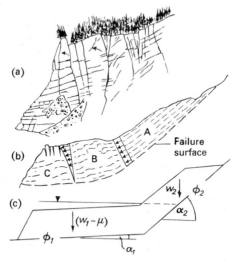

Fig. 13.10   (a) Sketch of rotation at front of slide (Muller). (b) General character of the
slide (after Muller); arrows show relative velocities of movement. (c) Section analysed by
Jaeger (1969); W = weight of rock; μ = uplift pressures; ϕ = angle of friction; α = dip
of sliding surface.

situated about 198 m below ground level. The front of the slide was
moving 8–10 cm per day and other parts of the mass at 3–5 cm per
day. Further, the eastern half of the slide was moving more slowly
than the western half, suggesting that the two were failing under

different conditions; the movement picture as a whole suggested that progressive failure and creep were occurring. The volume of the moving mass precluded all remedial measures other than those which would reduce the pressure of water within the slide. It was decided to drain the slope by adits and to lower the level of water in the reservoir, which by this time had risen during impounding to 85 per cent of its full height. The rate of lowering was such that excess water pressures would not develop in the slide, however, the reservoir level rose a further 10 m to 89 per cent of the final height before this drainage programme could begin. On November 4th, 7×10^5 m^3 of material slid from the toe of the slide into the reservoir in 10 minutes, and is known as the 1960 slip (Fig. 13.9,*c*). The stability of the slope was evidently closely related to the reservoir level. By January 1961 the reservoir level had been lowered by 50 m and from then until September 1961 it oscillated between that position and one about 15 m lower; ground-water levels hardly moved during that time. Movements which had pushed the toe of the slide into the reservoir for a distance of 25 cm to 100 cm gradually stopped, and the slide had apparently reached a condition of stability. Filling of the reservoir recommenced, and by February 1962 the water level had risen by 50 m to its former (1961) position. Rock movements were insignificant, and filling therefore continued until, in November 1962, the level had risen a further 50 m. Movements then increased but were still 'small', about 12 mm per day. Lowering was soon started again and in March 1963 the reservoir was back to its level of January 1962. Since that date the toe of the slide had travelled 125 cm further into the reservoir.

By March 1963 it had been noted that the greatest movement occurred when rock was flooded for the first time, and it was thought that if water levels were raised in stages the sliding mass would eventually reach equilibrium, or at least move so slowly that no serious problem would arise. Raising the water level began again in April 1963 and was accompanied by mass movement which did not exceed 2.5 mm per day, until the former highest level had been reached. The rate of movement then increased, though it was still less than one-third of that recorded when the valley was originally flooded at a lower level. Filling was temporarily halted in mid-July but recommenced in August as rock-movements had been less than those of 1962. At this time heavy rain fell; the level of the reservoir was the highest ever reached, though still below the approved top water level. The slow movement of the slide increased and lowering of the level begun at the end of September with the intention, based on earlier experience, of bringing the creep to a standstill. Lowering was carried out at a slow rate (15 m per week) and by October 9th the water level had dropped to where it had stood in November 1962 and June 1963. That night at 23.38 G.M.T. there was a violent failure that lasted a full minute. The whole of the disturbed mountain-side slid downhill with such momentum that it crossed the river gorge (99 m wide) and rode 135 m up the farther side of the valley (Fig. 13.9,*a*). More than 300×10^6 m^3 had moved, at a speed of about 24 m/s. This sliding lasted for 20 seconds, produced seismic shocks that were recorded throughout Europe, and sent a huge wave over the dam which, however, survived. The wave levelled five villages in the valley below and killed more than 1500 people.

The cause of the disaster is still not known with certainty, but Jaeger (1969) has proposed a theory which links many of the observed facts. He suggests that the failure occurred first in the upper part of the mass, A in Fig. 13.10,*b* (the 1960 tension scar), where the value of friction fell to very low levels. The zone nearer the gorge (C in the figure) was anchored at depth with rotation occurring in its upper levels. Jaeger simplifies the slide geometry to that given in Fig. 13.10,*c* and shows that the factor of safety for a creeping system of this character decreases with a decrease in the ratio of the angles of friction (ϕ_2/ϕ_1) whatever the values for the ratios W_2/W_1 and α_2/α_1. ϕ_2 must have been low to permit the observed deformations in zone A, hence ϕ_1, the friction on the 'seat' of the slide became critical to the overall stability. Increase in the uplift forces (μ) would decrease ϕ_1 and the periodic movements related to rises in reservoir level indicate that the factor of safety must have been hovering between 1.0 and some value above 1.0. The process continued until cohesive forces on the flatter lower section of the slide had been so reduced that rock there failed by rapid fracture. Investigations made after the failure support this theory. High up on the concave surface of sliding the failure was parallel to rock strata, whereas along the flatter part of the surface the failure cuts across the strata, suggesting a brittle fracture. The sudden reduction in cohesion permitted the accelerations that followed.

Vajont illustrates the culmination of a progressive failure that had probably been operating for thousands of years. It is possible that it would have occurred under natural conditions, but it is clear that conditions were aggravated by the reservoir construction. The disaster is a reminder of the possible consequences of interfering with natural systems which are poorly understood.

The Turtle Mountain slide   April 29, 1903. (References: McConnell and Brock, 1904. Daley *et al.*, 1911).

Turtle Mountain is a long, narrow, wedge-shaped ridge that forms part of the front range of the Canadian Rocky Mountains in South Alberta. Its peak rises 3100 feet above the valley of the Old Man River, and its eastern face overshadows the mining town of Frank. This face has a talus slope at 30 degrees which extends to 800 feet above the valley floor where it ends against a precipitous upper cliff.

The mountain is composed of Devonian and Carboniferous lime-stones which have been thrust east over the western limb of a syncline of Cretaceous shales, sandstones, and coals (Fig. 13.11). The lime-stones are an alternating sequence of contrasting beds, some massive

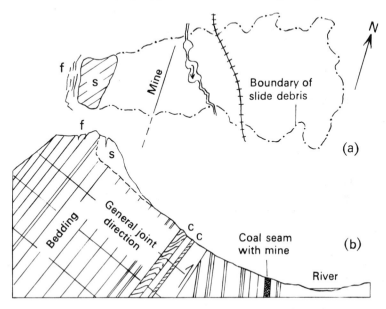

Fig. 13.11   Turtle Mountain Slide.   (a) Plan.   f = fractures; s = slide.   (b) Section. c = contorted zone. See Cruden and Krahn (1973) for a recent interpretation.

and coherent, others flaggy. They contain a 25 ft thick horizon of dark shale and two zones of tectonic deformation. One of these zones formed the toe of the slide which was to come. The general dip of the limestones is 50° to 65°, and two major joint sets are well developed, both are normal to bedding. The joints thus dip east, i.e. towards the valley, at 25° to 40°; the angle of friction along them is a little less than 32°. These angles are important for if the strata had been dipping at less than 50° the joints would have been steeper and would have promoted numerous small rock-falls; had they been dipping at more than 65° the joints would have been flatter and not provided surfaces of likely failure. The topography was also comparatively young, as the retreat of Pleistocene glaciers had removed lateral support from the glacially oversteepened valley sides. Thus in many ways Turtle Mountain was potentially unstable, and it was surprising that no slide resulted from the severe earthquake in 1901 that was centred in the Aleutian Islands.

A drift mine was opened in 1901 in the nearly vertical coal seam at the foot of the mountain. From the mouth of the mine, 27 feet above river level, a level gangway was driven for a mile along the strike of the coal. A second level was then dug, 30 feet below the first, for drainage and ventilation. Large chambers were opened up from the first gangway, each chamber being some 130 feet long and 14 feet wide and separated from its neighbour by pillars 40 feet long; these pillars contained manways 4–5 feet square. Most of the coal was stoped down into the chambers, where it was drawn off at chutes.

In October 1902 the miners noticed that the chambers were beginning to squeeze with noticeable severity, particularly between 1 a.m. and 3 a.m. Gangways were being continually re-timbered and manways, driven up to the outcrop, which could not be kept timbered were abandoned. Early in 1903 coal was being mined with unusual ease, and was mining itself over 1500 feet length of the working, bringing with it parts of the hanging wall. By April 1903 the mined chambers were up to 400 feet high; 245 000 cubic yards of coal had been removed, with extraction continuing at about 1000 tons per day.

Above ground level, the countryside was enjoying a warm spell of weather with temperatures up to 74°F, and April 28th in particular was a very warm day. However it was followed by a cold night with heavy frost.

At 4 a.m. on April 29th miners underground noticed that coal was breaking and running down the manways. The men took to the ladders to escape and, continually battered by falling coal, eventually reached the main level. A few minutes later the driver of a shunting engine on the surface heard the cracking of rocks on the mountainside. This was followed suddenly by a sound like an explosion, and 90 million tons of limestone fell 2500 feet from the peak on the mountain. The fall of rock hit the ground with a heavy thud that shook the valley, and was deflected into the air by a sandstone ridge just uphill of the coal outcrop, sending it over the outcrop and downwards into the Old Man River. But it did not stop. Instead, it scoured its way through the river and then slid for more than a mile over the rolling hills to the east, making a noise that resembled steam escaping under high pressure. It came to rest abruptly after climbing 400 feet up the opposite side of the valley, having covered in its wake an area of over 1 square mile. From eye-witness accounts it appears that the whole event took no longer than 100 seconds. The slide killed 70 people (Note 10).

As at Vajont, the slide at Frank illustrates the way in which many variables can become involved in such a movement. The geological structure of the slope and its glacial erosion made a failure inevitable. Mining, with the removal of rock from large underground chambers, hastened the event; the weather too probably played its part, as the warm spell had melted much snow so that melt water entered the limestone and increased the water pressure in joints. The failure of a 'key-zone' somewhere was probably the last of a chain of events that culminated in the great slide.

The Folkestone Warren slides, Kent. (Reference: Hutchinson, 1969) Folkstone Warren is a two mile stretch of naturally unstable coastline that lies between the old ports of Folkstone and Dover in south-east England. It is an area where landsliding is common. The first recorded slip at the Warren dates from 1765, but sliding may have occurred for centuries before that. The sea-cliff along the coast is cut into an irregular shelf of slipped material which has a width up to 1200 feet. This shelf gradually rises inland from about 50 feet above sea level at the cliff, to about 180 feet where it ends against the steep upper slope known as the High Cliff; the latter reaches a height of over 500 feet O.D. (Fig. 13.12*a*). In the 1840's it was proposed to carry the

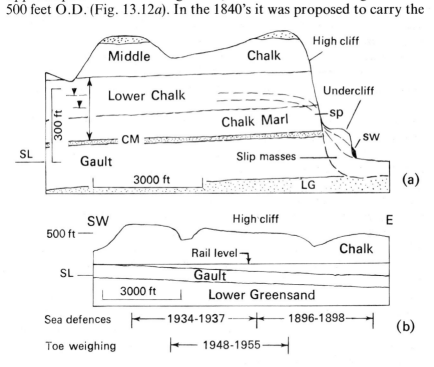

Fig. 13.12   (*a*) General section across Folkestone Warren area showing the main slip surface and maximum and minimum water levels. *sp* = spring; *sw* = sea wall; *SL* = sea level; *LG* = Lower Greensand; *CM* = Chloritic Marl. (*b*) View of Warren looking north (after Hutchinson).

London to Dover railway along the undercliff. The unstable nature of the ground was not then fully appreciated, and since its completion in 1844 the line has been disturbed by repeated movements of the ground on which it was founded. A large disturbance in December 1935 moved an area nearly 2 miles long and $\frac{1}{2}$ mile wide, almost the whole of the Undercliff (Plate 12b). The railway was pushed 165 feet towards the sea and lowered 20 feet, and the foreshore was heaved up by the emerging toe of the slide.

Some geological details of the Warren are shown in Fig. 13.12(a,b). The High Cliff is composed of Middle and Lower Chalk (see Chapter 8), with an impervious horizon, the Chalk Marl at its base. The latter lies above 10 feet of glauconitic sandy marl known as the 'Chloritic' Marl. Below this is the over-consolidated, fissured and jointed Gault clay, 140 to 160 feet thick. Beneath the Gault lies the Lower Greensand, a coarse yellow sand, highly permeable, containing calcareous and glauconitic horizons.

Three types of movement (Fig. 13.13) are termed by Hutchinson:

(i) M-slides: movements in multiple rotational landslips of the undercliff resulting in large seaward displacements; these movements seem to occur on non-circular surfaces.

(ii) R-slides: smaller features which are rotational and involve movement only in slipped masses close to the sea cliff.

(iii) Chalk falls of large masses of Chalk, which are commonly preceded by downward movements known as 'sets'; these have been associated with subsidences of up to 5 feet and can affect the Chalk for 60 feet from the edge of High Cliff.

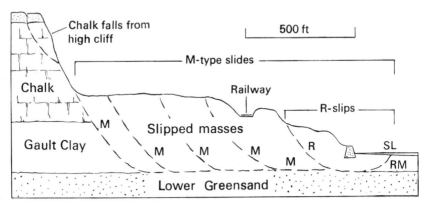

Fig. 13.13 Main types of landslide at Folkestone Warren (after Hutchinson).

Hutchinson suggests that the main controls promoting the above movements are the intensity of marine erosion along the toe of the slides and the influence of pore water pressure at potential slip surfaces. His table of landslide occurrence is reproduced as Table 13.2. Loading by heavy concrete structures placed on the foreshore to protect the toe, together with suitable drainage of the slipped mass, has reduced the incidence of sliding.

Table 13.2 Relation between Incidence of Landslides and Seasonal Variations in the Piezometric Level in the Slipped Mass of the Undercliff

| Month | Piez. Level in Undercliff | Incidence of Movement | | |
		R-Type	M-Type	Chalk Falls
	(feet)			
Jly	81			
Aug	79			
Sep	78			1
Oct	80			
Nov	87			
Dec	100	1 ~ 2 (1)	2	3 (2)
Jan	104	1 ~ 2 (1)	1	2 (2)
Feb	102	1 ~ 2 (1)		1
Mar	100	2		1
Apr	94			
May	89			1
Jun	87			

(1) One definitely, two possibly.
(2) Associated with M-type slips.

The hydrogeology of the area can be conveniently separated into two regimes: (1) ground-water in the slipped mass (Fig. 13.12a), and (2) ground-water in the Greensand. Water levels from the Greensand, whose piezometric surface is above the top of the sand, exhibit little seasonal fluctuation. The Greensand dips gently as shown in the figure. Ground-water in the slipped mass appears to be in hydraulic connection with the Chalk of the High Cliff, which is a good aquifer and supplies water to the slipped masses. Water levels in the latter, unlike the Greensand, fluctuate seasonally and are normally at their highest between December and March – the same period in which movement frequently occurs (Table 13.2).

Large movements have invariably involved the Gault clay, and attention has been directed to its shear strength. The clay is by no means uniform in its physical properties; two residual angles of shearing resistance (quoted by Hutchinson), in terms of effective stress, are as follows:

$\phi'_r = 12°$ Calcium-montmorillonite content = 20 to 34%
$\phi'_r = 19°$ Calcium-montmorillonite content   0 to 9%.

Calculations from analyses of slips in various zones of the Gault give an overall ϕ'_r of 16–17°. The undrained shear strength of the clay appears to be at least 20 tons per square foot. Residual angles of shearing for the Chalk are within the range 19 to 35°.

The cliffs are therefore composed of contrasting materials, and the following mechanism has been suggested by Hutchinson to explain the retrogression of the slips which has gone on for many years. Considering first the geological setting, the cliffs at Folkestone Warren lie on the northern limb of the Wealden anticline (Fig. 8.9), a structure of Oligocene or early Miocene age. Geological evidence suggests that the upper Gault was consolidated under an effective pressure of some 35 to 40 tons per square foot. Erosion since the formation of the anticline has left the Gault in an over-consolidated state, the present effective pressure on its upper surface being about 24 tons per square foot. Under such conditions the retrogression of the rear scarp of the landslip could have occurred in the manner shown in Fig. 13.14. The

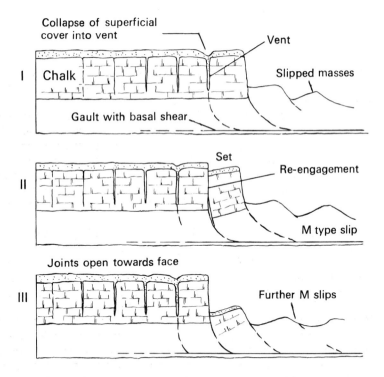

Fig. 13.14 Suggested mechanism for retrogression at Folkestone Warren (after Hutchinson).

over-consolidated Gault, with its content of active clay minerals and their strong diagenetic bonding will, during the unloading by erosion, have released little of the strain energy it accumulated during consolidation. If therefore, as is likely, an expansion potential exists in the Gault, its seaward portion (where marine erosion has removed lateral

support) could undergo expansion with the resulting development of a shear surface at or near the base of the formation. This
surface would progress inland as erosion ate further into the slipped
masses. The overlying Chalk could be put into tension as a result of
the seaward movement of the Gault, and vertical joints opened up.
The local load on the Gault is thus increased, and with reduced
lateral support to seaward the clay will fail and the slope subside.
After such subsidence, re-engagement of the Chalk could occur and
might be sufficient to temporarily stop further movement. Final
collapse would coincide with the next M-type slide, which would
allow the cycle of regression to be repeated.

The slides at Folkestone Warren therefore illustrate the way in
which geological history, mineralogy, ground-water, and erosion can
combine to produce large scale instability.

NOTES

1. The reader interested in the analyses of slopes will find a selection of
   views and theories in the selected bibliography at the end of the Chapter.
   The analyses cope with anisotropic materials, peculiar slope shapes,
   water pressures etc. but have difficulty incorporating the time dependency of some of the parameters used. It is important to realize that
   failure need not be a spontaneous event which occurs everywhere about
   some surface within a slope. Often it is initiated locally at points near
   the base of the slope, sometimes slowly other times quickly. This general
   process has been termed 'progressive failure' and was described by
   Terzaghi, with reference to the stability of clay slopes, in the following
   way:
   'The term 'progressive failure' indicates the spreading of the failure
   over a potential surface of sliding from a point or line toward the
   boundaries of the surface. While the stresses in the clay near the
   periphery of this surface approach the peak value, the shearing
   resistance of the clay at the area where the failure started is already
   approaching a much smaller ultimate value'.
   Hence a factor of safety for slopes is as time-dependent as the parameters from which it is calculated. Engineers therefore calculate a long
   term and a short term stability for slopes, where short term usually
   means 'for the duration of a contract' and long term 'for the engineering
   life of the scheme'. Two interesting papers concerning progressive failure
   are those by Bjerrum (1967) and Lutton (1971).

2. The parameters of immediate interest are normally those of cohesion
   and friction because they can be conveniently combined with values
   for the effective pressure normal to some plane and used in Coulomb's
   empirical law to predict the maximum resistance to shear (τ_f) on the
   plane. The expression used is

$$\tau_f = c' + (\sigma n - u) \tan \phi'$$

Where c' = apparent cohesion in terms of effective stress.

ϕ' = angle of shearing resistance in terms of effective stress.

σn = total pressure normal to plane being considered.

u = pore pressure at the point of interest.

Because the movement of many slopes suggests the presence of progressive failure it is often true to say that the characters which are sought, viz. c' and ϕ' as measured after some period of testing which is short by geological standards, are not all that are desired.

3. Flocculation is the coagulation of clay particles which form flocs, or clusters.

4. This only applies to situations where a clearly defined, coarse, anisotropic fabric is present. Very fine, isotropic fabrics such as those found in very broken rock and many arenaceous deposits will tend to have no directional characters and will support slopes of a uniform angle regardless of their orientation within the fabric. It must be remembered that fabrics cover a range from anisotropic to isotropic and that their behaviour will change accordingly. An interesting study of this phenomena has been made by Bray (1967).

5. A secondary effect of drainage in this situation would be the consolidation of the clay and consideration should therefore be given to the possible settlement of existing structures when such procedures are adopted.

6. The flow of water through the ground towards dewatering systems often carries with it fine material, which if transported through the pores will eventually enter the installations. Such a situation is undesirable on two accounts; firstly it means that internal erosion is occurring somewhere within the surrounding material, a situation which if left unchecked can result in subsidence and failure of the material, and secondly it fills up the dewatering system with sediment. This problem can be overcome in 3 ways:

1. By protecting the system with delicate meshes and slots which screen the flow and prevent material greater than a given size from entering the system.

2. By developing a graded filter around the installations, (this is further described in Chapter 16).

3. By placing a filter around the installations. These filters are composed of granular material, often sand, which has a grading that produces a network of pores which inhibits the transport of fine material above a certain size. The grading required to protect any situation can be ascertained by using Terzaghi's rule for filters. This simply states that a material satisfies the essential requirements for a filter if its 15% size (D_{15}) is at least 4 times as large as that of the coarsest layer of material in contact with the filter, and not more than 4 times as large as the 85% size (D_{85}) of the finest adjoining layer of material. Fig. 13.15 illustrates the filter requirements for the sand at Castle Hill. A beach sand from the Crumbles at Eastbourne proved satisfactory and was used.

7. Casagrande developed the notion of critical density (or critical void ratio) to explain the differences observed in the strength of granular

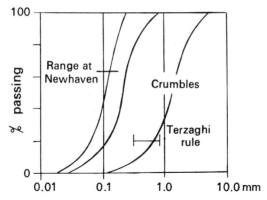

Fig. 13.15 The application of Terzaghi's Filter Rule.

material packed at differing void ratios; e.g. the difference in the strength of a sand when packed (a) densely and (b) loosely. According to the definition, granular materials which are less dense than their critical density, under some vertical pressure, collapse and become more dense on displacement and vice versa. The collapse of open fabrics can increase pore pressures and reduce the shear strength of the material, until such time as the pressures can be dissipated.

8.   The terminology involved in the description of *in situ* stresses tends to be confusing; an interesting account of it, and of the general character of the stresses, can be found in: Discussion of Theme 4, Lisbon Conference 1966; *Proc. 1st. Congr. Int. Soc. of Rock Mechanics*, **3,** 311–383.

9.   Old landslides and other periglacial features can often be seen from the air, and the study of air photos is often rewarding in such investigations.

10. The behaviour of the slipped mass after its detachment from the mountain is worth noting. Its great speed was assessed by witnesses at 90 m.p.h. Calculations suggest speeds of the order of 110 m.p.h., yet the slide was carrying blocks of limestone 20 to 40 feet long over hummocky ground, and it eventually climbed 400 feet. Its character was more that of a fluidized sheet than a typical flow; yet it was deflected by terraces and banks in its path and flowed along them rather than over them, sometimes changing direction by 90 degrees. It stopped suddenly after the rapid motion; the whole brief event suggested an explosion to those who heard it, and was followed by a noise like exhaust steam under high pressure. Men in the mine who had reached the main level and were making for the exit were blow off their feet by a blast of air. These and other points have been studied by Shreve (1968) who concludes that the slide was a highly lubricated moving mass; he puts forward other evidence from similar slides elsewhere, to suggest that the lubricant was nothing more than a layer of compressed air trapped beneath the debris. According to this theory a mass detaches itself, descends a slope, is launched into the air by some topographic feature, and traps on its descent a layer of air on which it rides at high speeds.

The vast area covered by the debris below Turtle Mountain points to the need for carefully assessing the likely extent of vulnerable ground below a high unstable slope.

REFERENCES

ASHBY, J. 1971. Sliding and Toppling modes of failure in models and jointed rock slopes. MSc. Thesis. London Univ. Imperial College.

BARR, D. 1970. Measurement of creep in a shallow slide-prone till soil. *Amer. Journ. Sci.* **269,** 467–480.

BEAUJOINT, M. and MARTIN, A. 1966. Observation of the behaviour of a natural slope. *Proc. 1st. Int. Cong. Int. Soc. Rock Mechanics, Lisbon,* 147–153.

BENSON, W. N. 1946. Landslides and their relation to engineering in the Dunedin District, New Zealand. *Econ. Geol.* **41,** 328.

BJERRUM, L. 1967. Mechanism of progressive failure in slopes of over-consolidated plastic clay and clay shales. *Proc. Amer. Soc. Civ. Engs., Jour. Soil Mechanics and Foundation Div.,* **93,** 1–49.

BJERRUM, L. 1968. Stability of rock slopes in Norway. *Norwegian Geotechnical Institute,* Publ. 79.

BRAY, J. 1967. A study of jointed and fractured rock. *Rock Mechanics and Engng. Geol.* **5.**

CHANDLER, R. 1969. The effect of weathering on the shear strength properties of Keuper Marl. *Geotechnique,* **19,** 321–334.

CHANDLER, R. 1971. (a) Creep movements in low gradient clay slopes since the late glacial. *Nature,* **229,** 399–400.

CHANDLER, R. 1971. (b) Landsliding on the Jurassic escarpment near Rockingham, Northamptonshire. Spec. Pub. No. 3, *The Institn. of Brit. Geographers.*

CLOSE, U. and MCCORMICK 1922. Where the mountains walked. *Nat. Geograph. Mag.* **41,** 445–464.

CRAWFORD, C. and EDEN, J. 1967. Stability of natural slopes in sensitive clay. *Proc. Amer. Soc. Civ. Engs., Journ. Soil Mech. and Foundations Div.,* **93,** 419–437.

CRUDEN, D. and KRAHN, J. 1973. A re-examination of the geology of the Frank Slide. Canadian *Geotechnical Jour.* **10,** 521–91.

DALY, R. A., MILLER, W. G. and RICE, G. S. 1911. Report of the Commission appointed to investigate Turtle Mountain, Frank, Alberta. *Canadian Dept. Mines* Memoir 27, Canadian Geol. Survey.

DROUHIN, G., GAUTIER, M. and DERVIEUX, F. 1948. Slide and subsidence of the hills of St. Raphael—Telemy. *Proc. 2nd. Int. Conf. Soil Mech.. and Foundation Engng. Rotterdam,* **5,** 104–106.

FIELD, J. 1966. Rock movements from load release in excavated cuts. *Proc. 1st. Int. Cong. Int. Soc. Rock Mechanics, Lisbon,* **2,** 139–140.

de FREITAS, M. and WATTERS, R. 1973. Some field examples of toppling failure. *Geotechnique,* **23,** No 4.

HENKEL, D. 1961. Slide movements on an inclined clay layer in the Avon Gorge in Bristol. *Proc. 5th. Int. Conf. Soil Mech. and Foundation Engng.* Paris, Vol. 2, p. 619–624. (For analyses of wedges see also Pauling, B. *Amer. Soc. Civil Eng. Proc. Soil Mech. and Foundations Div.,* 1970.)

HENKEL, D. 1966. The stability of slopes in the Siwalik rocks in India. *Proc. 1st. Int. Cong. Int. Soc. Rock Mechanics, Lisbon,* **2,** 161–165.

HOLLINGWORTH, S., TAYLOR, J. and KELLAWAY, G. 1944. Large scale superficial structures in the Northamptonshire Ironstone field. *Quart. Jour. Geol. Soc. London,* **100,** 1–44.

HOLM, O. 1961. Stabilization of a quick clay slope by vertical sand drains. *Proc. 5th. Int. Conf. Soil Mech. and Foundation Engng.* Paris, Vol. 2, 625–627.

HUTCHINSON, J. 1968. Mass movement. *Encyclopedia of Geomorphology,* p. 688–695. *Ed.* R. Fairbridge. Reinhold, London.

HUTCHINSON, J. 1969. A reconsideration of the coastal landslides at Folkestone Warren, Kent. *Geotechnique,* **19,** 6–38.

JAEGER, C. 1969. The stability of partly immersed fissured rock masses and the Vajont rock slide. *Civil Engng. and Pub. Works Review.* 1204–1207.

KARLSSON, R. and PUSCH, R. 1967. Shear strength parameters and microstructure characteristics of a quick clay of extremely high water content. *Proc. Geotech. Conf. Oslo,* **1,** 35–42.

KIERSCH, G. (1964 March). Vajont reservoir disaster. *Civil Engineering,* 32–39.

KOPPEJAN, A., WAMELEN, B. and WEINBERG, L. 1948. Coastal flow slides in the Dutch Province of Zeeland. *Proc. 2nd. Int. Cong. Soil Mech. and Foundation Engng. Rotterdam,* **5,** 89–96.

KOWALSKI, W. 1970. The interdependence between strength, softening, swelling, and shrinkage of Cretaceous marls and 'opokas' and their lithology. *Proc. 1st. Int. Cong. Int. Assoc. Engng. Geol. Paris,* **1,** 457–464.

LANE, K. 1969. Engineering problems due to fluid pressure in rock. *Berkeley Symp. on Rock Mechanics.*

LUTTON, R. 1971. A mechanism for progressive rock mass failure as revealed by loess slumps. *Int. Jour. Rock Mechanics and Mining Sci.,* **8,** 143–151.

MATSUO, S. 1957. A study of the effect of cation exchange on the stability of slopes. *Proc. 4th. Int. Conf. Soil Mech. and Foundation Engng., London,* **2,** 330–333.

MCCONNEL, R. and BROCK, R. 1904. The Great landslide at Frank, Alberta. *Canadian Parliament Sessional Papers,* **38,** No. 10, Sessional Paper No. 25, pt. 8. *Report Supt. Mines,* Appendix.

MORGENSTERN, N. and TCHALENKO, J. 1967. Microscopic structures in Kaolin subjected to direct shear. *Geotechnique,* **17,** 309–328.

MULLER, L. 1964. (a) The stability of rock bank slopes and the effect of rock water on the same. *Int. Joun. Rock Mechanics and Mining Sci.,* **1,** 475–504.

MULLER, L. 1964. (b) The rock slide in the Vajont valley. *Rock Mechanics and Engineering Geology,* **2,** 148–212.

MURAYMA, S. 1966. Swelling of mudstone due to sucking of water. *Proc. 1st. Int. Cong. Int. Soc. Rock Mechanics, Lisbon,* **1,** 495–498.

PRICE, N. 1966. Fault and joint development in brittle and semi-brittle rock. Section 1. (Pergamon Press, London).

RENGERS, N. 1970. Influence of surface roughness on the friction properties of rock planes. *Proc. 2nd. Int. Cong. Int. Soc. Rock Mechanics, Belgrade,* **1,** paper 1/31.

SEED, H. 1967. (a) Slope stability during earthquakes, p. 299–323.

SEED, H. 1967. (b) The Turnagain Heights landslide, p. 325–353. *Proc. Amer. Soc. Civ. Engs., Journ. Soil Mech. and Foundations Div.,* **93.**

SEED, H. 1969. Characteristics of rock motions during earthquakes. *Proc. Amer. Soc. Civ. Engs., Jour. Soil Mech. and Foundations Div.,* **95,** 1199–1218.

SHARP, J., *et al.* 1972. Influence of ground water on the stability of rock masses. *Trans. Sec. A. Inst. Mining and Metall. Trans.* **81,** Bull. No. 782. Jan.

SHREVE, R. 1968. The Blackhawk landslide. *Geol. Soc. Amer.* Spec. Paper. No. 108.

SKEMPTON, A. 1964. Long-term stability of clay slopes. *Geotechnique,* **14,** 75–102.

SKEMPTON, A. 1966. Some observations on tectonic shear zones. *Proc. 1st. Int. Cong. Int. Soc. Rock Mechanics, Lisbon,* **1,** 329–335.

STIMPSON, B. and WALTON, G. 1970. Clay mylonites in English Coal Measures: their significance in open-cast slope stability. *Proc. 1st. Int. Cong. Int. Assoc. Engng. Geol., Paris,* **2,** 1388–1393.

STRUILLON, R. 1966. Some aspects of rapid weathering of silicate rocks in temperate climates. *Proc. 1st. Int. Cong. Int. Soc. Rock Mechanics, Lisbon,* **1,** 303–306.

TERZAGHI, K. 1950. Mechanism of Landslides, in Application of geology to engineering practice. *Berkey, Vol. Geol. Soc. Amer.*

TERZAGHI, K. 1962. Stability of steep slopes on hard unweathered rock. *Geotechnique,* **12,** 251–270.

TERZAGHI, K. and PECK, R. 1962. Soil Mechanics in engineering practice. John Wiley & Sons; New York.

WARD, W. 1948. A coastal landslip. *Proc. 2nd. Int. Conf. Soil Mech. and Foundation Engng., Rotterdam,* **2,** 33–37.

WEEKS, A. 1969. The stability of natural slopes in S.E. England as affected by periglacial activity. *Quart. Jour. Engng. Geol. (Geol. Soc. London),* **2,** 49–61.

ZISCHINSKY, U. 1966. On the deformation of high slopes. *Proc. 1st. Int. Cong. Int. Soc. Rock Mechanics, Lisbon,* **2,** 179–186.

SELECTED BIBLIOGRAPHY

The analysis of slope stability is continually developing and any recommended publications are likely to become dated after a few years. The reader is therefore advised to consult an index for the following:

Géotechnique.
Amer. Soc. Civ. Engs. (Proc. Soil Mech. and Foundations Div.)
Int. Journ. of Rock Mechanics and Mining Sciences.
Proc. Int. Cong. of Int. Assoc. Rock Mechanics.
Proc. Int. Cong. on Large Dams.
Rock Mechanics and Engineering Geology.
Technical Publications of Norwegian Geotechnical Institute.
Trans. Inst. of Mining and Metallurgy.

Plate 13

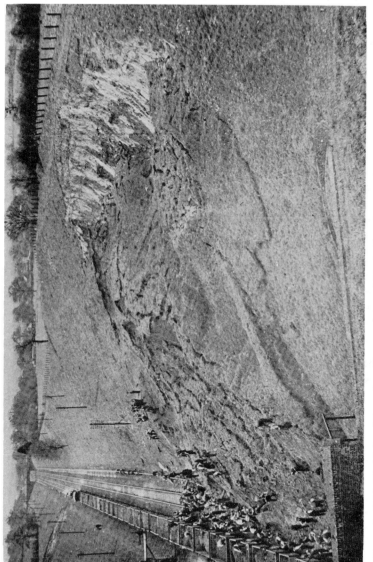

Sevenoaks slip in Weald clay. (Reproduced by permission of Mr. G. Ellson, Chief Engineer, Southern Railway)

Plate 14

1 Access

2 Future top water level

3 Buttresses

4 Excavation for key into hillside

5 Spread base for low bearing pressures

6 River channel

Farahnaz Pahlavi Dam, Iran. This is a concrete buttress dam with a maximum height of 107 metres. Completed in 1968, the dam is shown during construction in July 1965. (*Photograph by Sir Alexander Gibb and Partners*)

Three general books on the subject are:
Landslides and Engineering Practice, Highway Research Board Special Report 29. *Nat. Acad. Sci;* National Research Council Pub. 544, Washington D.C. Editor: E. Eckel. 1958.
Landslides and their Control, 1969 by Q. Zaruba and V. Menel. (Elsevier).
Rock Slope Engineering, 1974 by E. Hoek and J. Bray, The Institution of Mining and Metallurgy, London.

Attention is also drawn to the 'Rock Mechanics Information Service', offered by the Rock Mechanics Section of Imperial College of Science & Technology, London S.W.7.

14 Reservoirs and Dams

This chapter reviews some of the geological factors that are relevant to the successful construction and operation of reservoirs and dams. It has four sections, viz: Surface Reservoirs, Surface Dams, Ground Improvement, and Underground Reservoirs. Many geological investigations that are commonly used in this field of engineering have been described in Chapters 9 and 10, and ways in which geological conditions can affect movement in reservoir slopes and the leakage of water are referred to in Chapters 12 and 13. Chapter 17 discusses naturally occurring materials for the construction of dams and ancillary works.

SURFACE RESERVOIRS

Many surface reservoirs are now being sited in areas which contain quite severe geological problems and geological studies are becoming an increasingly important aspect of such schemes. The geological investigations which can be conducted are grouped here as follows; geographical considerations; sedimentation problems; water-tightness; and dam foundations.

Geographical considerations. Although these are not strictly of a geological nature, they often have a direct bearing on many geological problems which can occur in the near-surface layers of a catchment. Four areas of study may be important; general topography, slopes, vegetation, and soils; all tend to be inter-related. General topography includes the shape and size of the catchment, and the way in which these affect its run-off and infiltration characters. Topographical maps and air photographs are commonly used in such studies. Springs should be identified wherever possible, as they represent boundaries above which recharge can occur and below which there is discharge (Chapter 12). The slopes of a catchment should be surveyed so that existing and potential areas of instability are identified, especially in that part of the catchment which is later to be flooded; the disaster at Vajont (p. 387) is an example on a large scale of what can happen in particular conditions. Problems resulting from water waves generated by landsliding into reservoirs are described by Wiegel *et al.* (1970). Slopes are also important in a detailed study of run-off and erosion. The vegetation of a catchment is important if it inhibits excessive erosion. Some forms of vegetation can be troublesome; peat, for example, produces organic acids which, over a period of time, may cause a deterioration of the more soluble rocks within a reservoir, and even of the fabric of the dam itself. Soils are a controlling factor in areas where soil erosion is either anticipated or known, for their erosion produces sediment which progressively fills the reservoir (see England and Holtan, 1969).

Sedimentation problems arise when erosion occurs within a catchment. Under natural conditions a certain amount of the sediment carried by rivers will be transported from the catchment and deposited at sea (Chapter 2). However, when a valley is dammed all the river sediment will be deposited in the reservoir and may silt it up in a comparatively short time. Cartmel (1971) reports that the £300 million Tarbela Dam in West Pakistan has an estimated working life of only 50 to 60 years before silting reduces its useable storage to unacceptable levels. This dam impounds the upper reaches of the River Indus, which at that point carries an average annual sediment load of 430 million tons (437 Mg). Silt traps across the rivers discharging into the reservoir would increase its life to 100 years (Note 1).

The sediment yield of a catchment depends upon the production of transportable material and the presence of a transporting agent—geology affects both. Transportable material is produced mainly by weathering and erosion. No adequate standard of 'erodibility' has yet been defined which can be applied to catchments, but analytical methods based on the mechanical strength of the rocks involved, which can be applied to the evaluation of erosion rates, are given by Carson (1971) (Note 2). Surveys should therefore assess (a) the liability of a catchment to the denudation processes that operate, and (b) the make-up of the sediments that are produced. Gottschalk (1964) states that grain size is of primary importance in the design of sediment traps and in predicting the sediment distribution in a reservoir.

It is also necessary to consider the effect the reservoir will have upon the catchment. For example, seepage from below the spring-line can keep the surface of the ground moist, and render a soil more (or less) erodible than its drier counterpart above the spring-line. If the damming of a valley raises the water level in the sides of the valley, the level of the spring-line during periods of infiltration may also rise, so that the erodibility of the area can change from its original state (Fig. 14.1). The zone that will lie between the high and low water-levels of the reservoir will often be affected by wave action and should also be studied.

Movement of the transportable material depends on the strength of the transporting agent. As this is usually water, the infiltration

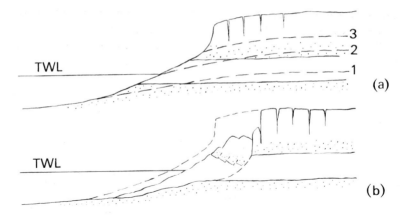

Fig. 14.1   Some results of impounding. (*a*) Elevation of water levels; 1 = original level, 2 = new level, 3 = new level resulting from infiltration. (*b*) Erosion; first of soil cover and then possibly of the rocks themselves by landsliding. TWL = top water level of reservoir.

characters of the ground should be considered, particularly its infiltration capacity (Chapter 12), as this determines whether water will remain on the surface. Impermeable areas can be expected to have natural drainage systems which are close to the surface. Such areas have concentrated, short-lived periods of run-off which can transport a wide range of grain sizes. The run-off from permeable areas is generally less powerful but a permeable catchment, by reason of its storage, is capable of maintaining a discharge during dry periods (pp. 342 and 364). Hence the total amount of sediment which is removed can be large.

In these studies, in order to identify the sediment which is on the move and possibly the areas from which it is derived, it is often useful to sample the load carried by a river. This helps to establish the form of erosion operating within the catchment. Sampling should cover a variety of flows including floods; flood deposits can be examined when the floods themselves cannot be sampled.

Water-tightness.   Some leakage may be expected from reservoirs, but in many cases the volume of water involved is so small, compared with the precipitation over the catchment, that no special counter-measures are required. The success of pumped storage schemes tends to be more sensitive to leakage than other schemes because the volume lost is wasted pumpage.

Two factors to be considered when assessing leakage are the permeability of the ground and the head of water operating; the latter is usually the more important. The total head of water at any point in the rocks around and below a reservoir is the sum of the elevation and pressure heads at that point (Chapter 12). When the total head at some point in the ground is greater than that in the reservoir and both are in hydraulic connection, there will be a flow of water from that point towards the reservoir. The reverse applies when the head in the ground is less than that in the reservoir, i.e. leakage will occur from the reservoir. Figure 14.2 illustrates the relationship. It

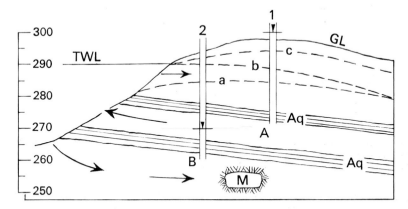

Fig. 14.2   Ground water movement in reservoir margins. GL = ground level; TWL = top water level of reservoir; 1 and 2 = piezometers; Aq = aquiclude; M = mine. a = original water level, b = water level resulting from impounding with the original water level at a; leakage from the reservoir would occur. No leakage would have occurred if the original water level had been at c; instead water would flow to the reservoir. Total head in aquifer A exceeds that in the reservoir and no leakage from the reservoir can occur. Total head in aquifer B is less than that in the reservoir and permits leakage from the reservoir.

should be noted that the permeability of the ground does not affect the above situation, although it does influence the rate of flow. Hence leakage cannot occur through even the most permeable ground if there is a flow of water from the ground to the reservoir. This has been illustrated by Kennard and Knill (1969) who describe how such a situation allowed the construction of a water-tight reservoir on cavernous limestones which contained old abandoned mine workings. Care is needed in isolating and measuring the piezometric levels and water-tables, and special attention should be devoted to identifying perched water-tables.

The direction of decreasing head beneath a dam is always down-stream because of the head of water impounded behind the dam. As a result, leakage commonly occurs beneath dams and if this is excessive it is controlled by a cut-off. The build-up of water pressure beneath the dam resulting from the presence of a cut-off must be considered, as it reduces the effective normal load on the dam foundation and can sometimes promote a major foundation failure. The importance of permeability to this pressure development is further discussed on p. 415.

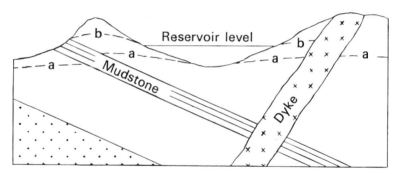

Fig. 14.3 Water-tight reservoir assured by sedimentary and igneous aquicludes. a = original water level; b = water level after impounding.

Permeability of the margins of a reservoir becomes important when leakage is taking place from the reservoir. Ideally, truly impermeable rocks, if advantageously oriented, can provide a barrier to leakage, as show in Fig. 14.3, but such a situation is rare. More common conditions are illustrated in Fig. 14.2 by the water levels a, b, and c in the upper aquifer. Water levels, resulting from impounding, which fall progressively away from the reservoir reflect the loss in head that is required to sustain flow through the aquifer. Vaughan (1969) illustrates the controls which permeability exerts in these circumstances. Total heads which are insufficient to stop leakage can be increased, if necessary, by artificially recharging water to the ground through wells. This would have the effect of raising the levels in Fig. 14.2 from a to c (Note 3).

Leakage from the floor of a reservoir cannot be controlled so easily, and permeable areas are usually grouted to prevent undue losses. Figure 14.4 illustrates two reservoirs in the Taf Fechan valley of Wales. Much of the upper (Pentwyn) reservoir, including the embankment, is situated on a small faulted outcrop of Carboniferous Limestone through which serious leakage occurred because of the low head within the formation. Much of the lost water was re-appearing further downstream, where it was brought to the surface by the

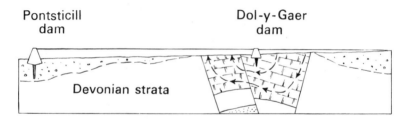

Fig. 14.4   The Taf Fechan Reservoirs, Pentwyn and Pontsticill.

presence of relatively impermeable Devonian sandstones and marls which lay beneath a shallow cover of glacial drift. Remedial measures failed to control the leakage sufficiently, and a second dam was constructed approximately $1\frac{1}{2}$ miles downstream to form the Pontsticill Reservoir. This impounded much of the water leaking from the upper reservoir.

The presence of a relatively impermeable layer in the floor of a valley need be no guarantee against leakage when the head of water in the rocks beneath the layer is much lower than that at the reservoir. Seepage will eventually reach these lower levels and may eventually cause the sealing layer to fail. The stratigraphy at a reservoir is there-

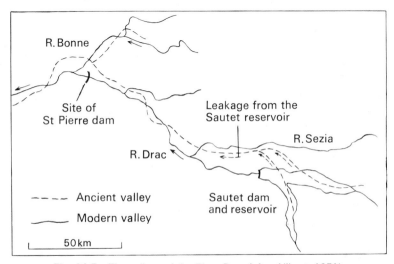

Fig. 14.5   The valleys of the River Drac (after Ailleret, 1951).

fore important. Special care should be taken to look for permeable horizons within superficial deposits. Buried valleys should also be sought especially in glaciated areas; they need not coincide with the position of existing river valleys. The valley of the River Drac, northern France (Fig. 14.5) illustrates such a situation. Leakage from the Sautet Reservoir flowed to the Sezia River and thence to the River Drac. A dam further downstream was constructed to form the St. Pierre Reservoir, but only after extensive field tests had confirmed that the permeability of the buried valley deposits at this point would not give rise to excessive leakage into the valley of the River Bonne.

The solubility of the rocks through which leakage water flows may also be important. The failure of the Macmillan Reservoir in the U.S.A. was due to water escaping through a gypsum layer which became widened by solution. Sivasubramanian and Carter (1969) describe how the solution of calcareous sandstones undermined the

Clubbiedean Dam in Scotland. Various formulas have been proposed for assessing the solubility of rock (see Douglas, 1968), although none are precise. The increase in leakage due to solution is not usually a problem, as that due to the removal of clayey joint- and fissure-fillings by internal erosion is more common. Such zones should be either protected with filters or sealed with grout.

SURFACE DAMS

The type of dam that is built, i.e. embankment, gravity or arch, within an allotted stretch of valley, is in part determined by the shape of the valley, the strength of the foundations and the availability of construction materials.

An *embankment dam* consists essentially of an impermeable core which is supported by permeable shoulders of soil and rock. The core is usually made from either rolled clay, when a suitable clay is available in sufficient quantities, or concrete. When clay is used it is normally flanked by filters and transition zones of permeable fill (Chapter 17). Examples of a clay core dam and a concrete core dam are described by Kennard (1963) and Paton (1956) (Note 4). The core is normally extended as a cut-off below ground level when seepage beneath the dam has to be controlled. These cut-offs may be very deep; that for the Ladybower Reservoir Dam (Hill, 1949) reached a maximum depth of 250 ft (76 m). The cut-off may also have to be extended beyond the ends of the dam and run into the sides of the valley if leakage around the ends of the dam is envisaged; the Butterly Dam in Huddersfield has lateral cut-offs which are $\frac{1}{4}$ mile (403 m) long.

A *gravity dam* is a massive, impermeable concrete monolith which has an essentially triangular cross-section; the Warragamba Dam in Australia (Nicol, 1964) and the Laggan Dam in Scotland (Naylor, 1936) are examples. Foundation conditions will determine whether a cut-off is required. A less massive form of gravity dam is the *buttress dam*. This consists of interlocked cellular buttress units, where each unit is a hollow gravity dam (Note 5). The Clywedog Dam is a typical example (Fordham *et al.*, 1970).

An *arch dam* is a relatively thin-walled, light weight, impermeable concrete shell which in plan is shaped as an arch. It may also be curved in the vertical direction to form a *cupola*. The Monar Dam is an example of an arch dam in the United Kingdom (Henkel *et al.*, 1964) and the Hongrin dam, Switzerland, of a cupola ('Water Power', 1970). Cut-offs can be extended from these dams if required. The geological conditions which favour any particular design can be appreciated from the character of the designs themselves.

Embankment dams, by virtue of the slopes required for their

stability, have a broad base and impose much lower stresses on the ground than concrete dams of similar height. Furthermore, their fill is plastic and can accommodate deformations, such as those associated with settlement, more readily than the rigid concrete dams. Embankment dams can therefore be built in areas where foundation rocks of high strength are not within easy reach of the surface. Their volume requires plentiful supplies of adequate building materials (p. 522) which should be located close to the site and be easily worked. Some problems involved in this respect are described by Hitchen (1968) (see reference 17 in Chapter 17).

The stability of gravity dams depends upon their weight, and they impose a high level of stress upon the foundation rocks, which must be of suitable strength. These dams are rigid structures and usually only the smallest differential movements can be safely tolerated in their foundations. Christiansen *et al.* (1971) illustrate the importance of the elastic and shear strength parameters of the rock mass for the assessment of foundations, and Zienkiewicz (1968) has shown how they are used in the analysis of foundation conditions. Adequate supplies of good quality aggregate should be close at hand for construction in concrete.

The stability of arch dams depends on the resistance offered to their lateral thrusts by the walls and floor of the site; for example, the thrusts involved at the Tachien dam, Central Taiwan (Formosa), are given by Zienkiewicz *et al.* (1970). The abutments and base of these dams are much thinner than those of gravity dams and impose high stresses upon the narrow zones of rock that are loaded.

All dams require good foundations, but these are rarely found and in many cases the selection of a dam site becomes an exercise in locating areas where the foundation rocks can be best improved. Most valleys have been subject to accentuated weathering and erosion, and the presence of unweathered rock and stable slopes tends to be the exception rather than the rule. Some valleys follow lines of structural weakness such as the crests of anticlines or lines of faults, so that the rocks are inherently broken and open. Numerous valleys were either cut or deepened when sea-level was lower than it is at the present (Chapter 8), so that they became sites of deposition when the sea-level rose and now contain thick deposits of alluvium. The glaciation of many valleys has added to this effect by the deposition of a variety of glacial deposits in them. Intercalated in these deposits may be layers of peat and solifluxion debris. The stress relief associated with the erosion of a valley can markedly affect the structure of its base and sides (Chapter 15 and account of the Coulee Dam, p. 419). Valleys are therefore sites where a variety of geological processes are likely to have been active, and an assessment of their

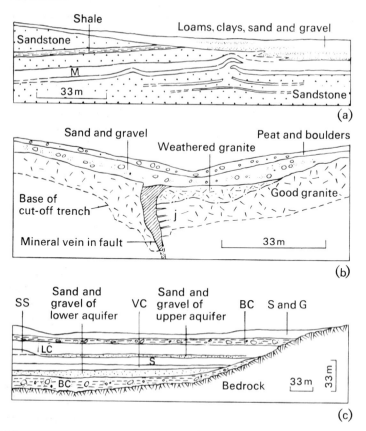

Fig. 14.6 Sections from dam foundations. (*a*) Weir Wood dam. Sussex. *M* = mudstone. Foundation strata deformed by valley bulge. (*b*) Fernworthy dam. Devon. *j* = open joints which yield water (after Kennard, J. and Lee, J., 1947). (*c*) Derwent dam. Northumberland and Durham. SS = sandy silt; *LC* = laminated clay; *VC* = varved clay; S = silt; BC = boulder clay; S & G = sand and gravel. Aquifers contained water under artesian pressure (after Ruffle, N., 1970).

potential foundation characters often requires a carefully considered geological survey. Figure 14.6 illustrates the geology of three dam sites. Information on the geology of many others is provided by Walters (1971), Wahlstrom (1974) and by the Proceedings of International Conferences on Large Dams.

Many dam sites which are unsuitable for one particular design can be used to support a dam of *composite design;* this may incorporate features of embankment, gravity and arch structures as required by the foundations. For example, a broad valley containing reasonably good rock but with a centrally placed deep zone of weathering may be dammed using gravity structures on its flanks which can support an arch that will span the centre, as at the Stithians Dam in Cornwall.

A broad valley having strong rocks on one side and a considerable thickness of weaker material on the other can be used for a gravity dam linked to an embankment; the Cow Green Dam in northern England is of this type (Kennard and Knill, 1969). The Glascarnock Dam in Scotland consists of a gravity dam linked to embankments on either side. The Fedaia Dam in Italy (Walters, 1971) contains both gravity and buttress sections. The Beni-Bahdel Dam in Algeria (see Walters, 1971) is a multiple arch dam where the arches are thrust against gravity buttresses. Many other instances exist, all illustrating that difficult foundations can be used if the geology is fully appreciated.

Ground-water is of great importance to the construction, efficiency and stability of any dam, and the pressures which may exist at depth should always be investigated. A study should also be made of the sum effect that the weight of the dam, the construction of a cut-off, and the head of water in the reservoir will have upon the development and dissipation of pore pressures. The permeability of the foundation rocks is important in this respect, and drainage may be needed to help dissipate the water pressures in ground of low permeability. Undesirable pore pressures can also exist in ground of high permeability, and here there is the added problem of seepage, causing excessive leakage beneath the dam and damage to the foundation strata. Seepage rates can be decreased by reducing the hydraulic gradient that exists beneath the dam. This is usually achieved by incorporating either an impermeable or a semi-permeable cut-off into the design of the dam foundation. The cut-off would be taken to some specified depth and usually sited in the region below the upstream face of the dam. An impermeable cut-off extending to an impermeable zone at depth will stop any leakage; such a situation is rarely achieved (Note 6). Many cut-offs are not completely impermeable because they allow water to flow either through them or beneath them. When flow-paths have to pass beneath a cut-off their length is increased and hence the hydraulic gradient along them is decreased. When flow is through a cut-off it is accompanied by a considerable loss in head. In both situations the hydraulic gradient is reduced and the pore pressures on the downstream side of the cut-off are lowered (Note 7). Cut-offs cannot be guaranteed to control pore pressures completely.

Bishop *et al.* (1963) describe the development of uplift pressures downstream of a cut-off during the impounding of the Selset Reservoir. The geology of the dam site consists of a gently-dipping series of shales, limestones and sandstones which are overlain by relatively impermeable boulder clay. The sandstones and limestones contained water under artesian pressure, which had to be reduced to ensure the stability of the embankment dam. It was thought that a cut-off would control these pressures, but it proved to be insufficient, and relief wells were installed downstream. The authors concluded that in planning a cut-off its effect on the ground-water hydrology of the valley should be considered. This was done with the Derwent Dam (Ruffle, 1970), an embankment founded on valley fill containing stratified deposits of boulder clay, varved clay, laminated clay, silty sand, silt, sand and gravel (Fig. 14.6). Sand and gravel horizons were confined at depth and contained artesian water which soaked the drillers when first encountered in the exploration bore-holes. The contractor had to pump from both the valley fill and the bed-rock in order to reduce water pressures sufficiently to enable the partial cut-off, which was used in this case, to be constructed. In addition, relief wells were required downstream of the dam to ensure its stability. An extensive system of vertical sand drains was also installed to assist the consolidation of the laminated clays so that their strength would progressively increase under the increasing load of the embankment.

Uplift pressures are thought to have been largely responsible for the failure of the Malpasset Dam in 1959. This was a cupola which was founded on severely broken and faulted rock. The thrusts imposed upon the foundations were concentrated over small areas, and such an intensity of loading can markedly reduce the *in situ* permeability of fissured rock (Chapter 12). It is believed that the loading at Malpasset may have been sufficient to produce a zone of much reduced permeability in the foundations, which would act as an underground dam and retard the dissipation of pore pressures at depth (Londe, 1967; and see Maury, 1970). These could eventually promote a deep failure in the foundations. It was known that the left side of the dam had moved some two metres downstream, prior to the catastrophic failure which later removed much of that part of the structure. Malpasset illustrates the importance of understanding the change in geological conditions which accompanies large engineering projects (Note 8).

Fluid pressures were also thought to have been responsible for the failure of the Baldwin Hills embankment in California. This was situated on a fault across which displacements later occurred and weakened the embankment, which rapidly failed. The displacements were thought to have resulted in part from the adjustment of fluid pressures in the region consequent to the pumping of oil from wells some miles away. A fuller account is given by Castle *et al.* (1973).

The damage caused by seepage should also be investigated, in addition to seepage pressures. The failure of the St. Francis gravity dam in California was attributed to the internal erosion and general weakening of the foundations which accompanied seepage beneath them. Two-thirds of this dam was founded on conglomerates with some shales and sands. When wetted these sediments virtually disintegrated into a gritty slurry so that the finer fractions could be removed from the rock (Ransome, 1928). Seepage through the conglomerate was noticed when the reservoir was first filled; it failed in the same month.

Earthquakes should be investigated in regions where they are either known or suspected to occur. Geological records giving their intensity in terms of one of the recognized scales (Chapter 1) are insufficient for design purposes, as actual values for ground accelerations are required. Typical precautions taken in Turkey to provide for such events are described by Üral (1967). Special attention is given to the geology of the dam site and the stability of slopes near the dam abutments and within the reservoir area itself. Any indication of an active fault, e.g. displacements of field boundaries, tracks, water adits, the presence of hot springs, are sought as these ae normally grounds for abandoning a potential site (e.g. Sherard *et al.*, 1974). A few dams in Britain have been damaged by weak earthquake shocks (Walters, 1971). Earth tremors have been recorded after the filling of large reservoirs and are generally thought to be related to stresses in the ground which arise from changes in ground-water conditions near the reservoir, and from the head of water in the reservior itself. Such events have been associated with the impounding by the Boulder Dam (Carder, 1945), and at the Kariba Dam on the Zambesi. Numerous incidents have recently been reported by Lane (1971), Housner (1969) and Judd (1974). Attention should also be given to the construction materials available in the area. These should allow suitable densities to be obtained in embankments, so as to resist 'liquification' from earth tremors (Casagrande, 1971), without giving rise to excessive problems of placing and compaction.

Construction materials are reviewed in Chapter 17. The location of suitable materials is an integral part of surveys associated with dams, as structures of any size should not be sited in areas that are remote from suitable supplies of raw materials. The type and volume of potentially useful rock should be assessed, together with the ease with which it can be won, transported and used; the design of a dam can often be tailored to suit both the foundation conditions and the available supplies. Much additional expenditure may be incurred if such supplies run out before the dam and its ancillary works are completed. The latter include diversion tunnels, spillway channels, access roads, haulage roads and other works. These may require geological investigation, which should be included in the general survey for the location of a dam.

Some existing sites The following short descriptions, in addition to those already mentioned, illustrate points discussed so far.

Vyrnwy Dam.   This was the first big masonry dam to be built in Britain (Note 9). It was constructed between 1887 and 1892 for the Liverpool water supply and is situated in North Wales at approximately 700 ft (213 m) above sea level. The area had been the site of an old glacial lake whose deposits of lacustrine clays, alluvium and peat covered the bedrock to an unknown depth. The engineer, G. Deacon (see Deacon, 1896), put down nearly 200 bore-holes in order to investigate the character of the buried rock surface. These revealed a rock bar at one point across the valley at a depth of 30 to 45 ft (13 m); upstream and downstream of this bar the solid rock surface was deeper. The bar probably represented the barrier of the former glacial lake and the dam was sited at this point, which also corresponded with a narrow part of the valley (Fig. 14.7). The upper

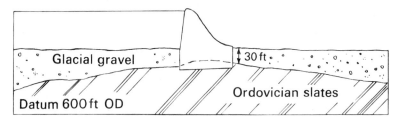

Fig. 14.7   Longitudinal section through the Vyrnwy dam site (after H. Lapworth).

6–7 feet of bedrock were in bad condition and were removed from the dam foundations. The final surface was washed with jets of water and scraped clean with wire brushes. No springs issued from the exposed foundation rocks, which were Ordovician slates that dipped upstream, but it was felt desirable to build rubble filled drains into the base of the dam. Building stone for the project was quarried about one mile from the site, and it is interesting to note that large quantities of stone were rejected because of their poor quarrying properties; 700 000 tons was tipped to waste in order to obtain the necessary rock for the masonry.

Haweswater Dam in Cumberland was built between 1934 and 1941 (Davis, 1940). It is a buttress dam, founded on Ordovician andesites and rhyolites which have *in situ* seismic velocities in excess of 18 000 ft/s (5484 m/s), and unconfined compressive strengths of 15 000 psi (1054 kg/cm^2) which is 35 times greater than the maximum stress that the buttresses impose on the ground. They are some of the best foundation rocks in Britain. The site of the dam was largely self-selecting, as the valley widens upstream and downstream from the point chosen. Bedrock was covered on one side of the valley by glacial drift to a maximum depth of 30 ft (9 m) and on the other side by a mass of boulders. The bedrock surface was explored with bore-holes and trenches. Exposures of rock were penetrated with percussive drills to ensure that they were not the upper surface of large boulders. All the superficial material was removed in a trench to expose the bedrock for foundations; rock excavation was confined, as far as possible, to the areas occupied by the buttress units of the dam, the natural rock surface being left undisturbed between them. The cut-off was excavated 5 feet deeper than the main excavation. A fault zone striking diagonally across the foundations was encountered; it dipped at 60° and contained much broken material but narrowed with depth. It was followed down in the cut-off trench to 45 ft (14 m) where it had diminished to a degree that was considered satisfactory. The foundations accepted very little grout. It was intended to face the dam with a dolerite which outcropped on both sides of the valley; but the rock was found to be too difficult to work and was a smaller mass than was first thought. Quarrying was restricted to areas below the top water level of the reservoir. Aggregate was also obtained from convenient local quarries in the Shap granite (see Taylor, 1951).

Coulee Dam is a concrete gravity structure 550 ft (167 m) high, situated on the Columbia River in the north-west of United States, and completed in 1941. Site investigations and geological mapping revealed that granite existed beneath a cover of glacial deposits which included river silts and gravels (p. 43). This cover varied in thickness from 50 to 300 ft (27 to 100 m). Some of the boreholes used for exploring the site were 36 inches (0.9 m) in diameter, and geologists and engineers could be lowered into the ground to inspect the foundations *in situ*. The granite was proved to be sound and adequate for bearing a gravity dam. Considerable difficulties were experienced in excavating the granite, because horizontal sheet jointing was encountered which did not appear to diminish in frequency with depth. It was eventually realized that these joints were being formed by the release of *in situ* stresses, following the unloading consequent on excavation. At the suggestion of the geologist, Dr. C. P. Berkey, excavation was stopped and the dam built, thus replacing the load. The ground was grouted to seal any other joints which may have existed at depth, and the foundations have since proved to be satisfactory. Aggregate for the concrete of the dam was obtained from suitable glacial gravels situated a mile from the site. Further information on the Coulee and neighbouring dams is provided by Berkey (1935).

Mangla Dam in West Pakistan is situated on the Jhelum River and forms part of the Indus Basin Scheme in the head-water tributaries of the River Indus. The scheme involves three embankment dams: Mangla which is two miles (3.2 km) long; Sukian, three miles (4.83 km) in length; and Jari, one-and-a-half miles (2.4 km). All are designed to withstand rapid drawdown, and earthquakes. They have been built so that they can be raised by 40 ft (12 m) at some future date. This meant that the cut-offs had to be designed with reference to the ultimate head of water. The geology of the area consists of an inter-bedded series of weakly cemented sandstones, stiff clays and silt-stones; all are of Miocene age and form part of the Siwalik Series. They are overlain by gravelly river terraces and flood plain deposits. It is estimated that between 4000 and 6000 feet (1830 m) of strata have been eroded at the Mangla site, leaving the foundation bedrock heavily over-consolidated (Binnie, 1967). The dam was designed to use the clay and sandstone excavated from the main and emergency spillways, the intake and tailrace, and other nearby structures. Gravel for the main dam was obtained from river gravel deposits and terraces. Cut-offs were required and difficulties were encountered with grouting, which tended to fracture the weak strata at moderate injection pres-sures (Little *et al.*, 1963). Excavations for the main works were well advanced when the site investigation on the Jari site revealed the presence of shear zones of low shear strength in many of the clay horizons. These were attributed to bedding plane slip during folding (p. 225). Re-examination of the sites at Mangla and Suki followed, and similar zones were found in many clays. This meant that the foundations were weaker than had first been thought, and remedial measures were required to ensure the stability of existing and future works. A toe weight was placed on the upstream side of the dam at Mangla; but construction of one section, an intake for diversion tunnels later to be incorporated in the hydro-electric scheme for the dam, was too far advanced to permit a change in design, and remedial measures there had to aim at reducing pore pressures in the bed-rock.

Farahnaz Pahlavi Dam (formerly known as Latiyan) was built between 1962 and 1967. It is situated on the Jaj-e-Rud 25 km N.E. of Tehran, Iran, and is a buttress structure with two abutment gravity blocks (Plate 14). The initial exploratory works were begun in 1960 at a time when the type of dam to be constructed (i.e. concrete or em-bankment) was undecided for lack of geological knowledge. The investigations had to establish rock types and quality, as well as structure. By 1961 it was realized that a buttress dam would be feasible. The foundation geology was complex and extensive ex-ploration was required to enable the final design to be completed, and an assessment made of the extent to which the strength of the rocks could be improved. Tunnels and adits totalling 915 m, in con-junction with several shafts and trenches and 110 bore-holes, were eventually used. Geophysical surveys were made at the surface and in the adits and tunnels; *in situ* shear strength, deformation and perme-ability tests, together with grouting trials, were also conducted. The geologists developed a classification based on estimates of the various geological characters which could together affect the engineering behaviour of the rocks, and the foundation area was mapped in these terms (Knill and Jones, 1965). As the investigations progressed it became apparent that the dam would have to be designed to cope with weak rock and variable conditions in the foundations. Much of the excavation was completed before any concrete was placed, so that a final review of the foundations and alignment could be made. The base of the buttresses was widened to spread the load (Plate 14), and an extensive system of subsurface drainage installed to strengthen the rock by controlling seepage and seepage pressures. Drainage holes were drilled from a concrete-lined drainage tunnel beneath the dam, sited just downstream of a multiple upstream grout-curtain. The left bank buttresses were placed on rock having a seismic velocity in excess of 200 m/s; the right bank buttresses, which were founded on more deformable material, were supported at their toe by a thrust-block founded on stronger quartzite which outcropped immediately downstream. The straight alignment of the dam was changed so that the right bank buttress thrust into the hillside, and consolidation grouting was used for strengthening and sealing the foundations. In addition, the dam had to withstand earthquakes, as the area lies close to one of the main seismic belts of the world. Excavated rock was only used for general fill, coarse and fine aggregate being obtained from a quarry opened in dolomitic limestone on the right bank. River gravels and sands were found to be potentially alkali-reactive and were rejected as a source for aggregate (Scott et al., 1967).

GROUND IMPROVEMENT

The numerous techniques which have been devised for improving ground can be applied to many engineering sites and their use is not restricted to the foundations of dams. They can be broadly divided into those which better the load-bearing properties of the ground, and those which control seepage.

The load-bearing properties of ground control its ability to withstand compressive and shear stresses. Tensile stresses could also be considered, but the design of most dams attempts to avoid the development of tensional stress in foundations. Compressive stresses promote a partial collapse of rock and soil structure by closing pores, fissures and other voids. Excessive settlement in the materials referred to by engineers as 'soils' mainly results from a re-arrangement of their weakly cemented mineral grains and can be avoided by partial consolidation of the ground prior to construction. Table 14.1 summarizes the methods currently employed and shows their average range of

Table 14.1   (after Mitchell, 1970)

Particle Size (mm)

>10	2	0.07	0.002	<0.0001
GRAVEL	SAND	SILT	CLAY	

```
– – ————Rolling and Pre-loading————————
  – ————Gravity drainage———————— –
  – ————Well-points with vacuum———— –
                           – – –Electro-osmosis—
————Vibro-flotation———————— – –
————Explosives———————— – –
————Grouts (see Table 14.2)———————— – –
              – – ————Chemical additives————
              – – ————Thermal treatment————
```

economic use. The geological factors relevant to each method are now briefly reviewed.

Rolling and pre-loading both apply a surcharge to the ground, the latter using water tanks, embankments and the like (Johnson, 1970a). Here it is important to know the distribution and thickness of layers of differing compressibility, if the anticipated settlements are to be obtained. A knowledge of the previous loading history of the site, and also its stratigraphical history, are relevant to design. The magnitude and rate at which settlement occurs is often related to the dissipation of pore pressures in the ground (Johnson, 1970b); this is important as pre-compression programmes are usually measured in terms of months rather than days. The geological factors involved in this aspect of pre-loading are similar to those for de-watering.

Gravity drainage and well-points (Note 10).   The permeability of the ground is a controlling factor here as are the variations in permeability which result from stratification (Rowe, 1968). Even thin layers of silt and sand are important, and their location and continuity within the ground should be established. They can be most variable, and geological advice should be sought when interpreting the site investigation reports of a large area. De-watering, besides being used for consolidation, is often employed in its own right for temporarily improving ground conditions by reducing pore-pressures and re-aligning, or reducing, seepage forces. As such it is used to assist in the excavation of trenches and in the stabilization of slopes. Usually a number of de-watering points are used, the total draw-down being the sum of the individual draw-downs around each installation. Most contractors have to de-water the foundations of dams in order economically to carry out their excavation: see, for example, Ruffle (1970) and instances in Chapters 13, 15, and 16.

In contrast to de-watering, some deposits, often of continental origin where deposition has involved little or no contact with water, can be consolidated by wetting. Loess is a well known example, although other deposits such as colluvium can behave in a similar manner (Fookes and Best, 1969). Such behaviour appears to result from the fragile bonding between the mineral grains; here, too, a knowledge of the mode of deposition and subsequent geological history is relevant to their 'pre-treatment'.

Electro-osmosis is a specialized technique sometimes used for the de-watering of fine-grained clayey materials. It is based on the fact that a flow of electricity through the ground can promote the movement of water. Because of this it has also been used for altering the direction in which seepage forces act when a control of these alone is sufficient to stabilize a slope or excavation. Further details and references to case histories are given by Mitchell (1970) (Note 11). The efficiency of this process can be gauged from easily measured geological factors:

1. the water content of the ground: the efficiency of the technique decreases with a decrease in water content;
2. the activity of the soil (p. 314, Note 9): the amount of water moved per hour per amp is greatest in inactive clays;
3. the chemistry of the pore water: in particular the presence of free electrolytes, which markedly retard the efficiency of the process in inactive clays.

Many side effects can accompany this method, e.g. the precipitation of secondary minerals around the cathodes, general desiccation of the ground from heat around the electrodes, the generation of low pH around the anodes: its use requires a fairly accurate knowledge of both the stratigraphy and the mineralogy of the area to be treated.

Vibroflotation can be used for increasing the density of loose granular deposits and works well in sands and fine gravels (Note 12). The best results are obtained in fairly coarse sands which contain little or no silt and clay, as the latter reduce the radius of ground which can be affected by the vibrating tool. Experience shows that the radius of influence decreases from approximately 7 ft (2 m) in clean sands to 2–3 ft (0.7 m) in sands containing more than 20% fines (Webb and Hall, 1968).

The geological features relevant to this operation are (a) the stratigraphy of the site, in particular the location and thickness of sedimentary layers of markedly differing grain size, and (b) the grain size of the deposits. Ground-water is not important to the process as the tool itself is jetted into the ground, but it may be of significance in later events because the columns produced by vibroflotation often act as drains, and can be paths up which water can migrate from horizons at depth. The method has been used to improve the ground at many engineering sites although its use in dam foundations appears to be limited. The United States Bureau of Reclamation (1948) describe its application in improving the foundations of the Enders Dam. Comparatively recently an adaptation of the method has been used to produce stone columns which can function as piles as well as drains (Luce, 1968).

Explosives are also occasionally used to increase the density of loosely compacted deposits. Their effect is similar to that produced by vibroflotation in that the detonation causes a liquification of the deposit; the escape of gas expels water from the ground. The charges are placed at some desirable depth below ground level, and major settlement occurs immediately after the blast. The method becomes a viable alternative to other techniques when large deposits have to be treated. The geological features relevant to this method are not known with certainty, but experience suggests that the best results are obtained in the same range of grain size that is suitable for vibroflotation. Two points of distinction are: (a) the level of the water-table is important and complete saturation of the ground is desirable for maximum consolidation; and (b) pockets and layers of fine material such as silt and clay can markedly reduce the efficiency of the method (Prugh, 1963). Variations in the cohesive strength of the sediment are therefore important. Hall (1962) describes the use of explosives in connection with the treatment of foundations for the Karnafuli Dam, Pakistan.

Grouting is the injection of a material of liquid consistency into rocks and 'soils' of any kind, where it sets after a period of time to become a permanent inclusion in pores and fissures (Plate 11b). The materials used for grouting fall into three categories:

(i) Suspensions of solid particles in water. Cement, clay and pulverized fly-ash slurries are examples; various mixtures can be used.
(ii) Emulsions, such as bitumen in water.
(iii) True solutions which form insoluble precipitates after injection; many of the so-called chemical grouts are of this type.

Grouts are commonly used for filling voids, as this reduces the permeability and the compressibility of the ground. They can also be used for increasing the cohesion between solid surfaces such as the contacts of mineral grains, bedding planes and joints, so improving the strength of rock in both tension and shear. The success of a grouting project is closely linked to the correct choice of grout, as each kind has its limitations (Bowen, 1975). Geology should affect this choice as the following factors are usually relevant to grouting projects:

(a) porosity, (b) flow of water through the ground, (c) permeability, (d) *in situ* stresses, (e) ground-water pressures, (f) ground-water chemistry, and (g) the variation of these characters over the site.

Porosity here includes all interconnected voids of any kind or origin. Pores and joints usually account for the majority of the voids, but other features may be present, e.g. solution channels, swallow holes, lava tunnels, underground workings, adits, wells, bore-holes. (See, for example, Khan *et al.*, 1970). General surveys of these features should enable the volume of grout needed to be estimated (Note 13). However, the accuracy of such an estimate depends on whether the volume of ground that will receive the grout is known. Two factors complicate this calculation: the permeability of the ground and the flow of water through it; although the two are related they are not synonymous.

When the flow of water through a rock mass is capable of transporting grout which is injected, it will markedly affect the efficiency of the programme. Normally this is not insuperable but care must be taken to establish the conditions which exist in materials having high field permeabilities. Special care is needed when chemical grouts are used (Karol, 1968).

Permeability is also important for another reason; it controls the time taken for a grout to migrate from its point of injection. As a rule injection becomes difficult when the permeability of the ground is less than 10^{-4} cm/s. Most grouts that are used are suspensions, and the size of the suspended particles should be related to the size of the voids through which they are expected to flow. The filter effect referred to on p. 400 is a useful guide, and experience has shown that for granular materials such as sands and gravels the ratio $D_{15 \text{ soil}}/D_{85 \text{ grout}}$ should exceed 25. In fissured rocks the problem is slightly different, and the width of the fissure has to be considered. As a guide, it is usually accepted that the ratio $D_{\text{fissure}}/D_{\text{max. grout}}$ should exceed 3. Experience shows that there is an upper limit to this ratio, as large quantities of grout have been lost from sites via open fissures and similar channels, which allowed it to travel considerable distances from the area of injection. Open structures in rocks should be noted, as the grouting procedures for them can differ from those used for filling openings of smaller dimensions. Table 14.2 provides a guide to the type of grout that is suitable for various geological situations.

Initial estimates of the 'groutability' of ground are often based upon the results of water tests; in particular pumping-in tests, in which water is pumped into the ground via a bore-hole. The dynamics of the test are analogous to those of grout injection (Note 14). The results are normally described in terms of Lugeon units, where one Lugeon = a flow of one litre/metre/minute at an excess pressure of 10 kg/cm^2.

Table 14.2   (Copyright N. G. Reid)*

Material	Grain Size	Grout
Rock		Cementicious grouts
Weathered rock		Cementicious grouts
Gravel (coarse)	20 mm	Sand filler grouts
Gravel (medium)	6	Fly ash filler grouts
Gravel (fine)	2	Clay filler grouts
Sand (coarse)	0.6	Bentonite filler grouts
Sand (medium)	0.2	Bentonite clay gel grouts
Sand (fine)	0.06	Chemical grouts (high viscosity)
Silt (coarse)	0.02	Chemical grouts (medium viscosity)
Silt (medium)	0.006	Chemical grouts (low viscosity)
Silt (fine)	0.002	Non-impregnating grouts
Clay		Non-impregnating grouts
Weathered clay		Non-impregnating grouts

* By permission of Geochemical Services Ltd., London.

Dam foundations are usually grouted if their absorption of water during such tests is greater than one Lugeon. Useful though these tests may be, they are no substitute for actual grouting tests. (See Little *et al*, 1963; and Geddes and Pradoura, 1967.) The rate at which grout can be injected into the ground generally increases with an increase in the grouting pressure, but there is a limit to this because excessive pressures will cause the ground to fracture and heave. The maximum pressure which can be safely used depends on the geology of the site. Parameters to note, in addition to the depth of the point of injection, are the cohesive strength and coefficient of friction of the ground, the *in situ* stresses and the ground-water pressures. As there is no simple relationship between these factors and the maximum grouting pressure that can be safely used, engineers employ hydraulic fracture tests to determine the optimum pressures which can be sustained (Note 15). This is better than using a rule-of-thumb such as '1 *psi* per ft. of cover', which assumes that the ground has no tensile strength and that the ratio of the principle stresses is unity: a change in either of these can allow grouting pressures to exceed overburden pressures. Although the geological parameters are rarely used to predict grouting pressures, a knowledge of their magnitude is helpful in understanding better the response of a site from test results.

The relevance of the chemistry of ground-water to the choice of grouts has been realized since the late 1870's, when cement grouts were first used (Glossop, 1960 and 1961). pH is taken as a guide to the aggressiveness of the ground-water. Super-sulphated cement grouts are now available for use where ground-water has a pH as low as 3.5.

These can resist the highest concentrations of sulphate usually encountered in natural ground-water, and can be used in areas subject to sea-water intrusion and acid water from peat. Another sulphate resisting grout is made from high alumina cement but this should not be used in waters containing caustic alkalies. As the chemistry of commercially available grouts becomes more sophisticated, so the chemistry of the ground-water becomes more important in controlling the choice of grout. For example, various agents are used for controlling the setting times of emulsions, but these will vary according to the ground-water chemistry. The suitability of clays for clay grouts is a function of ground-water chemistry, as this affects the viscosity of the grout and the strength of the resulting gel. Increased water temperatures in association with hot springs and other sources of heat can affect setting times, and appropriate grouts are required. It is therefore important to establish the physical and chemical characters of the ground-water at a site, and to consider the chemistry of the water that will pass through the grouted zone once the structure is built (Note 16).

The variation in natural parameters over a site can be important in a grouting project because it necessitates a tailor-made grouting programme. This may simply mean changes in the spacing of grout holes, but in some cases it may also require changes in the viscosity, setting time and grain size of the grout itself, depending upon the position of the injection hole within the geology of the site. Permeability is the feature most likely to change, especially in stratified deposits when the stratigraphy may be complicated, as in valley fill. (See Haffen, 1962; Ischy and Glossop, 1962). It should also be remembered that the direction of greatest permeability in many sediments lies parallel to their bedding, and that it can be much lower in directions across bedding. Ideally every geological horizon and structural unit should be considered in terms of its grouting peculiarities. An important factor in rocks is the orientation of joints and other discontinuities, and grout holes should be oriented so as to intersect the largest number of such discontinuities. Kreuzer and Schneider (1970) describe how such orientations can be obtained; good geological mapping of the ground is a fundamental requirement. Adits are often driven, especially in abutment areas, so that grouting operations can proceed from inside the ground (Soejima and Shidomoto, 1970).

Chemical additives are normally designed to react with the ground, rather than with themselves as occurs in many grouts. The stabilization of the slip at Kashio (p. 385) by the use of a solution of calcium salts is one example; the use of sodium chloride to control expansive clay minerals is another (Highways Research Board, 1966). The technique is, in general, considered only when no other method is likely to give a permanent result. This is because the most chemically active soils are usually clays, and their low permeability makes them difficult to treat (Note 17). Further, elaborate geological investigations are needed to ensure the success of chemical treatment because it largely depends on a detailed knowledge of the clay mineralogy.

Thermal treatment covers both freezing and heating of the ground at a site.

(1) *Freezing* is carried out by passing a refrigerant, usually brine but sometimes liquid nitrogen, through a series of tubes which are inserted into the ground. If the tubes are sufficiently close to each other, the cylinders of frozen ground which develop around them coalesce to form an underground wall of ice. The method temporarily improves the strength of weak ground while excavation is in progress, (Plate 15a) although it has been used as a permanent means of stabilization. Gordon (1937) describes how the freezing process was used to support an excavation on the east side of the Grand Coulee Dam (p. 419). Details and other case histories are given by Sanger (1968). Four geological factors affect the design and behaviour of freezing systems:

(i) The presence of ground-water: total saturation is not necessary and the system will work in rocks above the capillary fringe, but the presence of some water is necessary and the moisture content of each unit to be treated should be known.

(ii) The rate of ground-water flow: here it is the velocity of flow *in the pores* that is important, and not the average value obtained from the Darcy formula (p. 310) (Note 18). Careful measurements of *in situ* permeability and head are required, together with some value for the size of the pores, obtained from a grading analysis (p. 303). Experience shows that the maximum velocities that can usually be tolerated are between 3 and 4 feet (90–120 cm) per day.

(iii) The temperature of the rock or soil and, if special design calculations are required, its thermal conductivity. Here again, each geological unit should be investigated.

(iv) The ground-water chemistry: this may sometimes be significant in coastal and estuarine areas where the pore-water may have a high concentration of chlorides. The effect of tidal fluctuations should be noted, when present.

(2) *Heating* of the ground at a site to high temperatures (600 to 1000°C) is a process that has been developed and exploited mainly in eastern Europe. It is carried out by burning liquid or gas fuels under pressure in sealed bore-holes, and can result in a significant increase of strength, in some circumstances competing economically with piled foundations for large structures in thick sedimentary layers. It has been extensively used for improving ground prior to construction and for stabilizing landslides (Litvinov *et al.*, 1961). Geological

features of importance to this process are: (1) the presence of water; the method is best suited for dry or partly saturated materials, and is adversely affected by flowing water. (2) The mineral content of the rocks to be treated: the success of the process largely depends on the conversion of clay minerals; temperatures above 100°C drive off the adsorbed water on clay particles and so increase the strength of the mass. Between 400 and 600°C irreversible changes begin in many clay minerals which make them less sensitive to water. At temperatures greater than 900°C certain clays begin to fuse to a brick-like material. The heating process is therefore mainly used in soils which have an appreciable content of clay.

Ground anchors   These are either cables or rods which have been inserted into the ground and there anchored, so that they can withstand tensile forces applied to their exposed ends (Note 19). The tensions can vary with the application of the anchor, but two procedures are commonly used in engineering practice:

(i) to tension an anchor with a terminal nut which reacts against a prepared face bedded against the rock. This compresses a roughly cylindrical zone coaxial with the anchor. The method is used for tying retaining walls to the sides of excavations, for supporting excavations that would otherwise collapse, for pinning down structures against the uplift forces of ground-water, and for increasing the normal load on sliding surfaces. (See Brown *et al.*, 1971). The strengthening of a rock abutment on the left bank of the Kawamata Dam, Japan, with the aid of anchors is described by Lancaster Jones (1968); this thin walled arch dam is 400 feet (122 m) high. Anchors are also frequently used to support the strata surrounding underground works in hydroelectric schemes. Other applications include the anchoring of gravity dams; see for example Vallarino (1971); Gosschalk and Taylor (1970); and 'The Consulting Engineer' May 1970.

(ii) to place a terminal nut on the exposed end of an anchor and allow the deformation of the rock mass, as it expands against the nut, to tension the anchor. This is more a form of suspension, and although used successfully in the support of slopes and underground works, it is not considered here to be an example of ground 'improvement'.

The first method increases the compressive stresses in the zone around the anchor. If the resulting forces are oriented favourably, relative to surfaces of potential failure, they can be sufficient to prevent failure occurring. Figure 14.8 illustrates a simple situation. Many rock fabrics, from very fine to very coarse, can be improved in this way (Lang, 1961).

The success of an anchor therefore depends, first, on its length—it should be secured in stable ground (although this need not be the case when suspension is operating: see Obert and Duval, 1967); and

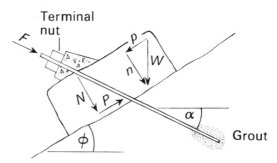

Fig. 14.8   Anchorage of a block in limiting equilibrium. Dip of sliding surface $= \phi =$ its angle of friction.

W = weight of rock per unit area;

$p = W \sin \phi$;

$n = W \cos \phi$.

F = weight per unit area applied via the anchor;

$P = F \cos (\phi + \alpha)$;

$N = F \sin (\phi + \alpha)$.

$$\text{Factor of safety for the block} = \frac{\tan \phi \, [\, W \cos \phi + F \sin (\phi + \alpha)\,]}{W \sin \phi - F \cos (\phi + \alpha)}$$

second, on its orientation, so that the resulting forces improve stability rather than promote failure. The design of ground-anchor systems in rock is facilitated if certain geological data are available:

(a) the position of sound rock and/or stable ground, which governs the length of the anchor;
(b) the dip and strike of discontinuities;
(c) the spacing of fractures in any given direction; and
(d) the strength, and in particular shear-strength, of the ground.

The last three items affect the spacing, tension, and orientation of an anchor. In very broken ground it is virtually impossible to design the system to strict geological controls and in such a situation an estimate of the isotropy of the geological structure may be of more help than details of numerous joints and fissures; see for example the *RQD* system (p. 247), and Goodman and Ewoldsen (1970). Geological advice should, however, be sought especially if secondary bolting (i.e. bolting in addition to that considered sufficient for general support) is anticipated. This problem will invariably arise if the structure of the ground and the range of rock-block sizes has not been adequately studied in advance.

Other geological factors affecting the *long term* improvement gained by an anchorage system include (i) change in permeability resulting from the compression of the ground; (ii) corrosion, which can arise from chemical factors; (iii) creep, which can develop during the life of an anchor; and (iv) surface weathering around the exposed terminal nut. Taking these in order: permeability can decrease as a result of compression, and drainage facilities can be provided, if necessary, when ground-water conditions are known. Corrosion usually results from either stray electrical currents in the ground, or aggressive ground-water. Geophysical investigation may be necessary to determine whether cathodic protection is desirable, and ground-water analysis will give information about pH and the content of sulphates, chlorides, and free CO_2. Creep is difficult to determine as it is so time dependent; the tension on an anchor decreases with time and this can be attributed either to failure of the anchor or to creep (Price and Knill, 1967). Severe difficulties in this respect have arisen in some fissured clays. Provision is normally made for the tension on permanent installations to be monitored and periodically adjusted. Weathering can be a minor nuisance when it promotes disintegration of the rock around a terminal nut, and rocks susceptible to this should be protected.

UNDERGROUND RESERVOIRS

Industrial development is now necessitating the storage of vast quantities of raw materials, such as oil, gas, and water. Solids can be stock-piled in tips as free-standing structures, but fluids must be retained either in a reservoir or in a container specially constructed. It is widely recognized that the storage of such materials at ground level is becoming increasingly difficult in developed areas as land becomes more and more at a premium. There is consequently a trend to use, wherever possible, any space that may be available below ground for the storage of fluids; underground reservoirs have therefore been developed. Such reservoirs, into which fluids may be injected and stored until required, may be the natural voids in rocks, e.g. the pores and fissures in an aquifer, man-made chambers such as worked-out mines, and cavities made specially for the purpose of storage. The subject is related in many ways to topics discussed earlier; e.g. aquifers (p. 350) can be natural reservoirs for oil and gas (p. 433) as well as water. Artificial recharge (p. 343) can be used to recharge any fluid to the ground including industrial waste (p. 495). These will not be discussed. Instead the emphasis of this section is on the geological factors that are pertinent to either the creation or utilization of reservoirs which have not been mentioned elsewhere. A few examples are given for illustration.

Storage of water. Cut-offs in aquifers have been constructed in some countries to retain water in certain geographical areas. These

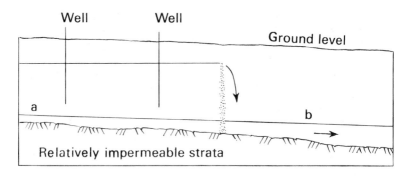

Fig. 14.9 An underground dam. a = original water level; b = unchanged water level downstream of the dam.

cut-offs act as underground dams, and prevent the natural sub-surface drainage of ground-water to lower levels in the aquifer, downstream of the area of infiltration. Their general effect is illustrated in Fig. 14.9. Wells can be installed to tap the reservoir and draw on the stored water when required. This system is generally employed in narrow aquifers such as valley fill, and is particularly effective when the aquifer overlies relatively impermeable strata. The advantages of such a dam are: (i) it requires no support, as that is provided by the ground; (ii) it need not be completely water-tight; (iii) it requires little or no maintenance; (iv) in the event of failure little or no damage would result; (v) it uses the storage potential of an aquifer, and the stored water is protected from evaporation losses; (vi) the risk of polluting the stored water is reduced, as all water reaches the reservoir by percolation and is thus filtered (Note 20).

Storage of oil.   Crude oil is now being stored in underground chambers that have been specially excavated for the purpose. Sites in salt domes are commonly used, and Lechler (1971) describes instances from Germany. Their popularity arises from the fact that evaporites are virtually impermeable to oil, and caverns can be excavated in them by leaching from specially designed boreholes. One design is shown in Fig. 14.10. This type of cavern has economic advantages in that it is cheaper to build than a surface structure, per unit volume of storage available; it requires little or no maintenance and is far less vulnerable. Disadvantages, however, arise mainly from geological problems which include: (1) the stability of the excavation (Chapter 15); pear-shaped caverns in evaporites are often chosen for their stability against collapse; (2) the solubility of the evaporite, which is important if a brine is used as the displacement fluid on the withdrawal of oil. Each 'turnover' results in enlarging the cavern, and vigilance is needed to ensure that the thickness of wall separating neighbouring caverns

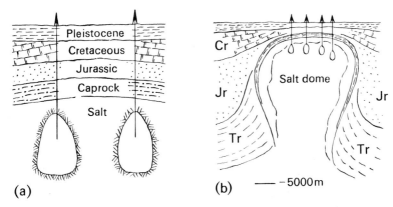

Fig. 14.10 Storage caverns in the Lesum dome, Germany. (*a*) General design. (*b*) Position of caverns within the salt dome. Tr = Triassic (after Lechler, 1971).

does not decrease beyond a prescribed limit; (3) injection pressures should not exceed the maximum injection pressure suitable to the ground (p. 426 and 495).

When salt domes are not available other chambers can be used. Morfeldt (1970) describes how large, unlined, underground excavations in hard unfissured rock have been made in Scandinavia for the storage of oil and gas. Similar installations operate in Finland (Kilpinen, 1970) and America. These caverns, unlike those in salt domes, are in hydraulic connection with the surrounding ground-water, and act as sinks to which ground-water will flow. Extensive migration of the oil or gas stored in them will therefore not occur as long as the specific gravity of the liquids is less than that of water. Problems associated with such systems arise, again, almost wholly from geological conditions and can be troublesome; when more than one cavern is used there is a risk of leakage from one to another, depending on how they are operated.

Storage of gas. The increasing use of natural gas as well as the products of oil refining has brought about a need for greater storage space to cope with fluctuating demands and peak loads. A simple storage method is to force gas through injection wells into porous aquifer horizons; ethane, propane, and butane are stored in different parts of the same horizon in the Borregos field, Texas (Turnbull, 1969). Each gas is injected into separate wells and saturates the rock in the immediate vicinity (Fig. 14.11). This is an elastic system, in that when gas is withdrawn its volume is replaced by ground-water. Between 70 and 80 per cent of the injected product is normally recoverable. Apart from problems attendant on injection (p. 495), difficulties of a geological kind may arise from contamination by natural gas already in the aquifer. If the geological conditions are

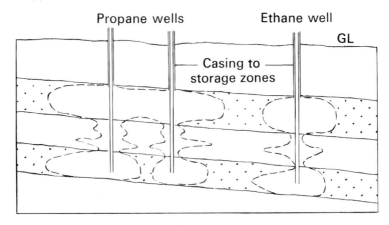

Fig. 14.11   Underground storage of gas. Two major storage horizons are shown. Zones storing gas are delineated.

suitable, this method of storage can be applied to large volumes of rock. Figure 14.12 shows how a simple anticlinal trap could be used. Many traps of this and other forms can be put to use if needed.

Another method is to store the gas in underground caverns that are ringed by frozen ground. A cylinder of sediment is frozen to the required depth and the unfrozen ground in the centre is removed; a gas-tight roof is then attached. Once the tank has been completed the temperature of the liquified gas will maintain its frozen walls (Fig. 14.13). Installations of this kind have been completed at Canvey Island, Essex, and were the first of their type in Europe. ('Consulting Engineer', 1967). Geological factors of special significance to these

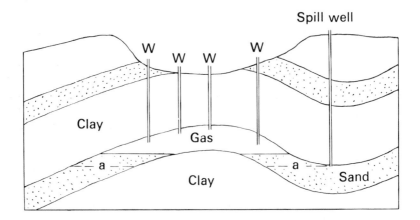

Fig. 14.12   Underground storage in anticlinal trap. a = limit of reservoir; w = injection and recovery wells linked to supply lines.

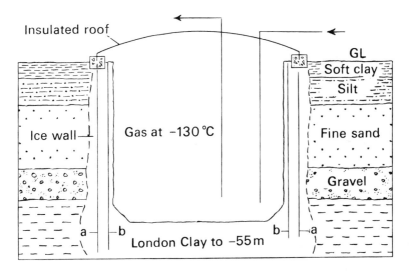

Insulated roof

GL

Soft clay

Silt

Ice wall

Gas at -130 °C

Fine sand

Gravel

a — b   London Clay to -55 m   b— a

Fig. 14.13   Gas chamber in frozen ground at Canvey Island. The chamber is 38.4 m (128 ft) deep and 39 m (130 ft) in diameter. a = outer ring of freezing tubes, b = the inner ring. The base of this chamber was frozen because the London Clay is underlain by sands containing artesian water pressures which would have caused the floor to heave. Had this not been the case only the walls would have been frozen.

structures are largely related to assessments of strength and insulation properties of the frozen material (Sanger, 1968; Corbett and Davies, 1966), Note 21. Various other designs for the underground storage of cryogenic fluids are given by Eakin *et al.* (1967).

Abandoned mines and underground cavities in hard, jointed rock have also been used for storage (Kelly, 1965, and Morfeldt, 1970). However, considerable design difficulties have to be overcome to make these projects work. The main problem is to keep the cooled gas in the cavities. If the gas temperature is about 0°C it must be stored at depth so that the ground-water pressure exceeds its vapour pressure at that temperature. The difficulty can be partly overcome by liquifying the gas, but at low temperatures of $-100°C$ or under, the surrounding rock begins to contract, so that fissures are continually opening. At the present time, unlined cavities of this type are generally unsuccessful.

Some geological conclusions. Four points of a general nature can be noted in connection with the extent to which geology controls the success of such schemes.

(1) The size of any scheme is controlled by the volume of storage available, which in turn depends on the geology of the site. Natural spaces in rocks (e.g. pores, joints, fissures) have been formed by past geological processes, and their usefulness may be limited by factors

such as cementation and overburden pressure. New excavations, especially if unlined, should be suited to the geological conditions.

(2) The successive stages of a scheme should be guided by the character of the site, especially when more than one cavern is to be formed.

(3) The usefulness of a scheme depends on the extent to which injection and abstraction rates can be maintained without damage to the installation. Disregard of geological controls may result in problems of subsidence, hydraulic fracturing, or pollution.

(4) The life of a scheme depends on the stability of the geological conditions which have been adapted. Caverns must neither collapse nor leak and porous reservoirs must continue to accept and yield the products stored within them.

NOTES

1. The use of silt traps is one of various ways in which the problem of reservoir sedimentation can be reduced. Many reservoirs which receive much silt have dams with specially constructed outlets, designed to permit the periodic flushing out of silt. It should be noted that silting of a reservoir progressively changes the profile of the valley floor, and if this is left unchecked it can promote sedimentation in the rivers upstream of the reservoir, a process called aggradation (in contrast to degradation). Care should be taken when reviewing the performance of old reservoirs in order to aid future design. It is commonly accepted, for example, that sedimentation in the Elephant Butte Reservoir on the Rio Grande was responsible for raising the river channel 13 ft (4 m) at San Marcial, 10 miles (16 km) upstream of the dam. Leopold *et al.* (1964) have demonstrated that many other factors contributed to this rise.

2. A more general approach based on the cohesiveness of the ground has been suggested by André and Anderson (1961). It is typical of many in this field of study, and illustrates that the erodibility of topsoils decreases with increased cohesion.

3. This may be needed if the impounded water is to stay in the reservoir, as may be the case with coastal pumped storage schemes, where sea water is pumped into inland reservoirs. Other methods of stopping leakage include cut-offs and grouting.

4. An excellent photograph of a concrete cut-off can be found in the 5th Edition of this book (Plate XV).

5. Gravity dams are normally made from high quality concrete that will withstand the great stresses at their heel and toe. However, the central part of the dam, in cross-section, is not stressed to the same degree and much of the potential strength of the concrete is wasted. The basic idea of the buttress dam is to reduce the total volume of concrete used, and to concentrate it in the areas where the stresses are highest.

6. An example is the Jari Dam in Pakistan (Binnie, 1967); the axis of the

dam followed the strike of the bedrock so that the silt core could be placed on a clay horizon which would then act as an impermeable cut-off.

7. Cut-offs can be made by excavating trenches and filling them with either clay or concrete; by driving sheet piling into the ground; or by grouting, using either liquid cement or some chemical filler. When a dam is built on very permeable ground, usually Recent sediments, a horizontal apron of clay or concrete can be extended some distance upstream from the dam: a most interesting example is described by Pira and Bernell (1967).

8. The hundreds of case histories that are now available show that grouting often fails satisfactorily to solve the problem of uplift pressures, and it is generally recognized that drainage is usually a far more efficient way of coping with this problem.

9. The term 'masonry dam' should strictly be applied only to those dams whose hearting and facing is of stone bedded in mortar: the Vyrnwy dam is such a structure. The term has since been loosely applied to dams containing large stones ('plums') bedded in concrete and faced with stonework. Many so-called masonry dams are in fact masonry-faced concrete dams.

10. The water flowing to a gravity well does so under the force of gravity, in response to the hydraulic gradient that is caused by lowering the level of water in the well. In fine-grained sediments such as fine sands and silts the permeability is so low that gravity alone is insufficient to overcome the frictional forces that retard flow. Pumping from these materials is usually ineffective, as the wells become drained before there has been any appreciable de-watering of the surrounding ground. In these situations a vacuum can be applied and well-points are normally used for the purpose. They consist of a perforated filtered intake section, the 'point', which is joined to an unperforated stem; diameters rarely exceed 4 in (10 cm). The well-point units are forced into the ground, usually by jetting but sometimes with percussion, until their point is at the required depth. Their stems are then linked to a common header pipe which connects them to a vacuum pump, situated so that the maximum vacuum head it can develop is not exceeded by the lift required to draw water from the points. (For further details see Leonards, 1962; also see Plate 11a).

11. The basic installation consists of metal conductors set at regular intervals in the ground. These are connected to an electric circuit so that one set acts as anodes and the other as cathodes. When current is switched on water will slowly migrate from the anodes towards the cathodes. Well-points are generally used as cathodes because their metal pipe acts as a conductor, and the well-point allows water migrating towards it to be discharged from the ground.

12. The process is begun by jetting a heavy vibrator some 15 in (38 cm) diameter and about 6 ft (2 m) long into the ground. Having reached some predetermined depth the jetting is stopped and the vibrator is started. Vibrations are transmitted to the ground as the tool is slowly

withdrawn to the surface. Extra gravel and sand is poured into the hollow which results as the column of ground around the tool settles to a higher density. Relative densities of at least 70% can often be obtained. (See reference Vibroflotation Foundation Co.).

13. In some cases it may be necessary to consider the material filling any fissures and large voids, as this may be of an undesirable nature, e.g. clay or sand, and have to be either treated or removed and replaced by grout. Treatment of this sort was required at the Kariba Dam. (Lane, (1963).

14. The test is conducted between two packers (Fig. 9.11), with the pressure of water in the test section between the packers exceeding that in the ground at that level. The amount by which it should exceed the existing water pressure is doubtful, but 10 kg/cm² is standard practice for a Lugeon Test (a packer test of this type). The accuracy of the permeability value obtained has been questioned, as the excess pressure can dilate fissures connected with the hole and allow unnatural flows to develop. Further, the distance between the packers, nominally 1 m, can also affect the flow rate and hence the value of permeability. Nevertheless the test does simulate grouting conditions and gives useful information.

15. The tests pump water into the ground at increasing pressures until fracture occurs. This is witnessed by a sharp rise in the rate at which water has to be injected to maintain the injection pressure. Optimum grouting pressures are therefore those which allow grouting to proceed at an economic rate, without disturbing the ground and so encountering excessive grout takes which will follow. A grouting system which takes account of flow rates and pressures has been described by Benko (1964).

16. The chemistry of reservoir waters is often different from that of the ground-water which existed prior to impounding. Note should therefore be taken of the ground-water chemistry of the catchment as a whole, particularly if this contains such features as peat, evaporites and mineral workings.

17. Electro-chemical injection has been developed to overcome this difficulty; the additive is carried through the ground by electro-osmosis. Materials having a permeability of 10^{-6} cm/s. have been successfully treated in this way. See Mitchell (1970) and Esrig (1968).

18. The velocity so calculated is obtained by dividing the discharge per unit time by the *total area* at right angles to the direction of flow, and not by the area of the pores alone. Hence the actual velocity of flow through the interstices is greater than the figure obtained. Farmer (1969) in a discussion of the leaching of cement from cast-*in situ* piles, describes how the real velocities can be assessed. The distinction applies equally to the success of grouting (p. 424). The use of tracers can be of great assistance in these investigations.

19. The kind of rock largely controls the choice of anchor, for example: *Medium to hard rock* favours the use of grouted anchors (cement or resin grouts), or dry anchors which expand against the sides of a bore-hole and transmit their load by friction.

In *soft granular material* grouted anchors with the anchorage embedded in either a bulb of grout, or multiple under-reamed anchorages, are commonly used.

Cohesive material can accept anchors similar to those used in granular materials, or embedded anchorages which bear against a column of grout in the bore-hole.

Further relevant points are: (i) fissures and cracks intersected by the bore-hole can result in loss of grout and the incomplete sealing of the anchor; and (ii) if maximum allowable grouting pressures are exceeded, the ground around the anchor is liable to fail (see p. 426).

20. The most serious disadvantage of such a scheme appears to be the question of its legality. The water rights of abstractors downstream of the dam must not be infringed. Once the reservoir has topped up it will overspill, and any additional surface water will pass downstream as if the dam did not exist.

21. This form of storage has been associated with many geotechnical problems, particularly fissuring of the ice wall and the subsequent leakage of liquified gas. See *New Civil Engineering*. 20 Nov. 1975.

REFERENCES

AILLERET, M. 1951. Estimation of Leakage from an Impounding Reservoir before Construction. *Jour. I. Water Eng.* **5**. See also *Jour. I. Water Eng.*, **12**, 1958 (Note after building of the reservoir).

ANDRÉ, J. and ANDERSON, H. 1961. Variation of soil erodibility with geology, geographic zone elevation, and vegetation type, in northern California wildlands. *Jour. of Geophys. Res.*, **66**, No. 8.

BENKO, K. 1964. Large Scale Experimental Rock Grouting for Portage Mountain Dam. *Trans. 8th. Int. Conf. on Large Dams*, Edinburgh, **1**, Paper 24.

BERKEY, C. 1935. Foundation conditions for Grand Coulee and Bonneville Projects. *Civil Engineering* (New York), **5**, No. 2.

BINNIE, G. 1967. The Mangla Dam Project. *Proc. Inst. Civ. Eng.*, **36**, 213.

BISHOP, A., KENNARD, M. and VAUGHAN, P. 1963. The development of uplift pressures downstream of a grouted cut-off during the impounding of the Selset Reservoir: in Grouts and Drilling Muds in Engineering Practice. Butterworth, London.

BOWEN, R. 1975. Grouting in Engineering Practice. Applied Science Publishers Ltd., England.

BROWN, G., MORGAN, E., DODD, J. 1971. Rock stabilization at Morrow Point Power Plant. *Amer. Soc. Civ. Eng.* (J.S.M. and F.D.), **97**.

CARDER, D. 1945. Seismic investigations in the Boulder Dam Area, 1940–44, and the influence of reservoir loading on local earthquake activity. *Bull. Seism. Soc. Amer.*, 35, No. 4.

CARSON, M. 1971. The Mechanics of erosion. Pion, Ltd., London.

CARTMEL, R. 1971. Construction work at Tarbela enters second stage. *Water Power*, **23**, No. 6.

CASAGRANDE, A. 1971. The liquification phenomena. *Géotechnique*, **21**, No. 3.

CASTLE, R., YERKES, R. and YOUD, T. 1973. Ground rupture in the Baldwin Hills—an alternative explanation: *Bull. Assoc. Engineering Geologists,* **10**, No. 1.

CHRISTIANSEN, J., THUN, J., and TARBOX, G. 1971. A new method for evaluating foundations, *Water Power*, **23**, March.

CONSULTING ENGINEER, THE. 1967. Underground liquid gas storage at Canvey Island, 31, No. 7.

CONSULTING ENGINEER, THE. May 1970. Ground anchors (a series of contributions).

CORBETT, R., DAVIES, C. 1966. Ground storage of LNG. Trans. Jour. Heat Transfer, *Gas World*, **164**, No. 4256.

DAVIS, D. 1940. The Haweswater Dam. *Trans. Liverpool Eng. Soc.,* **61**.

DEACON, G. 1896. The Vyrnwy Works for the water supply of Liverpool. *Jour. Inst. Civ. Eng.,* **126**.

DOUGLAS, I. 1968. Some hydrologic factors in the denudation of limestone terrains. *Zeitschrift Für Geomorphologie,* **12**.

ENGLAND, G. and HOLTAN, H. 1969. Geomorphic grouping of soils in watershed engineering. *Jour. of Hydrology*, **7**, No. 2.

EAKIN, B., KHAN, A. and ANDERSON, P. 1967. Below ground storage systems for LNG. *Trans. Jour. Heat Transfer*, **1**.

ESRIG, M. 1968. Applications of electro-kinetics to grouting. *Proc. Amer. Soc. Civ. Eng.* (J.S.M. and F.D.), **94**.

FARMER, I. 1969. Leaching of cement from a concrete pile section by groundwater flow. *Civ. Eng. & Pub. Wks. Review*.

FOOKES, P. and BEST, R. 1969. Consolidation characteristics of some late Pleistocene periglacial metastable soils of east Kent. *Q. Jour. Eng. Geol.,* **2**, No. 2.

FORDHAM, A., COCHRANE, N., KRETSCHMER, J. and BAXTER, R. 1970. The Clywedog Reservoir Project. *Jour. Inst. Water Eng.,* **24**, No. 1.

GEDDES, W. and PRADOURA, H. 1967. Backwater Dam in the County of Angus, Scotland. Grouted Cut-off. *9th. Int. Conf. on Large Dams*, Istanbul, 1967, **1**, Paper 16.

GLOSSOP, R. 1960 and 1961. The Invention and Development of injection processes (in two parts). *Géotechnique,* **10** and **11**.

GOODMAN, R. and EWOLDSEN, H. 1970. A design approach for rock-bolt reinforcement in underground galleries, in *Large Permanent Underground Openings*. (Scand. Univ. Books).

GORDON, G. 1937. Arch Dam of Ice stops slide. *Eng. News Record*, Feb. II.

GOSSCHALK, E. and TAYLOR, R. 1970. Strengthening of Mud Dam foundations using cable anchors. *Proc. 2nd. Cong. Int. Soc. Rock Mech., Belgrade,* **3**, Paper 6/11.

GOTTSCHALK, L. 1964. Reservoir Sedimentation, in *Handbook of Applied Hydrology*, by Ven te Chow. McGraw-Hill Book Co.

HAFFEN, M. 1962. Grouting deep alluvial fill in the Durance River Valley, Serre Poncon Dam, France. Engineering Geology Case Histories, No. 4, Ed. Kiersch. *Geol. Soc. America*.

HALL, C. 1962. Compaction of a dam foundation by blasting. *Am. Soc. Civ. Eng.*, (J.S.M. and F.E.), **88.**

HENKEL, D., KNILL, J., LLOYD, D. and SKEMPTON, A. 1964. Stability of the foundations of Monar Dam. *Int. Conf. on Large Dams, Edinburgh,* **1.**

HIGHWAY RESEARCH BOARD. 1966. Statement of the art of soil stabilization with sodium chloride. *HRB circular 19, Nat. Acad. Sci. U.S.A.*

HILL, H. 1949. The Ladybower Reservoir. *Jour. Inst. Water Eng.*, **3,** No. 5

HOUSNER, G. 1969. Seismic events at Koyna Dam. *Proc. 11th. Symp. Rock Mech.* (Amer. Inst. Mining, Metall. and Petr. Eng. Inc.)

ISCHY, E. and GLOSSOP, R. 1962. An Introduction to Alluvial Grouting. *Proc. Inst. Civ. Eng.*, **21,** March.

JAMES, L. 1970. Failure at Baldwin Hills Reservoir, Los Angeles, California. Eng. Geol. Case Hist. No. 6, Geol. Soc. Amer.

JOHNSON, S. 1970. (a) Pre-compression for improving foundation soils. *Proc. Am. Soc. Civ. Eng.*, (J.S.M. and F.E.), **96.**

JOHNSON, S. 1970. (b) Foundation Pre-compression with vertical sand drains. *Proc. Am. Soc. Civ. Eng.*, (J.S.M. and F.E.), **96.**

JUDD, W. 1974. Editor. Seismic effects of reservoir impounding. Proc. Internat. Colloquium, UNESCO, London 27–29 March 1973. Elsevier Pub. Co. Amsterdam.

KAROL, R. 1968. Chemical Grouting Technology. *Amer. Soc. Civil. Eng.*, (J.S.M. and F.E.), **94.** See also discussion in Vol. 95, Sept. 1969.

KELLY, R. 1965. PSC's experience with storing natural gas in an abandoned Colorado coal mine. *Gas,* **41,** No. 6.

KENNARD, M. 1963. The Construction of Balderhead Reservoir. *Civil Engineering and Public Works Review,* **58,** 633.

KENNARD, M. and KNILL, J. 1969. Reservoirs on Limestone, with particular reference to the Cow Green Scheme. *Jour. Inst. Water Eng.*, **23,** No. 2.

KENNARD, J. and LEE, J. 1947. Some features of the construction of Fernworthy Dam. *Jour. Inst. Water Eng.*, **1,** 11.

KHAN, S. and ALINAQUI, S. 1970. Foundation treatment for under-seepage control at Tarbela Dam Project. *Trans. 10th. Int. Cong. on Large Dams, Montreal.* **2,** Paper 60.

KNILL, J. and JONES, K. 1965. The recording and interpretation of geological conditions in the foundations of the Roseires, Kariba and Latiyan Dams. *Géotechnique,* **15,** 94.

KREUZER, H. and SCHNEIDER, T. 1970. The orientation of Grout curtains according to the systems of discontinuities in the Bedrock. *Proc. 2nd. Conf. of Int. Soc. Rock Mechanics, Belgrade,* **3,** Paper 6/1.

KILPINEN, M. 1970. Underground oil storage in Finland, TUNCON 70 Conf., Johannesberg.

LANE, R. 1971. Seismic activity at man-made reservoirs. *Proc. Inst. Civ. Eng.*, **50** (and see discussion in *ibid,* **51,** 1972).

LANE, R. 1963. The Jetting and Grouting of Fissured Quartzite at Kariba, in Grouts and Drilling Muds in Engineering Practice. Butterworth, London.

LANCASTER-JONES, P. 1968. Methods of improving the property of rock masses, in Rock Mechanics in Engineering Practice, Ed. Stagg and Zienkiewicz. (Wiley & Sons).

LANG, T. 1961. Theory and practice of rock-bolting. *Trans. Amer. Inst. Mining, Metall. and Petr. Eng.*, 220.

LAPWORTH, C. 1942. A concrete dam. *Water and Water Eng.*, No. 45.

LECHLER, S. 1971. Storage caverns in German salt-domes. *Petr. and Petrochem. Internat.*, 11, No. 12.

LEONARDS, G. 1962. Foundation Engineering. McGraw-Hill.

LEOPOLD, L., WOLMAN, M. and MILLER, J. 1964. Fluvial Processes in Geomorphology (see Chapter 11). Freeman and Co.

LITTLE, A., STEWART, J. and FOOKES, P. 1963. Bedrock grouting tests at Mangla Dam, West Pakistan, in Grouts and Drilling Muds in Engineering Practice. Butterworth, London.

LITVINOV, I., RZHANITZN, and BEZRUK, V. 1961. Stabilisation of soil for constructural purposes. *Proc. 5th. Int. Cong. Soil Mech. and Fndn. Eng.*, 2.

LONDE, P. 1967. Discussion of Theme 6. *Proc. 1st. Int. Cong. of Int. Soc. Rock Mechanics, Lisbon*, 3, 449–453.

LUCE, A. Jan. 1968. The strengthening in depth of weak soils by vibration at Tolka Quay. *Inst. Civ. Eng.*, Ireland.

MAURY, V. 1970. Distribution of stresses in discontinuous layered systems. *Water Power*, 22, No. 5/6.

MITCHELL, J. 1970. In-place treatment of foundation soils. *Proc. Amer. Soc. Civ. Eng.* (J.S.M. and F.D.), 96.

MORFELDT, C. 1970. Significance of ground-water at rock constructions of different types, in Large Permanent Underground Openings, Ed. Brekke and Jorstad, (Scand. Univ. Books).

NICOL, T. 1964. The Warragamba Dam. *Proc. Inst. Civ. Eng. (Australia)*, 27.

NAYLOR, A. 1936. The second stage development of the Lochaber Water-Power Scheme. *Jour. Inst. Civ. Eng.*, 5.

OBERT, L. and DUVAL, W. 1967. Rock Mechanics and the design of structures in rock. Chapter 20. (Wiley & Sons).

PATON, J. 1956. The Glen Shira Hydro-electric Project. *Proc. Inst. Civ. Eng.*, 5, Sept.

PIRA, G. and BERNELL, L. 1967. Järkvissle dam, an earthfill dam founded under water. *Int. Conf. on Large Dams*, 4, 213.

PRICE, D. and KNILL, J. L. 1967. The Engineering Geology of Edinburgh Castle Rock. *Geotechnique*, 17, No. 4.

PRUGH, B. 1963. Densification of soils by explosive vibrations. *Amer. Soc. Civ. Eng.*, Jour. of Construction Div., 89, No. C01.

RANSOME, F. 1928. Geology of the St. Francis Dam site. *Economic Geology*, 23.

ROWE, P. 1968. The influence of geological features of clay deposits on the design and performance of sand drains. *Proc. Inst. Civ. Eng.*, Paper 7058-S. (Supplementary Vol.)

RUFFLE, N. 1970. The Derwent dam: design considerations. *Proc. Inst. Civ.*

Eng., **45**. (See also in same volume papers on the construction and stability of the dam).

SANGER, F. 1968. Ground freezing in construction. *Amer. Soc. Civ. Eng.* (J.S.M and F.D.), **94**.

SCOTT, K., REEVE, W. and GERMOND, J. 1967. Farahnaz Pahlavi Dam at Latiyan. *Proc. Inst. Civ. Eng.*, **39**, 353.

SHERRARD, J., CLUFF, L., and ALLEN, C. 1974. Potentially active faults in dam foundations. *Géotechnique, 24,* No. 3.

SIVASUBRAMANIAN, A. and CARTER, A. 1969. Investigation and treatment of leakage through Carboniferous rocks at Clubbiedean Dam, Midlothian, *Scottish Jour. of Geol.*, **5**, No. 3.

SOEJIMA, T. and SHIDOMOTO, Y. 1970. Foundation Improvement of an Arch Dam by Special Consolidation Grouting. Vol. 3, Paper 6/6. *Proc. 2nd. Cong. of Int. Soc. Rock Mechanics, Belgrade.*

TAYLOR, G. 1951. The Haweswater Reservoir. *Jour. Inst. Water Eng.*, **5**.

TURNBULL, W. 1969. Storage of LPG in gas reservoirs. *Jour. Petrol. Tech.*, **21**, No. 5.

UNITED STATES BUREAU OF RECLAMATION, 1948. Vibroflotation experiments at Enders Dam. *U.S.B.R. Research and Geol. Div.* Earth Lab. Secn., EM 178.

ÜRAL, O. 1967. Design and construction of earthquake resistant dams in Turkey. *9th Int. Cong. on Large Dams*, **4**. Istanbul. Paper 18.

VALLARINO, E. and ALVAREZ, A. 1971. Strengthening the Mequinezna Dam to prevent sliding. *Waterpower,* **23**, March and April.

VAUGHAN, P. 1969. See discussion of Kennard and Knill, 1969. *Jour. Inst. Water Eng.*, **23**, No. 2.

VIBROFLOTATION FOUNDATION CO. Soil Compaction by Vibroflotation. Pittsburgh, Pennsylvania, U.S.A.

WAHLSTROM, E. 1974. Dams, Dam Foundations, and Reservoir Sites. *Developments in Geotechnical Engineering,* **6,** Elsevier Sci. Pub. Co. Amsterdam.

WALTERS, R. 1971. Dam Geology. Butterworth, London.

WATER POWER, 1970. The Hongrin-Léman development. May to Sept.

WEBB, D. and HALL, R. 1968. Effects of Vibroflotation on Clayey Sands. *Amer. Soc. Civ. Eng.*, Speciality Conf. on Placement and Improvement of Soils. Session IV, Mass. Inst. Tech., Aug. 1968.

WIEGEL, R., *et al.* 1970. Water waves generated by landslides in reservoirs. *Amer. Soc. Civ. Eng.*, Jour. of Highways and Harbours Division, **96**.

ZIENKIEWICZ, O. 1968. Continuous Mechanics as an approach to Rock Mass Problems. Chapter 8 in Rock Mechanics in Engineering Practice. Wiley and Sons.

ZIENKIEWICZ, O., TAYLOR, G. AND GALLICO, A. 1970. Three-dimensional finite element analysis of the Tachien arch dam. *Water Power,* **22,** May/ June.

15 Excavations

In this chapter the geological factors that commonly affect the making of excavations, and their stability, are discussed. There are five sections: Excavation methods for geological materials; Surface excavations; Shafts; Subsurface excavations (including tunnels); and The disposal of excavated material. Dredging has been excluded but see Selected Bibliography (p. 485) for references.

EXCAVATION OF MATERIALS

Geological considerations such as those discussed here should be taken into account when deciding upon equipment and methods to be used; costly mistakes can sometimes be made if the nature of the rock or soil to be excavated is disregarded. The five aspects dealt with are:

(1) drilling
(2) mechanical tunnelling
(3) blasting
(4) scraping and ripping
(5) the clearing of broken ground.

1. **Drilling** is a technology that covers many fields, of which geology only is considered here; readers interested in the whole subject are referred to McGregor's book 'The drilling of rock' (see references, p. 484). As a general rule, rocks can be thought of in terms of their influence on the life of a bit and its rate of penetration (Note 1). Experience has shown that the most important characters are (*a*) hardness of the mineral constituents, (*b*) toughness of the rock, (*c*) abrasiveness, and (*d*) rock fabric and structure. The first three terms, hardness, toughness and abrasiveness, although rather ambiguous, have become an established part of drilling terminology. The relative importance of the four factors varies according to the drilling system used, hardness and toughness usually being of more importance to percussive drilling than abrasiveness and structure. In rotary drilling hardness and abrasiveness are commonly more important than toughness and structure.

(i) *Hardness*, as far as drilling is concerned, can be considered as the ability of one material either to penetrate or scratch another. Geologists have a scale of hardness for minerals, the Mohs scale (p. 76).

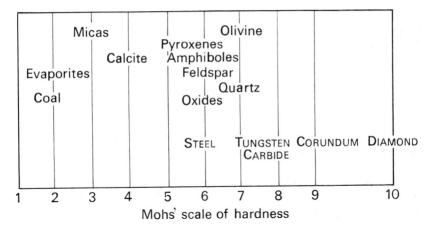

Fig. 15.1   Relative hardness of drill bits and geological materials.

Figure 15.1 shows the positions of certain rock forming minerals relative to commonly used drilling materials. All bits have to penetrate the material they are going to cut; and high pressure, hence strong bits, will be required to drill through hard materials. Two microscopic characters of rock are notable when assessing their drilling resistance; the mineral content, from which an average rock hardness in terms of the Mohs scale can be calculated, and the nature of the mineral boundaries, particularly their degree of interlocking and cementation. Both these factors are also important in determining toughness.

(ii) *Toughness* is essentially the resistance to fracture that comes from the cohesion between the mineral grains. For example, a calcareous sandstone has quartz grains cemented with calcite and is a relatively soft rock, whereas a quartzite consists of quartz grains cemented with silica and is a very tough rock. Interlocking of the grains, and the presence of mineral cleavage and fracture, also affect toughness, as do grain size and texture; it is common to find that coarse-grained rocks such as gabbros drill faster than their finer-grained equivalents, e.g. dolerites. It would appear therefore that 'toughness', as used in the drilling industry, describes the resistance to fracture that comes from the tensile strength of the rock. This is an important parameter because the efficiency of many drilling bits depends upon their ability to induce local tensile failures within the rock to be drilled (Note 2).

Lack of toughness, as found in many soft sediments and weathered rocks, is equally important because in such cases the teeth and flushing channels of the bits become caked and blocked, so reducing the rate of penetration. In these situations the nature of the drilling lubricant plays an important part, particularly in the recovery of core and the finished diameter of the hole. Most production drilling, outside the oil industry, uses either water or air; the latter usually gives the better core recovery, when required, and also better finished hole diameter— a matter which may be crucial for blast holes because the efficiency of any explosive depends upon its tightness in a hole.

(iii) *Abrasiveness* is a third term that has become an accepted part of drilling terminology. An abrasive material will scratch one that is softer, and abrasion results when materials of different hardness slide against each other. The problem of cutter wear tends to be more critical in rotary drilling than in percussive drilling (Note 3). Hardness, however, is not the only character that contributes to the abrasive qualities of a rock; toughness is also important. For example, a good bit in an abrasive yet weak material, such as poorly cemented sandstone, will produce large chips and suffer comparatively little wear; whereas a bit in tougher yet less abrasive material will probably produce smaller chips which have more contact with the bit and so produce greater wear. The continual blasting of bits by relatively unabrasive rock-dust can polish cutting edges as hard as tungsten and diamond, and so reduce their cutting power. A general guide to the abrasive characters of some commonly encountered rocks is given in Table 15.1.

(iv) *Fabric and structure* (and see p. 370) can be taken here to describe the network of discontinuities in a rock mass and the orientation of this network relative to vertical drill holes. An open discontinuity is a surface across which part of the energy controlling drill penetration is often wasted, either through interference with the flushing system or vibration of the bit. This occurs whilst penetrating new surfaces and can result in non-alignment of the hole. Rock-fabric is therefore a controlling factor in bit performance. In percussive drilling, for example, a four-point cross-bit usually gives a better performance in broken ground than a single chisel bit, because its four cutting faces have greater contact with the broken rock and give the bit less opportunity to stick in a single fissure or to wander from its prescribed course. The dimensions of the fabric also control the size of the fragments that can be recovered by coring tools, and it is usually found that recovery increases with increased core diameter. Drilling in broken ground may necessitate the use of some temporary wall support such as casing.

The presence of joints and other discontinuities that are filled with soft material can prove to be a source of delay, because the softer material can bind the bit, clog the flushing holes and permit the drill to wander off course. These problems may occur on a large scale in limestones where solution has riddled the strata with tunnels and

Table 15.1 Rocks Grouped by Certain Drilling Characters (after McGregor, Table 5, *q.v.*)

Igneous:
  Abrasive: Obsidian, rhyolite, aplite, felsite, granite (fine and coarse), pegmatite, quartz-porphyry, welded tuffs.
  Less abrasive: Basalt, dolerite, gabbro, andesite, diorite, syenite.
  Weathered igneous: decomposed rocks, e.g. kaolinized granite, serpentine, are usually ripped or dug, but drilling and blasting may sometimes be needed.
Metamorphic.
  Hard and abrasive: Quartzite,* hornfels, gneiss.
  Intermediate: Schist, marble.
  Softer: Slate, phyllite, schist, marble.
Sedimentary.
  Hard, siliceous: Flint, chert, quartzite, sandstone, ganister, quartz-conglomerate.
  Abrasive but less hard: Siltstone, pyroclastics, siliceous limestones, many sandstones.
  Friable, abrasive; friable sandstones and grits.
  Relatively hard, not abrasive: Limestones, mudstones, shales.
  Soft, not abrasive: Marl, mudstone, shale, chalk, coal, oolitic limestone.

* Usually the most difficult to drill of the common rocks.

opened many joints, often infilling them with sands and clays washed down from the surface. Such effects tend to be more severe in the older limestones which have been exposed to long periods of weathering.

The relationship between structure and the orientation of a drill hole can be an important factor in controlling both the rate of penetration and the alignment of the hole. Penetration is usually faster in holes which cross discontinuities at a steep angle than in holes which either follow discontinuities or cross them at a low angle. The latter course tends to jam the bit, encourage excessive wear, and lead to a deviation in the direction of the hole. Large deviation can result in the abandonment of a hole, which has to be back-filled and re-drilled on the correct course. Hence drilling down dip can be more difficult than drilling across dip, even though the latter may mean coping with different rock types. As a rule deviation tends to be greater in angled holes than in vertical holes, and tends to increase with distance from the rig (Note 4).

Anyone familar with rocks will know that their toughness, structure and other properties can often vary considerably and the choice of correct equipment may be difficult. Other problems can also affect drilling economics, such as headroom and manoeuverability of equipment. The experience of the drilling personnel is an important factor. Some guide-lines regarding the choice of equipment are given by McGregor (*loc. cit*) Fig. 11; simple tests for assessing the ease of drilling of rocks have been described in Chapter 10. It is generally advisable to consult manufacturers of drilling equipment before taking expensive decisions.

Augering.   This form of drilling can be briefly considered here. An auger consists essentially of a drag-bit cutting head which is connected to a spiral conveyor, or flight. As a tool it functions most efficiently in soft materials such as clays, soft shales, and weathering zones, and can produce holes with diameters up to 1.5 m (5 feet) or more. It can be used with casing in zones of weak or caving ground, and requires no lubricant such as air or water (Note 5). A major problem that may arise in augering is the presence of boulders, which make penetration very difficult if they are greater than one-third the diameter of the tool. Penetration rate and bit wear are largely governed by the characters noted previously, but the variability of the material (e.g. presence of boulders) and its toughness are usually more important than hardness, abrasion and structure.

2. **Mechanical tunnelling**   Many of the geological factors that influence drilling and auguring also affect the progress of mechanical tunnelling, commonly called 'mole digging'. This is in many ways an extension of drilling technology. The holes produced have a much greater diameter than those formed by single bits, and are often horizontal or inclined at a low angle. Its use as a constructional procedure is increasing, and it is now a viable alternative, in many materials, to the more conventional tunnelling techniques (Note 6).

The geological features which affect 'moles' are similar to those that influence other drilling methods as discussed above. In conventional tunnelling the drills have only to make small diameter holes; blasting does the rest, and as a result problems connected with rock hardness are usually subordinate to those of structure, which largely controls the question of support. In machine tunnelling progress is much influenced by hardness, toughness and abrasion, because the efficiency of the machines depends on their cutting-faces breaking *all* the ground. This is not to say that structure and hence support is unimportant, for slabbing and falls can occur at the face and from the arch immediately behind the machine, however, its shield can usually offer what support is needed near the cutting face, and additional support can be provided immediately behind the shield in the form of rings. Most delays, either from bit wear or slow rates of penetration, are due to hardness and toughness of the rock. It has been found that cutter costs for the machines vary almost directly as the compressive strength of the ground encountered. Standards of hardness, in terms of compressive strength, are approximately as follows:

Strength (lb/in.^{2})	Nature of rock
<10 000	Soft (10 000 (lb/in.^{2}) $= 700$ kg/cm^2)
10 000–20 000	Moderate hard
20 000–30 000	Medium hard
30 000–40 000	Hard
40 000–50 000	Very hard

These values are likely to change as cutter development improves the strength of bits.

It is important to know the character of the ground which is to be penetrated. Difficulties can develop when a tunnelling machine which has been designed for cutting sediments unexpectedly encounters a dyke, or when a machine designed for work in fissured clay hits the gravel fill of a buried channel. Geological homogeneity, at least in terms of strength, should therefore be considered when the use of such machines is proposed.

3. **Blasting**   As a rock-breaking process, blasting is commonly used to excavate tunnels, shafts, cuttings, trenches, foundations and quarries (Plate 15b). The explosives are employed in one of three ways: (1) down holes that have been drilled for blasting, (2) in shafts or headings that have been dug for blasting, (3) on the surface (plaster blasting). Numbers (1) and (2) are normally part of a primary blasting programme, while (3) is used in secondary blasting designed to break excessively large masses that have survived a primary blast.

The significance of geological factors for efficient rock-breaking by blasting is best appreciated from an understanding of the sequence of events which follows a detonation (Note 7).

In particular, the presence of 'free faces' in the rock is important; every joint, cleavage surface, or discontinuity of any kind provides a potential free surface which aids breaking. Hence the rock fabric is an important control in the blasting process; two fundamental geological observations can be made: (i) the orientation of the fabric (or structure of the mass), and (ii) the degree of contact across discontinuities in the rock.

(i) *Structure*.   Discontinuities act as free surfaces for reflection of the shock waves produced by an explosion and provide openings along which much energy is lost. The direction of greatest attenuation of the wave will be that which intersects most rock fractures, and the orientation of the rock-fabric therefore controls the directions of local breakage. Fissures are opened by the shock wave, and this relaxes the adjacent rock, allowing little tension to build up. As a result, the explosion opens existing cracks but produces few new fractures. Also the loss of gases along existing fractures results in a reduction of pressure which would otherwise have been available for assisting the formation of new fractures. Thus, blasting can result simply in loosening large masses of rock which then have to be broken by secondary blasting; this can be a considerable drain on the margin of profit of a contract.

In theory, therefore, blast-holes should run nearly parallel to main fissures in the rock, so as to intersect as few as possible. In practice this is rarely achieved, and the best compromise should be sought. Rock-structure also plays a part in determining the sequence of shot firing and the timing of detonations.

(ii) *Openness* of discontinuities is important, for open fractures cannot transmit tensile stresses unless there is some connection between them. Openings can therefore be stress traps. However, it is unlikely that the walls of fractures have to be in total contact, or even touching, for them to be able to transmit shock waves, as the accelerations following a detonation can cause surfaces that are reasonably close to touch.

A situation may develop where blasting simply follows existing fractures, giving a very uneven contour to the resulting faces, and producing large blocks which require secondary blasting. Considerable advances in overcoming these problems have been made with a blasting system called pre-splitting (Note 8).

Three further points should be mentioned in connection with primary blasting: *in situ* stresses, rock types and ground vibrations. *In situ* stresses affect the propagation of shock waves and hence the whole process of fragmentation. In surface excavations they are usually so small that they can be ignored, but in underground works they can be large and exert an influence upon the choice of blasting techniques. Severe overbreak can occur when these stresses are neglected.

Rock types which blast well are those that are reasonably strong. There is little point in blasting weak rock, which can normally be excavated with less trouble by other means. The rock must be capable of transmitting the energy for some distance away from the blast-holes. Unless this occurs a blast will only pulverize the rock adjacent to the holes, leaving the column between the holes unfractured. Perhaps the most difficult materials to assess from this point of view are fragmental rocks. One example was the blasting of a Devonian conglomerate that contained rounded boulders of quartzite, 0.6 m (1 to 2 feet) in diameter, set in a matrix of coarse sand cemented by silica. On blasting, the cohesion of the conglomerate broke and the quartzite boulders were hurled like cannon balls into the surrounding area. Cohesion is clearly important, and rocks having a high tensile strength, i.e. good cohesion, will hinder the initiation of cracks. Such rocks usually have an isotropic mineral fabric and interlocking grains. Unweathered gabbros, dolerites and hornfels are typical examples. Anisotropic fabrics as found in schists and slates usually have a cohesive strength which varies with direction, and this in turn accentuates the anisotropy of crack generation and propagation that occurs in rocks with well defined fabrics. Simple tests for assessing the blasting character of rocks are mentioned in Chapter 10.

It is often necessary to consider the ground vibrations that will result from blasting. Significant geological factors which control their intensity appear to be the compressional seismic velocity of the rock and its density. Some rocks will damp vibrations more efficiently than others, and hence the seismic velocity of a rock can be used as a guide to its transmissive properties. Geology plays a small, but important, part in the overall problem; a short review of the subject is given by Ambraseys and Hendron (1968).

4. **Scraping and ripping**   Under this heading are included techniques that are commonly used for removing fairly soft material such as clay, silt, shale, sand, weathered rock and top-soil. Scrapers, rippers, dozers, graders and excavators are the machines commonly employed, and experience indicates that they work best in ground that has a seismic velocity lower than 1000 m/s. The threshold for most of this equipment occurs in materials with seismic velocities in excess of 2000 m/s. No hard and fast rules can be made, but Table 15.2 is a guide to machine capabilities.

The geological features that are important to the efficiency of these

Table 15.2   Comparison of Machine Performance in Fairly Soft Material

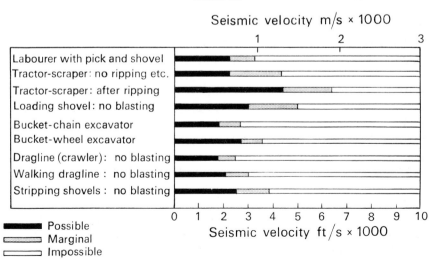

techniques are: (1) the *in situ* strength of the rock or soil and its range of variation, particularly in cohesion and frictional strength; (2) density of the material and its adhesion properties; (3) the extent to which (1) and (2) vary with moisture content; (4) abrasiveness of the material; (5) bulking of the broken material. Many of the comments made above on percussive and drag-bit drilling are applicable to the performance of the cutting edges of machines considered here. Further, the intact strength of the deposit governs the type and size of machines required, since they must be supported by the load-bearing capacity of the ground. Care should be taken in using heavy equipment for clearing ground that is underlain by weak material, such as peat. Many glacial and flood-plain deposits contain soft horizons interbedded with firmer strata. Efficient traction for these vehicles also relies on the strength of the ground.

Adhesion is a non-technical term that describes the extent to which the material sticks to the machinery rather than 'to itself' (cohesion). Clays, for example, can be difficult in this respect, and if adhesion is excessive may influence the choice of machines and cutting edges.

Moisture content affects the weight of the material to be moved as well as its strength. The Plasticity Index (p. 313) is an important parameter in earth-works and can be used as a guide to the planning of equipment and contract schedules. Alternatively, desiccation can be a problem, as cohesive materials, including clays and clay-rich sediments tend to harden on drying. A dry, clean sand, with little or no cohesion, can influence traction and cause wear on moving parts.

Bulking is the increase in volume which occurs when material is broken and results in a decrease in the overall density of the moved mass. The amount of bulking is therefore equal to the porosity of the broken mass and will rarely be less than 20 per cent of the original volume. Table 15.3 gives general values for the bulking of common materials.

Table 15.3 Amount of Swell as % of Original Volume

Clean sands and gravel	10–15
Top soil	10–25
Sandy clays and silts	10–35
Weathered rock	20–40
Clayey sands and gravels	25–50
Fissured clay	30–60
Dry clay	30–70
Shales and soft beds	40–80
Hard rocks (poorly blasted)	50–100

5. Clearing of broken ground The contractural term *mucking* is often used to describe the clearing of broken ground from an advancing tunnel face, and can also be used to cover all loading and moving operations that accompany excavation. Significant geological factors involved are bulking; abrasion, and the flow properties of the broken ground. The last is important when gravity flow through chutes and hoppers is used, and is governed by the size of fragments, by the frequency of these sizes, and by the friction and strength of the material.

The first point can be considered in terms of a grading curve (p. 303) because granular mixtures having a uniform particle size flow with greater ease than those of varying particle size. Kvapil (1965) has grouped the various gradings into four categories:

1. Mixtures containing pieces of essentially the same size and shape.
2. Mixtures containing pieces of essentially the same size but having different shapes.
3. Mixtures containing pieces of different size and shape.
4. Mixtures containing fine material, dust and clay.

In this grouping 'fine' materials are those whose average diameter is less than 10 cm (4 inches). The degree of mobility of these groups decreases from 1 to 4, with group 4 usually having a much lower mobility than group 3. This is because much of the clogging that hinders the flow in group 4 is due to cohesion from the dust and clay which, when moist, sticks to the larger fragments and fills the cavities between them, giving the mass a cohesive strength. Kvapil relates these grades to the inclination (β) of a chute, or hopper base, as follows:

Grade 1: $\beta = 85°-90°$; Grade 2: $\beta = 70°-85°$;
Grade 3: $\beta = 55°-70°$ and Grade 4: $\beta = 40°-55°$.

The angle β is related to the friction of the system in two ways: first, to the internal friction of the material (ϕ), and second, to the friction between the material and the chute (ϕ_w). Table 15.4 gives a series of general values that can be used as a guide to design requirements; actual tests should be conducted where possible.

The tendency of some materials to chip and disintegrate with transport results in a change in their grading as they travel through the system. This always increases the percentage of fines, and hence decreases the mobility of the mass, a point to be noted in the planning stage of any contract. The matter is well discussed by Kvapil (1965), and Reisner *et al.* (1971), with reference to other authors.

SURFACE EXCAVATIONS

Surface excavations enter many aspects of ground engineering, as in cutting, quarries, open cast mines, docks and foundations. Most of

Table 15.4   Angles of Friction (after Kvapil)

Type of Material	Wall Material	Angle of Friction along Wall (ϕ_w)
Limestone, dolomite, marble	Steel	30–40°
	concrete	33–43
	wood	37
Granite, gneiss	steel	31–42°
	concrete	35–42
Iron ores	steel	33–42°
	concrete	36–43
	wood	40
Sandstone	steel	32–42°
	concrete	34–42
Shale, mudstone	steel	28–40°
	concrete	29–42

them present two basic problems: maintaining the stability of the walls and floor, and keeping the excavation dry. For the contractor much depends upon the way material is excavated and this has already been considered (p. 444). The present section deals mainly with the geological factors that are important to the design of stable excavations; de-watering is more fully discussed in Chapter 14. Two other related topics are also considered here, viz. quarries and bored foundations.

The geological aspects of slope stability, described in Chapter 13, apply to most excavations. It should be remembered that *in situ* stresses can influence the angle of stable slopes in large excavations (see papers by Merrill and Wisecarner, 1967; and Piteau and Jennings, 1970). Methods for stabilizing excavated slopes are all designed to restore the equilibrium that has been disturbed, either by increasing the shear strength of a slope or decreasing the shear stresses within it. Rock bolting and grouting are examples of the former, and drainage (Plate 11a) and regarding of the latter. Methods for retaining slopes range from wire cages placed over the face, to cribs and more elaborate retaining structures. These topics are beyond the scope of this text, but a useful review of the subject is given by Barron *et al.* (1970).

Floor stability is important in any large excavation as it affects the stability of the slopes, the movement of plant and the final cost of the contract. Movement is normally described with reference to the vertical as being either heave or subsidence.

Heave   This is commonly the result of one or more of the following: (i) excess water pressures at depth; (ii) *in situ* stresses other than those associated with ground water; (iii) swelling ground.

Upward pressures in excess of overburden pressures can develop in areas of high hydraulic gradient e.g. as through an impermeable boundary. Two such boundaries are the natural aquiclude and the man-made cut-off. The former is illustrated in Fig. 15.2(*a*).

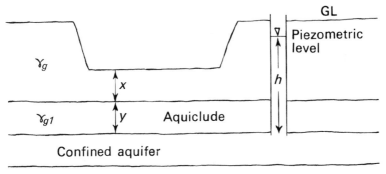

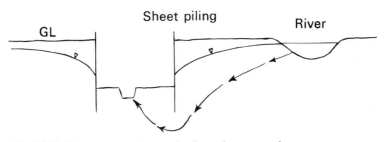

Fig. 15.2   (*a*) Uplift pressures acting on the floor of an excavation.

γ_g = density of strata of thickness x below excavation.
γ_{g_1} = density of strata of thickness y below excavation.

(*b*) Example of man-made cut-off.

A confined aquifer contains water under a pressure equal to $\gamma_w h$, where

γ_w = density of water
h = pressure head as measured at the base of the aquiclude.

This acts as an uplift pressure against the base of the aquiclude and is counteracted by the weight of strata above. If rock is removed by excavation a level will be reached when the overburden pressure will equal the uplift pressure, sometimes called the 'limit line for heave'. If the floor of the excavation shown in Fig. 15.2 were at that level, then

$$(\gamma_g \times x) + (\gamma_{g_1} \times y) = \gamma_w h.$$

When excavation is continued beyond this level the floor will eventually break and the pit will be flooded. As mentioned on p. 351 piezometric levels fluctuate with time and it is therefore normal practice either to design to the highest piezometric level, or to install pumps that drain the aquifer and prevent the level exceeding a given height.

Figure 15.2(b) illustrates a situation involving a man-made cut-off where the hydraulic gradient represents the loss in head sustained by flow from the river to the excavation. This generates a seepage force (p. 356) which is normally less than the intergranular forces due to the weight of the saturated strata. If, however, the excavation is deepened, the length of the flow path is decreased and the difference in head on either side of the cut-off is increased. The hydraulic gradient is therefore steepened. When, as a result, this self-weight of the strata is insufficient to counteract the upward seepage forces, failure will occur and the floor by the cut-off will fail. Analyses of these situations can be found in any good text on Soil Mechanics. The geological features to note are: (i) the position of hydrogeological boundaries, (ii) the distribution of head on either side of the boundaries, (iii) the density of the materials, and (iv) their permeability. In situ stresses other than those associated with ground-water are usually a direct consequence of the geological history of the area, as described in Chapter 13. Many that are encountered in soil engineering are derived from the overburden, the removal of which results in unloading and subsequent heave.

The rebound that occurred in an excavation some 35 × 118 m and 38.4 m deep (115 × 388 × 126 feet) in alluvial fan deposits is described by Bara and Hill (1967). The deposits, 38 m (125 feet) thick, consisted of lenticles of sand, silt and clay. Below them was a lacustrine deposit of highly plastic montmorillonitic clay, some 12 m (40 feet) thick, which overlay a series of highly consolidated gravels, sands and silts. Two water-tables existed at the site, one in the alluvium and one in the gravels. The area was de-watered in advance of excavation, with the drainage wells penetrating the upper half of the clay. Laboratory tests had indicated that both the alluvium and the clay would rebound on unloading (the nature of such tests is described in Chapter 10), and the site was therefore instrumented. The excavation was completed in five months and heave developed; there were no observed discontinuities in the final excavation surface. The heave at the centre of the excavation is illustrated by Bara and Hill (Figs. 8 and 9).

There are occasions when the pressures at depth, particularly the lateral pressures, are greater than would be expected from the overburden. The behaviour of the floor of an ore quarry in Canada is described by Coates (1964). The ore was contained in Pre-Cambrian beds which were overlain by up to 53 m (160 feet) of thickly bedded

Ordovician limestone and this was covered by 3 to 5 m of glacial till. The pit, some 300 × 600 m (1000 × 2000 feet) was begun and shipping of the limestone commenced. When these operations had reached a depth of 15 m (50 feet) a crack some 150 m (500 feet) long appeared in the floor. Within a few minutes the rock on either side of the crack had risen 2.4 m (8 feet), producing an elongated dome approximately 150 m long and 30 m wide. The upheaval was attributed to the presence of high horizontal stresses.

As stated in Chapter 13, the geological factors that are significant in such a situation are: the magnitude of the *in situ* stresses and their direction, their susceptibility to change and the presence of potential zones of stress relief. Two methods adopted by engineers for coping with this kind of problem are pinning the floor down with bolts, anchors and similar ties, and de-stressing the ground with small trenches and localized blasting.

Water pressures and other *in situ* stresses account for most of the problems associated with the heave of excavation floors. But difficulties can arise when excavating materials that swell. This phenomenon normally occurs in sediments, clays and shales in particular, which are susceptible to absorbing water. Excavations in these materials should therefore be protected from surface water.

The upward movement of ground below an excavation is a possibility which should be considered in the design stage of a contract. The depth to which ground may be disturbed can be considerable, and care should be taken when excavating above underground installations. One example of the problems that can arise is described by by Measor and Williams (1962), with reference to the movement of an underground railway below a deep excavation on the south bank of the River Thames in London.

Subsidence The majority of subsidence problems that occur during excavation, as distinct from construction, result from the presence of underground voids. These can be either man-made, e.g. mines, or natural. The ground most susceptible to natural cavities is limestone (p. 25); drilling should always be employed to verify the condition of the strata at depth. Enormous solution cavities were encountered in dolomites during the sinking of No. 1 Dreifontein shaft in the Rand, S. Africa. Smaller features from the limestones of S. Wales are mentioned by North (1952). Volcanic terrain can also contain large cavities in the form of lava tunnels (p. 133), and these too can collapse. Large construction work on cliff tops has revealed a rather unusual source of cavities at depth, the tension scars within large rock-slides. If the slides are old it is normal to find the upper portion of the tension cracks filled with debris that has fallen from the walls of the crack. In the floor of an excavation this appears as rather loose, broken ground. When disturbed this material can collapse into the cavity which often remains open to some depth. Once again, drilling is the only sure way of checking the continuity of the ground.

Large voids are an obvious hazard, but small voids, e.g. pores, can also collapse, particularly if they contain water under a pressure that is reduced by de-watering operations. As already mentioned (p. 352 and p. 376), de-watering can result in consolidation and subsidence, and care should be taken to ensure that de-watering operations do not damage either the property or the ground in the neighbourhood of the excavation. The geological factors associated with de-watering are described on p. 422.

Quarrying This is a process that requires the profitable development of a surface excavation and the planning of a quarry conveniently illustrates the role of geology to surface excavation contracts. Five points should be considered:

    (i) the size and shape of the pit; this will depend on the rock structure and on the stable angle of the quarry faces;

   (ii) the condition of the rock, e.g. in relation to weathering and ground-water (Chapters 12 and 14);

  (iii) the method of excavation (see earlier this chapter);

  (iv) the handling of broken material, and equipment used in the pit (p. 453); and

   (v) the disposal of spoil (p. 476).

Concerning (i) above, in many quarries, it is possible, in the planning stage, to choose the direction in which to drive the working faces. Geological structure is important, and some of its effects upon the stability of slopes are illustrated in Fig. 13.2. It is often an advantage to orientate the working faces parallel to the direction of dip. Figure 15.3 illustrates some hazards of working in the direction of dip.

A very different type of excavation that is commonly employed in engineering practice is the bored foundation. Here the most common problem is wall support so that the hole does not cave in. Groundwater is usually the cause of most collapses, particularly when present in poorly cemented granular material. Lenses of waterbearing sand and gravel in deposits such as clays are often troublesome in this respect. Before embarking upon the bores it should be ascertained that they are either suitable for the ground in question, or that adequate provision has been made for casing the holes.

SHAFTS

A shaft is the link between the surface and a sub-surface excavation. Many of the problems associated with shafts arise in connection

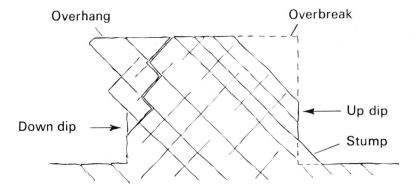

Fig. 15.3   Diagrams to illustrate quarrying conditions in dipping strata.

with either wall stability or ground-water; the two are often complementary and control to a large extent the type of shaft-sinking method that is used. The diameter of many shafts is so large that failure can occur in rock by slabs sliding out along discontinuities that have been intersected by the shaft. The problem is more acute in weak granular materials such as poorly cemented gravels, sands and silts. The following index has been found useful in assessing the nature of the ground when little or no experience of its behaviour in a shaft is available (T. Atkinson, personal communication).

$$S = \frac{c(1 + \sin \mu)^2}{0.1h(1 - \sin \mu)}$$

where S = depth independent index for stability of unsupported shaft walls,

c = cohesion (Kp/cm^2) as measured either in the field or in the laboratory,

μ = angle of friction = $\tan \phi'$ (see Note 2 of Chapter 13),

h = depth, in metres, of point of investigation.

The dimensions of this index $(Kp/cm^2/m)$ should be ignored.

As a guide, values of S between 0 and 1 indicate unstable ground that is likely to flow into the excavated space, even if dry. Sinking will require special methods including pre-treatment of the ground.

$S = 1.0–1.5$ rather unstable ground, stability very sensitive to water. Sinking best conducted in short sections. Special methods needed if large inflows occur.

$S = 1.5–2.0$ reasonably stable ground, allowing sinking in short to medium length sections.

$S = 2.0–4.0$ sound rock that will permit sinking to proceed in long sections.

Weak zones should if possible be identified prior to shaft sinking. Structures such as faults and shatter zones are usually easy to locate as their extent means they are normally intersected by the many exploratory bore-holes that precede the shaft. Limestone, however, can contain solution cavities which are less easy to locate, and often contain water or a mixture of water and fine sediment. The dolomites of South Africa are riddled with such cavities which, when intersected by shafts, discharge a liquified mud which is highly mobile and lethal if it entraps workmen. Water filled fissures and cavities do not have to be intersected to be dangerous. If the thickness of rock separating

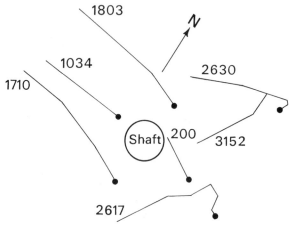

Fig. 15.4 Plan of borehole deflections with depth of holes given in feet.

them from the shaft is insufficient to withstand the head of water within them it will fail and burst into the shaft.

Dangers are always involved when unexpected, water-bearing fissures are encountered, and efforts should be made to locate them during the site investigation. A point that is easily overlooked in this connection is the alignment of exploratory holes. Long drill-holes may wander off course (see p. 447), so that the geology in the area of the proposed shaft can be either completely missed or misinterpreted. Figure 15.4 illustrates the bore-hole deflections recorded around a shaft: most holes wander in the same direction, and all have been influenced by north-west dipping structures. Holes drilled for ground improvement, e.g. grouting, freezing, chemical injection, etc., can also deviate from their prescribed direction and so miss areas that should be treated. It is interesting to note that the profitable sinking of shafts usually requires preliminary treatment of the ground ahead of the advancing face; many shaft sinkers use pilot holes that are always 15 m (50 feet) or so ahead of the working face.

The kind of geological information commonly required for the successful excavation of shafts will therefore concern (i) rock types, particularly the characters pertinent to drilling, blasting and the subsequent removal of broken material; (ii) ground-water, its location and movement; (iii) geological structure; the transmissive properties of the ground and other characters affecting the movement of water and the success of ground improvement techniques such as grouting and freezing; (iv) shear strength of ground. Other factors become important as depth increases, e.g. *in situ* stresses, geothermal gradient, etc. Most of these factors are also important to subsurface excavations, as discussed in the following section.

SUB-SURFACE EXCAVATIONS

Problems involving geology that are presented by sub-surface excavations can be summarized as follows: (1) general stability of the excavation; (2) effects of *in situ* stresses; (3) contents of pores and fissures; (4) excavation techniques; (5) finished profile; (6) effects at ground level.

1. **General stability**, when assessed, largely determines the amount of support that will be required below ground level. It is controlled by the effect that magnitude and orientation of *in situ* stresses have upon the excavation (Jaeger, 1975–76). An assessment of their effect is a most complicated task, if possible at all, and many engineers therefore base their general assessment of stability on some measure of the quality of the rock mass. The RQD system (see p. 274) is commonly employed. Deere (1968) has found that construction difficulties in tunnels can be related to RQD values in the following way:

0–25 %   Very poor rock, found usually in weathered zones and shear zones. Good support will be required as squeezing and ravelling ground can be expected.

25–50 %   Poor rock, again requiring extensive support.

50–75 %   Fairly good rock, that presents the problem of deciding which support system is the best to use. These RQD values usually indicate blocky ground containing broken seams along which blocks and slabs can slide into the excavation. In these conditions sets spaced at frequent intervals are often favoured. Some engineers specify rock bolts, because the ground is essentially blocky. However, rock bolting programmes can be difficult to cost because the intensity of primary drilling and bolting has to be estimated; secondary drilling and bolting can sometimes be very expensive.

75–100 %   Normally good rock, requiring the occasional support usually with bolts.

Rock Type	Rock Quality	Unconfined Compressive Strength, psi	Rock Quality Designation	Probable Bridge Action Period	Support Requirements
A	Solid, high strength, widely spaced joints	16 000–32 000	>90%	Tens of years	None
B	Massive, medium strength	8000–16 000	>75%	Month to years	None to very light
C	Foliated or stratified, high strength moderately jointed	16 000–32 000	>60%	Weeks to months	Light
D	Blocky, medium strength, closely jointed	8000–16 000	>50%	Days to weeks	Light to moderate
E	Blocky and seamy, low strength, closely bedded and jointed	4000–8000	>40%	Hours to days	Heavy
F	Very blocky and seamy, low strength	4000–8000	>25%	Minutes to hours	Heavy
G	Squeezing and swelling, very low strength, highly fractured and sheared	<4000	<25%	Seconds to minutes	Very heavy, full shield

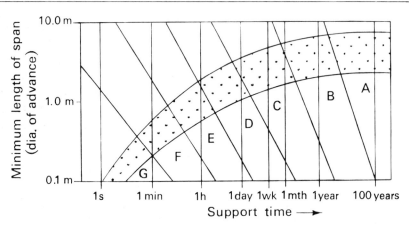

Fig. 15.5 A rock classification based on the time a certain advance will stand without support (after Stini and Lauffer).

For an extension of this system see Stini and Lauffer (1965), Fig. 15.5. Others exist (see Bieniawski, 1973—ref. on p. 293, and Cartney 1977).

Classifications of this kind are usually based upon information obtained from drilling along the line of a tunnel, and although helpful, they can easily obscure an assessment of the condition of ground that has not been sampled. In heterogeneous ground it is unwise to classify the character of large areas on the basis of isolated investigations (e.g. Robinson *et al.*, 1974). Bad ground is likely to be found in excavations close to weathered zones, e.g. below the normal weathering profile, near shear zones and solution cavities. Weak and often saturated zones can suddenly be encountered in young sediments.

In circumstances such as deep tunnelling or large excavations the general assessments of the above kind should at some stage be superceded by more specific design criteria, which require an appreciation of the presence and character of the *in situ* stresses.

2. **In situ stresses** should be known if some attempt is to be made at designing either a system of supports or an excavation that will support itself. Inadequate design in this respect can result in violent failures, variously called rock bursts (or popping or spalling, depending upon their severity), a related phenomenon, called 'bumps', the deformation and sometimes closure of an excavation in squeezing and swelling ground, and the development of rock-falls.

Rock bursts are the sudden detachment of masses of rock from the sides of an underground excavation. Failures of this type can release hundreds of tons of rock with explosive violence, and produce ground motions that can be detected at seismic stations many miles away. They result from the development of static stresses that exceed the static strength of the ground around the excavation, and their frequency and severity tend to increase with depth. Some of the most spectacular rock bursts have been recorded from deep gold mines in the Witwatersrand, South Africa; from Ontario, Canada; and the zinc mines of Idaho, U.S.A. Most rock bursts occur below 2000 feet (609 m) of cover, although minor bursts have been experienced in quarries less than 1000 feet (304 m) deep. Some mines, however, can operate at depths much greater than 2000 feet and have virtually no rock bursts.

It appears that these failures depend not only on horizontal and vertical stresses but on the way in which they are released. Mining management that permits controlled or partly-controlled de-stressing of the ground is an important contribution to solving the rock burst problem; an excellent outline of the subject is given by Obert and Duval (1967). The rocks that appear most susceptible to failure are hard, brittle and very strong. Those that fail most violently usually have an unconfined compressive strength that is greater than 20 000 lb/in.^{2} (1400 kg/cm^2) and a modulus of elasticity that exceeds 5×10^6 lb/in.^{2} (3.5×10^5 kg/cm^2). Rocks such as unweathered dolerites, and quartzites, are more susceptible to 'bursting' than sediments. In general, hardness, grain size and rock structure are important factors. The violence of failure in any stress field seems to increase with increase in hardness, decrease in grain size, and proximity to discordant structures such as dykes and faults.

Popping is a similar but less violent form of failure that occurs in more elastic materials, e.g. porous sandstones. The sides of an excavation can be seen to bulge before failure of the deformed slab occurs, and detached slabs can rarely, if ever, be fitted back on to the surface from which they have popped—indicating that the areas so affected have been subjected to considerable stress and subsequent strain.

Spalling tends to occur in fractured rock (the fractures can be cleavage planes or larger features such as joints), where the rock fragments are kept in contact by the *in situ* stresses. Such a rock mass can bulge, to a certain extent, as a sheet. Collapse of this sheet occurs when a key block either fails or detaches itself from the mass. 'Sloughing' is a term used when the detached sheets remain in a relatively intact state, e.g. as in the failure of pillars in salt mines and other evaporate deposits.

Rock-falls take place when there are no *in situ* stresses to keep rock fragments in contact. The stability of the excavated surface comes from the jamming of the detached interlocked fragments; when they fail to become jammed a rock-fall occurs. A fall of very broken ground, or cohesionless sand, that tends to occur soon after exposure in headings is called ravelling.

Bumps are shocks that originate at points behind excavated faces. They are usually audible reports and often generate strong, but local, seismic shocks which may be sufficient to dislodge material from the walls and roof of the excavation. They are thought to come from the sudden displacements in surrounding rock, as might be caused by small movements along a fault. The shearing of overlying strata is thought to be a common cause of bumps in coal mines.

Squeezing ground moves into the excavation whenever it is not restrained by some support; walls and roof can cave in, and floors can heave (Plate 16b). There are various reasons why ground will squeeze, and one of the most common is failure under excessive overburden pressure. However, there are occasions when other *in situ* stresses are responsible, e.g. failure under high horizontal pressures, as often occurs in overconsolidated clays. Squeezing deforms the ground and the fabric of fissured materials such as shales, fissured clays, and intensely fractured rock (e.g. cleaved schists), and often permits the surface of the excavation to flake. In clays and shales this can be attributed to their exposure to air, and care should be taken to ensure whether or not it is a sign of squeeze. The softer materials, e.g. soft clays, weathered and crushed rocks, usually squeeze, and ground affected in this way is normally too weak to support an arch. The ground pressure is usually exerted on all sides of the excavation and it is for this reason circular tunnel cross-sections are commonly adopted. As mentioned above, ground pressures in hard rocks generally produce some sort of spalling or popping. Squeezing as a phenomenon is associated essentially with soft materials. One condition which gives rise to squeezing is swelling ground which expands into an excavation. Although there are various definitions the term is normally applied to any material that increases its volume by more than 2 per cent when immersed in water. Swelling is always accompanied by a migration of water into the expanding fabric, and most of this water appears to come from the surrounding ground and not from the atmosphere within the excavation. Argillaceus materials such as shales, mudstones and over-consolidated clays, particularly those containing expansive minerals such as montmorillonite, are more susceptible to swelling than other rock types. Swelling in ground that does not contain expansive material is usually the result of the removal of support, as this permits a relaxation of the surrounding strata. Laboratory tests can be used to assess the extent to which swelling can occur with any reduction in load (p. 308).

Local swelling can occur when limited and relatively small amounts of expansive material are encountered. Proctor and White (1946) recorded that during the driving of a tunnel through the Alps, layers of shale below the floor of the tunnel were disturbed and allowed water to enter thin interbedded seams of anhydrite. This hydrated to gypsum and expanded, raising the tunnel floor for several years at about 25 cm (10 inches) a year. Difficulties that have resulted from the presence of pyrrhotite in Swedish rocks are described by Martna (1970). Crystals of this mineral may be disseminated throughout a rock, often a carbonaceus or argillaceus deposit, or in seams in faults and fissures. Pyrrhotite is stable either when surrounded by water low in oxygen, or in dry air, but decomposes when exposed to oxygen-rich water (i.e. most flowing ground-water) or to moist air. Decomposition can therefore start when the water level is lowered during construction, or when previously dry rock is wetted. The weathering is gradual and usually produces solutions of sulphuric acid and iron sulphate; these attack concrete and steel. The problems which arise can be considerable and pyrrhotite-bearing rocks should be avoided wherever possible.

Brekke and Selmer-Olsen (1965) record some problems arising from seams of expansive montmorillite clay in Norwegian excavations. These can promote large falls and can occasionally result in considerable amounts of clay flowing into a heading. Such zones require careful investigation, because when dry they may be sufficiently strong to give no difficulty during construction; hence their effect on the long-term stability of an excavation can easily be underestimated.

The pressures that squeezing ground will develop on supports normally increase with time, but at a decreasing rate as the ground adjusts itself to new conditions. The increase in pressure can be large, particularly in over-consolidated clays, and supports that were sufficient when installed can fail as the ground pressure on them increases. This applies to any support system that restricts the relief of residual stresses, whether in hard or soft materials. In contrast, active tectonic stresses can fluctuate with time. Normally they gradually build up until some failure occurs, and then decrease rapidly. However, the decrease may be temporary, as stresses will re-develop until similar relief again occurs. Two further points should be mentioned in connection with *in situ* stresses: the shear stress around portals, and creep.

Portals often pass through the weathering profile where a mantle of mechanically weak material is present. The depth of this mantle varies according to the rock type and climate, but can reach depths of several metres in zones of tropical weathering. It may become involved in downhill movements and contain sliding surfaces on which movement has already occurred. Such zones therefore require a careful survey in order to secure lasting stability for the portal.

Creep.   Field measurements have confirmed that *in situ* rock can exhibit varying degrees of transient and steady-state creep (Chapter 2), which in certain rocks can promote a gradual deformation of pillars and the general profile of an excavation. The rocks affected are generally those which undergo plastic deformation under the compressive stresses which exist in the crust. More elastic rocks would fail by spalling, popping or bursting. Stress measurements on the surfaces of excavations cut into massive non-elastic materials, such as evaporites, show that the level of stress decreases with time because of creep. Hence it can be mistakenly assumed that if the excavation surfaces do not fail when they are made they will remain stable and free from spalling, etc. This, however, is not the case, for failure in such rocks is found to occur after a period of time, ranging from days to many years, and to be closely related to the total strain, or the rate of strain, developed in them. As the total strain increases, either at a constant or a decreasing rate, so the incidence of spalling and sloughing will also be liable to increase.

Most engineering problems associated with *in situ* stresses are really problems that result from the release of the stresses, and it should be remembered that the stresses encountered need not be a simple function of depth. Table 15.5 gives some data obtained at certain underground sites in the Snowy Mountains Scheme, S.E. Australia.

At the Tumut sites in the steep V-notch valley of the River Tumut, stress values were influenced by topography. All the sites in the

Table 15.5   Natural Stresses at Localities in the Snowy Mountains Scheme (after Moye)

Location	Thickness of Cover (feet)	Compressive Stress (lb/in²) v h_1 h_2			Ratios h_1/v h_2/v v/g		
Tumut 1 Power Station	1200	1800	1500		0.83		1.3
Tumut 2 exploratory tunnel	600	1500	1900	1800	1.3	1.2	2.1
Eucumbene-Tumut tunnel	1100	1040	2700		2.6		0.81

v = vertical compression,
h_1 = horizontal compression,
h_2 = horizontal compression in direction perpendicular to v and h_1,
g = compression due to weight of rock vertically above site.
Note: in the Eucumbene-Tumut tunnel, average h_1 was 2.6 times v, which in turn was 0.8 times g. The horizontal stress was thus about 10 times greater than that due to the weight of overlying rock (assumed to be elastic, isotropic, and having a Poisson's Ratio of 0.2).

Snowy Mountains were affected by the tectonic history of the region (see p. 382), as well as the present level of tectonic activity, and illustrate the need for making actual measurements of the state of stress in rocks at depth.

3. **Contents of pores and fissures**   Pores, fissures and other voids in rock are potential reservoirs for liquids and gases. Water is the medium most commonly encountered, and may present considerable hazards, especially in rocks that contain well-developed interconnected fissure systems. The karstic characters of massive limestones (see p. 26) are important, as the cavernous nature of the rock can provide it with a very large reservoir potential and the ability to transmit water at high velocities. The greatest ground-water hazard in underground work is the presence of unexpected water-bearing zones, and it is important that the position of hydrogeological boundaries should be known. Sometimes an impermeable boundary can be used as underground dam and so turned to the engineers' advantage; this should only be attempted after a detailed investigation has been made of the ground and its response to the de-stressing that follows excavation.

A reminder of the problems that constantly trouble even experienced miners occurred in 1968 with the flooding of the gold mine at West Dreifontein in South Africa. At this mine, in the Wonderfontein Valley, the gold-bearing beds are overlain by some 1260 m (4000 feet) of cavernous water-bearing dolomites. Both the gold-bearing rocks and the dolomites are cut through by near vertical dykes about 58 m (190 feet) in width and spaced about 200 m (660 feet) apart. They strike across the valley, dividing it into a series of compartments. In order to reduce the likelihood of water suddenly draining from the dolomite into the mine, as can happen when fractures behind the hanging wall of a stoped area meet others which connect with the dolomite, the compartment in which the mine is situated was de-watered. The neighbouring compartment was not, and a difference of head developed across the dyke that separated the two. Failure of a hanging wall near this dyke occurred on 26th October, 1968 and water poured into the mine at an uncontrollable rate. The threat of total flooding was averted, however; the incident is described by Cousens and Garrett (1969).

Water flows are less predictable than water pressures, which are nearly always a simple function of the head of water above the point of interest, and can be very high, especially in confined aquifers (see Chapter 13). Such pressures should be considered when designing the thickness of divides that will separate an aquifer from a mine, and if possible should be measured.

Gases.   A variety of gases, some dangerous, can be encountered in underground works; the most common are:

(a) Carbon dioxide (CO_2). Being 1.5 times as heavy as air, it tends to accumulate at the bottom of shafts and drifts, and can be lethal. It is normally produced by the slow oxidation of coal or the decay of organic compounds, and may accompany magmatic processes in volcanic areas. The gas is aggressive to concrete and corrodes steel.

(b) Carbon monoxide (CO) is 0.97 the weight of air and tends to rise. It is far more toxic than CO_2. Normally associated with coal-bearing rocks, it is a common constituent of coal dust explosions, and is often accompanied by methane.

(c) Methane (CH_4) is almost half the weight of air and, like CO, will rise. This very mobile gas can readily migrate from its point of origin, so that strata above the source zone can become contaminated (Note 9). The gas is combustible and forms a highly explosive mixture with air. Normally it originates from bitumens, e.g. from oil and coal. Recent sediments such as alluvium and flood plain deposits can also contain pockets of the gas produced by the decay of organic sub-stances. For this reason it has long been called 'Marsh gas'. Shales and clays with organic matter have been known to release methane. The presence of the gas normally requires strict control over excavation processes because of the hazard of fire.

(d) Hydrogen sulphide (H_2S) is 1.7 times as heavy as air and is highly toxic. It is also explosive when mixed with air. The gas may come either from the decay of organic substances or from magmatic activity; it is readily absorbed by water. Water entering excavations in H_2S-rich rocks is often accompanied by the gas, which is aggressive to concrete.

Other gases which have been encountered below ground include sulphur dioxide, hydrogen, nitrogen and nitrous oxide. The gases appear to be stored in pore spaces and considerable volumes of gas can be stored under pressure. For example, 1 cm^3 of coal may liberate 0.25 cm^3 or more of gas. However, much of the gas that enters an excavation probably comes from fissures and it is therefore possible, in certain situations, to ventilate a gas-rich rock by drilling holes ahead of the working face and promoting fracturing by shot firing. But even with these precautions fatal accidents can occur. (Note 10).

Gas bursts are due to the sudden liberation of gas under pressure, and are usually associated with a violent failure of the rock. Coal mines are understandably the excavations most susceptible to this form of failure, and recent investigations indicate that 'coal bursts' (rock-bursts in coal seams or pillars) are connected with the release of pore gases. When such bursts occur, many hundred of tonnes of coal can disintegrate into a coal-dust laden cloud of methane. The report on the explosion at the Hampton Valley Colliery, Lancashire, describes the dangerous consequences of this phenomenon (Stephenson, 1962). Patching (1962) and Price (1972) give explanations that have been forwarded to explain its nature.

The temperature of liquids and gases found in the ground is a function of the geothermal gradient at the location in question. Temperature gradually increases with depth (see p. 3) but the rate of increase varies from place to place. The average temperature gradient is normally taken as 1°C per 30 m increase in cover, but this should be checked for deep excavations because significant variations can exist. Some examples of temperatures encountered in four Alpine tunnels are listed in Table 15.6.

Working conditions deteriorate markedly once the ground temperature exceeds 25°C (77°F). Other factors which influence the temperature at depth are the thermal conductivity of the rock, its structure (conductivity is usually greatest in directions parallel to stratification), and the presence of circulating meteoric water. Gases may also provide a certain amount of thermal insulation. During the construction of the Great Apennine Tunnel the ground temperature suddenly increased in a clay shale from 27°C to 45°C, and exceptionally to 63°C after an inrush of methane (Szechy, 1966). High temperatures were also recorded in the Tecolote Tunnel through the Santa Ynez Mountains, near Santa Barbara, California. This tunnel encountered ground-water at a temperature of 65°C (U.S.B.R. Technical Record, 1959).

Table 15.6   Some Underground Temperatures in Alpine Tunnels

Location	Approximate Geothermal Gradient °C per 30 m	Maximum Temperature °C
Simplon Tunnel	0.8	55–56
St. Gothard	0.64	30–31
Mont Blanc	0.39	25–30
Ricken	1.1	23–25

In general it was found that the geothermal gradients were steeper beneath valleys than below higher ground.

4. **Excavation techniques** These can now range from the simplicity of a pick and shovel to the sophistication of an atomic explosion. (For the latter see *Proc. 1970 Symp. on Eng. with Nuclear Explosives*). The successful choice of excavation technique depends partly on the geological conditions mentioned above, and partly on the shape and purpose of the excavation. The excavation techniques in common use are grouped here according to geological conditions in which they are normally employed, viz. hard rock, moderately hard to firm ground and loose ground. Mention is also made of some problems encountered with the use of compressed air.

Hard rock: here techniques are centred around either blasting or mechanical cutting using moles (see p. 449 and p. 448). Blasting is now commonly used for breaking (either breaking-in or full-face blasting), for pre-splitting and for cutting (Note 11). Blasting and mechanical cutting are not normally used together; the former is still favoured in ground that contains zones of differing hardness, e.g. schists cut by dolerite dykes.

When blasting alone is used the excavation is advanced in a series of benches, the most forward of which forms a drift, e.g. at the crown, sides, centre or invert of a tunnel. This allows for variations in rock-mass quality to be handled easily and for the poorer quality rock to be supported where necessary. Full-face excavations should be used with caution where rocks are variable, even though all may be classed as 'hard rock'. The problem of water is usually solved by one or more of the following methods, depending upon the hydraulic head encountered: (i) pumping from the excavation area or complete dewatering; (ii) grouting; (iii) use of compressed air within the excavation.

Moderately hard to firm ground: here techniques are mainly designed

Plate 15

(a) Excavation in artificially frozen ground. Note the
freezing tubes and unsupported excavation in sand overlying
darker peat and more sand; all are frozen. (*Photograph by
Foraky Ltd.*)

(b) A heading blast. (*Photograph by Nobel's Explosives Co. Ltd.*)

Plate 16

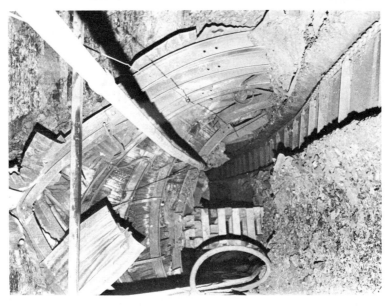

(b) Partial closure of an originally horizontal main road. (*Courtesy of the National Coal Board, London*)

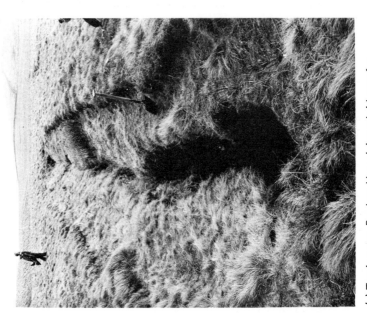

(a) Tension scar. Produced by mining subsidence; for explanation see Fig. 15.7. (*Courtesy of the National Coal Board, London*)

to cause the least possible disturbance to the ground and allow speedy erection of supports. Correctly designed moles are normally suited to these conditions and can be used for extensive tunnel work. When irregularly shaped excavations are required various mining methods are adopted, and in these situations the safety of excavating and the life of unsupported arches both increase as the size of cross-section decreases. Headings therefore play an important part in this type of ground, and their arrangement, sequence of advance and support can be designed with reference to the geological conditions. Pilot headings are normally essential. Unexpected sources of additional expenditure arise from the sudden collapse of the working face or some part of the roof, or from excessive overbreak (p. 472). In some respects moderately hard or firm ground is the most difficult to excavate, because geological conditions can vary from good to bad within any one job, and the techniques used should be flexible. In some cases it may be economic to improve the ground before excavating it (see Chapter 14). A good account of tunnelling in this type of ground is given by Szechy (1966).

Loose ground has little or no cohesion: sands and gravels are typical examples. Support is vital and usually supplied by a shield (Note 12); and also see Plate 11b. Intrusion of the ground into the face of the shield is a hazard, and the *in situ* stresses likely to be encountered should be calculated or measured before a shield is used. Hoods can be added to shields in order to reduce the chance of such intrusions but the shield then becomes more difficult to steer. In very weak ground it is common practice to blank off part of the working face with bulkheads, allowing the ground to enter the shield through restricted openings. A remarkable example of this was the driving of the southern tube of the Lincoln Tunnel in the Hudson River silt. Here, all but 0.5 to 0.8 % of the shield face was sealed so that 70–80 % of the silt was pushed around the shield as it penetrated the ground, the remainder being admitted into the shield. This method, however, increases lateral pressures on adjacent structures, such as other tunnels or foundations, and cannot always be used. Shields have been built with mechanical excavators at their working face so that they virtually become a type of mole. They normally cut a full face, and although their rate of progress exceeds that of a manually driven shield, they can be dangerous and become useless if they penetrate very weak or heterogeneous ground. A bulkhead shield is generally used in running ground, as found in many recent silts, silty sands and soft clays. Stratification should be noted because it provides the working face with a variety of materials at any one time. If the properties of the strata are markedly variable it may be necessary to position the excavation so that the optimum use can be made of the shield design. When the direction of stratification is at an oblique angle to the long axis of the shield it often tends to send the shield off its prescribed course as it is jacked forward.

Compressed air. Despite the operational and medical problems that are associated with its use, compressed air has long been employed to counteract water pressures and so assist the excavation of water-bearing weak ground (Note 13). Certain geological data should be known before compressed air is used, including (1) the water pressure to be counteracted (for medical reasons this should not exceed certain levels); (2) the permeability of the ground with respect to air; (3) the density of the ground above the excavation.

With this information the excavation can be designed to prevent the compressed air from blowing out at the face of a tunnel. If this occurs, there is a drop in the internal pressure of the tunnel, which may permit an inrush of water and loose sediment. Blow-outs are frequently lethal to personnel at the working face and the bigger the diameter of the tunnel the greater will be the chance of their occurrence (Note 14).

Three techniques are employed to overcome the problem of blowouts: (i) decrease the permeability of the overburden, (ii) increase the load above the tunnel, and (iii) increase the length of the flow path for gas escaping from the tunnel. The first method is perhaps the most frequently used and a typical example of its application was provided during the construction of the Dartford–Purfleet tunnel beneath the Thames in 1957. Here a mixture of clay-chemical grout was injected into the river gravels of the Thames, so reducing their permeability. Temporary freezing of the ground can also be employed, if for some reason this is preferred to permanent grouting. (See Anon. *Engineering News Record*, 1957).

The two remaining methods are often used together. A blanket of material is deposited above the line of the tunnel, and this can only be done when space is available, as above some tunnel sections which pass beneath rivers. The blanket may be permeable or impermeable material. An impermeable blanket, such as clay, would not only increase the load above the tunnel but would also promote a lengthening of flow paths for the air escaping from the tunnel, for the air would have to flow laterally to the edge of the blanket before it could escape.

5. **Finished profile** of an underground excavation. Perhaps the most common problem here is that of overbreak. This is the removal of material beyond the designed boundary of the excavation; often it cannot be avoided, especially in strata where discontinuities permit rock to fall. Overbreak generally occurs in excavations made without some kind of shield. Unless care is taken, excessive amounts of rock

can be removed; this has to be handled and the space so formed back-filled. One example where this happened was in the Lochaber pressure tunnel, a 24 km (15 miles) long tunnel connecting the reservoir at Loch Treig to a pipe-line on the western slopes of Ben Nevis. A large part of the tunnel was cut, by blasting, through schists and when it crossed the strike of the steeply dipping foliation a good cross-section resulted, in close agreement with the outer ring of drill holes. When, however, it became aligned to the strike of the foliation the rock breakage was more irregular and considerable overbreak resulted. Further details are given in the papers by Peach (1929) and Halcrow (1930). Since this tunnel was constructed in the 1920's considerable advances have been made in drilling and blasting which have greatly reduced the incidence of overbreak. Nevertheless, the problem remains and its solution depends largely upon an appreciation of the geological conditions in the rock to be penetrated. These are illustrated in Fig. 15.6, which shows:

(a) Strike of rocks parallel to axis of excavation.

Here large scale, continuous, and often planar surfaces can form much of the roof, and difficulties may arise when the dip (α) is less than 15° or greater than 70°, for at these inclinations joints will allow loose blocks to drop from the roof; *in situ* stresses can play an important part in determining the final dimensions of the excavation. In some rocks small discontinuities can also make the walls difficult to control on the up-dip side of the excavation, when α is greater than 25°–30°.

(b) Strike of rocks at right angles to axis of excavation.

When α is less than 15° or 20° the situation is similar to (a), with large scale surfaces accounting for much of the roof, and joints permitting local overbreak. Walls tend to be reasonably easy to control and overbreak should be small. When α is greater than 20° the roof gives less difficulty, although severe falls can occur between weak layers and form chimneys in the roof instead of long gashes along the crown of the excavation, as can occur in case (a).

(c) Strike of rocks oblique to axis of excavation.

This condition lies between (a) and (b), and when $\beta = 90°$ the situation is equivalent to (b) with $\alpha = 90°$; when $\beta = 0°$ the situation is the same as (a) when $\alpha = 90°$.

The finished profile is not only important to the contractor but also to the design engineer, especially when the excavation is for an unlined hydraulic tunnel. Here the roughness of the tunnel walls generates resistance that affects the total head in the tunnel and hence its discharge. The rock structure and the irregularity of the finished profile are both significant, and design criteria for such excavations have been proposed by Wright (1971).

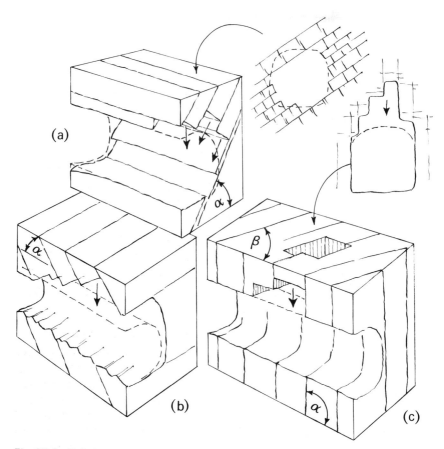

Fig. 15.6   Relations between overbreak and structures in tunnelling. See text for explanations of (a), (b), and (c).

Effects at ground level   Underground excavations that are not fully supported invariably result in subsidence at ground level (Note 15); this, in its simplest sense, is the lowering of the ground surface. However, the strains involved normally have both vertical and horizontal components of movement (Fig. 15.7). The most obvious consequence of subsidence is the damage that occurs to buildings, but there can be many other effects that are less obvious though no less important. For example, the permeability of the ground can be markedly changed; areas liable to flooding adjacent to rivers can be increased. The amount of subsidence that occurs is generally controlled by the thickness of material removed; the support given to the excavation; the depth and area of the workings; the shape of the excavation and its orientation with respect to *in situ* stresses; and the nature of the surrounding ground.

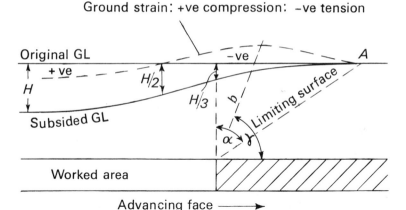

Ground strain: +ve compression: -ve tension

Fig. 15.7   Typical subsidence resulting from the extraction of a level seam. α = angle of draw (measured from the horizontal in some countries); γ = angle of break; H = maximum subsidence. In the typical cycle of ground movement, tension occurs behind the limiting surface and may be associated with the opening of cracks at ground level (Plate 16a) Tensile strain reaches a maximum at the surface of break (*b*). Later the strains become compressive and close cracks. An excavation advancing beneath point A would subject it first to tension and then to compression.

The first indication of impending subsidence is usually seen in the excavation, as a roof sag. This process, sometimes called subsurface subsidence, may continue, unless arrested, as the area of the excavation is increased, until the roof either fails or meets the floor. Surface subsidence usually starts soon after this subsurface process begins.

The nature of the neighbouring ground is important and can control the type of subsidence that occurs. Well stratified deposits of reasonable strength usually offer a certain amount of beam support, and this tends to contain the zone of breaking rock which develops around an excavation. The rock will fall into the unsupported excavation and may eventually fill it, giving some support to the overlying ground. Even so, the depth of surface subsidence is commonly 80–90% of the thickness of the material removed (Note 16). When the geology does not provide this type of support the ground may continue to break and a front of broken rock will travel up towards ground level. It is sometimes described as caving, and can result in the appearance of holes at the surface; it develops most readily in rock that is either weak or broken by fractures across which there is little or no cohesion.

Faults, dykes and other steeply dipping structures can laterally restrict a zone of caving and are boundaries to be noted when investigating such a subsidence (see Boyum, 1961). Very occasionally the roof of a caving system will collapse as a unit, rather like a plug. (The collapse of plugs of cohesive fissured soils into near surface excavations is a similar phenomenon.) Here too vertical or near vertical fissures normally define the limits of the plug. This type of subsidence, unlike the other forms, can occur almost instantaneously, and if large, can send a blast of air through the excavation below.

When definite lateral boundaries do not exist the areal extent of subsidence becomes largely controlled by the angle of draw (Fig. 15.7). Many theories exist for predicting this angle but none are entirely satisfactory, for they cannot cope with the variety of possible geological conditions. The angle is measured from field subsidence and in Coal Measure strata of the United Kingdom it normally lies between 35° and 38°, but this is not always the case. In Holland, where the Coal Measures are overlain by weaker rocks than are found in Britain, the angle of draw is around 45°. Better agreement has been found between the predicted and measured angle of break (Fig. 15.7), which defines a 'surface of break' that joins the areas of maximum tensile strain. Whether or not breaking actually occurs depends upon the strains involved. Some fairly typical measurements are as follows:

	Angle of Break
Weak and generally loose strata	60°–40°
Cohesive materials of clay type	60°
Sands and sandy soils	45°
Shales of medium strength	60°
Hard sandstones and similar rocks	85°

THE DISPOSAL OF EXCAVATED MATERIAL

The disposal of unwanted material is a problem that most large excavating concerns have to solve. For centuries the normal method of disposal has been to tip the material at ground level so as to form a mound or tip, and the unsightly consequences of this procedure are visible in most mining districts of the world. However, a growing awareness of the danger that tips present to those who live near them, and to the scenery as a whole, is promoting a body of opinion that may ultimately curtail this form of disposal in certain areas. (See also Refuse Disposal, Chapter 16). Nevertheless, tipping is likely to continue for many years and this, together with the heritage of tips that will be passed on, makes the whole subject of tip stability one which cannot be neglected. It is briefly considered in this section.

The stability of a tip depends not only on the strength of the tipped spoil but also on the strength of its foundations. The geological characters that are pertinent to foundation stability can be of critical importance, as was tragically illustrated by the disaster at Aberfan in Wales; see also Bishop, 1973.

Aberfan is a small mining village situated in the valley of the River Taff. The village itself extends up the western slope of the valley, from its floor at approximately 420 feet above sea level to about 500 feet O.D. From there the slopes rise at 1 in 4 to 900 feet, steepening to 1 in 3 up to 1000 feet and thereafter flattening towards a summit line at about 1400 feet. Coal waste had been tipped on these slopes since 1914, and by 1966 seven spoil heaps had been formed ranging in height from 60 to 100 feet. All lay between the 650 and 1200 feet contours.

The rocks of the valley are of Carboniferous age, and consist in the main of well jointed sandstones separated from each other by varying thicknesses of mudstone. The mudstones commonly contain thin seams of coal, the coal being underlain by a clay. The dip of the strata is towards the valley at about 1 in 12. The sandstone is not particularly permeable, but its jointing renders it pervious and it is a good aquifer. The mudstones and clays, on the other hand, are relatively impermeable and behave as aquicludes. As a result springs issue from the sandstone-mudstone junctions and feed streams which flow down the slopes to the River Taff. Two well defined 'spring lines' existed, one between 800 and 850 feet, the other around 1300 feet. The lower spring zone, therefore, passed beneath the complex of tips, which, as it happens, also covered two stream courses. The tips themselves lay mainly on a sandstone outcrop but this, like most of the outcrops on the valley slope, is not exposed because of the cover provided by superficial deposits. These deposits extend over the whole slope and, being of a clayey character, are relatively impermeable: their thickness generally varies from 5 to 20 feet. The tips were therefore founded on slopes which were covered with a clay that impeded, although not entirely prevented, the drainage of water from the sandstone aquifers. Water had long been a problem below these tips and in 1944 part of Tip 4 failed and slid some distance down the slope: it overlay a stream course. Another tip, No. 5, partially failed between 1947 and 1951; it too overlay a stream course. Tip 7, which caused the disaster in 1966, lay above a stream course and a zone of springs. The tips placed on saddles between the streams have not been so affected.

Tip 7 was begun in 1958 and had been the site of considerable tipping; slips had been known to occur within it. By 1966 its down-slope face was approximately 130 feet high and tipping was still in progress. On the morning of 21st October, 1966 the tipping gang noticed that the crest of the tip had settled 10 feet over a distance of 30 to 40 feet from its edge; this settlement had taken place overnight. An hour later the settlement had increased to 20 feet, and about 40 minutes later the toe of the tip began to move forward. Suddenly a wave of liquified tip material appeared at the toe of the mass and moved as a flow-slide very rapidly down the hillside carrying with it drier material that had come from higher in the tip. This slide engulfed part of the village, killing 144 people and injuring others. The amount of material transported was estimated as about 140 000 cubic yards. At the enquiry that followed (see Report of the Tribunal and a Selection of Technical reports) a detailed analysis of the failure was presented and the following points emerged:

1.   Between 1962 and 1963 tailings had been tipped on to the heap (Note 17). Its toe also had gradually encroached upon and then covered one of the hillside springs, so that by September 1963 back-sapping (defined as the removal of material from the toe of the tip by an issue of water, leading to intermittent slips of progressively increasing size) had become well advanced.

2.   A considerable slip had occurred in the latter part of 1963, generating a sliding surface on which all subsequent slide movements seem to have taken place.

3.   By 1964 back-sapping and associated slipping had already removed the upper part of the mantle of relatively impermeable drift, so facilitating the discharge of ground water from the sandstone aquifer beneath.

4.   Tensional strains associated with mining subsidence had increased the separation of joints in the sandstone aquifers, giving them an increased storage, greater perviousness and higher rates of infiltration. Further, compressional strains had closed joints to the south of the tip, decreasing the perviousness of the aquifers and hindering their drainage.

5.   On the night prior to the disaster the water pressures in the ground beneath the tip and in the lower part of the tip had, as a result of recent rain, been able to reactivate movement on the pre-existing failure surface. This was accompanied by a 10 foot settlement at the crest.

6.   Shearing continued, and the lower, most saturated, part of the tip was brought to a critical condition for flow-sliding to begin.

7.   In the slide that followed an area of drift was stripped from the bedrock and so permitted the release of water that was stored in the sandstones. A strong spring resulted which promoted a 'mud-run' that carried further material down the slope and into the village.

The events at Aberfan reaffirm that the geological characters important to tip stability are essentially the nature of the ground, including its slope, and the presence of water. Tips become unmanageable if placed on ground that cannot support them. A general guide to the performance of various materials is given in Table 15.7. Further, tips should not be placed over zones of natural discharge without the provision for ground-water control and if necessary, drainage.

Since the Aberfan disaster many organizations have revised their

codes of practice regarding tipping and a more enlightened approach
now exists. (See National Coal Board Technical Handbook, 1970).
Nevertheless, the engineer should remember that codes of practice in
these matters should not over-ride common-sense; geology does not
follow a code of practice.

Table 15.7

Ground	Safe Pressure in tonnes/m²	(tons/ft²)	Height of Tip Producing this Pressure in m	(ft)[a]
Alluvial deposits Soft clay Moist clay and sand Made ground	0.04–0.09	0.5–1.0	3.66–7.33[b]	11–22
Firm-stiff clay	0.19–0.28	2.0–3.0	15–22.33	45–67
Clayey coarse sand	0.19–0.37	2.0–4.0	15–30	45–90
Dry sand	0.28	3.0	22.33	67
Coarse sand and gravel	0.37–0.56	4.0–6.0	30–44.66	90–134
Compact gravel	0.46–2.79	5.0–30.0	28 and upwards[c]	112

[a] Assumes a unit weight of 45 kg/283 × 10^{-4} m³ (100 lb/ft³).
[b] No tipping should occur on sloping ground.
[c] Attention should be given to the state of the rock mass.

NOTES

1. There are now a number of ways of making a hole in geological ma-
    terials, e.g. jetting with water, breaking with electrical currents,
    melting with thermal lances, however, conventional drilling remains the
    method by far the most commonly used. Although there are many
    varieties of drill the majority can be divided into two general classes,
    viz. percussive and rotary. Normally in percussive drilling a chisel-
    shaped bit is repeatedly struck against the rock so as to form a hole. Its
    primary aim is to fracture and pulverize the rock to a fine debris which
    can be flushed from the hole by the returning current of air. In rotary
    drilling the aim is either to cut or crush and grind the material with
    tough blades or points, which are rotated against the rock under load.
    Drilling machines have combined both rotary and percussive methods
    so that bits designed to both crush and shear are now available.

2. Toughness can therefore influence the choice of bit. For example, dragbits and other non-coring bits that are used in tough rock are robust, having strong shoulders and small flushing grooves of sufficient size to permit the finely comminuted debris to leave the cutting area. Bits used in soft rocks do not need to be so strong and can have wider flushing grooves so that the cuttings produced, which are normally larger than those from tougher rocks, can be flushed away without further comminution. Similar differences can be seen in the Tricone bits, those for weak rock have large, sharp pointed, widely spaced teeth and those for tough rock have smaller, blunter, more closely spaced teeth.

3. Gauge wear is also a problem because it reduces the diameter of the bit and results in a tapering hole which will tend to jam any new bit that is introduced.

4. Deviation of coring drills can lead to an incorrect appreciation of subsurface geology, see Fig. 15.4. For example, a vertical drill-hole intersecting a steeply dipping fault-zone may be deflected some way along the zone before continuing on through the rock. The core recovered from this hole would indicate a thicker fault-zone than was actually present. Other situations can similarly occur: the interaction of soft shale horizons or weathered dykes are two examples. Where alignment is important for correct interpretation it is often advisable to have the holes surveyed while they are being drilled. Instruments used for this range from the simple, yet reliable, clockwork Pajari apparatus to the more sophisticated electronic pendulum dipmeters. Surveys of this kind are commonly needed in holes longer than 30 m.

5. Augers are frequently used for drilling through thick overburden prior to hard rock drilling, for drilling soft rocks, and for excavating large holes for cast piles and similar structures. They are also being increasingly used for small scale jobs such as pipe conduit and drainage installation, where horizontally mounted augers are used to drill small diameter holes for distances of 60 m without disturbing the ground surface.

6. Conventional tunnelling techniques, at the time of writing, can be grouped according to use, viz those for soft ground and those for hard ground. In soft ground, such as clays, pneumatic spades and similar tools can be used literally to hand-dig the working face. In hard ground the standard procedure is to drill, blast, and clear; this has reached a high degree of development—as witnessed by the shaft sinkers in South Africa—and can cope with the hardest of rocks. Rates of advance in excess of 15 m (50 feet) per day can be achieved by competitive crews. However, it suffers from the disadvantage of having the sequence of work punctuated by blasting, still a relatively uncontrolled operation which can result in weakening the surrounding rock, and overbreaking the design profile of the excavation. Nevertheless, the system, like hand digging, is flexible, mobile, and relatively inexpensive. New tunnelling machines weigh tens of tonnes, consume several hundred horsepower, are not particularly mobile, and are very expensive to build, but they

do offer a continuous and controlled means of tunnelling and are profitable when used in the right situations (Yardley, 1970).

7. Most commonly used explosive is in a solid form that can liberate energy in an extremely fast reaction so that on detonation it is transformed almost instantaneously into a gas of high pressure, sometimes in excess of 10 000 atmospheres. This reaction sends a shock wave through the rock at a velocity of around 3000–5000 m/s. This wave accelerates the ground around the blasthole in an outward radial direction causing a zone of intense deformation immediately around the hole, and circumferentially orientated tensile stresses in a zone 1–2 borehole diameters away from the hole. When these exceed the tensile strength of the ground they produce a series of radial cracks which propagate outwards at approximately a quarter of the speed of the compressive wave front. When the front meets a free face, such as a joint, or the face of an excavation, it is reflected back into the rock mass so generating a tensile stress. This reflected wave is most important because it can produce spalling at the free face and assists the propogation of the radial cracks which it meets on its return journey. In the meantime the radial cracks extend back into the blasthole and become filled with the gases generated by the explosion. The gas pressure once in the cracks increases the tensile stress at their tip and so permits them to extend even further from the hole. Maximum acceleration at free faces normally occurs soon after these cracks arrive and fragments are usually ejected from the faces themselves. (See references on the mechanics of blasting in Langefors and Kihlström, 1967.)

8. In pre-splitting a number of holes are drilled relatively close together along the desired line of break to the full depth of the cut. The holes are then charged and stemmed so that on detonation the interaction between adjacent charges is so strong that the effects of the geological fabric are largely smothered. This interaction induces a failure surface that links the holes, and forms a smooth surface that follows the excavation boundary rather than geological surfaces, as would be the case with normal blasts. The technique is proving versatile and popular. (See Kutter and Fairhurst (1967) and Paine *et al.* (1961).)

9. The natural gas found below the North Sea contains small amounts of ethane, butane, and propane, but is essentially a methane gas. It originates from the decay of organic remains trapped within the Carboniferous rocks of the area some 9000 feet below present sea level, but is drawn off from strata ABOVE the Carboniferous to which it has migrated, notably the porous Permian Sandstone known as the Rötliegendes and the Bunter sandstone (Triassic). The gas (sp. gr. 0.62) has been trapped in these sandstones because above them lie impermeable layers of shale and evaporite which have halted its upward migration. The best traps are antiformal structures (see p. 222, Chapter 7). The North Sea area contains some of the largest gas fields discovered to date in the world. The first important strike, in what is now known as the Groningen gas field, was made at Slochteren in Holland in 1959. In October 1965 gas was found beneath the North Sea: by

the mid-1970s the whole of the United Kingdom should be served by this natural gas; the gas is non-toxic.

10. Care should be taken when gas is unexpectedly encountered in site investigation boreholes. The first indication of its presence is often the bubbling which occurs in the returning water. Some odour may be present and a gurgling noise may also be heard coming from the hole.

11. 'Cutting' is the creation of 'cut holes' from which the break-out of the remaining cross-section can be started. They are normally blasted, and the success of the operation depends greatly upon the relationship of the spacing, size, and detonation of the charges to the tensile strength of the rock.

12. A shield, in its simplest form, is a rigid steel cylinder that is driven in advance of a permanent lining to support the ground and protect those at the working face. The forward edge of the shield is heavily reinforced and driven into the ground ahead of the working face. When material is excavated manually it is usually from a number of gantries, and can start from one or more at any one time depending upon the stability of the ground. Permanent pre-fabricated lining segments are assembled in the trunk of the shield, and the key segment for any ring is located when the shield has passed the forward edge of the ring. The key segment expands the ring so that it fully occupies the cavity left by the shield. Any spaces left between the permanent lining and the ground are normally filled with grout. The shield is usually propelled forward and steered by hydraulic jacks mounted around its perimeter. The cutting edge of these machines should if possible be designed for the ground in question particularly with respect to its angle of friction and degree of hardness.

13. One big difficulty in shield tunnelling through weak water-bearing ground is the exclusion of water; when this enters the shield it usually carries with it suspended material from the ground, thus facilitating subsidence at the surface. Compressed air can be used to counteract the water pressure and so relieve the problem. An airtight bulkhead is installed either in the trunk of the shield, so that only the area in the vicinity of the working face is pressurized, or, more usually, at some point behind the shield in a section of the tunnel where the lining is complete. Compressed air is avoided with the use of a Bentonite Shield. This is a conventional full faced rotary digger which has a circulating system of bentonite slurry that is kept in contact with the cutting face. This supports the face, renders the walls impermeable and reduces the caulking requirements. Cuttings from the face are carried away in the return slurry (see the New Civil Engineer, 1972).

14. This is because a tunnel in water-bearing strata will have a pressure at its invert which exceeds that at its crown. Hence the air pressure required to balance the water pressure at the base of the tunnel will be γd greater than at the crown of the tunnel (where γ = density of water and d = diameter of tunnel). Blow-outs should be anticipated if γd approaches either the overburden pressure above the working face or a level near to that. Contractors commonly use pressures which balance

the upper two-thirds of the working face and drain the water that enters the lower third, thus reducing the danger of blow-outs.

15. Subsidence is generally associated with the excavation of solids, e.g. ores, coal, evaporites, however, ground-level lowering cun also occur with the extraction of pore fluids, e.g. water, oil, and gas.

16. Sometimes the beam support permits a downward separation of strata about bedding surfaces, but is sufficient to stop them from breaking. As a result a dangerous situation is created, for such ground would undergo severe settlement if overstressed by surface structures.

17. 'Tailings' are the waste materials from mines, and if incorporated in tips they can significantly reduce their stability. However, when handled properly they can be safely incorporated into engineering structures (see Pettibone and Kealy, 1971).

REFERENCES

AMBRASEYS, N. and HENDRON, A. 1968. Dynamic behaviour of rock masses. Chapter 7 in Rock Mechanics in Engineering Practice. *Ed.* Stagg and Zienkiewicz. J. Wiley and Sons.

ANON. 1957. River Gravel Solidified for British Tunnel, *Engineering News Record.* Nov.

ATKINSON, T. 1971. Selection of open-pit excavating and loading equipment. *Trans. Inst. Mining & Met.* Section A,80.

BARA J. and HILL, R. 1967. Foundation rebound at the Dos Amigos pumping plant, J.S.M. and F.E. *Proc. A.S.C.E.*, **93**, 153.

BAILEY, J. and DEAN, R. 1966. In Failure and Breakage of Rock. *Proc. 8th. Symp. Rock Mech. Amer. Inst. Mech. Eng., Minneapolis.*

BARRON, K., COATES, D. and GYENTE, M. 1970. Artificial support of rock Slopes. Res. Rep. R 228, *Dep. Energy, Mines & Resources* (Mines Branch), Ottawa, Canada.

BISHOP, A. 1973. The stability of tips and spoil heaps. *Quart. Jour. Engng. Geol.*, **6**, 335–377.

BOYUM, B. 1961. Subsidence Case Histories in Michigan Mines. *4th. Rock Mech. Symp., Penn. State Univ., U.S.A.*

BREKKE, T. and SELMER-OLSEN, R. 1965. Stability problems in underground construction caused by montmorillonite carrying joints and faults. *Engineering Geol.*, **1**, 3–19.

CARTNEY, S, 1977. The ubiquitous joint method. Cavern design at Dinorwic Power Station. *Tunnels and Tunnelling,* **9**, 54–7.

COATES, D. 1964. Some cases of residual stress effects in engineering work. in State of stress in the earth's crust. pp. 679–688. *Ed.* Judd. (Amer. Elsevier Pub. Co.).

COUSENS, R. and GARRETT, W. 1969. The flooding at the West Dreifontein Mine. *9th. Commonwealth Mining and Metallurg. Congr., London.*

DEERE, D. 1968. Indexing rock for Machine Tunnelling in Rapid Excavations, pp. 32–38 Soc. Mining Eng., Am. Inst. Mining Met. & Petr. Eng., New York.

HALCROW, W. T. 1930. The Lochaber Water-Power Scheme. *Proc. Inst. Civ. Eng.*, **231**, 1930–31.

JAEGER, C. 1975–76. Assessing problems of underground structures. *Water Power and Dam Construction.* Dec. 75, Jan 76.

KVAPIL, R. 1965. Gravity flow of granular materials in Hoppers and Bins in Mines. Part II, Coarse Materials. *Int. Jour. Rock Mechanics and Mining Sci.*, **2**, 277–304, and Part I, **2**, 35–41.

KUTTER, H. and FAIRHURST, C. 1967. The roles of stress wave and gas pressure in pre-splitting. *9th. Symp. on Rock Mechanics, Colorado School of Mines.*

LANGEFORS, O. and KIHLSTRÖM, B. 1967. The modern techniques of rock blasting. J. Wiley and Sons, 2nd. Edn.

MCGREGOR, K. 1967. The drilling of rock. C. R. Books, Ltd. (A. Maclaren Co.), London.

MARTNA, J. 1970. Engineering problems in rocks containing pyrrhotite, in Large Permanent Underground Openings. *Proc. Int. Symposium, Oslo. Ed.* T. Brekke and F. Jörstad, Published in Scandinavian Univ. Books, Oslo.

MEASOR, E. and WILLIAMS, G. 1962. Features in the design and construction of the Shell Centre. *Proc. Inst. Civ. Eng.*, **21**, London. (Discussion in **24**, 1963).

MERRILL, R. and WISECARNER, D. 1967. The stress in rock around surface openings, in Failure and Breakage in Rock. *Ed.* Fairhurst, C. *Amer. Inst. Min. Metall. Petr. Eng.*, 337–350.

MOYE, D. 1964. Rock Mechanics in the investigation and construction of Tumut 1 Underground Power Station, Snowy Mountains, Australia, pp. 123–154. Engineering Geol. Case Histories, No. 3 *Geol. Soc. Amer.*

NATIONAL COAL BOARD. 1970. Tech. Handbook for Spoil Heaps and Lagoons.

NEW CIVIL ENGINEER. 1972. Newly proved Bentonite Shield takes tunnels into new ground, cuts costs by a third. (21st. Sept.)

NORTH, F. 1952. Some geological aspects of subsidence not due to Mining. *Proc. S.W. Inst. Eng.*, **67**, 127.

OBERT, L. and DUVAL, W. 1967. Rock Mechanics and the design of structures in rock. (Chapter 19 in particular). J. Wiley and Sons.

PAINE, R., HOLMES, D., and CLARK, H. 1961. Controlling overbreak by Pre-splitting. *Int. Symp. on Mining Res., Univ. of Missouri*, **1**, 179–209. Issued by Pergamon Press, 1962.

PATCHING, T. H. 1962. Investigation related to the sudden outburst of coal gas. *Proc. Rock Mech. Symp., McGill Univ., U.S.A.*

PEACH, B. N. 1929–31. The Lochaber Water-Power Scheme and its Geological Aspect. *Trans. Inst. Mining Eng.*, **78**.

PERSSON, P., et al. 1970. The Basic Mechanisms in Rock Blasting. *Proc. 2nd. Congr. Int. Soc. Rock Mechanics, Belgrade*, Paper 5/3.

PETTIBONE, H. and KEALY, C. 1971. Engineering Properties of Mine Tailings. *Proc. Amer. Soc. Civ. Eng.*, **97** (Sept.).

PITEAU, D. and JENNINGS, J. 1970. The effects of Plan Geometry on the stability of natural slopes in rock. Kimberley area of South Africa. *Belgrade (ibid)*, Paper 7/4.

PRICE, N. 1972. Report from the Department of Mining and Metallurgy, Univ. of Minnesota, U.S.A.

PROCTOR, R. and WHITE, T. 1946. Rock Tunnelling with steel supports. Commercial Shearing and Stamping Co., Youngstown, Ohio.

REISNER, W., ROTH, M. and COLIJN, H. 1971. Bins and Bunkers for handling bulk materials. **1.** No. 1. Series on Rock Mechanics and Soil Mechanics. *Trans. Tech. S.A.* CH-4711, Switzerland.

Report of Tribunal appointed to inquire into the Aberfan disaster on Oct. 21, 1966. H.M.S.O. London, 1967. And Technical Reports submitted to the Tribunal, items 1–7 and 8. H.M.S.O. 1969.

ROBINSON, C. *et al.* 1974. Engineering Geologic, Geophysical, Hydrologic, and Rock Mechanics investigations of the Straight Creek Tunnel Site and Pilot Bore, Colorado. United States. Geol. Survey. Prof. Paper 815.

STEPHENSON, H. S. 1962. Report on the Explosion at Hampton Valley Colliery, Lancashire. Ministry of Power, H.M.S.O. See also the Specialists' Reports, Ministry of Power (Chief Scientists' Division). Safety in Mines Research Establishment. Hampton Valley Report No. 4. Examination of affected area with special reference to the nature and spread of the explosion.

STINI, J. and LAUFFER, H. 1965. in Functional Rock Classification. *Rock Mech. Symp. IVA Swedish Acad. Eng. Sci., Stockholm.* Brochure 142, p. 124.

SZECHY, K. 1966. The Art of Tunnelling. Akademai Kiado., Budapest.

TAYLOR, R. K. 1968. Site investigations in coalfield—the problem of shallow mine working. *Q. Jour. Eng. Geol.,* **1,** 115.

U.S.B.R. 1959. Technical Record of the Design and Construction of the Tecolote Tunnel. United States Bureau of Reclamation, Denver, Colorado.

WRIGHT, D. E. 1971. The hydraulic design of unlined and lined invert rock tunnels. Construction Industry Research and Information Association, Report 29, London.

YARDLEY, D. (ed.). 1970. Rapid Excavations, Problems and Progress. *Proc. Tunnel and Shaft Conf., Minneapolis.* Published by *Am. Inst. Mining, Metallurg. and Petrol. Eng. Inc.,* New York, 1970.

SELECTED BIBLIOGRAPHY

Mainly terrestrial environments

American Symposia on Rock Mechanics (from 1958, annually) *Amer. Inst. Mech. Eng.*
Bulletins of the Colorado School of Mines. (Quart. Journal).
Int. Jour. Rock Mechanics and Mining Sciences. Pergamon Press.
Tunnels and Tunnelling. Pub. of British Tunnelling Soc.
Quarry Managers' Jour. Pub. of Inst. of Quarrying, London.

Mainly aquatic environments

Terra et Aqua. Pub. of Internat. Assoc. of Dredging Companies. The Hague.
Offshore Engineer. A publication of Inst. Civ. Engineers. London.
Proc. Waterways and Harbours and Coastal Engineering Division. Amer. Soc. Civil Engineers.

16 Development and Redevelopment

The development of an area is usually accompanied by the utilization of natural resources and the alteration of geological regimes. Attitudes towards the environment in general now dictate that the success of such schemes depends, in some measure, on the extent to which the ensuing changes are in keeping with the general sense of conservation. This means that the geology of such areas must be more fully appreciated than ever before, and that geological investigations should extend beyond the boundaries of the developed area so that the effects of the development on neighbouring districts can be estimated. In this Chapter a variety of situations are described where geology is important to the success of development. Such a review is by no means comprehensive, and a wider treatment of the subject has been provided by Knill (1970), Scheidegger (1975) and others. Four topics related to development in general are discussed under the following headings: Infiltration and Run-off; Water supply; Pollution; and Valley development.

INFILTRATION AND RUN-OFF

Infiltration is the natural process by which water enters the ground (Chapter 12). Obviously if the ground is progressively covered by urban development the volume of water it will receive from infiltration will decrease. However, the area of such development is usually only a small part of the intake area for an aquifer; of far greater significance is the area of ground that is farmed. Ploughing and tilling for example will tend to disturb the delicate structure of the surface soil layers and usually decrease their permeability. This can often be restored only by returning the ground to its original state, and in this respect the practice of rotation farming is of value. The continual cultivation of land which cannot regain its infiltration characters within the period of a growing season should be avoided. One example of the effect changing agricultural policies can have upon the hydrogeological regime comes from farming districts in east Gloucestershire, which lie on Jurassic limestones. Old records giving the water levels in wells and position of springs indicate that there has been a gradual drop in the level of the water-table since 1860. As far as can be ascertained, rainfall has not decreased, and the abstraction of water from the ground has not increased; indeed the only widespread change that has occurred is the swing from pastoral to arable farming.

Changes in infiltration are normally accompanied by changes in run-off, and these can be most marked in areas where extensive alterations have been made to the character of the topsoil or to its vegetation (Pereira, 1962). This can occur on a large scale when urban growth and industry enter an area, and it is not uncommon for run-off to increase from a gradual process to something much quicker. In many cases both the rate and quality of run-off are altered. Walling and Gregory (1970) describe the changes which accompanied construction in a small catchment. They noted that the suspended load in run-off from the catchment had been generally increased by a factor of ten, and sometimes by a hundredfold; major increases usually accompany large development programmes. The geological factor which most affects infiltration and run-off is the stability of the surface layers. This is partly a function of their grain size, mineral composition and structure, and partly a function of the angle of slope. Careful mapping of slope angles and soil types is therefore required, particularly in relation to their susceptibility to erosion. A general review of the whole subject is provided by Chow (1964).

WATER SUPPLIES

Water can be provided from either surface or sub-surface sources. In many countries both are used in an integrated programme that aims at achieving the best use of each. However, both are dependent upon the storage capacity of the catchment. For example, a catchment composed of impermeable rocks, such as slate or clay, will have little, if any sub-surface storage that is of use for water-supplies. Run-off will occur soon after precipitation begins and regular water-supplies will only come from storing this run-off with an impounding structure. Furthermore, the amount stored must be sufficient to maintain an adequate supply during periods when run-off is not making good the volumes drawn off. A permeable catchment such as those in many limestone regions is capable of storing water below ground; but the water in most aquifers is flowing and areas above sea level will eventually drain to some base-level (Chapter 12) if not periodically re-charged. Hence storage, in water supply problems, means volume with respect to time. Quantitative assessments of water-supply must therefore include the element of time in the components of the hydro-logical cycle. Normally this is described in terms of hydrological seasons which may or may not comprise a calendar year, depending upon the climate of the area in question. The seasons are essentially periods when precipitation is either effectively or ineffectively recharging the storage of the catchment. Initial studies for water-supply therefore take account of meteorological data, in particular the rainfall and evaporation for the area. Figure 16.1 illustrates the annual distribution of rainfall in the British Isles (Note 1). By comparing this

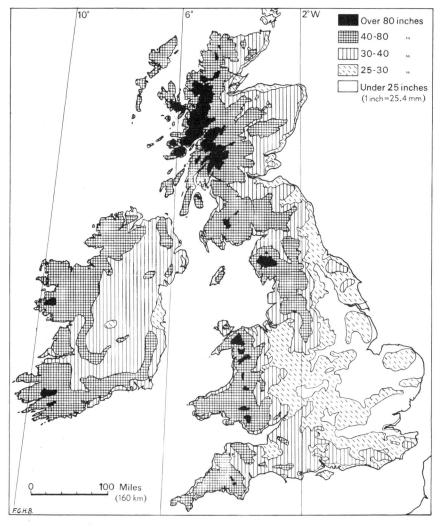

Fig. 16.1   Mean Annual Rainfall of British Isles for the 35-year period 1916–1950.

distribution with that of the rocks in the British Isles (Fig. 8.7) it can be seen that the greatest rainfall happens to occur in areas where the rocks have low permeability and little storage, e.g. the metamorphic and igneous rocks of N.W. Scotland, the Lake District, and Wales. As a result water in these areas is stored at the surface in reservoirs, many of which are used to generate hydro-electric power, e.g. the Cruachan Scheme (p. 270). The outcrops of the country's best aquifers (Permo-Triassic sandstones, Jurassic and Cretaceous limestones) which lie mainly to the south of a line joining the River Tyne to the Exe estuary, receive lesser amounts of rainfall. Great Britain is therefore in the unfortunate situation of having its greatest concentration of population, and hence its greatest demand for water, in the south-east, where the least rainfall is received. As a result severe problems of water-supply have developed (Water Resources Board Report, 1966). A simple review of the meteorological data needed for the assessment of water resources is provided by Ward (1967). From the meteorological data it is possible to determine in which months rainfall is in excess of evaporation and capable of restoring the soil moisture deficits to zero (Headworth, 1970). After that, further rainfall will replenish the storage available until such a time when evaporation dominates the climatic regime. A certain amount of this rainfall is often lost to the catchment as run-off, and when this occurs it should be measured and its volume subtracted from the effective rainfall. Records for run-off in the United Kingdom can be obtained from the Surface Water Year Book. What remains can be assumed to have penetrated the ground and replenished its storage. This will be witnessed by a rise in the level of water in wells and boreholes (Fig. 12.5). The volumes of water involved are obtained by multiplying the depth of effective rainfall by the area of the catchment.

In making calculations for catchments, two catchment areas must be considered; that for ground-water and that for surface water. The surface water catchment is defined by the topographical divide which joins the ridges of hills within the catchment and separates basins of surface drainage. This boundary frequently lies directly above the ground-water divide, which joins the ridges of the water-table, and separates basins of ground-water drainage. However, significant differences in the areas of the two catchments can occur in any one region. For example the River Itchin in Hampshire has a ground-water catchment that is 20 per cent greater than its surface catchment (Fig. 16.2). In this situation it is necessary to consider the infiltration which enters the ground-water catchment that lies beneath the neighbouring surface catchments. Finally a study must be made of the natural depletion of stored water which occurs with the drainage of ground-water, either to rivers and lakes, or to the sea. The former can

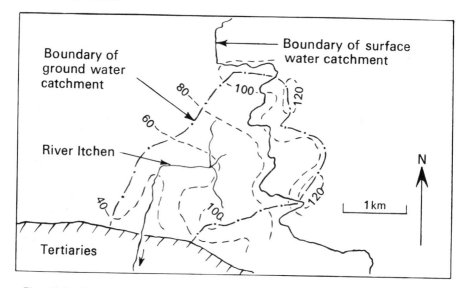

Fig. 16.2 The catchments of the River Itchen (after Headworth). The boundary of the surface water catchment separates the drainage basins of the rivers Wye and Rother, to the east, from those of the Meon, Itchen, and Test to the west; only the River Itchen is shown. The northern part of these catchments is of Chalk, but this is covered in the south by Tertiary sands and clays. The ground-water contours that are shown are in metres above sea level and are the levels measured in October 1969.

be gauged by analysing the ground-water component of river run-off, the latter by constructing flow nets for that part of the catchment which discharges ground-water towards the coast. Electrical analogues of the type described by Rushton (1970) are commonly used to assist in such studies.

Having obtained the quantities involved in these fundamental processes it is possible to quantify the ground-water balance as defined by the general Hydrological Cycle, i.e.

Precipitation = Run-off + Evapo-transpiration + Infiltration

This can be considered as

Input = Output + Changes in Storage

Normally a more detailed listing of processes needs to be made and that suggested by Farvolden (1967) is typical of many used. He writes the Hydrological Cycle as follows:

$$\text{Precipitation} = SW_r + GW_u + ET \pm \Delta SW_s \pm \Delta GW_s \pm \Delta SM_s$$

where   Precipitation = Total input of water into the system,
   SW_r = Surface water run-off from the system,
   GW_u = Ground-water underflow from the system,

$ET =$ Total evapo-transpiration loss from the system,

$\Delta SW_s =$ Net changes in surface water storage,

$\Delta GW_s =$ Net changes in ground-water storage,

$\Delta SM_s =$ Net changes in soil moisture storage.

The ground-water storage available for supply in an aquifer, at any time, can be assessed by multiplying the effective porosity of the aquifer (Note 2) by its saturated volume. How this volume of water is used is a matter of management. For many years it has been considered good practice to budget the available supply on the basis that no more is removed from an aquifer than is received by it. Any withdrawals which exceeded this budget were considered to over-develop the aquifer and if permitted to continue would progressively deplete its storage. This policy does not use the bulk of the stored water within an aquifer and in certain cases it can be a wasteful form of ground water management. For example, the water level in some aquifers in southern England becomes so high, during the winter season of recharge, that much of the precipitation which falls over them is soon lost to the sea as run-off.

More recently it has become accepted that the stored water in the ground should be used more fully so that a larger volume of ground becomes de-watered during the summer season, when supplies are in greatest demand. The volume of potential storage so produced can then be used to house some portion of the winter precipitation which usually runs to waste. This form of ground-water management therefore controls not only the ground-water but also the river run-off and so helps to alleviate the risk of winter floods. Such a scheme has been initiated by the Thames Conservancy; water is pumped from the ground in the summer and discharged into rivers so as to augment their summer flows. Burrow (1971) describes the conjunctive use of ground-water and surface reservoirs.

Having assessed the volumes that can be pumped it is then necessary to select sites from which these volumes are to be taken. Sites are normally chosen on the basis of field permeability, which should be as great as possible; the proximity of hydrogeological boundaries, which should be carefully investigated; and the regional direction of ground-water flow, which should ideally be towards the area that is pumped; for a general review see Walton (1970). Water is usually pumped from either wells or adits; Plate 12a (facing p. 377).

Wells These are normally sunk by rotary or percussion methods, although hand digging may be economical in certain areas. Casing is often installed in the upper levels of a well to prevent polluted surface water from entering the supply. Screens may or may not be required at depth, depending upon the ability of the ground to maintain an unsupported hole. Careful geological investigations should be conducted when screens are required in gravels and sands so that the size of slots in the screens are correct for the grading of the deposit. This may vary within any one hole, according to the strata encountered, so that the well screen may have different size openings along its length. Screen apertures which allow the entry of too great a quantity of material into the well will clog the well, damage the pumps, and can result in the subsidence of the ground around the well. Samples for grain size analysis are therefore required for the design of such filters (see Efficiency of Well Screens Report 1969). The chemistry of the ground-water should also be analysed so that the screen can be made from the most suitable material to resist corrosion and incrustation. Factors which commonly cause this are a low pH coupled with low alkalinity and a high content of free carbon dioxide, the presence of hydrogen sulphide, sulphur dioxide or similar gases and the presence of organic acids. An ordinary mineral analysis will not supply all the information required, as a measure of the dissolved gases is also necessary, particularly in connection with incrustation, which eventually results from the presence of dissolved carbon dioxide.

Many wells need to be developed in order to reduce the draw-down which occurs at the well for a given discharge. This can be done by increasing the permeability of the rock around the well; the methods commonly used to achieve this are flushing, surging, jetting, acidization, shot-firing and hydraulic fracturing (Note 3). Any of these, if used incorrectly, can ruin the well. Experienced personel should be employed for well development. In some cases the permeability around the well is increased by surrounding the well screen with a gravel pack, which is designed to suit both the ground and the screen (see Efficiency of Well Screens Report, Anon, 1969).

Adits Conventionally dug by hand, these tend to be either horizontal or gently inclined towards a central shaft which can be several metres in diameter. Many old shafts of this nature have been unexpectedly encountered during site investigations in the London Basin. Adits are usually oriented to intersect as many fissures as possible (Plate 12a), and can be periodically dammed along their length to provide a certain storage. The ganats (qanats) found in many arid and semi-arid regions are essentially adits which tap the upper levels of the water-table. They are normally lined with tiles and stones when support is required. The general problems of water supply in arid regions are reviewed by Hills (1966).

A horizontal collector of a different type is the Ranney Collector. This is a series of well screens which are jacked horizontally from a central shaft so as to radiate into the ground. They can be developed by flushing, jetting, and surging and are ideally suited for obtaining large supplies from stratified alluvium. General guidelines to the construction, maintenance, and operation of wells and adits are provided by the Manual of British Water Engineering Practice (1969), and by Huisman (1972).

The quality of water obtained from aquifers can vary considerably, and is essentially a function of the chemistry of the infiltrated supply and the mineral characters of the ground. Table 16.1 gives the chemistry of typical ground-water, but these values should only be taken as a guide. Nevertheless, significant departures from the values quoted normally indicates some 'peculiarity' within the aquifer. The analysis of water is referred to in Chapter 10. Recommended standards for human consumption and for some industrial processes vary from country to country, but the general recommendations given in the Manual of British Water Engineering Practice and by the World Health Organization are almost universally accepted norms.

Artificial recharge   In many countries the demand for water that arises in an area may exceed the natural replenishment to such an extent that the aquifers have to be artificially recharged. This can be accomplished in a number of ways, the most common of which use either wells or basins. The efficiency of all methods depends largely upon the quality of the water being recharged, which should be chemically and biologically clean so that the pores and fissures of the intake surfaces do not become clogged. In this respect special care should be taken to exclude suspended material and entrapped air from the recharge water. The chemistry of the water should also be similar to that already in the aquifer. All these factors are difficult to control and the efficiency of most recharge systems decreases with time. A comprehensive review of the subject was presented at the Artificial Ground-water Recharge Conference in 1970: (Water Research Assoc. 1970).

POLLUTION

Whether or not a substance constitutes a pollutant depends largely upon the situation in which it is liberated. The waste from a clay pit, for example, would probably not be considered a pollutant if it were discharged into disused clay pits, but the reverse would be true if it was poured into rivers. Thus the disposal of waste material need not result in pollution and for this reason the substances which can cause pollution are not considered here. Instead, the emphasis is given to the description of some commonly used disposal methods which involve geological factors and can possibly pollute existing geological

Table 16.1   Analyses of Typical Groundwaters (All Values are as Parts per Million)

	SiO_2	Al	Fe	Mn	Ca	Mg	Na	K	HCO_3	CO_3	SO_4	Cl	F	NO_3	PO_4	pH Min	pH Av	pH Max
Igneous																		
Acid	37.0	0.1	0.3	0.02	28.0	6.1	11.6	2.3	142.0	0.0	15.7	5.0	0.2	3.8	0.10	6.3	7.2	7.7
Basic	41.0	0.1	0.6	0.06	26.0	14.3	7.7	3.1	202.0	0.0	12.1	22.6	0.2	6.5	0.03	5.6	7.5	8.2
	20.0	0.05	0.1	0.01	27.0	3.1	6.1	1.6	66.0	0.0	31.1	4.4	0.1	2.4	0.03	7.2	7.5	7.7
Sedimentary																		
Arenac.	23.0	0.1	0.5	0.04	49.0	17.7	19.1	2.8	252.0	2.0	69.0	9.8	0.4	8.6	0.02	6.2	7.3	9.2
Argillac.	27.0	0.8	1.6	0.06	110.0	51.2	179.0	5.7	330.0	2.8	96.6	121.0	0.6	4.1	0.02	4.0	6.7	8.6
Limestone	13.0	0.1	0.4	0.05	71.0	19.1	12.9	2.2	228.0	0.0	8.8	9.7	0.2	8.9	0.02	7.0	7.5	8.2
Dolomite	14.0	0.2	0.4	0.07	62.0	43.7	13.7	1.1	272.0	0.7	35.0	6.9	0.5	6.3	0.00	7.4	7.7	8.2
Metamorphic																		
Quartzite	10.0	0.07	0.7	0.10	33.0	12.4	6.5	3.3	119.0	0.0	37.5	5.9	0.2	1.7	0.02	6.5	7.1	7.4
Marble	10.0	0.15	0.1	0.03	52.0	10.0	3.0	1.2	192.0	0.0	10.7	4.7	0.1	0.05	0.00	7.6	7.7	7.9
Slate	16.0	0.02	0.5	0.10	39.0	6.0	14.7	3.3	143.0	0.0	40.1	7.8	0.3	4.7	0.03	5.2	7.1	8.0
Schist	16.0	0.02	0.5	0.10	39.0	6.0	14.7	3.3	143.0	0.0	40.1	7.8	0.3	4.7	0.03	5.2	7.1	8.0
Gneiss	33.0	0.1	0.5	0.07	39.0	24.2	29.5	2.5	219.0	0.0	34.3	34.5	0.9	5.8	0.00	5.8	7.1	8.1
Drift																		
Igneous	39.0	0.00	0.2	0.01	18.0	6.2	25.7	3.7	220.0	0.0	74.1	32.6	0.4	5.4	0.00	6.1	7.2	7.9
Sedimentary	19.0	0.00	0.2	0.00	62.0	20.4	53.2	3.9	269.0	0.6	71.1	22.4	0.2	8.6	0.00	7.4	7.8	8.4

regimes. The movement of many pollutants normally follows that of ground-water flow and can thus be predicted by the distribution of hydraulic head within the ground. Pollutants will not seep from disposal sites if there is no hydraulic gradient away from the sites themselves. For a comprehensive review see Water Research Assoc. 1972.

Injection wells These are wells which serve as conduits for pumping fluids into the ground; many old wells in abandoned oil fields are being used for this purpose. The supply to these wells must be as clean as possible, and free from suspended material, biological matter, and entrapped air; these can clog pores and fissures at the well and so reduce its recharge performance. The waste is often chemically different from the ground-water which it displaces, and this is the cause of most pollution that is associated with such schemes. If possible the waste should be made chemically inert before injection, so that it does not promote unfavourable reacting with either the ground-water or the geological formation. The clay content of the ground is usually the most reactive solid phase and its mineral content should be carefully studied. Most of the investigations are completed in the laboratory and experiments should be conducted at temperatures and pressures that are comparable with those at the level of injection. General reviews of the subject are provided by Bernard (1957) and Warner (1968). The extent to which such pollution will affect the ground is a function of the regional and local ground-water flow, which should be studied in both the horizontal and vertical directions.

Injection wells have been known to cause earth tremors and both their siting and operation may need to be based on detailed geological studies. The cause of such tremors is thought to be the increase in ground-water pressures which result from the introduction of fluid under pressure. This increases the pore pressures in the ground and so decreases the effective stress which operates across any surface. Local quakes are thought to occur when the normal stress on surfaces subject to shear stresses is reduced to the point where shear failure can occur. The mechanics of this system are described by Hubbert and Rubey (1959).

Poolen and Hoover (1970) discuss the earth tremors which accompanied the operation of a 12 000 ft (4000 m) deep disposal well in Colorado. Figure 16.3 illustrates the history of the well which, by reason of the tremors it was causing, was eventually abandoned.

Waste pits Waste can be tipped either on the ground or into pits, both are among the most commonly used methods for disposing of solid and liquid matter. Suitable tipping sites include low-lying land, marsh, foreshore, quarries, subsidence areas, ground to be

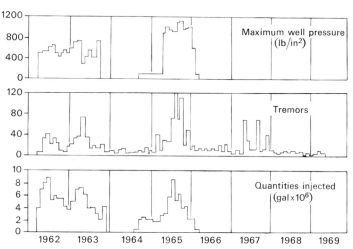

Fig. 16.3 Correlation between injection and earthquake history for the Rocky Mountain Arsenal Well, Colorado (after Poolen and Hoover).

reclaimed, undulating ground to be levelled, valleys, cuttings, and ravines to be filled. The main geological factor to consider in these operations is usually ground-water, for tipping can easily pollute ground-water supplies. Two types of pit are recognized; a wet pit where material is dumped into a flooded pit whose base lies below the water-table, and a dry pit which is founded above the highest level of the water-table. Both can give rise to pollution, although the greatest risks are associated with wet pits. This is because they are foreign bodies within the ground, and the water in them eventually becomes incorporated into the local system of ground-water flow. As a general rule, such pits should be filled with inert material, or be situated in areas where there is little or no ground water-flow, or be placed in ground which is neither an existing nor a potential aquifer for water supply. Clay pits are an example of the last point, but care should be taken to assess whether flow exists between such ground and neighbouring aquifers. When reports describe ground as being 'impermeable' they usually mean that it has a very low permeability (Note 4).

Dry pits do not interfere so directly with the ground-water regime and offer a little more protection from pollution, because of the filtration which occurs when water drains through the ground from the pit to the water-table. The efficiency of this filtration depends on the permeability and porosity of the ground. Truly permeable materials such as sands and gravels are excellent filters for bacterial and organic pollutants, whereas pervious material such as jointed rock offers little if any filtration. Pits built on fissured ground should be given a proper foundation which may entail filling all the fissures with sand, cement or some impermeable fill such as clay, and providing the base of the pit with a drainage system which discharges into a sewer. Such a method is currently used when tipping in Chalk quarries in Surrey (Summer, 1964). Few natural materials filter chemical pollutants, and as far as is known no near-surface geological deposits will prevent the downward migration of gases such as carbon dioxide and methane, both of which are generated in certain refuse tips. The travel of polluted water from a pit to an aquifer is more fully discussed in the report on the Disposal of Toxic Wastes (1970).

Care should also be taken with pits that cross spring lines. These can not only erode the base of the pit but can also flush any polluted infiltrate out of it. Permeable foundations, such as a gravel base linked to a piped collector system would ensure a degree of control over these pits. A full review of the subject is provided by Bevan (1967).

Coastal pollution   A form of pollution which commonly arises in coastal regions is the intrusion of sea water into a fresh water aquifer. The density of sea water is 0.025 greater than that of fresh water, so that under static conditions a head of 41 units of fresh water can be balanced by a head of 40 units of sea water. This is described as the Ghyben-Herzberg relationship, after its discoverers and is illustrated in Fig. 16.4(a); this is an idealized situation as conditions are rarely static and fresh water is normally flowing out to sea. This flow tends to deflect the interface between fresh and salt water seawards. Tidal fluctuations also mix the sea water with the fresh to produce a zone of brackish water rather than a sharp interface. Tidal fluctuations can cause the water levels in coastal wells to rise and fall as illustrated in Fig. 12.5.

If a well is sunk close to the coast and pumped so as to develop a cone of depression, a corresponding rise will result in the level of the salt water interface, as shown in Fig. 16.4(b). This progressively affects the aquifer and if pumping continues will eventually pollute the well. Instances of such pollution have occurred on most coast lines and along many estuaries, e.g. along the Mersey, Thames, Humber, Clyde, Tyne, and Severn. The correction and subsequent management of coastal aquifers is considered in most textbooks on ground-water hydrology (see Chapter 12).

DEVELOPMENT IN VALLEYS

Ideally a river system should be considered as one hydrological unit that extends from its headwaters to its estuary. The geological problems associated with the development and re-development of such

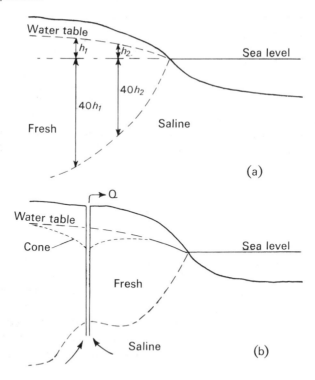

Fig. 16.4 Sea water intrusion in coastal aquifers. (*a*) Natural conditions, (*b*) conditions near a pumped well. For explanation see text.

a unit can be linked with its cross-sectional characters. Three basic situations can exist: (i) steep sided, often narrow valleys; (ii) broad valleys which are the sites of substantial alluvial deposits; and (iii) estuaries.

Steep sided valleys have slopes which are often unstable, and it is these that frequently control the pattern and extent of development which takes place. For example, the heads of the valleys of the South Wales coalfield are narrow and steep. Many slopes have failed and many may still be failing. Since mining was established in the 1860's much of the ground has been affected at one time or another by de-watering and subsidence. Local residents believe that shallow mining disturbed many potentially unstable slopes, but few appear to have totally failed before the pit closures of the 1920's and 1930's— possibly because the ground was drained by mining operations. Mining recommenced after the industrial slump and continues at the present; will further slips develop when this eventually ceases? Other problems associated with the re-use of these valleys were reviewed by the 1969 Conference at the Institution of Civil Engineers. The stability of slopes is considered in Chapter 13.

Broader valleys usually contain alluvium that provides a flat valley floor and a natural site for development. However, much of the alluvium is likely to be saturated and some may have a high permeability. Ground-water is usually the more difficult of the engineering problems to cope with here, particularly with respect to foundations although the prediction of the settlement of structures founded on materials of differing strength can also be difficult. It is often best to avoid the need for deep excavations in saturated gravelly alluvium. Piles, if used, should be chosen with care, particularly if the alluvium is dense and has been re-worked either naturally or by earlier construction. Re-working involves the disturbance of pre-viously deposited material, and usually results in clay and silt-sized particles being removed from the sediment, thus increasing the permeability of the ground. Much alluvium in developed valleys has been seriously re-worked by the trenching required for laying sewers and other piped services. In these operations the material which has been excavated to form the trench is normally used to re-fill the trench, but it is rarely replaced at the same density at which it existed before, so that the infilled trench may act as a vertical drain of high permeability. If many of these exist in an area they can completely alter the natural pattern of ground-water flow; de Freitas and Fookes (1970) have described a situation where back-filled trenches of this kind allowed the discharge from fractured sewers to damage newly placed cast-*in situ* piles. Small streams are often culverted and placed below ground in re-development schemes. In some instances this procedure can seal them from the ground they are supposed to be draining, and can result in local flooding. Water supplies from alluvial deposits are usually polluted if no adequate sanitation facilities are provided in the region.

The majority of the important valleys in north-west Europe and North America have already been altered to such an extent that the original character of their alluvial fill, in highly developed areas, is becoming of diminishing significance to local engineering projects. When investigating such areas great care is required, for the geologist has now to interpret not only the remnants of an original natural situation, but also the character and possible extent of one that is completely man-made. Gray (1972) describes ground-water con-ditions which now exist in the superficial deposits of the Thames Flood Plain, London where the underground railways are definite zones of discharge and have a widespread effect upon the overall pattern of ground-water flow through the deposits. Permeable river embankments are points of recharge to, and discharge from, the surrounding ground.

Problems of stability can also arise in the gentle slopes of broad valleys (e.g. cambering p. 383) and attention should be directed to both the solid and drift geology of the slopes, (see for example Chandler, 1972).

Estuaries are a meeting point for river and sea. The valleys associated with them are usually broad and contain a variety of sediments, fresh-water, estuarine, and marine. Foundation design in these materials requires the able application of soil mechanics, and the advances in this subject have now made many estuaries potential

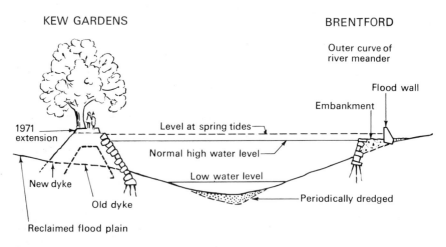

Fig. 16.5   Section through the River Thames, West London. Vertical scale approximately twice the horizontal.

areas for future development. Unfortunately, their very position makes them vulnerable to invasion by the sea. For example, in February 1938 at Horsey in Norfolk, a violent storm breached the line of sand dunes which filled the old silted estuary of the 12th century 'Hundred Stream', and caused disastrous floods. In January 1953 a severe storm swept down the North Sea and flooded many of the estuary areas between the Humber and the Medway. The lowlands of Holland were also flooded. Flood control has therefore become an integral factor in the development of river areas (see Chapter 2). Initially it was restricted to canalizing and strengthening river sides with embankments, and straightening their courses, all of which was included in the general term 'river training.' Figure 16.5 illustrates some of these operations.

More ambitious schemes, however, are now usually required to protect the estuary areas satisfactorily. In 1918, two years after a disastrous flood in Holland, the Dutch Government decided to dam the Zuider Zee with an embankment dam 30–40 km long. This outer dam was completed in 1932, transforming the Zuider Zee into a fresh-water lake. The success of the polder reclamations which followed has given great impetus in north-west Europe to the use of barrages for protecting estuaries. Several are now proposed for the U.K., including Morecombe Bay, the Severn estuary, and the Wash. London's new airport may be built in the outer reaches of the Thames estuary on reclaimed land protected by embankments. Schemes of this magnitude can affect the hydrology of large areas. For example, ground-water levels in the Thames Flood Plain will be raised as a result of the new flood barrier which will cross the upper reaches of the Thames estuary. This could in turn affect existing drainage facilities in the area, and subject many basements and foundations to positive pore water pressures for perhaps the first time in many years.

The geological evaluations that will be required to solve the problems associated with re-development will always neccessitate reference to existing geological records, and it should be remembered that routine geological maps, memoirs, and reports are rarely produced with either the engineer or the developer in mind. They are normally statements of fact (see Chapter 11), and it is for the user to interpret them as best he can.

NOTES

1. **Rainfall statistics**   Detailed records of rainfall in Britain have been kept since about 1860, when G. J. Symons began to organize the recording of rainfall data; local records go back much farther. The statistics show that the average rainfall for the whole of England, taken over a period of thirty-five years (1916–1950) is 33.8 (858 mm), and for Great Britain 41.9 inches (1063 mm) per year.

The rainfall for a particular locality may vary considerably from the average for the whole country: thus Greenwich has an average of 24.5 inches (620 mm) per year, while a mountain station such as Borrowdale in Cumberland receives as much as 140 inches (3556 mm).

Maximum intensities of rainfall generally occur over short periods of time, as in the case of thunderstorm rain. A typical example is the fall in 24 hours of 9.56 inches (243 mm) at Bruton, Somerset, on June 28, 1917. A fall of 9.00 inches (229 mm) on August 15, 1952, was recorded at Longstone Barrow, Exmoor, and led to disastrous floods at Lynmouth.

Measurements of rainfall are made by means of a gauge, which usually consists of a metal cylinder 18 inches (457 mm) long and 5 inches (127 mm) in diameter, containing an inner cylinder in which the rain is collected through a funnel. The rain collected is decanted into a measuring glass. At less accessible moorland stations, a larger type of gauge (such as the 'Bradford') is used, in which the amount of water collected is measured by inserting a graduated rod.

The rainfall recorded in any given period may be shown on a map, on which are plotted the observations at a large number of stations, and lines of equal rainfall (*isohyetal* lines) are drawn at intervals by interpolation. Maps of this kind, together with rainfall statistics, are issued annually in the Meteorological Office publication known as *British Rainfall*.

Of the rain that falls on the earth's surface, part returns direct to the air as water vapour by evaporation and by transpiration from plants; part flows away in streams and rivers; and part enters the soil and subsoil and percolates to the rocks beneath. The amount of evaporation is controlled by the air temperature and humidity, by air currents, and by the temperature of the ground on which the rain falls.

Measurements of evaporation have been made with water in tanks, and with pans filled with soil and buried so as to be level with the ground surface; evaporation has also been estimated from records of water levels, and of intake and outflow at reservoirs. Measurements from the tank at Camden Square, London, date from 1885 and give a value for the average annual evaporation there of 15.5 inches (394 mm).

According to H. L. Penman (1948)[1] the evaporation loss from *continuously wet* bare soil is 0.9 times that from an open water surface exposed to the same weather conditions. Much detailed work has been done on the evaluation of evaporation and transpiration losses since the Rothamsted Station's Reports of 1907–09. Penman (1950) has shown that evaporation in Great Britain ranges from about 12 inches (304 mm) p.a. in the north to 21 inches (533 mm) in the south-west; values of 17 or 18 inches (432 or 457 mm) hold for much of southern England.[2]. As the rainfall increases broadly from east to west (Fig. 16.1), many different combinations throughout the range of rainfall and evaporation exist; one extreme is the low rainfall and high evaporation in the south-east of the country, where percolation is in consequence very low. By contrast, some areas have high rainfall and low evaporation.

Measurements of percolation have been made at certain localities by means of 'percolation gauges.' A section of the soil and sub-soil, either a cube or cylinder about 3 feet (1 metre) in depth (but sometimes larger), is enclosed in a metal container which is sunk nearly flush with the ground. From its base a drain leads off the percolating rainwater. Records obtained with such gauges at Rothamsted are published annually in *British Rainfall*. These records are consistent among themselves. Varying results can be obtained, however, since infiltration is affected by the nature of the surface on which the rain falls, e.g. whether it is bare soil, or grass-covered, or supporting crops of different kinds. The setting up of a

[1] Penman, H. L. 1948. Natural Evaporation from Open Water, Bare Soil, and Grass, *Min. Proc. Roy. Soc.*, **193A**, 120.

[2] Penman, H. L. 1950. Evaporation over the British Isles, *Q. J. Roy. Met. Soc.*, **76**, 372. (Reprinted in *Journ. I.W.E.*, **8**, 1954, 415).

See also: Lapworth, C. F. 1965. Evaporation from a Reservoir near London. *Journ. I.W.E.*, **19**, 163.

gauge also involves re-packing the excavated rock and soil into the gauge, and it is difficult to reproduce exactly the natural conditions before the ground was disturbed. A further criticism of percolation gauges is that they give data for only small areas; variations in geology may be present in a formation throughout its extent, and these are not allowed for when local gauge data are assumed to apply to a whole outcrop; see p. 342 and 364.

Percolation is therefore often estimated by assuming a value for evaporation (and for soil-moisture deficit, if available) and deducting this together with the flow of surface streams (expressed as depth of rain per annum in inches or millimetres) from the annual rainfall.

Careful observations extending over long periods have been made for areas of Chalk grass-land with no surface streams in south-east England. It has been shown that at Compton, West Sussex, the annual evaporation is nearly constant and almost independent of variations in rainfall; it amounts to 17.4 inches (440 mm)—averaged over 50 years—out of an average annual rainfall of 37.6 inches (955 mm).[3] Again, by plotting values of average yearly rainfall (R) against percolation (P), for several stations on the Chalk of south-east England, the following formula has been derived:[4] $P = 0.9R - 13.5$ measured in inches, (or $P = 0.9R - 340$, in millimetres). The data used in this instance are obtained from records of gauges, and from measurements of water levels in wells and the flow of springs. It will be seen that this formula, when applied to the Compton figures, gives a result in agreement with that found there independently. It appears, therefore, that for grassed Chalk areas in south-east England, where annual rainfall varies between 35 and 40 inches (889 and 1016 mm), evaporation accounts for between 17 and 17.5 inches (431.8 and 444.5 mm).

2. The volume of water released by an aquifer is commonly described by the coefficient of storage, this being the ratio of the volume of water derived from storage, from a vertical column of unit area extending the full saturated thickness of the aquifer during a reduction in head equivalent to a unit fall in water level, to the unit volume of the aquifer. It is usually expressed as a decimal, and for water-table conditions where water comes from gravity drainage will commonly be in the range 10^{-2} to 10^{-1}. In confined conditions where the water is mainly coming from the compaction of the aquifer the range will be between 10^{-5} and 10^{-3}.

3. 'Flushing' pumps clean water into the hole and flushes out sediment and drilling debris that is lodged on the walls of the hole.

'Surging' cyclicly pushes water into the walls of the hole and draws it back out so that material blocking pores and fissures is dislodged and drawn into the hole. This is then baled out. Surging can be achieved by using either plungers or compressed air; it is usually more effective than flushing.

'Jetting' blasts the sides of the hole with jets of water which break the

[3] Thomson, D. Halton 1947. *Journ. Inst. W.E.*, 1, 39. (See also 1931, *Trans. Inst. W.E.*, 36, 176; and 1938, 43, 154.)

[4] Lapworth, C. F. 1963. *Journ. Inst. W.E.*, 17.

cake of drilling debris that may be plastered to its sides. The hole is pumped while jetting is in progress so that the dislodged material can be withdrawn.

'Acidizing' injects acid into the hole in order to dissolve the debris blocking fissures and pores. The hole is then pumped to remove all dissolved material and clean the ground of acid. The method is often used to develop wells in limestones.

'Shot firing' disturbs the ground around a hole by means of an explosion, with the intention of increasing the fracturing in the immediate vicinity and increasing the permeability of the ground.

'Hydraulic fracturing' is generated by pumping water into a hole that is sealed at the surface and building up a water pressure within the hole which opens fissures in the ground. The method is not commonly used for water supply wells.

4.   An interesting source of pollution which occurred through an apparently impermeable zone is described by Knill, M. (1970). Two water supply wells were sunk through the Chalk and Gault of North Kent into the Lower Greensand from which they drew their supply. The Greensand is underlain by an 'impermeable' clay of Jurassic age. The wells produced good quality water for a number of years, when suddenly droplets of oil appeared in their discharge. Chemical analysis suggested that it was a weathered crude oil. After an exhaustive investigation it was concluded that natural oil in the Jurassic strata below the base of the well had managed to seep up through the clays below the sands and had entered the wells.

REFERENCES

ANON. Efficiency of well screens and gravel packs. 1969. *Jour. Inst. Water Eng.*, **23.**

BERNARD, G. 1957. Effect of reactions between interstitial and injected waters on permeability of reservoir rocks, in *Symp. on water flooding*, Bull. 80, Illinois State Geol. Surv.

BEVAN, R. 1967. Notes on the Science and Practice of the Controlled Tipping of Refuse. *Inst. of Public Cleansing*, London.

BRITISH RAINFALL. An Annual Publication produced by the Meteorological Office.

BURROW, D. 1971. Conjunctive use of water resources. *Jour. Inst. Water Eng.*, **25,** No. 7.

CHANDLER, R. 1972. See Chapter 13 references.

CHOW VENT-TE, 1964. Handbook of applied Hydrology. McGraw-Hill Book Co. Ltd.

DISPOSAL OF TRADE WASTES (REPORT). 1970. *Min. of Housing and Local Govmnt.* H.M.S.O. London.

DISPOSAL OF TOXIC WASTES, 1970. Technical Committee Report. Dept. of Environment. H.M.S.O.

FARVOLDEN, R. 1967. Methods of study of the ground-water budget in N. America. Contribution No. 126, Dept. Geology Univ. West Ontario, Canada.

de FREITAS, M. and FOOKES, P. 1970. Groundwater problems associated with ground engineering in the re-worked alluvium at High Wycombe. *Proc. 1st. Int. Conf. of Int. Assoc. Eng. Geol., Paris,* **2.**

GRAY, D. and FOSTER, S. 1972. Urban influences upon ground water conditions in Thames Flood Plain Deposits in Central London. *Phil. Trans. R. Soc. Lond.,* **A272.**

HEADWORTH, H. 1970. The selection of root constants for the calculation of actual evaporation and infiltration for Chalk catchments. *Jour. Inst. Water Eng.,* **24,** No. 7

HILLS, E. 1966. Arid Lands. A geographical appraisal. (Chap. 5, Water supply, use and management.) Methuen & Co.

HUBBERT, M. and RUBEY, W. 1959. Mechanics of fluid-filled porous solids and its application to overthrust faulting. *Bull. Geol. Soc. Amer.,* **70.**

HUISMAN, L. 1972. Groundwater Recovery. Macmillan.

INSTITUTION OF CIVIL ENGINEERS. 1969. Conf. on Civ. Eng. Problems of the S. Wales Valleys. *Proc. I.C.E.*

KNILL, J. 1970. Environmental Geology. *Proc. Geol. Assoc.* **81,** 529.

KNILL, M. 1970. Oil pollution of a Greensand source. *Jour. Inst. Water Eng.,* **24,** No. 7.

MANUAL OF BRITISH WATER ENGINEERING PRACTICE. 1969. Compiled by Inst. Water Engineers. 4th Edition. W. Heffer & Sons.

PEREIRA, H. 1962. Hydrological effects of changes in land use in some East African catchment areas. *E. African Agric. & Forestry Jour.* Special Issue, **27.**

POOLEN, H. and HOOVER, D. 1970. Waste disposal and earthquakes at the Rocky Mountain Arsenal, Derby, Colorado. *Jour. Petr. Tech.* August.

RUSHTON, K. and HERBERT, R. 1970. Resistance network for three-dimensional unconfined ground-water problems with examples of deep well dewatering. *Proc. Inst. Civ. Engs.,* **45,** March.

SCHEIDEGGER, A. 1975. Physical aspects of material catastrophies. Elsevier, Amsterdam.

SUMMER, J. 1964. Technical developments in refuse collection and disposal. *Jour. Inst. Munic. Engrs.*

SURFACE WATER YEAR BOOK. An annual publication produced by the Department of the Environment. H.M.S.O.

WALLING, D. and GREGORY, K. 1970. The measurement of the effects of building construction on drainage basin dynamics. *Jour. of Hydrology,* **11.**

WALTON, W. 1970. Ground-water Resource Evaluation. McGraw-Hill.

WARD, R. 1967. Principles of Hydrology. McGraw-Hill.

WARNER, D. 1968. Subsurface disposal of liquid industrial wastes by deep well injection, in Subsurface Disposal in Geological Basins. *Amer. Assoc. Pet. Geol.* Mem. 10.

WATER RESEARCH ASSOCIATION. 1970. Artificial Ground Water Recharge Conference. Proc. of Conf. given at Reading, England.

WATER RESEARCH ASSOCIATION. 1972. Conf. on ground-water pollution, Reading.

WORLD HEALTH ORGANISATION. 1963. Geneva. International Standards for drinking-water.

17 Materials of Construction

Introduction   Any source of materials, geological or otherwise, which is to be used in engineering construction must satisfy certain requirements which are fundamentally economic. The source should be economically sited so that the cost to the consumer in respect of transport charges is kept to a minimum. The geological conditions at the source should also be economically favourable; for example, overburden which would have to be removed before the extraction of the material should not be excessive; and the concentration of the required material in a deposit should be such as to exclude high costs in separating it from other material with which it may be associated. Geologically, the rock itself should be suitable for the purpose for which it is to be used: this may be a matter of grain size, as with sands and gravels; or it may be a question of purity in mineral materials, or the quality of rock for a particular purpose. Other factors to be taken into account include the method of working, and transport to the construction site. This chapter deals only with geological materials that occur naturally; manufactured or treated materials such as burnt shale, slag and coke breeze for aggregates are not discussed further.

In a general sense the materials used for engineering purposes can be broadly divided into three groups—those required for their mineral characteristics, those required for their structural properties, and water. The first group would contain such materials as the clays used for bricks, pottery, and as fillers; asbestos and gypsum; ore minerals; fuels, including coal products; bitumen; the raw materials for making cement. Any investigation of these materials should include an assessment of their quality as well as their volume.

Building stones, aggregates, and the like would more naturally come into the second group, and their suitability would be assessed not only on their volume and general quality, but also on rock-mass properties, e.g. size and shape of blocks. Thirdly, the importance of water in the construction industry lies mainly in its ability to mix satisfactorily with cement. The chemical characters of a supply would therefore be considered as well as the volume available.

It is convenient to divide the subject matter of this chapter into the following sections, which deal with specific requirements for different types of construction:

  (i)   aggregates
  (ii) building stones
 (iii) road stones and ballast
 (iv) materials for embankment dams
  (v) cements and fillers
 (vi) water used in construction
(vii) materials for bricks, boards, and glass.
Ornamental stones are not discussed.

AGGREGATES

Concrete is of great importance as a structural material and special attention has been paid to the nature of the aggregates employed. The term *aggregate* includes gravels, crushed stone, sand, and other materials which are mixed with cement and water in making concrete. In the United Kingdom the specifications for concrete aggregates and building sands are set out in British Standards 882 and 1201 (see list of references on p. 529). Aggregates for road-stone and sub-base materials are covered by B.S. 882. The selection of a suitable aggregate is of prime importance because it forms over three-quarters of the volume of the concrete. The presence of an undesirable mineral in a particular rock may be a source of weakness, and may give rise to aggressive chemical reactions when in contact with the cement used; these can progressively weaken the concrete. Fragments of rocks which decompose readily, or swell on the addition of water, can also affect the strength of the resulting mix. Such constituents, if present, can be detected by proper tests in advance of the adoption of a rock for use as an aggregate; the study of thin sections of the aggregate with the microscope can often yield important information.

It is usual to describe aggregates as *coarse* or *fine* according to their particle size, the dividing line between the two being taken as 4.8 mm (formerly $\frac{3}{16}$ inch). 'Coarse aggregate' is material mainly retained on à 4.8 mm sieve and 'fine aggregate' is material which mainly passes a 4.8 mm B.S. sieve. (For particulars of Test Sieve aperture sizes see B.S. 812, 1967, and B.S. 410).

Trade Groups. To overcome difficulties of nomenclature which arise from the numerous petrographical names for rocks, aggregates are classified under 11 Trade Groups, as follows:

Name of Trade Group	Rocks included in Group (or other materials).
(1) Artificial	Slag, crushed brick, breeze, etc.
(2) Basalt	Basalt, andesite, porphyrite, dolerite, hornblende-schist.
(3) Flint	Flint, chert.

Name of Trade Group	Rocks included in Group (or other materials).
(4) Gabbro	Gabbro, diorite, norite, peridotite, picrite, serpentinite.
(5) Granite	Granite, gneiss, granodiorite, pegmatite, quartz-diorite, syenite.
(6) Gritstone	Grit, agglomerate, arkose, breccia, conglomerate, greywacke, sandstone, tuff, pumice.
(7) Hornfels	All contact-altered rocks except marble.
(8) Limestone	Limestone, dolomite, marble.
(9) Porphyry	Porphyry, aplite, felsite, granophyre, microgranite, quartz-porphyry, rhyolite, trachyte.
(10) Quartzite	Quartzite, ganister, siliceous sandstone.
(11) Schist	Schist, phyllite, slate, severely altered rocks.

Any particular stone offered for aggregate needs to be checked as being chemically and structurally suitable before it is accepted; this is important if a new source is being opened up. Rock from an existing quarry can vary in type or quality (e.g. fresh or weathered), and checks on samples should be carried out at regular intervals.

Sands Naturally occurring sands are commonly used as fine aggregate, and in terms of particle size the grains of suitable deposits lie between the limits 2 mm and 0.06 mm (see definitions in Chapter 5, p. 177). Sands have been derived from the weathering of rocks of many kinds, and individual grains are generally single mineral particles (coarse grains may be an exception to this). Only the more resistant minerals survive weathering processes and form sand deposits, and of them *quartz* is the most abundant because of its resistance to abrasion, decay, and solution. Quartz therefore forms the bulk of most sands; feldspar persists for a shorter time and is found in some quickly sedimented deposits. Muscovite, a stable mica, persists through long periods of weathering and transport. Other minor constituents of sands are mentioned in Chapter 5.

Most of the deposits used for fine aggregate therefore have a relatively simple mineral composition. The grains may, however, have coatings which colour the deposit, such as red and yellow iron oxides. These, unless the grains are thickly coated, are not objectionable; if necessary they can be largely removed by washing the sand.

The grading in a deposit depends on the conditions under which it was formed: wind-blown or eolian sands are very uniformly graded; water deposited sands have a wider variation in grade size (see Fig. 17.1, where a No. 52 sieve represents an aperture size of 0.3 mm and a No. 7 sieve 2.40 mm).

Gravels The harder, more resistant rocks form the greater part

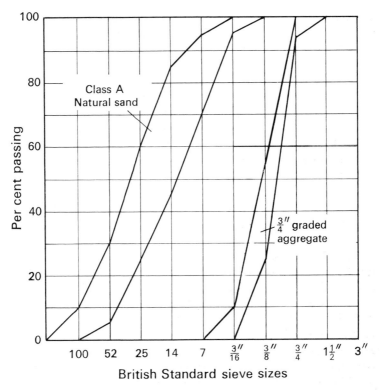

Fig. 17.1   Grading limits for a natural sand and a $\frac{3}{4}$ inch (19 mm) coarse aggregate.

of most gravels, because they withstand the wear and tear of weathering and transport (see Chapter 2). Common constituents of gravels are therefore materials such as flint, chert, vein quartz, quartzite, granite, gneiss, and in some deposits schist, slate, sandstone and limestone. Certain flint gravels of Tertiary age in England consist of about 98 per cent of rounded flint pebbles together with 2 per cent of vein quartz or quartzite. In the Thames Valley area such deposits are worked extensively for use as aggregate for concrete, e.g. 'Thames Ballast' from the River Colne (Note 1). On the other hand, in an area of metamorphic rocks the river gravels are more varied in composition; one deposit from the River Tay, Scotland contained approximately 72 per cent of gneiss (with some schist), 16 per cent of vein quartz or quartzite, and 12 per cent granite. Occasionally softer rocks bulk large in a deposit, as in the Dodder River gravels near Dublin, where over 80 per cent of the deposit was limestone, the remainder being a mixture of chert, vein quartz, granite and gneiss. Such a deposit would have been formed after a short distance of transport, so preserving the softer (limestone) material. Sandstone pebbles are also found in some gravels but are relatively short-lived unless the sandstone has a siliceous cement. Schist and slate fragments, because of their shape, tend to split into small flat-sided particles.

Glacial deposits such as outwash gravels and eskers (see Chapter 2) are one of the main sources of gravel and sand in Great Britain and elsewhere; they are frequently mixtures of sand and pebbles, i.e. they contain a variety of grain sizes. Raised beach deposits are another source, as at Scottish and other localities. When glacial debris has been moved and re-worked by rivers and deposited in their valleys such deposits are termed fluvio-glacial; they commonly provide useful supplies in accessible localities.

It may be necessary to treat or 'process' a gravel deposit before its use as aggregate by *screening*, *crushing*, or *washing*. Screening is used to separate larger from smaller fragments, e.g. in a deposit where pebbles are embedded in a matrix of sand or where pebbly layers and sandy layers are intercalated. The requisite proportions of pebble and sand grades for a particular concrete can then be obtained by the mixing of measured quantities. If pebbles are very smooth and rounded, a light crushing may be necessary to provide partly angular fragments with rough surfaces, which give better adhesion when mixed with cement than the smoother pebbles. Thus, light crushing improves some aggregates. Again, when the surfaces of fragments are coated with matter, e.g. silt, iron oxides which would inhibit good adhesion with cement, the coatings can be removed by washing to provide a clean aggregate. The treatment for a particular deposit varies according to the characters of the materials in it, the objective being to secure an aggregate which is composed of clean fragments with surfaces sufficiently rough to give a good bond with cement (Note 2).

Crushed aggregates   A large output of aggregate, both coarse and fine, is obtained from quarried rock which is broken down by crushing. It then consists of mainly angular fragments which have a rough surface; the crushed stone is graded into suitable sizes and stock-piled at the quarry. Aggregates are then made up by blending the necessary grades of material from the stock-piles of different sized fragments to give a mix of any desired proportions. The angular shape of the fragments enables denser concrete to be made, if required, than with more rounded fragments. During the crushing a certain amount of dust is inevitably produced; a high proportion of dust in the product can render it unsuitable, and this is important especially in connection with fine aggregates. The extraction of dust during the crushing process is therefore carried out as part of the production process. Many kinds of rock, both igneous and sedimentary (e.g. limestone), provide crushed aggregate material.

Specifications for crushed stone aggregates in the United Kingdom are given in British Standard 882. The grading limits for a coarse and a fine aggregate are illustrated in Fig. 17.1 (for a description of the method of plotting grading curves see Chapter 10). In the figure two points lie on an ordinate for a particular particle size: these are the maximum and minimum allowable percentages specified. The grading curve for a particular aggregate, produced by blending of crushed stock grades, must lie between the two curves, i.e. within the limits of the specification. The left-hand pair of curves indicate the limits set for a natural sand (fine aggregate) in a similar way.

Crushed rock for coarse aggregate and natural sand for the fine aggregate are used at many sites. However, in view of difficulties in obtaining supplies of natural sand in some areas, the use of the finer fractions from crushed rock-material instead of sand is being investigated. The crushed rock 'fines' have characteristics that differ from those of natural sands; preliminary results of tests have shown that concrete of satisfactory strength is produced when crushed rock fines are used in this way, but the properties of the concrete vary according to the aggregate used (Note 3).

Light-weight aggregates.   The vesicular rock *pumice* (p. 158) is used as an aggregate in making low density concrete, such as slabs for interior walls. The pumice has a high proportion of void space, and slabs made with it also have sound insulating properties. Its specific gravity is between 0.3 and 0.9. Pumice from the Lipari Islands and from New Zealand localities has been extensively used in the past.

Other light-weight materials are coke-breeze (sp. gr. about 1.0) and crushed brick (sp. gr. about 1.4); these are manufactured aggregates of relatively low cost, and are extensively employed in the building industry.

Shape of fragments   In British Standard 812 the shapes of particles are described by six terms:

(1) *rounded*, water-worn or rounded by attrition, e.g. beach gravels.
(2) *irregular*, partly rounded by attrition or naturally, sub-angular, e.g. glacial gravels, river gravels.
(3) *angular*, possessing well defined edges and roughly planar faces, as in crushed rock and scree.
(4) *flaky*, material having a small thickness relative to other dimensions, as in fragments derived from laminated rocks such as slate and schist.
(5) *elongated*, fragments (not flaky) whose length is considerably larger than their other dimensions.
(6) *flaky and elongated*, having length much greater than the width, and width considerably larger than thickness.

The *surface texture* of fragments can also be described by non-geological terms:

glassy, characteristically with conchoidal fracture;
smooth, water-worn, or with smooth fracture as in fine-grained or laminated rocks;
granular, fracture surface showing rounded grains, as in sandstone;
rough, the rough fracture of fine- or medium-grained rock containing no easily visible crystalline constituents;
crystalline, containing easily visible crystalline constituents;
honeycombed, having visible pores or cavities.

The above terms enable a surface to be described by visual examination of hand specimens. They do *not* refer to petrographical textures as seen in thin sections with a microscope.

Undesirable constituents   The presence of fragments of flat or elongated shape in an aggregate, such as shale, slate or schist, in anything but very small amounts, can be deleterious to the resulting concrete. In particular, shale fragments have low strength and high water absorption; their area: volume ratio is several times that of a spherical fragment of equal volume, and this greater area allows much higher water absorption. Slate and schist have a tendency to split preferentially parallel to their flat surfaces, and their strength is weaker in that direction than at right angles to it. Other harmful constituents or impurities in aggregates include clay, coal, hydrated Fe-oxides, sulphides and sulphates, salt and organic material. They are listed in B.S. 882, where also tests for their determination and limiting amounts are prescribed. The amount of clay in a fine aggregate, for example, should not exceed 3 per cent of the total weight.

Reference should also be made to B.S. 812 for details of tests for the determination of clay, silt and dust in aggregates, and for water absorption, organic impurities, and moisture content; mechanical properties such as crushing strength and resistance to abrasion, and methods of sampling an aggregate, are also described (Note 4).

Minerals which react with cements.   In 1938 and the early 1940's certain concrete structures made with high-alkali cements showed deterioration from cracking caused by internal expansion. One road bridge failed in this way a year after completion (Stanton, 1940). The reason for the deterioration of the concrete was eventually traced to the presence of certain mineral constituents in the aggregates used, in particular opal and chalcedony; fragments of volcanic rocks, tuffs, and certain linestones, which were present in the aggregates, were also suspected of contributing to the process. It was shown by tests that when high-alkali cement is mixed with water an alkaline solution is produced which reacts with opal, and a gel begins to form which continues to absorb water from its immediate surroundings. This alkaline silica gel swells and sets up pressures which distend the paste surrounding the reactive particles, and the swelling may eventually lead to the cracking of raw concrete when the internal pressure exceeds its strength. Mortars made with opal, chalcedony, and chert fragments showed rapid expansion during the first six or eight months. Cracks are formed first on a microscopic scale; they then become enlarged and may eventually extend to the surface, where the gel may be exuded. This was observed in several concrete structures. Detection of the reactive constituents is readily made by microscopic examination of an aggregate; opal or chalcedony may be present in the vesicles of lavas such as rhyolite and andesite, or filling cracks in limestones. An account of the laboratory investigations which led to an understanding of the process of deterioration is given in the Berkey Volume, 1951 (Note 4).

Another method of avoiding deleterious cement-aggregate reaction is to replace a part of the cement by a pozzolan (p. 527). The pozzolan is finely ground and reaction takes place between it and the alkalies in the cement; the alkalies are thus used up and the products of reaction are spread throughout the concrete (Note 4).

Low-grade aggregates Recent studies have been made of relatively low quality materials which may be proposed for use where good quality aggregate is not available at an economic price. Difficulties of supply have arisen in many developing countries, where the choice of materials is limited, in connection with road construction. Among the samples of low-grade aggregate from overseas which have been tested are lateritic gravels from Sierra Leone and Uganda, soft limestone from Nigeria, gravelly soils from Kenya, and crushed coral limestone from the Bahamas (Note 5).

Methods of studying unsound rock include the use of thin sections on which point counts of secondary minerals are made, using a petrographical microscope. Chloritic decomposition products, after e.g. pyroxenes in altered dolerites, are readily identified and their amount estimated. It was found that weathered dolerite gravels from Lesotho, which were studied by this method, showed high percentages of secondary minerals in point counts, generally from 33 to 80 per cent and over. Other tests were made, e.g. for soundness and shrinkage. Attention is also directed to the stripping of overburden down to fresh rock below weathered rock, especially important in the opening up of new quarries. When a quarry is producing, any weathered rock going through the crushers is largely reduced to 'fines', and these can be separated by passing the crushed rock over a screen, a process known as scalping.

BUILDING STONE

Natural building materials can be conveniently classified into three groups, namely dimension stone (including 'armour' stone), roofing, and ornamental stone. The choice of natural stone for use in engineering construction depends on its cost and availability at a site, and on properties of the particular material such as strength, durability and appearance.

Dimension stone   The cost of rock for dressed stonework is controlled by the ease (or difficulty) of quarrying and the kind of finish required—rough, fine-axed, or polished, and on the distance from the site to which it is to be supplied. The first factor is a geological matter, and geological advice on it may be valuable, especially when a new quarry is to be opened, in order to obtain suitable rock. Distance from a site, if considerable, may put up haulage costs to a prohibitive extent—unless no alternative source is easily available. Strength can be determined by testing (see Chapter 10), and durability assessed by observing the extent of weathering in natural exposures of the rock. The appearance of a stone in a structure depends on textural and mineral features, and on its resistance to weathering; in general, uniformity of colour and texture are desirable. Proper selection of the stone at the quarry is of vital importance.

The main types of rock which yield good quality stone in large enough sizes for structural work are granite, limestone, and some varieties of sandstone. The properties of each of these rocks are now discussed, with examples.

(i) *Granite.*   The term here includes rocks which are strictly granite and also other granitic rocks such as granodiorite. All have a high content of quartz and feldspar (see Chapter 4) and usually a small amount of mica (muscovite or biotite), and are coarse grained. Jointing in these rocks is widely spaced so that large blocks can be taken out of a quarry and split down to the requisite sizes for a particular construction (Note 6).

The strength of a rock depends on the way in which its component minerals are held together. In granitic rocks (and many other igneous rocks) the minerals interlock, giving high strength values. The range of crushing strength for most granites is from about 96×10^6 N/m^2 to 240×10^6 N/m^2 (14 000 to 35 000 lbs/sq. inch), and their specific gravity from 2.55 to 2.75 (see Table 17.1). Many exceed the higher value in crushing strength. But taking the lower value, and allowing a factor of safety of 5, a block of granite 0.5 m^2 ($5\frac{1}{2}$ sq. ft.) in superficial area would carry a load of 7620 kg ($7\frac{1}{2}$ tons). The rock is therefore eminently suitable for heavy constructional work; and in addition it is a very durable material. Because of these properties many dock

Table 17.1   Some Properties of Structural Stone

Rock	Crushing Strength		Specific Gravity	Absorption % by Weight
	10^6 N/m²	lbs/sq. inch		
Granite	103–240	15 000–35 000+	2.55–2.75	0.09–0.3
Limestone (Oolitic)	14–62	2 000–9 000	2.0–2.4	
Limestone (Carboniferous)	c.82	c.12 000	2.5–2.7	
Sandstone	34–103	5 000–15 000+	c.2.0–2.7	2.0–12.0
Slate			2.7–2.88	0.1–0.55

walls have been constructed in granite in the past, an example being the dock and harbour works at Colombo, Ceylon (built from 1884 onwards in Cornish granite).

At the present time the use of granite as dimension stone has been partly superseded by its use as a facing material for concrete structures. Facing slabs of granite, which may be some 10 cm thick, provide a permanent finish of good appearance and are very resistant to weathering agents, including acids in the atmosphere of large cities. They can be given a 'fine-axed' finish, or in some situations a more expensive finish by polishing. Bridge parapets and balustrades, for instance, have been constructed in polished stone-work. The cost of using granite in this way has to be set against the permanency and aesthetic value imparted by it to the structure.

The low porosity of granite ($< 1\%$) gives it a low *absorption* value (see Table 17.1); absorption is the measure of the quantity of water which a rock will absorb when immersed, and is measured by the difference between the dry weight (W_1) of a sample and the weight after soaking (W_2), expressed as a percentage of the dry weight, i.e. $100(W_2 - W_1)/W_1$. A low absorption value in turn means that the rock has a high resistance to the disintegrating effects of frost.

The extremes of temperature to which a stone may be subjected in a structure are due to frost and fire, the latter being a hazard which will only arise under exceptional circumstances, if at all. A steep temperature gradient is produced at the surface of stonework subjected to fire, and outer parts of the stone tend to flake off. The two main minerals of granite, quartz and feldspar have different coefficients of expansion, and this assists the flaking process by the production of textural strains in the rock (p. 35).

(ii) *Limestone.*   Although most limestones have a similar mineral composition, consisting mainly of calcite (or calcite and dolomite) with subordinate amounts of other minerals, they show great variety because of differences in texture, porosity and fossil content. Relatively few limestones, therefore, have acquired a reputation for being sound structural material; they must also occur in beds of sufficient thickness to yield dimension blocks of suitable size. Among them are certain oolitic limestones from Jurassic formations in England, and the somewhat heavier grey limestones of Carboniferous age. Some data for these two varieties are given in Table 17.1.

As a first example, the cream coloured oolitic limestones of the Portland Series in Dorset have a uniform texture and weather evenly over a long period of time (see Note 12, Chapter 5). They are 'free-stones', that is, without bedding planes in a considerable thickness of the sediment (4.5 metres in the case of the Whit Bed), and have been worked at intervals for over 300 years (Note 7). The main joints (north–south and east–west) are widely spaced, so that large blocks of the stone can be hoisted out of the quarries for splitting and trim-ming to size. Before this is done, however, the overburden which may be up to 18 metres (60 feet) thick above the building stone horizon has to be removed. And since the quality of the stone varies, even in a single quarry, much rejected material is accumulated. The selected blocks are marked in the quarry before delivery to a site (Note 8). Under present day conditions the use of Portland Stone as facing material for buildings in ferro-concrete and repairs to the stonework of existing buildings has increased.

Another freestone which yields high quality material is worked at the Clipsham quarries, Rutland, in a bed of limestone 9.4 metres (31 feet) thick. The Clipsham Stone is an even-grained shallow-water deposit, not oolitic but composed of calcite fragments, and is part of the Lincolnshire Limestone of Jurassic age. From this thick stratum large blocks are cut out in the quarry and then split by means of wedges. As at Portland, much rock (over 80 per cent) is rejected because of the need for strict selection of first-class material. But part of the reject stone is crushed and marketed as lime for agriculture. The Houses of Parliament in London, re-built in 1948–49, were faced with Clipsham Stone, which lends itself to shaping for architectural detail; and material from the same source has been used for repair work on Canterbury Cathedral and York Minster.

Grey Carboniferous Limestone from the Pennine area was in the past worked extensively at many localities, and used for domestic and municipal buildings, railway viaducts, and bridges. This use has now been superseded, but it has left its distinctive mark on many towns and villages in northern England and adjoining areas. A large pro-duction of crushed Carboniferous Limestone is maintained at the present time for use in road construction (see p. 521).

Decay of stonework in cities. The presence of carbon dioxide and sulphur dioxide in the atmosphere of cities results in the formation of weak solutions of these gases in the rain-water, giving carbonic acid and sulphurous acid. The effect of the former is to dissolve away slowly the surface layers of a limestone, as noted in Chapter 2.

Sulphurous acid, however, reacts with the $CaCO_3$ of the rock to form the compound $CaSO_4$, which on hydration becomes crystalline gypsum, occupying a greater volume. A sulphate skin is thus formed on the surface of stonework except where the products of chemical reaction are washed away, as in parts of a building exposed to much rain; the sulphate skin gradually splits off and falls away, a process known as *exfoliation*. This and other kinds of decay are described in detail in 'The Weathering of Natural Building Stones' (Schaffer, R. J., 1972, facsimile reprint, Building Research Establishment, Dept. of the Environment, London). The rate of decay which formerly affected buildings in London and elsewhere has now been reduced by the formation of smokeless zones.

(iii) *Sandstones.* Variation among sandstones arises mainly from the uniformity or otherwise of the sand particles which make up the rocks, and from the kind of cement present (see Chapter 5, p. 182). The strength of the rock depends largely on the cohesive strength of the cement and on its amount; the best structural stones in this group are those with siliceous or ferruginous cements. The rock should also have a relatively low porosity. Water absorption is governed by the porosity; it has a wide range among sandstones (Table 17.1), and the resistance of the stone to frost damage varies accordingly. Sandstones, like limestones, are bedded rocks, and when the rock is used in construction it is important that the bedding direction in the blocks should be horizontal. If the sandstone is placed in a structure with the bedding vertical (called 'face bedding') there is a tendency for the rock to flake off, and the resulting patches of spalled rock are a source of weakness as well as a disfigurement.

Sandstone should not be used with limestone in adjacent courses of stonework; a solution of calcium sulphate (derived from the limestone and from sulphur in the atmosphere) can percolate into the surface layers of a sandstone course, and the growth of gypsum crystals in the sandstone may lead to rapid decay of the rock. Also, dense mortar should be avoided (e.g. Portland cement); in general mortar for stonework should have a low lime content.

As examples of sandstones used in modern construction in Great Britain the Pennant Sandstone from South Wales and the Bunter Sandstone of South-west Lancashire may be cited. The rocks of the Pennant Series, in the Coal Measures of South Wales, include massive feldspathic and micaceous sandstones; some varieties with angular quartz grains are locally known as grits. The Pennant Sandstones in places reach a thickness of more than 304 metres and are much quarried for building stone (dimension blocks).

The Triassic sandstones are on the whole softer than the Carboniferous rocks, but their red colour is used to good effect in many public buildings. Keuper sandstone is worked in the Wirral peninsula, near Liverpool, and red Bunter sandstones in West Lancashire, as at Woolton (see Note 6, Chapter 5). The thick beds at some localities allow large sizes of dimension stone to be obtained. In Southern Scotland, red Permian sandstones in Ayrshire and near Dumfries are similarly worked for building stone (see p. 257).

Flagstones (p. 183) provide material for facing slabs, set with the bedding vertical, as exterior cladding to large steel-framed buildings and in other situations. Instances include the Elland Flags of Yorkshire, and in Ireland the Moher Flags from the coast of Co. Clare.

Selection of a quarry site for dimension stone. Factors which should be taken into account include:

(i) absence of excessive overburden;
(ii) availability of stone of suitable quality and uniformity;
(iii) spacing of joints and bedding surfaces such that blocks of economical size are obtainable;
(iv) adequate drainage of quarry site, and haulage distance to the site of works which is not prohibitive.

The use of black powder (a mixture of charcoal, KNO_3 and sulphur) for blasting is common practice in the working of dimension stone; it burns more slowly than other explosives and the rock is 'pushed' from the quarry face rather than shattered.

Armour stone consists of large blocks of stone placed around the base of dock walls or breakwaters to take the force of waves which would otherwise impinge on the masonry. Many kinds of hard rock can be used, but they should possess neither directional characters which allow them to split after repeated impacts, nor easily weathered minerals. Granite is usually eminently suitable; hard limestones and well-cemented sandstones are also used. An example of a rock which failed because of the presence of a weak mineral in it was the coarse-grained gneiss from Shai Hills, Nigeria, used as armour stone at marine works on the coast. Heavy blocks of gneiss began to break up after a relatively short time, and investigation showed this to be due to a small content of the mineral prehnite, which was detected in thin sections cut from the rock. Prehnite is a white or greenish secondary mineral, a hydrous calcium silicate, sometimes classed with the zeolites because of its occurrence. Although only present in small amount in the gneiss its rapid weathering weakened the rock, resulting in lower resistance to the impact forces which the stone had to withstand.

Roofing stone   The production of *slate* for roofing has been much reduced by the introduction of manufactured products such as clay tiles, concrete tiles, and other materials. But in some areas the use of slate still continues on a smaller scale, both for roofing and as vertical courses where weatherproofing is needed. Slate is a low-grade metamorphic rock, formed under the influence of high pressures which have resulted in the growth of oriented minerals in the rock mass, as discussed in Chapter 6, giving a dense rock (Table 17.1). Because of its texture (or fabric) the rock splits preferentially along parallel planes, and from a block cut out in the quarry slates are split off in the cleavage direction and trimmed to size. The properties of slate which give it a commercial value are listed on p. 211, and the main localities where slates are produced in the United Kingdom are given there. Much dark grey slate of Lower Carboniferous age has been worked in Co. Cork, and Ordovician slates at Killaloe and other Irish localities. In Norway the Caledonian orogeny gave rise to belts of slate in the areas affected by the compression (p. 252). In the U.S.A. slates of Ordovician age are produced in Pennsylvania and Vermont from the slate belts of the Appalachians.

Certain thin-bedded limestones or sandstones have been used for roofing in the past, such as the Collyweston 'Slate' (Inferior Oolite) of the Cotswolds and the Stonesfield 'Slate' near Woodstock, Oxford. This material is now little used, but gives character to the stone buildings of the Cotswold area. They are not true slates.

In the Auvergne district of France thin plates of phonolite (*q.v.*) have been employed for roofing over a long period of time.

ROAD-STONE AND BALLAST

Many forms of road construction are employed at the present time, varying according to the conditions and materials available, and this large subject is discussed in works of a specialized character; a general reference is given on p. 530 in Note 9 and later in Note 24.

The *surfacing* of a road is the uppermost layer, which rests on a base and sub-base. Depending on the kind of base used, the surfacing may consist of crushed rock of small grade size (chippings) with a bituminous binder; or it may be the top layer of a concrete base; or under some conditions overseas it may consist of cemented stone or stabilized gravel. The first of these, i.e. crushed rock with bituminous binder, is discussed below. Much research has been carried out on other road materials, and Reports of the Road Research Laboratory in Great Britain, and the American Society for Testing and Materials in U.S.A. should be consulted.

A road surface has to withstand abrasion and resist impact. Any rock used in a surfacing should be fresh and strong enough to remain uncrushed under the roller; the fragments should preferably be angular, and without oxidized coatings on their surfaces. In addition, the higher speeds which are common on modern roads require that the 'polishing' properties of the stone should be low: certain rocks tend to give a slippery road surface in rain more rapidly than others. For igneous and metamorphic rocks, good resistance to polishing was found when minerals of different hardness were present in a stone; medium grain size rather than fine was favourable. In sedimentary rocks the presence of harder minerals set in a softer matrix also gave good resistance to polishing (Note 12).

The crushed stone itself forms a large part of the surfacing, and tests of physical properties have been used for many years in Britain and America to assess in advance the performance of a particular material. They include tests for crushing strength, abrasion, impact, attrition (Deval and Los Angeles tests), aggregate crushing strength, water absorption, and polishing properties. They are described in British Standard 812 and in corresponding American Standards (Note 10).

Rocks of the Basalt Group (see Trade Group nomenclature, p. 507) give some of the best results from the tests, together with those of the Granite and Hornfels Groups. Particular types of stone within each Group are especially favourable. In the polishing test the Gritstone Group has shown superior results. Examples of rocks in these Groups are now discussed.

(i) *Dolerite* (Basalt Group). Many dolerite quarries in the United Kingdom yield high quality material; in the North of England and in Scotland the term *whinstone* is used for any dark, hard dolerite or basalt (typified by the Whin Sill, including *tholeiites, q.v.*). Tests on fresh dolerites from Northumberland showed high impact and low attrition values, low water absorption, and crushing strengths in the range 27 000–37 000 lbs/sq. inch $(186 \times 10^6 – 255 \times 10^6$ N/m^2). When the rock is partly weathered the test values are less favourable but individual dolerites may be suitable for use in special circumstances (Note 11). In *olivine-dolerite* the secondary mineral serpentine may be present as a decomposition product. The alteration of pyroxene to chlorite is another sign of incipient decomposition. Rocks of finer-grained texture than dolerite, e.g. basalt, may be useful material; but very fine-grained types tend to break with a splintery fracture, which is less desirable in a surfacing stone. The type of fracture is readily observable in a sample. Some andesites (included in the Basalt Group) have been used without difficulty, but glassy types of andesite are generally unsuitable.

(ii) The coarser-grained rocks of the Gabbro Group are, in

general, weaker than the dolerites; they may find uses as larger sized material for hard-core.

(iii) *Granite*. Rocks of the Granite Trade Group include true granites (with a relatively high quartz content) and also diorites and syenites with little or no quartz. The bond between a quartz-rich rock and a tar or bitumen binder is poorer than with a more basic rock. Fine-grained granites, however, and felsites, do not suffer so much from this disadvantage. (See Chapter 4, p. 140, Tinto porphyrite).

(iv) *Hornfels*. Many rocks of this type from contact aureoles are tough and possess a high crushing strength; one is on record as reaching 57 000 lbs/sq. inch (393×10^6 N/m^2). Hornfelses derived from shale by metamorphism are often suitable for use as roadstone; they are also extensively used as ballast, an English example being the crushed rock from Meldon Quarry, Okehampton, in the aureole of the Dartmoor granite, which has supplied ballast for the Southern Region railways over many years.

(v) *Gritstone*. Tests made with a wheel apparatus for giving a laboratory estimate of wear on rock chippings have shown that rocks of the Gritstone Group have a low 'polishing coefficient' (Note 12). They are therefore used for surfacing roads where high speeds are maintained. The Group includes sandstones, grits, quartzites, greywackes and pyroclastics. Samples from Basalt Group and Granite Group rocks also behave well in this respect and are second to the Gritstones. The least satisfactory are limestones and flint; the latter is only used on secondary roads.

(vi) *Limestones*. A high production of hard limestone is maintained in the United Kingdom, especially at many quarries in the Carboniferous Limestone outcrop (e.g. in the Mendips, South Wales, Derbyshire, and the North of England). Most of the rock is crushed at the quarry, graded, and mixed with tar or asphalt for delivery to sites as a coated aggregate. The bonding between the limestone and the coating is good (*cf*. granite, above), and the product is extensively used for macadam surfaces. A 'wearing course' of chippings (hard rock or gravel) is spread on the macadam. Another use of limestone in some areas is for *granular base* waterbound macadam. An aggregate of mixed grades, including fines, is laid to form the road base, or the fines can be omitted and added after laying the stone, by vibrating them into the voids of the base using a vibrating roller. The incorporation of vibration in the laying process gives a well interlocked texture, and the surface is sealed by tar immediately after laying. The road base is permeable, and adequate drainage is essential.

Studies of limestones have been made using dyed resins with which the rock is impregnated; the resin strengthens the limestone, and when thin sections are cut for petrological examination the void spaces in the rock are clearly delineated. Early investigations with this method were carried out by the Building Research Station at Garston, England (Note 13). More recently a group of soft limestones from Jamaica has been studied in this way, and their engineering properties (as revealed by laboratory tests) have been correlated with the proportion of voids in the stone. Strength increases as the content of voids decreases; the voids themselves are affected by the kind and proportion of fossils present in the limestones. (Micropores are present, for example, within the ooliths of an oolitic limestone, and macropores lie in the calcite cement between the ooliths; both are revealed by the stained resin filler) (Note 13).

Petrological examination can thus help considerably in the selection of suitable roadstone material among the limestones. The correlation between petrological details and engineering properties is greater for the soft limestones examined than for harder roadstones of the United Kingdom (Note 14).

MATERIALS FOR EMBANKMENT DAMS

These include permeable and impermeable fill, and clays for core and cut-off. Rocks of many kinds have been used for *rock-fill dams;* the broken rock can be tipped loosely and compacted largely under its own weight, giving a permeable structure. Hand-packed rubble is often used to form a layer at the upstream face, parallel to the slope of the embankment, with an overlying facing of concrete or masonry. Sometimes an additional facing is used, as at the Salazar Dam, Portugal, where a flexible steel membrane with expansion joints was placed on a thin layer of concrete, which in turn rested on masonry above the sloping face of the rock-fill. This arrangement reduced leakage through the facing almost to zero. The Loch Quoich Dam in Scotland, built in 1955 and the first rock-fill dam in Great Britain, is an embankment 304 metres long and 36.6 metres high.

An *earth dam* (embankment) is usually constructed of clay or shale, suitably placed in position in successive layers, with a core-wall which acts as an impermeable barrier along the centre line of the embankment (p. 412). The core-wall may be of puddled clay, for which a good glacial clay kneaded and worked into place, or rolled on, has been used at many sites. If a suitable clay is not readily available a concrete core-wall can be used, below which a cut-off wall extends vertically downwards to extend the flow path of impounded water below the dam (Chapter 14). The embankment itself is commonly constructed of layers of material which are rolled into position. Vibration during the rolling process can be used to give a well-compacted material. Coarse armour stone or 'rip-rap' (broken stone), may be placed on the outer slopes for protection. In some structures a gradual change from 'earth' to rock-fill has been employed, with successive zones of sand followed by fine gravel and coarse gravel between the 'earth' and rock-fill; these act as a filter and prevent penetration of the 'earth' into the rock-fill. By the use of an earth core with rock-fill shoulders the cross-sectional area of the embankment is reduced, compared with an earth dam formed entirely of one kind of material (Note 15).

During the construction of the Usk Dam (30.5 metres high) for Swansea, South Wales, pore-water pressure recorders buried in the earth bank showed unusually high pressure values. This was confirmed by water levels in tubes which were driven vertically into the then uncompleted structure. The pressure was relieved by incorporating layers of coarse materials of higher permeability into the dam embankment.

At the Balderhead Dam, near Middleton-on-Tees in Yorkshire, which was completed in 1965, a rolled fill construction was employed, with shale for the embankment obtained locally and a central core of boulder clay. The embankment is 3000 feet (914 m) long and 157 feet (47 m) high. Carboniferous shale from the Yoredale sequence was extensively tested before being used in construction, and at the site was compacted with a vibrating roller. Piezometers placed in the shale fill gave information about pore-pressures, which were low, and settlement gauges recorded maximum settlements of about 2 feet (0.6 m). A rip-rap facing of dolerite, from quarries in the Whin Sill, was placed on the upstream face of the dam.

Before construction of the Balderhead Dam similar shale which had been used 25 years earlier at the Burnhope Dam was investigated by means of a trial pit sunk on the downstream side of the Burnhope embankment. It was found that this material was unweathered 25 years after it had been placed in position. Evidence from the spoil tips of a 50 year old tunnel, driven in the shales between the Balder valley and the Lune valley, was also in agreement in the matter of the slow weathering of the material. The experience at Balderhead showed that fresh Carboniferous shale could be satisfactory as a fill when excavated and placed rapidly by modern machinery (Note 16).

Kainji Dam. By way of illustrating the materials which may be available for use in a large structure overseas, data for the *Kainji Dam* site in Northern Nigeria may be summarized. This large dam on the River Niger, some 55 miles (89 km) north of Jebba, consists of a central section of concrete gravity form flanked by two rock-fill wings. The whole structure has a length of over two miles (3.2 km), apart from ancillary works, and was completed in 1968. It is sited on igneous and metamorphic rocks including Pre-Cambrian granites and gneisses, which were covered by medium to coarse quartz sand and silty alluvium at the river, to depths up to 79 feet (24 m). The naturally occurring materials needed for construction were (i) rock for rock-fill, rip-rap, and coarse aggregate; (ii) sand, for concrete and drainage blankets; (iii) earth for earth-fill.

(i) Granite and granodiorite from the site, when tested, gave compressive strengths from 25 000 to 40 000 lbs/sq in (172 × 10^6–275 × 10^6 N/m^2). These hard rocks were available from the excavation for the dam itself, and to supplement them a quarry was opened in hard and massive hornblende-granite-gneiss not far from the dam. In all, a total of 7 × 10^6 cu yd (5.3 × 10^6 m^3) was estimated to be available from these sources; coarse aggregate for concrete was produced by crushing.

(ii) Deposits in the bed of the Niger, and as sand-banks adjacent to the river banks and to islands, provided quartz sand of suitable quality. Grading tests were carried out on this material and showed 80 to 85 per cent of sand (fine, medium, and coarse), with coarse silt not more than 5 per cent, and the rest fine gravel. Some 5 × 10^6 cu yd (3.8 × 10^6 m^3) of this deposit was available from one of the two main channels of the river.

(iii) For earth-fill, borehole samples of soils obtained from the Saddle area north-east of the main site were shown, after testing, to include sandy silty clays and clayey silts with some fine gravel. These were divided into two categories, namely material having a clay content of over 14 per cent and usable as core material, and that having less than 14 per cent clay. The figure of 14 per cent is arbitrary, but it is often at about this limit that a soil begins to show the properties of what is usually called clay as distinct from more granular material. Between 13 and 14 million cu yd (10 to 10.7 × 10^6 m^3) of material suitable for earth-fill could be obtained from the Saddle area on the left bank of the river.

The availability and amount of the above materials governed the design of the different sections of the main dam. In addition, a large power-house building 1800 feet (548 m) in length was constructed, in concrete made with crushed rock aggregate. The surveys for materials and other investigations are described by C. S. Hitchen (Note 17).

Sasumua Dam.   An earth dam 34 m high was completed in 1956 at the Sasumua site, Kenya, 64 km NNW of Nairobi. Deep weathering in nearly horizontal lavas and tuffs at the site had reduced parts of these rocks to a residual soil, known as the *Sasumua clay*. This relatively impermeable material was used as a fill for the greater part of the earth dam, and was found to possess unusual properties which were of considerable engineering value during construction.

The abnormal properties of the Sasumua clay were: (i) a plasticity index much lower than that of a normal clay of similar liquid limit; (ii) a higher angle of internal friction and higher permeability than for a normal clay of equal liquid limit; (iii) variable results for the Atterburg tests and mechanical analysis, depending largely on the chemical which was used as a dispersing agent. Because of these features, the mineral composition of the clay was investigated and found to be as follows:

Halloysite, 58.9; kaolinite, 3.9; gibbsite, 9.1; goethite, 15.9; quartz, 6.4; other constituents, 5.8 (per cent). The halloysite crystals had the form of small tubes about $\frac{1}{2}$ micrometre in length. Particle size measurements combined with the mineralogical data indicated that most of the clay fraction of the weathered lavas is present as clusters of clay mineral particles; these clusters in effect form porous grains with rough surfaces. As a result, the 'clay' has a low plasticity index, and has engineering properties which are characteristic of relatively coarse-grained soils. Water is contained in the voids of the clusters as well as between them, giving a high value of liquid limit because the material consists largely of these porous aggregates of minute halloysite crystals.

The behaviour of the dam after construction, in respect of (i) compaction of the clay fill, and (ii) seepage from the reservoir, was very satisfactory. Other earth dams made of similar material (weathered volcanic rocks) had been built in Java (Tjipanoendjang, 1927) and in Australia (the Silvan Dam, 1931). Samples of the materials from these localities, when investigated at the time of the building of the Sasumua Dam, also showed a high content of halloysite (76.4 per cent, and 80–90 per cent respectively). The mineral was present as minute spongy aggregates, as at Sasumua. It appears, therefore, that such 'clays', derived from the weathering of lavas and tuffs, can be as good as normal clays of high quality and low liquid limit, for the construction of earth dams (Note 18).

CEMENTS AND FILLERS

The geological raw materials used in making Portland Cement are limestone (as a source of calcium) and a deposit containing silica and alumina, such as clay or mud. Thus, alluvial mud and chalk are used at some localitiès in Great Britain where both are available; and the Chalk Marl, a clayey limestone which occurs at the base of the Chalk formation (see p. 263) is worked at localities in the Chalk outcrop. Certain argillaceous limestones of the Lias near Rugby, known as 'cement-stones', have been employed for the manufacture of cement.

Where river mud and chalk are used, as in the Thames and Medway area, the two constituents are mixed and ground in correct proportions, either wet or dry. A fine slurry is produced and is pumped into inclined rotating kilns, where it becomes dried. As it moves forward into a higher temperature, chemical combination takes place with the formation of calcium and aluminium silicates in the form of small clinker. After cooling, the clinker is ground to an extremely fine powder. During the grinding, an additive such as gypsum may be introduced, to retard the setting time of the cement, and thus allow for satisfactory mixing and placing of the concrete. The proportion of lime (CaO) to silica plus alumina (calculated as chemical equivalents) should not exceed 2.85 (Note 19).

The main constituents present in Portland Cement are tricalcium silicate, dicalcium silicate, tricalcium aluminate, and an alumino-ferrate $4CaO \cdot Al_2O_3 \cdot Fe_2O_3$. Minor constituents include Mg, Ti, P, and alkalies; the latter, Na_2O and K_2O, generally amount to less than 1.5 per cent of the whole. High-alkali cements have a greater alkali content than 1.5 per cent, and have been shown to give rise to deterioration of concrete through reaction with certain constituents of aggregates (p. 512).

High-alkali cement, with Na_2O and K_2O content up to 3.5 per cent, is used where a concrete is required to gain strength rapidly, or where concrete foundations are placed in certain chemically aggressive soils. A sulphate-resisting cement is made by the addition of a very small quantity of tricalcium aluminate to normal Portland cement. This additive can combine chemically with sulphates in solution, which may be absorbed by e.g. a concrete foundation in an aggressive soil. The amount of tricalcium aluminate to be added is calculated according to the method set out in B.S. 4027 (Note 19).

Cement made from septaria (calcareous nodules) in shale, obtained at the Isle of Sheppey, gives a concrete which sets rapidly under water and is often called 'hydraulic cement'. (*Cf.* hydraulic lime, which also hardens under water). The nodules contain about 20 per cent silica, 15 per cent alumina, and calcium carbonate; the CO_2 is driven off when the material is calcined.

A rise of temperature takes place during the setting of Portland Cement, and this may affect mass concrete in large structures such as dams. *Low-heat cements* have therefore been developed from pozzuo-lana (poorly consolidated tuff deposits, p. 134), and from artificial pozzolans which are now manufactured. The pozzolan is mixed with Portland Cement, to a specification set out in B.S. 1370 (Note 19). As well as less heat being evolved during the setting, the low-heat cement has a higher resistance to chemical deterioration than normal Portland Cement.

WATER USED IN CONSTRUCTION

The *hardness* of water is due to dissolved substances; without them a pure water is *soft*. The chief salts in solution which cause hardness are the carbonates and sulphates of calcium and magnesium; carbonates produce 'temporary hardness', which can be removed by boiling and by other means, and sulphates 'permanent hardness', which can be removed by chemical treatment. In addition chlorides may be present and in smaller amounts nitrates, fluorine, and metal ions such as Fe and Mg. Hardness is expressed in parts per million (by weight).

A typical water from an upland source is relatively soft, with a total hardness around 20 p.p.m. and a pH of 6 or less. By contrast, ground-water from calcareous rocks may have a hardness of 50 or 60 p.p.m. or more, present as calcium bicarbonate (see Chapter 2, p. 25). Soft waters have a solvent action on lead, and attention should be given to this aspect if lead pipes are involved. The concentration of hydrogen ions in water, or pH value, is a measure of the degree of acidity (pH less than 7) or alkalinity (pH more than 7) (Note 20).

The sulphate content of ground-water, expressed as sulphur trioxide (SO_3), varies from zero to over 0.5 per cent (or 5000 p.p.m.). The sulphate may be derived from selenite (gypsum) crystals in a sediment, or from pyrrhotite (*q.v.*), or from H_2S in solution; such water is aggressive to concrete. When the concentration of sulphide ions, in water to be used for mixing with cement for concrete, exceeds about 0.12 per cent (1200 p.p.m.), consideration should be given to the use of sulphate resisting cement (see B.S. 4027).

Where connate water has been trapped in an aquifer, a high sulphate content may be due to concentration in the original sea-water in which the sediment was deposited (Note 21).

Ground-water containing sulphate in solution is a main cause of corrosion in iron pipes, and in other metals (Note 22).

MATERIALS FOR BRICKS, BOARDS, AND GLASS

Brick clays   Clays are essentially mixtures of clay minerals (hydrous aluminium silicates) with some silt and fine sand, and are plastic when wet (see Chapter 5). The extent to which shrinkage occurs when a clay is dried depends on the amount of water held between the mineral particles (or within their lattices), and can vary considerably. Clays suitable for brickmaking should not show a shrinkage of more than 30 per cent (by volume). A small proportion of minor constituents in the clay, such as Na, Ca and Mg, which assist fluxing at a low temperature, are desirable; a little iron imparts a red or brown colour to the brick. Organic carbon in a grey clay is burnt out during the kilning of the brick; if ferrous sulphate is present in the clay it combines to form sulphates of alkalies, magnesium or calcium. An excess of sulphate in clay leads to deterioration in the resulting brick.

During firing in the kiln much water is driven off from the clay, and the remaining water enters into chemical combination with alumina, silica, iron and alkalies to form a silicate glass. The production of sufficient glass to bind together the unfused particles is a main result of the firing process.

Marine clays are extensively used in the United Kingdom for brickmaking, and include the Lower Lias grey clays in the Cheltenham district, the Oxford Clay in Bedfordshire and Northamptonshire, and the Weald Clay of Sussex (see Chapter 8). Brickearths in the Lea Valley, near London, were used to make bricks known as London Stocks.

Boards   The group of fibrous minerals known as "asbestos" in commerce have important fire-resistant properties; they are incombustible and infusible, and have low heat conductivity. They can be separated easily into fibres, which are used in a range of fireproof materials including asbestos board (in flat or corrugated sheets), roofing cements and tiles, roofing shingles, and fire-resistant packing. The longer fibres can be woven into textiles. The bulk of the world's commercial production consists of chrystotile asbestos (p. 150) which occurs as cross-fibres in narrow veins traversing serpentinite and other ultrabasic rocks. The largest production comes from the Thetford district of Quebec, Canada, and the Central Urals, U.S.S.R.

'Plaster of paris' is employed in the building industry in the form of plaster-board, wallboard, and other materials. It is made from the mineral gypsum (p. 125) which, when heated to a temperature between 110° and 120°C, loses more than half its water of crystallization and is converted into a white powder, plaster of paris. This can absorb water and then sets to a relatively hard mass. Kaolin, an inert substance (p. 117) is used as a filler in certain wall-plasters.

Another geological constituent which may be mentioned here is the ground-up mica scrap that is used as a backing for rolled asphalt roofing to prevent sticking.

Glass   An important use of quartz is in the manufacture of glass of various kinds, and in silica brick, which is used for the linings of furnaces. The quartz is obtained mainly from silica sand deposits. Glass is a non-crystalline transparent substance which is a mixture of fused silicates; it behaves as a rigid elastic solid. To its main ingredient, silica, is added a proportion of calcium carbonate and sodium carbonate or sulphate. The mixture melts over a range of temperature from about 1200 to 1500°C, and can then be poured; the tank-furnace in which the melting is carried out is lined with fire-clay. The choice of sand is important; many sands contain a little iron, which gives a green colour to the glass, and the Fe-content should be kept as low as possible. Glasses with special properties, e.g. resistance to extreme temperature changes, are made by the addition of particular substances such as borates and phosphates. (Note 23).

NOTES AND REFERENCES

1. For a discussion of the geological characters of gravels see The Geology of sand and gravel deposits in Great Britain by Knill, D. C. 1963. (*Cement, Lime, and Gravel*, October, 1963).
2. These properties are discussed in B.S. 882 and 1201 (1965), Aggregates for Concrete (published by the British Standards Institution, 2 Park Street, London, W.1); and B.S. 812, Methods for sampling and testing of Mineral Aggregates, Sands and Fillers, is also relevant.
3. See A survey of crushed stone sands for concrete, by Teychenne, D. C. 1967, *British Granite and Whinstone Federation Journal*, **7,** No. 1.
4. Corresponding American specifications and tests are set out in publications of the American Society for Testing and Materials, Philadelphia, Pa. See 1970 Book of A.S.T.M. Standards, part 10, Concrete and Mineral Aggregates.
    An account of cement-aggregate reaction by McConnell, D. *et al.* is given in the Berkey Volume, Applications of Geology to Engineering Practice, *Geol. Soc. Amer.*, 1950, p. 225. See also Engineering Geology Case Histories Nos. 1–5, 1964, p. 83. (*Geol. Soc. Amer.*, Eng. Geol. Division). And Stanton, T. E., 1940, in *Am. Soc. Civ. Eng. Proc.* **66,** 1781.
5. RRL Report, LR 293, by Hosking, J. R. and Tubey, L. W. 1969, Research on low-grade and unsound aggregates. (Road Research Laboratory, Crowthorne, Berkshire, England).
6. The traditional method of splitting a large block of granite, which may weigh 15 or 16 tons when moved by quarry tackle, is known as 'feather and tear'. A line of shallow holes (4 inches deep) is drilled in the upper face of the block, the line being parallel to the *rift* direction in the granite. Rift is the direction in which the rock splits most readily, and is probably due to the presence of oriented microscopic structures in certain crystals (*cf.* S-joints in Fig. 7.24). About six holes are needed to split a block four feet long. Two steel strips, or 'feathers', are inserted in each hole and between them a wedge. The wedges are struck in succession and the block splits vertically along the line of holes.

In the Creetown granite, S.W. Scotland (sheet N.S. 16) biotite crystals in the rock form an arch of flow in the central part of the dyke-like intrusion. The easy-split direction is nearly horizontal and at this quarry is called the *reed* (*cf.* rift, above). A second parting, the *hem*, is vertical and parallel to the length of the intrusion; and cross-joints which run nearly at right angles to the reed and hem provide a third direction of splitting. The structure is described in The Kirkmabreck Granodiorite, near Creetown, South Galloway by Blyth, F. G. H. 1955. *Geol. Mag.*, **107**, no. 4, p. 321–328. Stone from this locality has been used in dock walls at Liverpool and elsewhere.

7.   The first recorded use of Portland Stone in London was in 1619, for the Banqueting Hall at Whitehall designed by Inigo Jones.

8.   For a description of the method of working Portland Stone see Schaffer, R. J. 1932, in *Proc. Geol. Assoc.* (London), **43**, p. 225: Portland Stone, its geology and properties as a building stone.

9.   Road Research Laboratory (Ministry of Transport) Reports include the following:

No. 47, Roadmaking materials in Basutoland, by Beaver, P. J. 1966.

LR 90, The frost susceptibility of soils and road materials, by Croney, D. and Jacobs, J. C. 1967.

LR 328, The use of cement-stabilised Chalk in road construction, by Pocock, R. G. 1970.

See also: A.S.T.M. Standard C.290 and C.292, Freeze-thaw testing. And British Standard 1924, Methods of test for stabilised soils; also B.S. 1984, Gravel aggregates for surface treatment on roads.

General Reference: Soil Mechanics for Road Engineers (revised edition), published by Road Research Laboratory. (H.M.S.O., London).

10.   See also D.S.I.R. Spec. Rep. no. 3, Roadstone, by Phemister, J. C., Markwick, R. and Shergold, F. A. 1946.

11.   A discussion is given by Sabine, P. A., Morey, F., and Shergold, F. A., 1954, in *Jour. Appl. Chemistry*, **4**, 134.

12.   See Maclean, D. J. and Shergold, F. A. 1958, The polishing of roadstone in relation to the resistance to skidding of bituminous road surfacing. D.S.I.R. Tech. Paper 43.

Also Knill, D. C. 1960, in *Jour. Appl. Chem.*, **10**, 28. And Knill, D. C. Petrographical aspects of the polishing of natural roadstones, 1960, *Quarry Manager's Jour.*, **44**(6), p. 215.

It has recently been shown by means of test surfaces, placed in roads and subjected to traffic, that the frictional resistance of chippings when wet (called 'skid-resistance' values) correlates with the polished stone coefficients found in the laboratory tests. The latter, therefore, give a reliable measure of the relative extent of polishing of different stones. See Maclean, D. J., A study of the mechanism governing the polishing of stone in road surfaces, 1968, *Q. Jour. Eng. Geol.* **1**, 135.

The blending of aggregates of different polishing characteristics with the aim of obtaining a desired level of skid resistance is discussed by Williams A. R. and Lees, G. in Topographical and petrological variation of road aggregates and the wet skidding resistance of tyres,

1970, *Q. Jour. Eng. Geol.* **2**, pt. 3, p. 217. This account gives illustrations of the effects of weathering on the micro-topography of polished limestones and other rocks, and full references up to 1968.

13. For the resin impregnation and other methods see Allman, M. and Lawrence, D. 1972. Geological Laboratory Techniques. Blandford Press, London, 336pp.

   Data for the Jamaican limestones is quoted from A study of the petrology of some soft limestones from Jamaica in relation to their engineering properties, by Tubey, L. W. and Beaver, P. J. 1966, RRL Report LR 21.

14. See, for example, Geotechnical properties of the Great Limestone in Northern England, by Atterwell, P. *Engineering Geology*, 5, 1971.

15. The necessary grading for the filter material is discussed by Terzaghi, K. and Peck, R. 1962, in Soil Mechanics and Engineering Practice, p. 50. (Wiley and Sons, New York.) See Note 6 of Chapter 13. Also Penman, A. and Charles, J. 1975. The quality and suitability of rockfill used in dam construction. Building Research Station, C.P.87/75.

16. See Kennard, M. F., Knill, J. L. and Vaughan, P. R. 1967. The geotechnical properties and behaviour of Carboniferous shale at the Balderhead Dam, *Q. Jour. Eng. Geol. (London)*, **1**, 3.

17. Materials surveys and investigation for the Kainji Dam project, Northern Nigeria, by Hitchen, C. S. 1968. *Q. Jour. Eng. Geol.*, **1**, 75. (Geol. Soc. London publication).

18. See the account by Terzaghi, K. 1958, Design and performance of the Sasumua Dam, *Proc. Inst. C.E.*, **9**, 369.

19. British Standards dealing with cement include the following, which are referred to in the text: B.S. 12, Portland Cement, ordinary and rapid hardening. B.S. 1370, Low heat Portland Cement, and B.S. 4246 (1968). B.S. 4027, Sulphate resisting Portland Cement. B.S. 4248, (1968), Supersulphated Cement. For a general account see Lea, F. M. 1970. The Chemistry of Cement and Concrete, 3rd. edition, Edward Arnold.

20. In water, some of the H_2O molecules are dissociated into hydroxyl ions (OH^-) and hydrogen ions (H^+). The concentration of these two components is very small and is governed by the Law of Mass Action. From this it can be shown that the product of the hydroxyl and hydrogen ion concentrations is a constant, the value of which is 10^{-14} at 21°C. The concentrations are expressed in terms of gram ions per litre. In pure water these concentrations are equal; that is, 10^{-7} at 21°C. The index -7 is the logarithm of the number, to base 10, and with a change of sign 7 is conveniently used to state the concentration of the hydrogen ions, denoted by the expression pH. Thus water of pH 7 is neutral; if less than 7 it is acid, and if greater than 7 alkaline.

21. An example of connate water with a high concentration of sulphate in solution, found in a confined synclinal area in the Chalk of East Anglia, is described by Woodland, A. W. 1946, in Chemistry of Chalk Groundwater in East Anglia (Essex and Suffolk); *War-time Pamphlet* No. 20, Pt. 1. (Institute of Geological Sciences, London.)

22. See Romanoff, M. Underground Corrosion, Circular 579 (Sections 3 & 4), *Nat. Bureau of Standards* (U.S. Dept. of Commerce, Washington D.C.). See also Hem, J. 1970, Study and interpretation of the chemical characters of natural waters. *U.S. Geol. Surv.* Water Supply Paper no. 1473.
23. See B.S. 2975 (1958), Sand for making colourless glass.
24. Fookes, P. 1976. Road geotechnics in hot deserts. *Journ. Institution of Highway Engineers,* **23,** 11–23.

Appendix
SOURCES OF GEOLOGICAL INFORMATION

by Joan E. Hardy, Library Information Officer, Imperial College

I **Libraries**

The following list of libraries specializing in publications on the earth sciences includes: Nos. 1–17, *Government libraries;* 19–29, *professional societies;* 30–36, *national libraries.* A rapid loans service is obtainable from the National Lending Library through any of the other libraries in the United Kingdom.

*1. Institute of Geological Sciences, Exhibition Road, London SW7.
      Exeter: Hoopern House, Pennsylvania Road, Exeter EX4 6DT.
      Leeds: Ring Road, Halton, Leeds LS15 8TQ.
      Edinburgh: 19 Grange Terrace, Edinburgh EH9 2LF.
      Belfast: 20 College Gardens, Belfast BT9 6BS.

2. Soil Survey of Great Britain
      England and Wales: Rothamsted Experimental Station, Harpenden,
      Scotland: Macaulay Institute for Soil Research, Craigiebuckler,
             Aberdeen AB9 2QJ.

3. Ordnance Survey of Great Britain
      England and Wales: Romsey Road, Maybush, Southampton
         S09 4DH. *Also:* Air Photographs Officer, Air Survey Branch
      Scotland: 43 Rose Street, Edinburgh 2.
      N. Ireland: Ladas Drive, Belfast BT6 9FJ.

4. National Institute of Oceanography, Wormley, Godalming, Surrey.

5. Department of the Environment
      (*includes former* Ministry of Housing and Local Government *and*
      Ministry of Public Buildings and Works *and* Ministry of Transport)
      2 Marsham Street, London SW1.
      Also: Map Library, Air Photographs Officer

6. Ministry of Agriculture, Fisheries and Food
      3 Whitehall Place, London SW1.
      Maps: Great Westminster House, Horseferry Road, London SW1.

7. Department of Trade and Industry
      Thames House South Library (*formerly* Ministry of Power Library)
      Millbank, London SW1P 4QJ.
      Also: Mining Record Office, Safety & Health Division
              Senior District Inspectors of Mines & Quarries

* Available to the public for reference only.

8. Directorate of Overseas Surveys (Overseas Development Administration of the Foreign and Commonwealth Office)
Kingston Road, Tolworth, Surbiton, Surrey.
Land Resources Division: 8th Floor, Tolworth Tower, Surbiton.

*9. Public Record Office, Chancery Lane, London WC2.

10. Scottish Development Department
Air Photographs Officer, York Buildings, Queen Street, Edinburgh 2.

*11. Public Records Office of Northern Ireland, Law Courts Building, May Street, Belfast 1.

12. Admiralty Chart Establishment
Hydrographic Department, Ministry of Defence, Taunton, Somerset.

13. Building Research Station, Bucknall's Lane, Garston, Watford, Herts.

14. Hydraulics Research Station, Wallingford, Berks.

15. Transport and Road Research Laboratory, Crowthorne, Berks.

16. Water Research Association, Medmenham, Marlow, Bucks.

17. Water Resources Board, Reading Bridge House, Reading, Berks.

18. Nature Conservancy, 19 Belgrave Square, London SW1.

19. Geological Society of London
Burlington House, Piccadilly, London W1V OJU.

20. Institute of Petroleum, 61 New Cavendish Street, London W1M 8AR.

21. Institution of Civil Engineers, Great George Street, London SW1.

22. Institute of Mining & Metallurgy
44 Portland Place, London W1N 4BR.

23. Institution of Water Engineers, 6–8 Sackville Street, London W1.

24. Royal Geographical Society, Kensington Gore, London SW7.
(Map Library is open to the public)

25. Geologists' Association
Library at: University College, Gower Street, London WC1E 6BT.

26. Edinburgh Geological Society
     Library at: University of Edinburgh, Department of Geology,
     Grant Building, West Mains Street, Edinburgh 8.

27. Liverpool Geological Society
     Department of Geology, The University, Liverpool 3.

28. Royal Geological Society of Cornwall
     West Wing, Public Buildings, Penzance, Cornwall.

29. Yorkshire Geological Society
     Library at: Brotherton Library, The University, Leeds LS2 9JT.

*30. British Museum (Natural History)
      Cromwell Road, London SW7.

*31. National Lending Library, Boston Spa, Yorks.

*32. Science Museum Library, South Kensington,
      London SW7 5NH.

*33. National Library of Wales, Aberystwyth, Cardiganshire.

*34. National Library of Scotland, George IV Bridge, Edinburgh 1.

*35. Ulster Museum, Stranmillis Road, Belfast 9.

36. National Coal Board
     Hobart House, Grosvenor Place, London SW1.
     Plans Record Office: Edinburgh
     Opencast Executive: Harrow, Middlesex.
     Mining Research Establishment: Ashby Road, Stanhope Bretby,
     Burton-on-Trent.

II Geological literature (mainly of the British Isles)

Sources of information are published in *handbooks and glossaries,
reviews and conference papers,* and in *journals.* The World List is the
most comprehensive. *Abstracting journals* for geology, and *catalogues*
of Government publications and *research reports,* are followed by
individual summaries and *map lists.*

HANDBOOKS AND GLOSSARIES

INFORMATION retrieval for soil engineers; report of the Committee on
Information Retrieval, Soil Mechanics Division. *Journal, Soil Mechanics
and Foundations Division, ASCE* **93,** No. SM5, Proc. Paper 5471, Sept.
1967 Part 2, 1–180.

KAPLAN, S. R. 1965. A guide to information sources in mining, minerals, and
geosciences. New York, Interscience.

MACKAY, JOHN W. 1973. Sources of information for the literature of geology: an introductory guide. London, Geological Society.

MASON, B. 1953. The literature of geology. New York, American Museum of Natural History.

WARD, D. C. et al. 1972. Geologic reference sources. New edition. Metuchen (NJ), Scarecrow Press. (1st edition, University of Colorado Press, 1967).

WOOD, D. N. (editor) 1973. Use of earth sciences literature. London, Butterworths.

WOODLAND, A. W. 1968. Field geology and the civil engineer. *Proc. Yorks. geol. Soc.*, **36,** No. 4, 531–578.

SCIENTIFIC research in British universities and colleges. (Annual). 3 vols. London, HMSO.

AMERICAN GEOLOGICAL INSTITUTE. 1972. Glossary of geology. New ed. Washington (DC), the Institute.

CHALLINOR, J. 1967. A dictionary of geology. 3rd ed. Cardiff, University of Wales Press.

NEDERLANDSCH GEOLOGISCH MIJNBOUWKUNDIG GENOOTSCHAP. 1959. Geological nomenclature. Edited by A. A. G. Schieferdecker. Gorinchem, J. Noorduijn, (English-Dutch-French-German).

NELSON, A. and NELSON, K. D. 1967. Dictionary of applied geology; mining and civil engineering. London, George Newnes.

ARKELL, W. J. and TOMKEIFF, S. I. 1953. English rock terms chiefly as used by miners and quarrymen. London, Oxford University Press.

BRITISH STANDARDS INSTITUTION. 1964. Glossary of mining terms. Section 5: Geology. (British Standard 3618).

FAIRBRIDGE, R. (editor) 1968. Encyclopaedia of geomorphology. New York, Reinhold.

RUNCORN, S. K. (editor) 1967. International dictionary of geophysics. 2 vols. Oxford, Pergamon.

KOTTLOWSKI, F. E. 1965. Measuring stratigraphic sections. New York, Holt, Rinehart & Winston.

DIRECTORY of quarries and pits. 1969. 20th ed. London, The Quarry Managers' Journal Ltd.

REVIEWS AND CONFERENCE PAPERS

Earth Science Reviews Amsterdam, Elsevier, 1966–

AHRENS, L. H. et al. (editors). Physics and chemistry of the earth, a progress series. New York, Pergamon, 1956–

Reviews in engineering geology New York, Geological Society of America, 1962–

INTERNATIONAL GEOLOGICAL CONGRESS. Proceedings *and* Reports.

WORLD PETROLEUM CONGRESS. Proceedings.

INTERDOK: directory of published proceedings, Series SEMT. New York, Interdok Corp., 1965–

GEOLOGICAL SOCIETY OF LONDON. 1962. List of periodicals currently taken by the Library.

ULRICH'S international periodicals directory. 1971. 2 vols. 14th ed. New York, Bowker.

WORLD list of scientific periodicals published in the years 1900–1960. 1963–65. 4th ed. 3 vols. London, Butterworths.
BRITISH union catalogue of periodicals: new titles. Quarterly 1966– (keeps the *World List* up-to-date).

JOURNALS

Nature London, Macmillan, 1869–
New Scientist London, New Science Publications, 1956–
Science Washington (DC), American Association for the Advancement of Science, 1883–
Scientific American New York, Scientific American Inc., 1845–
Engineering Geology California, Association of Engineering Geologists, 1964–
Engineering Geology Amsterdam, Elsevier, 1965–
Geographical Journal London, Royal Geographical Society, 1893–
Geological Journal Liverpool, Liverpool Geological Society, 1964– (formerly *Liverpool and Manchester Geological Journal* 1951–63, *Proceedings Liverpool Geological Society* 1859–1950 and *Journal of Manchester Geological Society* 1925–50).
Geological Magazine Cambridge, Cambridge University Press, 1864–
Géotechnique London, Institution of Civil Engineers, 1948–
Institute of British Geographers, Transactions London, the Institute, 1935–
International Journal of Rock Mechanics and Mining Sciences Oxford, Pergamon Press, 1964–
Journal of the Geological Society London, the Society, 1971– (formerly *Quarterly Journal of the Geological Society* 1845–1970).
Journal of Geology Chicago, University of Chicago Press, 1893–
Journal of Hydrology Amsterdam, North-Holland, 1963–
Journal of the Soil Mechanics and Foundations Division, American Society of Civil Engineers New York, the Society, 1956–
Proceedings of the Geologists' Association Colchester, Benham & Co., 1859–
Proceedings of the Ussher Society Camborne, the Society, 1962–
Proceedings of the Yorkshire Geological Society Leeds, the Society, 1839–
Quarterly Journal of Engineering Geology London, the Geological Society, 1967–
Rock Mechanics Berlin, Springer-Verlag, 1969– (formerly *Rock Mechanics & Engineering Geology* 1963–68).
Scottish Journal of Geology Edinburgh, Oliver & Boyd, 1965– (formerly *Transactions Edinburgh Geological Society* 1866–1963 and *Transactions Geological Society of Glasgow* 1868–1963).
Transactions of the Royal Geological Society of Cornwall Penzance, the Society, 1814–
Welsh Geological Quarterly Cardiff, Geologists' Association, South Wales Group, 1966–

ABSTRACTING JOURNALS

Bibliography and Index of Geology New York, Geological Society of America, 1967– (formerly *Bibliography and index of geology exclusive of North*

America 1934–66 and *Geological literature added to the Geological Society's library* 1895–1935. Incorporating *Geophysical Abstracts* and *Abstracts of North American geology* 1972–).

Engineering Index New York, Engineering Index Inc., 1885–

Geodex: Soil Mechanics Information Service California, Geodex International, 1968–

Geographical Abstracts Section A; Geomorphology. Norwich, Geo-Abstracts, 1966–

Geomechanics Abstracts Oxford, Pergamon, 1973– (formerly *Rock Mechanics Abstracts* 1970–72).

Geotechnical Abstracts Essen, International Society for Soil Mechanics & Foundation Engineering, 1970–

CATALOGUES AND RESEARCH REPORTS

GREAT BRITAIN. *Her Majesty's Stationery Office*, Government publications. London, HMSO. (Monthly and annual issues).

GREAT BRITAIN. *Her Majesty's Stationery Office*, Government publications; Sectional list 45—Institute of Geological Sciences. London, HMSO.

U.S. BUREAU OF MINES, 1960. List of publications issued by the Bureau of Mines from July 1, 1910 to January 1, 1960 with subject and author index. Washington (DC), Government Printing Office. (Annual supplements).

U.S. GEOLOGICAL SURVEY, 1964. Publications of the Geological Survey 1879–1961. Washington (DC), Government Printing Office. (Annual supplements).

INDIVIDUAL SUMMARIES AND MAP LISTS

DUMBLETON, M. J. and WEST, G. 1971. Preliminary sources of information for site investigations in Britain. Crowthorne (Berks), Road Research Laboratory (RRL Report LR 403).

GEOLOGICAL SOCIETY ENGINEERING GROUP WORKING PARTY. The preparation of maps and plans in terms of engineering geology; a report. London, Geological Society.

BASSETT, D. A. 1967. A source-book of geological, geomorphological and soil maps for Wales and the Welsh Borders (1800–1966). Cardiff, National Museum of Wales.

GEOLOGICAL SURVEY OF GREAT BRITAIN. 1937. List of memoirs, maps, sections, etc. published by the Geological Survey of Great Britain and the Museum of Practical Geology to 31 December 1936. London, HMSO.

INSTITUTE OF GEOLOGICAL SCIENCES. List of geological maps (in print). London, HMSO. Bookshop: 49 High Holborn, London W.C.1 *or* Edward Stanford Ltd., 12 Long Acre, W.C.2.

Old Series 1″ to 1 mile maps for England and Wales, Scotland and N. Ireland (uncoloured).

New Series 1″ to 1 mile (1:63 360) solid and drift for England and Wales and Scotland; with memoirs.

    1:25 000 with memoirs

    1:10 560 showing boreholes and brief borehole data

    1:253 440 and 1:625 000

Hydrogeological maps.
Economic and coalfield maps and memoirs, including borehole and shaft records with the depth and character of the beds.
ORDNANCE SURVEY. Map catalogue (annual) *and* Publication Report (monthly).
BUNTING, B. T. (editor and compiler) 1964. An annotated bibliography of memoirs and papers on the soils of the British Isles. Part 1. London, Geomorphological Abstracts.
SOIL SURVEY OF GREAT BRITAIN. The Soil Survey. Harpenden (Herts), Rothamsted Experimental Station.
(includes series: 1:25 000 for selected areas, 1:63 360 for most areas, 1:10 560 are not published)
INSTITUTE OF GEOLOGICAL SCIENCES. Water supply memoirs *and* Well catalogue series. London, HMSO.
WATER RESOURCES BOARD. 1970. Groundwater yearbook 1964/66. London, HMSO.
WATER RESOURCES BOARD and SCOTTISH DEVELOPMENT DEPARTMENT. The surface water year book of Great Britain. London, HMSO, 1961–
ADMIRALTY. Catalogue of Admiralty charts and hydrographical publications. NP131 (complete edition); NP91 (Gt. Britain & Ireland).
DUMBLETON, M. J. and WEST, G. 1970. Air-photograph interpretation for road engineers in Britain. Crowthorne (Berks), Road Research Laboratory. (RRL Report LR 369).

BRITISH—SI EQUIVALENTS

Length:	1 inch	25.4 mm
	1 foot	0.3048 m
	1 mile	1.609 km
Area:	1 in.^{2}	645.2 mm^2
	1 ft^2	0.0929 m^2
	1 acre	4047 m^2
	1 mile^2	2.590 km^2
Volume:	1 in.^{3}	16.39 × 10^3 mm^3
	1 ft^3	0.028 m^3
	1 yard^3	0.7646 m^3
	1 gallon	4.546 litres
Mass:	1 lb	0.4536 kg
	1 ton	1.016 kg
Density:	1 lb/ft^3	16.02 kg/m^3
Force	1 lbf	4.448 N
Pressure:	1 lbf/in.^{2}	6895 N/m^2
Temperature:	1 deg F	5/9 deg. C

Index